ALAN TURING: THE ENIGMA
THE BOOK THAT INSPIRED THE FILM
THE IMITATION GAME

Andrew Hodges is Tutor in Mathematics at Wadham College, Oxford University. His classic text of 1983, since translated into several languages, created a new kind of biography, with mathematics, science, computing, war history, philosophy and gay liberation woven into a single personal narrative. Since 1983 his main work has been in the mathematics of fundamental physics, as a colleague of Roger Penrose. But he has continued to involve himself with Alan Turing's story, through dramatisation, television documentaries and scholarly articles. Since 1995 he has maintained a website at www.turing.org.uk to enhance and support his original work.

D0093875

TO THEE OLD CAUSE!

The dedication, epigraphs and epitaph, are taken
from the *Leaves of Grass* of Walt Whitman.

'Alan Turing was by any reckoning one of the most remarkable Englishmen of the century. A brilliant mathematician at Cambridge in the '30s, Turing discovered that his was precisely the kind of intelligence needed by Britain during the war and became the presiding genius at Bletchley Park, the boffin centre which cracked the German Enigma code. (A character in McEwan's *The Imitation Game* was loosely based on him.) There he became obsessed by the notion of machine intelligence and was, in effect, the father of the modern computer. Mistrust and bureaucracy, however, frustrated many of his plans after the war, when Turing was to discover that though he was the master of his own sphere, politically he remained as he was in 1941 – a servant. A homosexual, Turing found his own morality and scientific ideas increasingly at odds with the values of the state which he served. Eventually, he committed suicide. Andrew Hodges's book is of exemplary scholarship and sympathy. Intimate, perceptive and insightful, it's also the most readable biography I've picked up in some time'

Richard Rayner, *Time Out*

'Researched and written extraordinarily well. It is a first-class contribution to history and an exemplary work of biography'

Nature

'Life and work are both made enthralling by Hodges, himself a scientist'

Sunday Times

'This rather shadowy figure has now finally been lifted into the light of day . . . it has to be said that Andrew Hodges has put together an extraordinary story'

Sunday Telegraph

'This book has a great deal to offer: clear technical descriptions set against their backgrounds; the story of a man largely at odds with the system he lived in; and the puzzle of Alan Turing himself'

Times Higher Education Supplement

'Andrew Hodges, in this fine biography *Alan Turing: The Enigma*, brings Turing the thinker and Turing the man alive for the reader and thus allows us all to share in the privilege of knowing him'

Financial Times

'This is not a book to be argued about. It is a book to be read'

New Scientist

'A major work at any level. Recommended'

Personal Computing World

'An almost perfect match of biographer and subject.... [A] great book.'

Ray Monk, *Guardian*

'A captivating, compassionate portrait of a first-rate scientist who gave so much to a world that in the end cruelly rejected him. Perceptive and absorbing, Andrew Hodges's book is scientific biography at its best.'

Paul Hoffman, author of *The Man Who Loved Only Numbers*

'A remarkable and admirable biography.'

Simon Singh, author of *The Code Book* and *Fermat's Enigma*

ANDREW HODGES

Alan Turing: The Enigma

The Book That Inspired the Film
The Imitation Game

Princeton University Press
Princeton and Oxford

Published in the United States by Princeton University Press
41 William Street, Princeton, New Jersey 08540
press.princeton.edu

First published by Burnett Books Ltd
in association with Hutchinson Publishing Group 1983
Unwin Paperbacks edition 1985
Reprinting 1985 (twice), 1986, 1987 (twice)

First published by Vintage in 1992

This edition published in the United Kingdom in 2014 by
Vintage
Random House, 20 Vauxhill Bridge Road,
London SW1V 2SA

A Penguin Random House Company

Library of Congress Control Number: 2014952514
ISBN: 978-0-691-16472-4
Printed on acid-free paper. ∞

Printed in the United States of America
1 3 5 7 9 10 8 6 4 2

Contents

List of Plates

Foreword

Is a mind a complicated kind of abstract pattern that develops in an underlying physical substrate, such as a vast network of nerve cells? If so, could something else be substituted for the nerve cells – something such as ants, giving rise to an ant colony that thinks as a whole and has an identity – that is to say, a self? Or could something else be substituted for the tiny nerve cells, such as millions of small computational units made of arrays of transistors, giving rise to an artificial neural network with a conscious mind? Or could software simulating such richly interconnected computational units be substituted, giving rise to a conventional computer (necessarily a far faster and more capacious one than we have ever seen) endowed with a mind and a soul and free will? In short, can thinking and feeling emerge from patterns of activity in different sorts of substrate – organic, electronic, or otherwise?

Could a machine communicate with humans on an unlimited set of topics through fluent use of a human language? Could a language-using machine give the appearance of understanding sentences and coming up with ideas while in truth being as devoid of thought and as empty inside as a nineteenth-century adding machine or a twentieth-century word processor? How might we distinguish between a genuinely conscious and intelligent mind and a cleverly constructed but hollow language-using facade? Are understanding and reasoning incompatible with a materialistic, mechanistic view of living beings?

Could a machine ever be said to have made its own decisions? Could a machine have beliefs? Could a machine make mistakes? Could a machine believe it made its own decisions? Could a machine erroneously attribute free will to itself? Could a machine come up with ideas that had not been programmed into it in advance? Could creativity emerge from a set of fixed rules? Are we – even the most creative among us – but passive slaves to the laws of physics that govern our neurons?

Could machines have emotions? Do our emotions and our intellects belong to separate compartments of our selves? Could machines be enchanted by ideas, by people, by other machines? Could machines be attracted to each other, fall in love? What would be the social norms for machines in love? Would there be proper and improper types of machine love affairs?

Could a machine be frustrated and suffer? Could a frustrated machine release its pent-up feelings by going outdoors and self-propelling ten miles? Could a machine learn to enjoy the sweet pain of marathon running? Could a machine with a seeming zest for life destroy itself purposefully one day, planning the entire episode so as to fool its mother machine into 'thinking' (which, of course, machines cannot do, since they are mere hunks of inorganic matter) that it had perished by accident?

These are the sorts of questions that burned in the brain of Alan Mathison Turing, the great British mathematician who spear-headed the science of computation; yet if they are read at another level, these questions also reveal highlights of Turing's troubled life. It would require someone who shares much with Turing to plumb his life story deeply enough to do it justice, and Andrew Hodges, an accomplished British mathematical physicist, has succeeded wonderfully in just that venture.

This biography of Turing, painstakingly assembled from innumerable sources, including conversations with scores of people who knew Turing at various stages of his life, provides a picture as vivid as one could hope of a most complex and intriguing individual. Turing's was a life that merits deep study, for not only was he a major player in the science of the twentieth century, but his interpersonal behavior was unconventional and caused him great grief. Even today, society as a whole has not learned how to grapple with Turing's brand of nonconformism.

Hodges's rich and engrossing portrait is not the first book about Turing; indeed, Turing's mother, Sara Turing, wrote a sketchy memoir a few years after her son's death, presenting an image of him as a lovable, eccentric boy of a man, filled with the joy of ideas and driven by an insatiable curiosity about questions concerning mind and life and mechanism. Although that little book has some merits and even some charm, it also whitewashes a great deal of the true story. Andrew Hodges explores Turing's mind, body, and soul far more deeply than Sara Turing ever dared to, for she wore conventional blinders and did not want to see, let alone say, how poorly her son fit into the standard molds of British society.

Alan Turing was homosexual – a fact that he took no particular pains to hide, especially as he grew older. For a boy growing up in the 1920s and for a grown man in the subsequent few decades, being homosexual – especially if one was British and a member of the upper classes – was an unmentionable, terrible, and mysterious affliction.

Atheist, homosexual, eccentric, marathon-running English mathematician, A. M. Turing was in large part responsible not only for the concept of computers, incisive theorems about their powers, and a clear vision of the possibility of computer minds, but also for the cracking of German ciphers during the Second World War. It is fair to say that we owe much to Alan Turing for the fact that we are not under Nazi rule today. And yet this salient figure in world history has remained, as the book's title says, an enigma.

In this biography, Andrew Hodges has painted an extraordinarily detailed and devoted portrait of a multifaceted man whose honesty and decency were too much for his society and his times, and who brought about his own downfall. Beyond the evident empathy that Hodges feels for his subject, there is another level of depth and understanding in this book, one that makes all the difference in a biography of a scientific figure: scientific accuracy and clarity. Hodges has done an admirable job of presenting to the lay reader each idea in detail, and most likely this is so because, as is obvious to a reader, he himself is passionately intrigued by all the ideas that fascinated Turing.

Alan Turing: The Enigma is a first-rate presentation of the life of a first-rate scientific mind, and given that this particular mind was attached to a body that had a mind of its own, the full story is an important document for social reasons as well. Alan Turing would probably have shuddered had he ever suspected that the tale of his personal life would one day be presented to the public at large, but he is in good hands: it is hard to imagine a more thoughtful and compassionate portrait of a human being than this one.

Douglas Hofstadter

Preface

On 25 May 2011, the President of the United States, Barack Obama, speaking to the parliament of the United Kingdom, singled out Newton, Darwin and Alan Turing as British contributors to science. Celebrity is an imperfect measure of significance, and politicians do not confer scientific status, but Obama's choice signalled that public recognition of Alan Turing had attained a level very much higher than in 1983, when this book first appeared.

Born in London on 23 June 1912, Alan Turing might just have lived to hear these words, had he not taken his own life on 7 June 1954. He perished in a very different world, and his name had gone unmentioned in its legislative forums. Yet even then, in its secret circles, over which Eisenhower and Churchill still reigned, and in which the names of NSA and GCHQ were spoken in whispers, Alan Turing had a unique place. He had been the chief backroom boy when American power overtook British in 1942, with a scientific role whose climax came on 6 June 1944, just ten years before that early death.

Alan Turing played a central part in world history. Yet it would be misleading to portray his drama as a power play, or as framed by the conventional political issues of the twentieth century. He was not political as defined by contemporary intellectuals, revolving as they did round alignment or non-alignment with the Communist party. Some of his friends and colleagues were indeed party members, but that was not his issue. (Incidentally, it is equally hard to find money-motivated 'free enterprise', idolised since the 1980s, playing any role in his story.) Rather, it was his individual freedom of mind, including his sexuality, that mattered – a question taken much more seriously in the post-1968, and even more in the post-1989, era. But beyond this, the global impact of pure science rises above all national boundaries, and the sheer timelessness of pure mathematics transcends the limitations of his twentieth-century span. When Turing returned to the prime numbers in 1950 they were unchanged from when he left them in 1939, wars and superpowers notwithstanding. As G. H. Hardy famously said, *they are so*. Such is mathematical culture, and such was his life, presenting a real difficulty to minds set in literary, artistic or political templates.

Yet it is not easy to separate transcendence from emergency: it is striking how leading scientific intellects were recruited to meet the existential threat Britain faced in 1939. The struggle with Nazi Germany called not just for scientific knowledge but the cutting edge of abstract thought, and so Turing's quiet logical preparations in 1936–38 for the war of codes and ciphers made him the most effective anti-Fascist amongst his many anti-Fascist contemporaries. The historical parallel with physics, with Turing as a figure roughly analogous to Robert Oppenheimer, is striking. This legacy of 1939 is still unresolved, in the way that secret state purposes are seamlessly woven into intellectual and scientific establishments today, a fact that is seldom remarked upon.

The same timelessness lies behind the central element of Alan Turing's story: the universal machine of 1936, which became the general-purpose digital computer in 1945. The universal machine is the focal, revolutionary idea of Turing's life, but it did not stand alone; it flowed from his having given a new and precise formulation of the old concept of algorithm, or mechanical process. He could then say with confidence that *all* algorithms, all possible mechanical processes, could be implemented on a universal machine. His formulation became known immediately as 'the Turing machine' but now it is impossible not to see Turing machines as computer programs, or software.

Nowadays it is perhaps taken rather for granted that computers can replace other machines, whether for record-keeping, photography, graphic design, printing, mail, telephony, or music, by virtue of appropriate software being written and executed. No one seems surprised that industrialised China can use just the same computers as does America. Yet that such universality is possible is far from obvious, and it was obvious to no one in the 1930s. That the technology is *digital* is not enough: to be all-purpose computers must allow for the storage and decoding of a program. That needs a certain irreducible degree of logical complexity, which can only be made to be of practical value if implemented in very fast and reliable electronics. That logic, first worked out by Alan Turing in 1936, implemented electronically in the 1940s, and nowadays embodied in microchips, is the mathematical idea of the universal machine.

In the 1930s only a very small club of mathematical logicians could appreciate Turing's ideas. But amongst these, only Turing himself had the practical urge as well, capable of turning his hand from the 1936 purity of definition to the software engineering of 1946: 'every known process has got to be translated into instruction table form...' (p. 409). Donald Davies, one of Turing's 1946 colleagues, later developed such instruction tables (as Turing called programs) for 'packet switching' and these grew into the Internet protocols. Giants of the computer industry did not see the Internet coming, but they were saved by Turing's universality: the computers of the 1980s did not need to be reinvented to handle these new tasks. They needed new software and peripheral devices, they needed greater speed and storage, but the fundamental principle remained. That principle might be described as the law of information technology: all mechanical processes, however ridiculous, evil, petty, wasteful or pointless, can be put on a computer. As such, it goes back to Alan Turing in 1936.

That Alan Turing's name has not from the start been consistently associated with praise or blame for this technological revolution is due partly to his lack of effective publication in the 1940s. Science absorbs and overtakes individuals, especially in mathematics, and Alan Turing swam in this anonymising culture, never trying to make his name, although frustrated at not being taken seriously. In fact, his competitive spirit went instead into marathon running at near-Olympic level. He omitted to write that monograph on 'the theory and practice of computing', which would have stamped his name on the emergent post-war computer world. In 2000 the leading mathematical logician Martin Davis, whose work since 1949 had greatly developed Turing's theory of computability, published a book[1] which was in essence just what Turing could have written in 1948, explaining the origin of the universal machine of 1936, showing how it became the stored-program computer of 1945, and making it clear that John von Neumann must have learnt from Turing's 1936 work in formulating his better-known plan. Turing's very last publication, the *Science News* article of 1954 on computability, demonstrates how ably he could have written such an analysis. But even there, on terrain that was incontestably his own discovery, he omitted to mention his own leading part.

Online search engines, which work with such astonishing speed and power, are algorithms, and so equivalent to Turing machines. They are also descendants of the particular algorithms, using sophisticated logic, statistics and parallel processing, that Turing expertly pioneered for Enigma-breaking. These were search engines for the keys to the Reich. But he asked for, and received, very little public credit for what has subsequently proved an all-conquering discovery: that all algorithms can be programmed systematically, and implemented on a universal machine. Instead, he nailed his colours to the mast of what he called 'intelligent machinery', but which came to be called Artificial Intelligence after 1956. This far more ambitious and contentious research programme has *not* developed as Turing hoped, at least as yet. Why did Turing go so public on AI, and make so little of himself as an established maestro of algorithms and the founder of programming? Partly because AI was for him the really fundamental scientific question. The puzzle of mind and matter was the question that drove him most deeply. But to some extent he must have been a victim of his own suppressed success. The fact that he knew so much of the algorithms of the secret war, and that the war had made the vital link between logic and electronics, cramped his style and constrained his communication. In his 1946 report his guarded allusion to the importance of cryptographic algorithms (p. 418) reflects an inhibition that must have infected all that came later.

Only after thirty years did the scale and depth of wartime cryptanalysis at Bletchley Park begin to leak out, allowing a serious assessment of Alan Turing's life to be attempted. This point coincided with the break-out of cryptology theory into an expanding computer science, with a reassessment of the Second World War in general, and with the impact of 1970s sexual liberation. The 1968 social revolution, which Turing anticipated, had to happen before his story could be liberated. (Even so, the change in UK vetting and military law came only in the 1990s, and a legal principle of equality was not established until 2000. 'Don't ask don't tell' ended only in 2011, showing how the issues of chapter 8 have remained literally unspeakable in the US military.) Alan Turing's story shows the first elements of this liberating process in the Norway of 1952, since the men-only dances he heard about (p. 599) were probably organised by the fledgling Scandinavian gay organisation. In addition to the gay-themed novels mentioned on p. 613, Norman Routledge recalled in 1992 how Turing expected him to read André Gide in French. One regret, voiced in note 8.31, is that his letters to Lyn Newman did not survive. Their content can be guessed from what in 1957 she wrote to a friend: 'Dear Alan, I remember his saying to me so simply & sadly "I just can't believe it's as nice to go to bed with a girl as with a boy" and all I could say was "I entirely agree with you – I also <u>much</u> prefer boys."' This interchange, then confined to a discreet privileged circle, could now be a TV chat show joke, with a happy resonance of the repartee of his famous imitation game. But Alan Turing's simple openness came decades too early.

It is not difficult to imagine the hostility and stigma of those days, for such hatred and fear is still, whether in Africa, the Middle East or the United States, a major cultural and political force. It is harder now to imagine a world where persecution was not just asserted but taken as an unquestionable axiom. Alan Turing faced the impossible irony that his demand for honesty ran up against the two things, state security and homosexuality, which were the most fraught questions of the 1950s. It is not surprising that it proved impossible to contain them in a single brain. His death left a jagged edge in history, something no one (with the extraordinary exception of his mother) wanted to talk about. My fusion of these elements into a single narrative certainly encountered criticism in 1983. But *nous avons changé tout cela:* since then, his life and death have been as celebrated as those of any scientific figure. Hugh Whitemore's play *Breaking the Code,* based on this book and featuring leading performers, pushed at the envelope of public acceptability. It made Alan Turing's life a popular story in 1986, reinforced by a television version in 1997. By that time the Internet had transformed personal openness. In a curious way, Turing had anticipated this use of his technology, already hinted at in the risqué text-messaging of his imitation game. The love letters created by the Manchester computer (p. 601), and his message about the Norwegian youth, rendered as a nerdy computer printout (p. 608), suggest a Turing who would have relished the opportunity for electronic communication with like-minded people.

In 2009 the British prime minister, Gordon Brown, made a statement of apology for Turing's trial and punishment in 1952–54, framed by a wider vision of how the values of post-war European civil society had been won with his secret help. This statement was enlisted through a popular web-based petition, something impossible in 1983, but already then being mooted as the sort of thing the 'mighty micro' could bring about. My own comments (p. 608) in the concluding Author's Note about future revision of printed text reflected this mood. And indeed from 1995 onwards my website has supplied updating material. In this light it is surprising that such a long volume has remained continuously in print since 1983. But perhaps one thing that a traditional stack of paper still makes possible is an immersion in storytelling, and this time-consuming experience was one I certainly supplied.

As narrator I adopted a standpoint of a periscope looking just a little ahead of Alan Turing's submerged voyage, punctuated by just a few isolated moments of prophecy. The book bears in mind that what is now the past, the 1940s and 1950s, was once the completely unknown future. This policy required an unwarranted confidence that readers would wade through the pettier details of Alan Turing's family origins and early life, before being offered any reason for supposing this life had any significance. But it has had the happy outcome that the text has not dated as do texts resting on assertions about 'what we know now'. So although so much has changed, the story that follows can be read without having to subtract 1983-era comment. (Of course, this is not true of the Notes, which now show what sources were available in 1983, but do not indicate a guide to 'further reading'.)

After a further thirty years, how would I reassess Alan Turing's pure scientific work and its significance? My book made no attempt to trace the legacy of Turing's work after 1954; that would be far too large a task. But naturally, the expansion of scientific discovery continually forces fresh appraisals of Turing's achievement. His morphogenesis theory, since 2000 more actively pursued as a physico-chemical mechanism, would now require more material on the various different approaches and models. As another example, Turing's strategy of combining top-down and bottom-up approaches to AI, and the neural nets he sketched in 1948, have acquired new significance. There has been a parallel explosion in quality and quantity of the history of science and technology since the 1970s, with many detailed studies of Turing's papers. The centenary year of 2012 saw a climax of new analysis from leading scientific figures. Alan Turing's work is accessible as never before, and topics that attracted scant attention in 1983 are now the subject of lively debate.

But I would not take a radically different point of view. My division of the book into Logical and Physical was already radical, reflecting a rejection of conventional description of him as a pure logician, and portraying him as always, and increasingly, involved in the nature of the physical world. This fundamental perception could now be asserted with even greater confidence. He came to the ideas of 1936 with an unusual knowledge of quantum mechanics, and this is now a more interesting connection, for since the mid-1980s quantum computing and quantum cryptography have become important extensions of Turing's ideas. Likewise, the renewed interest in quantum mechanics in Turing's last year, whose significance was correctly signalled with a supersized footnote (p. 645) could now be linked more closely with his 1950 and 1951 arguments about computers and minds. These issues have arisen sharply since 1989, when Roger Penrose[2] discussed the significance for minds of the uncomputable numbers Turing had discovered. Penrose himself suggested an answer which related Turing machines to a radical new view of quantum mechanics. Writing now, I would draw more attention to what is now called the physical Church–Turing thesis. Did Turing consider that the scope of the computable includes everything that can be done by any physical object? And what would this mean for his philosophy of the mind? In this light, Church's 1937 review (p. 157) of Turing's work has more importance than I noted. Turing's decisive shift of focus to what *could* be done by algorithms, as stated on p. 138, I would now move from 1936 to 1941 (at p. 266). Turing's argument about infallibility (p. 454) would deserve more analysis, as also his use of 'random' elements, and a number of general statements about thinking and doing in my text. But sharper sensitivity to these questions would bring out few if any new answers; it would only make more acute the questions about what Turing really thought.

Much more positive detail could now be given regarding his secret wartime work. Even in the 1992 preface to the Vintage edition, new material could be given from the third volume of

F. H. Hinsley's official history of British Intelligence. But since the mid-1990s, raw American and British documents on Second World War cryptanalysis have been officially released, and it has been possible to elucidate the internal story with far more details than Hinsley allowed. What has emerged has only enhanced the quality and significance of Bletchley Park work, and of Turing as its chief scientific figure. The park itself is now a famous visitor attraction, though its lesson, that reason and scientific methods were the heroes of the hour, has not really caught on.

These documents show how on 1 November 1939 Turing could announce 'the machine now being made at Letchworth, resembling, but far larger than the Bombe of the Poles (superbombe machine)'. That prefix 'super' dramatised the advance that my explanation (p. 229) was unable, for lack of supporting narrative detail, to highlight as the crucial breakthrough. Turing's own 1940 report on the Enigma-breaking methods clarified how he made this advance, called 'parallel scanning'. All of this is now working physically in the rebuilt Bombe at the Bletchley Park Museum. In addition to the document release, members of the original cryptanalytic team have written fully about the technical work, such as the details of the bigram tables which made the Naval Enigma so much more challenging, and the statistical Banburismus method. The super-fast bombes, the break into the Lorenz cipher, and the now-famous Colossus are all open to study, a great deal being due to the inspiring work of the late Tony Sale. The description in this book is now unnecessarily hazy. On the other hand, there was no room for any more codebreaking technicalities in the book, and the reader will not be seriously misled by its summary.

In particular, these revelations have only reinforced the significance of the 'Bridge Passage' between the logical and the physical, Turing's top-level liaison visit to the United States in the winter of 1942–43. His report of 28 November 1942 from Washington, now released, documents the difficult and anomalous position he faced, including an initial confinement to Ellis Island (p. 305). He was not overawed by the US Navy: 'I am persuaded that one cannot very well trust these people where a matter of judgment in cryptography is concerned.' Something that I had heard only as rumour in 1983 has been confirmed: on 21 December a train brought Turing to Dayton, Ohio, where the US Bombes were under construction. There is also more revealed on his initiation into the most secret US speech encipherment technology. There is more on his response to it, the Delilah speech scrambler – an interim report dated 6 June 1944, and a later complete description. As a precursor of the mobile phone, this belongs to the future, whilst the Enigma was a mediocre adaptation of 1920s mechanical engineering. This new material only underlines that in the post-war period, Turing had a unique knowledge of the most advanced American technology, as it emerged from victory in 1945.

This fact draws further attention to the question of what he did for GCHQ after 1948. In the 1992 preface I floated the suggestion that this might have been connected with the now famous Venona problem of Soviet messages. But there has been no comparable release of GCHQ or other secret documents on 1948–54, which might indicate the nature of his work. Richard Aldrich's recent history of GCHQ[3] opened by saying that 'Today it is more important than ever – yet we know almost nothing about it.' We know more now, from Edward Snowden, about the work whose foundation stone was laid by Alan Turing. No one could miss seeing how much it has to do with the power of the universal machine. And it is hard to believe that Turing played no part in giving secret advice about the potential of computing in the early days of the Cold War. As I wrote in the 1992 preface, who else could have done this?

Answers to such questions were strikingly absent from the British government statement of 24 December 2013 when, in response to a demand from illustrious figures, a posthumous Royal Pardon was granted to Turing in respect of the conviction for 'gross indecency' of 31 March 1952. There must have been memoranda within government detailing the prospect of their top scientific consultant going on public trial for a crime which rang all the alarm bells of Security. Hugh Alexander must have reported on his participation in the trial. As observed in note 8.17 the question arises as to whether the Foreign Office had an influence on Turing being treated with hormones (then seen as a soft option) instead of going to prison. But no such papers appeared; nor indeed were they asked for.

The pardon captured the public imagination and was received with elation as a fairy-story Christmas present. But its principle was less elevated: it conceded the plea made in Turing's defence at his trial in 1952, that he was a national asset. It was an act of pulling rank which, in 1952, Hugh Alexander's advocacy had not been able to achieve, but sixty years later, dressed up with the formal magic of monarchy, was widely applauded.

Queen Elizabeth's reign had begun with Turing's arrest, giving extra piquancy to the medieval language of the Royal Pardon, but those unfamiliar with the decorative aspects of the British constitution should appreciate that this was simply an executive action by the government. Its language recognised Turing as exceptional in his service of the state, and thus re-emphasised the central question of Turing's relationship with that state. But even by 1954, the state that really counted was on the other side of the Atlantic. What did American authorities know of what transpired in 1951–52, and how did they react, given that Turing had explicitly been granted special access to US secrets? Was he vetted under US-inspired rules in 1948? Did he knowingly flout the new rules when he started cruising the Manchester milk bar in 1950? Did British authorities convey the reality of Turing's sex tourism in Europe in 1952–53? What demands, threats and surveillance did Turing have to deal with as a result? None of this was addressed.

The petitioners for the pardon explicitly said Turing was *sui generis*, and that a pardon would create no precedent to apply to anyone else. It was granted on that exceptional basis. So it left Arnold Murray, who was charged in exactly the same way as Turing, unpardoned: no reference was even made to whether he was still alive (he was not). Readers of this book will see that Turing himself took great interest in the background and character of this vulnerable young man, writing a short story based on his breaching of class barriers. It is hard to believe that Turing would have been high-minded enough actually to object if the trial had been stopped on the grounds of his rank, and the whole thing hushed up. On the other hand, it is also hard to believe that he would rejoice in an exception being made for himself while so oppressive a law was enforced on thousands of others.

In 1950, Turing had written a description of what would now be called 'the butterfly effect', ending with a man being killed by an avalanche, and when writing his short story, the events of 1951–52 might have appeared to him in just this light. We now have a further glimpse of the chance events which precipitated the crisis. Amidst that scene on the Oxford Road, as described on p. 540, was an eighteen-year-old lad on weekend leave from the Navy. He recognised and greeted Alan Turing in the milk bar – not as a mathematician, but as a champion amateur runner. This young man, Alan Edwards, later noticed Turing connect with Arnold Murray. An athlete himself, of keen intelligence, and very clear about his homosexual identity, Alan Edwards would have made a far more suitable boy. But, by the cussedness of human nature, Alan Turing was not his type. Not because he was too old, but because he was too similar, being lithe and fit.

Another witness has emerged to the importance of running in Turing's life, even after his competitive days were over. This is Alan Garner, famous for *The Owl Service* (1967). In 2011 he told a story that he alone knew. He had been Alan Turing's training partner; they had run perhaps a thousand miles together through Cheshire country lanes in 1951–52. Garner was seventeen in 1951, and a sixth-former at Manchester Grammar School, studying classics. The meeting arose in that year, as fellow athletes spotting each other on the road. From the start, Garner felt himself treated as an equal, something he could appreciate and cope with because of this school's distinctive ambience (a culture that yet another Alan has evoked in *The History Boys*). He was also just about to become a serious competitive young sprinter. Their disparate long- and short-distance strengths were compatible with an equal pace over a run of several miles. Equality also was found in banter full of word play and scurrilous humour. It came as no surprise to Garner when Turing asked him if he thought intelligent machinery was possible. After running silently for ten minutes, along Mottram Road, Alderley Edge, he said no. Turing did not argue. 'Why learn classical languages?' Turing asked, and Garner said, 'You have to learn to use your brain in a different way': the kind of answer that Turing would have appreciated.

Their chat kept away from the personal: it was focused on sustaining the six or seven miles of running. But once, probably late in 1951, Turing mentioned the story of Snow White. 'You too!' said Garner, amazed. For he connected immediately with a singular event from his childhood. It was his first cinema outing when five years old. *Snow White and the Seven Dwarfs* had terrified him with the image of the poisoned apple. Turing responded with immediate empathy. 'He used to go over the scene in detail, dwelling on the ambiguity of the apple, red on one side, green on the other, one of which gave death.' Their shared trauma – as Garner saw it – remained a bond.

The training extended into 1952 and overlapped with the period of Turing's trial. Turing never spoke of what he was undergoing and somehow Garner only heard the news late in 1952, when he was warned by the police not to associate with Turing. Garner was very angry at this, and at what he learnt had happened, and he never had the least sense of having been approached in any predatory way. And yet, inevitably, it ended sadly. Alan Garner painfully recalls seeing Turing for the last time in 1953, as a fellow passenger on the bus from Wilmslow to Manchester. Being with his girlfriend, Garner found it too difficult to say anything appropriate and so he pretended not to have noticed his presence. This incident, so redolent of the fiction and film of final teenage years, was soon followed by Garner's departure to National Service, during which he heard of Turing's death. Alan Garner revealed nothing of this for sixty years.[4]

Alan Turing would naturally have delighted to see a lad from a very ordinary Cheshire village background showing such curiosity and intellectual ambition. But it is as though he also saw something extra in Garner, sensing a writer of the future who would combine modernity and mythology. The story of the apple is like a glimpse of the Jungian analysis he went in for after 1952, of which we know virtually nothing. It is also striking to know that when Turing saw *Snow White* at its Cambridge release in 1938 (p. 189) a five-year-old boy was reacting in parallel, and one day would share it. The year 1938 was Turing's year of choice: he chose to return from America and chose active engagement with war rather than pure mathematics. He accepted secrecy and the death of innocence. That the apple had already figured in a suicide plan (p. 164) must have made the film scene an intense (and as Garner saw it, traumatic) image. His analyst, Franz Greenbaum, might have been the perfect confidant in working out such conflicts, but state secrecy would have made it impossible for Turing to convey the true seriousness of his situation. His total isolation in 1954 is virtually impossible to conceive of in today's world.

It will be surprising if any more such witnesses emerge, but further personal documents do exist and may eventually become available. Meanwhile, this Preface ends with a few gems of writing which surfaced too late for the 1983 book, but were included in the 1992 preface. They are given again here.

A cosy continuity between King's College, Cambridge, and the pre-war codebreaking establishment is evoked by some brief letters placed in the King's archive in 1990. 'Dilly Knox, who is my boss, sends you greetings,' wrote Turing on 14 September 1939 to the Provost, John Sheppard. 'It is always a joy to have you here,' wrote back the Provost, encouraging him to visit. The economist J. M. Keynes, who looked after the question of Turing's fellowship for the duration of the war, also knew the older generation of codebreakers (and indeed had apparently enjoyed an intimate relation with the 'boss'). These connections lend further colour to my description (p. 148) of how in 1938 Turing's interest in ciphers could have been transmitted to the British government, thus making possible his fateful appointment.

The following account, which in 1983 was only available in Polish, also concerns the early months of the war.[5] It settles the question raised in note 4.10 as to whether Turing was the personal emissary who took the new perforated sheets to the Polish and French cryptanalysts. Indeed he was: there is no mistaking his voice in this account of their farewell supper.

> In a cosy restaurant outside Paris staffed by Deuxième Bureau workers, the cryptologists and the chiefs of the secret decryptment centre, Bertrand and Langer, wished to spend an evening in a casual atmosphere free of everyday concerns. Before the dishes ordered and the choice wine selected for the occasion had been served, the attention of the diners was drawn to a crystal flower glass with flowers, placed on the middle of the tablecloth. They were delicate rosy-lilac flowers with slender, funnel-shaped calyces. It was probably Langer who uttered their German and then their Polish names: 'Herbstzeitlose ... Zimowity jesienne ...'

> This meant nothing to Turing, as he gazed in silence at the flowers and the dry lanceolate leaves. He was brought back from his reverie, however, by the Latin name, *Colchicum autumnale* (autumn crocus, or meadow saffron), spoken by mathematician-geographer Jerzy Różycki.

> 'Why, that's a powerful poison!' said Turing in a raised voice.

> To which Różycki slowly, as though weighing each word, added: 'It would suffice to bite into and suck at a couple of stalks in order to attain eternity.'

For a moment there was an awkward silence. Soon, however, the crocuses and the treacherous beauty of the autumnal flowers were forgotten, and an animated discussion began at the richly laid table. But despite the earnest intention of the participants not to raise professional questions, it proved impossible to get completely away from Enigma. Once again, there was talk of the errors committed by German operators and of the perforated sheets, now machine- rather than handmade, which the British sent in series from Bletchley to the Poles working at Gretz-Armainvillers, outside Paris. The inventor of the perforated sheets, Zygalski, wondered why their measurements were so peculiar, with each little square being about eight and a half millimetres on a side.

'That's perfectly obvious,' laughed Alan Turing. 'It's simply one-third of an inch!'

This remark in turn gave rise to a dispute as to which system of measures and currency, the traditionally chaotic British one or the lucid decimal system used in France and Poland, could be regarded as the more logical and convenient. Turing jocularly and eloquently defended the former. What other currency in the world was as admirably divided as the pound sterling, composed of 240 pence (20 shillings, each containing 12 pence)? It alone enabled three, four, five, six or eight persons to precisely, to the penny, split a tab (with tip, generally rounded off to a full pound) at a restaurant or pub.

The dark tone of Turing's knowledge of poisonous plants, arising unexpectedly in the midst of secret work and mathematical banter, recalls the manner of his death. The shock of that event is vividly portrayed by another first-hand account, that written by Turing's housekeeper Mrs Clayton on the night of Tuesday 8 June 1954:

My dear Mrs Turing

You will by now have heard of the death of Mr Alan. It was such an awfull shock. I just didn't know what to do. So I flew across to Mrs Gibson's and she rang Police & they wouldn't let me touch or do a thing & I just couldn't remember your address. I had been away for the weekend and went up tonight as usual to get his meal. Saw his bedroom light on the lounge curtains not drawn back, milk on steps & paper in door. So I thought he'd gone out early & forgot to put his light off so I went & knocked at his bedroom door. Got no answer so walked in, Saw him in bed he must have died during the night. The police have been up here, again tonight for me to make a statement & I understand the inquest will be Thursday. Shall you or Mr [John] Turing be coming over[?] I feel so helpless & not able to do anything. The Webbs removed last Wed. & I don't know their new address yet. Mr & Mrs Gibson saw Mr Alan out walking Mon. evening he was perfectly all right then. The weekend before he'd had Mr Gandy over for the weekend & they seemed to have had a really good time. The Mr & Mrs Webb came to dinner Tues. & Mrs Webb had aftern[oon] tea with him Wed. the day she removed.

You do know you have my very deepest sympathy in your great loss & what I can do to help at this end you know I will continue to do so.

Yours respectfully, S. Clayton

This account indicates how the police took charge of the house immediately, leaving open the possibility that there was information in official hands not made public at the inquest. It is now in the archive at King's College.

The police also feature in two valuable letters written by Alan Turing himself to his friend Norman Routledge, and now also in the archive. The first, undated, must be from early 1952:

My dear Norman,

I don't think I really do know much about jobs, except the one I had during the war, and that certainly did not involve any travelling. I think they do take on conscripts. It certainly involved a good deal of hard thinking, but whether you'd be interested I don't know. Philip Hall was in the same racket and on the whole, I should say, he didn't care for it. However I am not at present in a state in which I am able to concentrate well, for reasons explained in next paragraph.

I've now got myself into the kind of trouble that I have always considered to be quite a possibility for me, though I have usually rated it at about 10:1 against. I shall shortly be pleading guilty to a charge of sexual offences with a young man. The story of how it all came to be found out is a long and fascinating one, which I shall have to make into a short story one day, but haven't the time to tell you now. No doubt I shall emerge from it all a different man, but quite who I've not found out.

Glad you enjoyed broadcast. J[efferson] certanly was rather disappointing though. I'm rather afraid that the following syllogism may be used by some in the future

> Turing believes machines think
> Turing lies with men
> Therefore machines do not think

Yours in distress, Alan.

The allusion to the traditional syllogism about Socrates, who drank the hemlock, is an extraordinary piece of black humour. (It also stands as a superb example of how Turing himself fused the elements of his life.) The opening of the letter is perhaps equally remarkable in its absurdly off-hand description of six years of crucial wartime work, and in its inexplicable statement that the work had not involved any travelling.

The second is dated February 22, and must be from 1953:

My dear Norman

Thanks for your letter. I should have answered it earlier.

I have a delightful story to tell you of my adventurous life when next we meet. I've had another round with the gendarmes, and it's positively round II to Turing. Half the police of N. England (by one report) were out searching for a supposed boyfriend of mine. It was all a mare's nest.

Perfect virtue and chastity had governed all our proceedings. But the poor sweeties never knew this. A very light kiss beneath a foreign flag under the influence of drink, was all that had ever occurred. Everything is now cosy again except that the poor boy has had rather a raw deal I think. I'll tell you all when we meet in March at Teddington. Being on probation my shining virtue was terrific, and had to be. If I had so much as parked my bicycle on the wrong side of the road there might have been 12 years for me. Of course the police are going to be a bit more nosy, so virtue must continue to shine.

I might try to get a job in France. But I've also been having psychoanalysis for a few months now, and it seems to be working a bit. It's quite fun, and I think I've got a good man. 80% of the time we are working out the significance of my dreams. No time to write about logic now!

Ever, Alan

The style is a reminder that whilst Turing's plain-speaking English might be compared with that of Orwell or Shaw, it also had a strong element of P. G. Wodehouse. Both letters perhaps indicate a state of denial about the seriousness with which those in charge of the nosy 'sweeties' would contemplate his Euro-adventures.

Alan Turing used logarithms of betting odds as the key to the work he had done for the 'racket' of cryptography, and his sustained fascination with probability is illustrated by that reference to a one-in-ten chance of being caught. In his 1953 stoic humour there is a link with innocent Anti-War undergraduate days of twenty years earlier, when he analysed Alfred Beuttell's Monte Carlo gambling system. While the tectonic forces of geopolitics ground away, Alan Turing dodged his way through as a nimble, insouciant, individual. The lucky streak did not last for ever.

As well as supplying *addenda,* this Preface must also confess to *corrigenda.* Inevitably, a number of errors are perpetuated by reprinting this text. Here are some examples. Note 2.11 severely understates the significance of Turing's work on normal numbers and of his friend David Champernowne's 1933 contribution. It seems possible that Turing's study of such infinite decimals suggested his model of 'computable numbers'. Note 3.40 on Turing's work on the Skewes number is inaccurate: his incomplete manuscript actually dates from about 1950 when he briefly resumed this work, and corresponded briefly with Skewes. Audrey *née* Bates (p. 505) did more interesting and substantial work than is suggested; her Master's thesis involved representing Church's lambda calculus on the Manchester computer, an advanced idea which was never published. This sharpens the point made on the footnote on that page, concerning how Turing failed to turn his vision for programming and logic into the creation of a lively school of research and innovation. One clue to the problems he faced comes from her recollection that 'Max Newman made the immortal statement that "there is nothing to do with computers that merits a Ph.D."' Further additional and corrective material may be found on www.turing.org.uk.

The curious cocktail of topics in this Preface is also offered as an aperitif for the story itself, inviting the reader to travel back over more than a century, and to enter the world of 1911. In making that journey as author, I had the peculiar experience of living a previous life. The strangeness is now doubled since I am as far now removed from that Reagan era as it was from Eisenhower's. The landscape has changed: the science-fiction *2001* in my Author's

Note has become well-trodden history, and Turing's scientific 'least waste of energy' now has a more urgent meaning. But the Victorian roots I drew upon, one English, one American, need no revision or apology. I chose a setting with the binary classicism of Lewis Carroll's mathematical chessboard, on which Alan Turing was the pawn. But I also imbued it with Whitman's romance of the 'history of the future'. These dreams from the nineteenth century still speak to the crimes and follies of the twenty-first.

1 Martin Davis, *The Universal Computer* (Norton, 2000).

2 Roger Penrose, *The Emperor's New Mind* (OUP, 1989).

3 Richard J. Aldrich, *GCHQ: The Uncensored Story of Britain's Most Secret Intelligence Agency* (Harper, 2010).

4 *The Observer*, 11 November 2011: 'My Hero, Alan Turing'.

5 W. Kozaczuk, tr. C. Kasparek, *Enigma . . .* (Arms and Armour Press, 1984). The original Polish text was published in Warsaw, 1979.

PART ONE

The Logical

1 Esprit de Corps

Beginning my studies the first step pleas'd me so much,
The mere fact consciousness, these forms, the power of motion,
The least insect or animal, the senses, eyesight, love,
The first step I say awed me and pleas'd me so much,
I have hardly gone and hardly wish'd to go any farther,
But stop and loiter all the time to sing it in ecstatic songs.

A son of the British Empire, Alan Turing's social origins lay just on the borderline between the landed gentry and the commercial classes. As merchants, soldiers and clergymen, his ancestors had been gentlemen, but not of the settled kind. Many of them had made their way through the expansion of British interests throughout the world.

The Turings could be traced back to Turins of Foveran, Aberdeenshire, in the fourteenth century. There was a baronetcy in the family, created in about 1638 for a John Turing, who left Scotland for England. *Audentes Fortuna Juvat* (Fortune Helps the Daring) was the motto of the Turings, but however brave, they were never very lucky. Sir John Turing backed the losing side in the English civil war, while Foveran was sacked by the Covenanters. Denied compensation after the Restoration, the Turings languished in obscurity during the eighteenth century, as the family history, the *Lay of the Turings*[1], was to describe:

Walter, and James and John have known,
Not the vain honours of a crown,
    But calm and peaceful life –
Life, brightened by the hallowing store
Derived from pure religion's lore!
And thus their quiet days passed by;
And Foveran's honours dormant lie,

Till good Sir Robert pleads his claim
To give once more the line to fame:
Banff's castled towers ring loud and high
To kindly hospitality
And thronging friends around his board
Rejoice in TURING's line restored!

Sir Robert Turing brought back a fortune from India in 1792 and revived the title. But he, and all the senior branches of the family, died off without male heirs, and by 1911 there were but three small clusters of Turings in the world. The baronetcy was held by the 84-year-old eighth baronet, who had been British Consul in Rotterdam. Then there was his brother, and his descendants, who formed a Dutch branch of Turings. The junior branch consisted of the descendants of their cousin, John Robert Turing, who was Alan's grandfather.

John Robert Turing took a degree in mathematics at Trinity College, Cambridge, in 1848, and was placed eleventh in rank, but abandoned mathematics for ordination and a Cambridge curacy. In 1861 he married nineteen year old Fanny Boyd and left Cambridge for a living in Nottinghamshire, where he fathered ten children. Two died in infancy and the surviving four girls and four boys were brought up in a regime of respectable poverty on a clerical stipend. Soon after the birth of his youngest son, John Robert suffered a stroke and resigned his living. He died in 1883.

As his widow was an invalid, the care of the family fell upon the eldest sister Jean, who ruled with a rod of iron. The family had moved to Bedford to take advantage of its grammar school, where the two elder boys were educated. Jean started her own school, and two of the other sisters went out as schoolteachers, and generally sacrificed themselves for the sake of advancing the boys. The eldest son, Arthur, was another Turing whom fortune did not help: he was commissioned in the Indian Army, but was ambushed and killed on the North-West Frontier in 1899. The third son Harvey[2] emigrated to Canada, and took up engineering, though he was to return for the First World War and then turn to genteel journalism, becoming editor of the *Salmon and Trout Magazine* and fishing editor of *The Field*. The fourth son Alick became a solicitor. Of the daughters only Jean was to marry: her husband was Sir Herbert Trustram Eve, a Bedford estate agent who became the foremost rating surveyor of his day. The formidable Lady Eve, Alan's Aunt Jean, became a moving spirit of the London County Council Parks Committee. Of the three unmarried aunts, kindly Sybil became a Deaconess and took the Gospel to the obstinate subjects of the Raj. And true to this Victorian story, Alan's grandmother Fanny Turing succumbed to consumption in 1902.

Julius Mathison Turing, Alan's father, was the second son, born on 9 November 1873. Devoid of his father's mathematical ability, he was an able student of literature and history, and won a scholarship to Corpus Christi College, Oxford, from where he graduated with a BA in 1894. He never forgot his early life of enforced economy, and typically never paid the 'farcical' three guineas to convert the BA into an MA. But he never spoke of the miseries of his childhood, too proud to moan of what he had left behind and risen above, for his life as a young man was a model of success. He entered for the Indian Civil Service, which had been thrown open to entry by competitive examination in the great liberal reform of 1853, and which enjoyed a reputation surpassing even that of the Foreign Office. He was placed seventh out of 154 in the open examination of August 1895.[3] His studies of the various branches of Indian law, the Tamil language and the history of British India then won him seventh place again in the Final ICS examination of 1896.

He was posted to the administration of the Presidency of Madras, which included most of southern India, reporting for duty on 7 December 1896, the senior in rank of seven new recruits to that province. British India had changed since Sir Robert left it in 1792. Fortune no longer helped the daring; fortune awaited the civil servant who could endure the climate for forty years. And while (as a contemporary writer put it) the district officer was 'glad of every opportunity to cultivate intercourse with the natives', the Victorian reforms had ensured that 'the doubtful alliances which in old days assisted our countrymen to learn the languages' were 'no longer tolerated by morality and society.' The Empire had become respectable.

With the help of a £100 loan from a family friend he bought his pony and saddlery, and was sent off into the interior. For ten years he served in the districts of Bellary, Kurnool and Vizigapatam as Assistant Collector and Magistrate. There he rode from village to village, reporting upon agriculture, sanitation, irrigation, vaccination, auditing accounts, and overseeing the native magistracy. He added the Telugu language to his repertoire, and became Head Assistant Collector in 1906. In April 1907 he made a first return to England. It was the traditional point for the rising man, after a decade of lonely labour, to seek a wife. It was on the voyage home that he met Ethel Stoney.

Alan's mother[4] was also the product of generations of empire-builders, being descended from a Yorkshireman, Thomas Stoney (1675–1726) who as a young man acquired lands in England's oldest colony after the 1688 revolution, and who became one of the Protestant landowners of Catholic Ireland. His estates in Tipperary passed down to his great-great-grandson Thomas George Stoney (1808-1886), who had five sons, the eldest inheriting the lands and the rest dispersing to various parts of the expanding empire. The third son was a hydraulic engineer, who designed sluices for the Thames, the Manchester Ship Canal and the Nile; the fifth emigrated to New Zealand, and the fourth, Edward Waller Stoney (1844–1931), Alan's maternal grandfather, went to India as an engineer. There he amassed a considerable fortune, becoming chief engineer of the Madras and Southern Mahratta Railway, responsible for the construction of the Tangabudra bridge, and the invention of Stoney's Patent Silent Punkah-Wheel.

A hard-headed, grumpy man, Edward Stoney married Sarah Crawford from another Anglo-Irish family, and they had two sons and two daughters. Of these, Richard followed his father as an engineer in India, Edward Crawford was a Major in the Royal Army Medical Corps, and Evelyn married an Anglo-Irish Major Kirwan of the Indian Army. Alan's mother, Ethel Sara Stoney, was born at Podanur, Madras, on 18 November 1881.

Although the Stoney family did not lack for funds, her early life was as grim as that of Julius Turing. All four Stoney children were sent back to Ireland to be educated. It was a pattern familiar to British India, whose children's loveless lives were part of the price of the Empire. They were landed upon their uncle William Crawford, a bank manager of County Clare, with two children of his own by a first marriage and four by a second. It was not a place for affection or attention. The

Crawfords moved to Dublin in 1891, where Ethel dutifully went to school each day on the horsebus, crushed by a regimen that permitted her a mean threepence for lunch. At seventeen, she was transferred to Cheltenham Ladies College, 'to get rid of her brogue,' and there she endured the legendary Miss Beale and Miss Buss, and the indignity of being the Irish product of the railway and the bank among the offspring of the English gentry. There remained a flickering dream of culture and freedom in Ethel Stoney's heart and for six months she was sent, at her own request, to study music and art at the Sorbonne. The brief experiment was vitiated by the discovery that French snobbery and Grundyism could equal that of the British Isles. So when in 1900 she returned with her elder sister Evie to her parents' grand bungalow in Coonoor, it was to an India which represented an end to petty privation, but left her knowing that there was a world of knowledge from which she had been forever excluded.

For seven years, Ethel and Evie led the life of young ladies of Coonoor – driving out in carriages to leave visiting cards, painting in water-colours, appearing in amateur theatricals and attending formal dinners and balls in the lavish and stifling manner of the day. Once her father took the family on holiday to Kashmir, where Ethel fell in love with a missionary doctor, and he with her. But the match was forbidden, for the missionary had no money. Duty triumphed over love, and she remained in the marriage market. And thus the scene was set, in the spring of 1907, for the meeting of Julius Turing and Ethel Stoney on board the homebound ship.

They had taken the Pacific route, and the romance was under way before they reached Japan. Here Julius took her out to dinner and wickedly instructed the Japanese waiter to 'bring beer and keep on bringing beer until I tell you to stop.' Though an abstemious man, he knew when to live it up. He made a formal proposal to Edward Stoney for Ethel's hand, and this time, he being a proud, impressive young man in the 'heaven-born' ICS, it was successful. The beer story, however, did not impress his future father-in-law, who lectured Ethel upon the prospect of life with a reckless drunkard. Together they crossed the Pacific and the United States, where they spent some time in the Yellowstone National Park, shocked by the familiarity of the young American guide. The wedding took place on 1 October 1907 in Dublin. (There remained a certain edge to the relationship between Mr Turing and the commercial Mr Stoney, with an argument over who was to pay for the wedding carpet rankling for years.) In January 1908 they returned to India, and their first child John was born on 1 September at the Stoney bungalow at Coonoor. Mr Turing's postings then took them on long travels around Madras: to Parvatipuram, Vizigapatam, Anantapur, Bezwada, Chicacole, Kurnool and Chatrapur, where they arrived in March 1911.

It was at Chatrapur, in the autumn of 1911, that their second son, the future Alan Turing, was conceived. At this obscure imperial station, a port on the eastern coast, the first cells divided, broke their symmetry, and separated head from heart. But he was not to be born in British India. His father arranged his second period of leave in 1912, and the Turings sailed *en famille* for England.

This passage from India was a journey into a world of crisis. Strikes, suffragettes, and near civil war in Ireland had changed political Britain. The National Insurance Act, the Official Secrets Act, and what Churchill called 'the gigantic fleets and armies which impress and oppress the civilisation of our time,' all marked the death of Victorian certainties and the extended role of the state. The substance of Christian doctrine had long evaporated, and the authority of science held greater sway. Yet even science was feeling a new uncertainty. And new technology, enormously expanding the means of expression and communication, had opened up what Whitman had eulogised as the *Years of the Modern*, in which no one knew what might happen next – whether a 'divine general war' or a 'tremendous issuing forth against the idea of caste'.

But this conception of the modern world was not shared by the Turings, who were no dreamers of the World-City. Well insulated from the twentieth century, and unfamiliar even with modern Britain, they were content to make the best of what the nineteenth had offered them. Their second son, launched into an age of conflicts with which he would become helplessly entangled, was likewise to be sheltered for twenty years from the consequences of the world crisis.

He was born on 23 June 1912 in a nursing home in Paddington,* and was baptised Alan Mathison Turing on 7 July. His father extended

---

* Warrington Lodge, now the Colonnade Hotel, Warrington Crescent, London W9. His baptism was at St Saviour's Church, immediately across the road.

his leave until March 1913, the family spending the winter in Italy. He then returned to take up a new posting, but Mrs Turing stayed on with the two boys, Alan a babe in arms and John now four, until September 1913. Then she too departed. Mr Turing had decided that his sons were to stay in England, so as not to risk their delicate health in the heat of Madras. So Alan never saw the kind Indian servants, nor the bright colours of the East. It was in the bracing sea winds of the English Channel that his childhood was to be spent, in an exile from exile.

Mr Turing had farmed out his sons with a retired Army couple, Colonel and Mrs Ward. They lived at St Leonards-on-Sea, the seaside town adjoining Hastings, in a large house called Baston Lodge just above the sea front. Across the road was the house of Sir Rider Haggard, the author of *King Solomon's Mines*, and once, when Alan was older and dawdling along the gutter in his usual way, he found a diamond and sapphire ring belonging to Lady Haggard, who rewarded him with two shillings.

The Wards were *not* the sort of people who dropped diamond rings in the street. Colonel Ward, ultimately kindly, was remote and gruff as God the Father. Mrs Ward believed in bringing up boys to be real men. Yet there was a twinkle in her eye and both boys became fond of 'Grannie'. In between lay Nanny Thompson, who ruled the nursery which was the boys' proper place, and the governess of the schoolroom. There were other children in the house, for the Wards had no fewer than four daughters of their own, as well as another boy boarder. Later they also took in the Turing boys' cousins, the three children of Major Kirwan. Alan was very fond of the Wards' second daughter Hazel, but hated the youngest Joan, who was intermediate in age between him and John.

Both Turing boys disappointed Mrs Ward, for they scorned fighting and toy weapons, even model Dreadnoughts. Indeed, Mrs Ward wrote to Mrs Turing complaining that John was a bookworm, and Mrs Turing loyally wrote to John chiding him. Walks on the windswept promenade, picnics on the stony beach, games at children's parties, and tea before a roaring fire in the nursery were the most that the Ward environment had to offer in the way of stimulation.

This was not home, but it had to do. The parents came to

England as often as they could, but even when they did, that was not home either. When Mrs Turing returned in spring 1915, she took the boys into furnished and serviced rooms in St Leonards – gloomy places decorated by samplers embroidered with the more sacrificial kind of hymn. By this time Alan could talk, and proved himself the kind of little boy who could attract the attention of strangers with precocious, rather penetratingly high-pitched comments, but also a naughty and wilful one, in whom winning ways could rapidly give way to tantrums when he was thwarted. Experiment, as with planting his broken toy sailors in the ground, hoping they would grow afresh, was easily confused with naughtiness. Alan was slow to learn that indistinct line that separated initiative from disobedience and resisted the duties of his childhood. Late, untidy and cheeky, he had constant battles with his mother, Nanny and Mrs Ward.

Mrs Turing returned to India in the autumn of 1915, saying to Alan, 'You'll be a good boy, won't you?', to which Alan replied 'Yes, but sometimes I shall forget!' But this separation was only for six months, for in March 1916 Sahib and Memsahib together braved the U-boats, wearing lifebelts all the way from Suez to Southampton. Mr Turing took his family for a holiday in the Western Highlands, where they stayed in a hotel at Kimelfort, and John was introduced to trout fishing. At the end of his leave, in August 1916, they decided not to risk travelling together again, but instead to separate for the next three-year period. Alan's father returned to India, and his mother resumed a double exile at St Leonards.

The Great War had remarkably little direct impact on the Turing family. The year 1917, with the mechanised slaughter, the U-boat siege, the air raids, the appearance of America and the Russian revolution, set up the pattern which was to be the newborn generation's inheritance. But it had no private meaning except in keeping Mrs Turing in England. John was packed off to a preparatory school called Hazelhurst near Tunbridge Wells in Kent in May of that year, and thereafter Mrs Turing had only Alan about her. Church-going was one of her favourite pastimes, and in St Leonards she adopted a certain very high Anglican church, where Alan was dragged every Sunday for the communion service. He did not like the incense, and called it 'the church with the bad smells'. Mrs Turing also pressed on with her water-colours, for which she enjoyed a definite talent.

She took Alan out on her sketching parties where, with big eyes and sailor hat, and with quaint expressions of his own like 'quockling' for the screech of seagulls, he delighted the lady art students.

Alan taught himself to read in about three weeks from a book called *Reading without Tears*. He was quicker, however, to recognise figures, and had an infuriating habit of stopping at every lamp post to identify its serial number. He was one of those many people without a natural sense of left and right, and he made a little red spot on his left thumb, which he called 'the knowing spot'.

He would say that he wanted to be a doctor when he grew up – an ambition that would have been agreeable to the Turings, for his father would approve of the fees, and his mother of the distinguished clients and the practice of good works. But he could not learn to be a doctor on his own. It was time for some education. And so in the summer of 1918 Mrs Turing sent him to a private day school called St Michael's, to learn Latin.

George Orwell, who was born nine years earlier but likewise to an ICS father, described himself [5] as from 'what you might describe as the lower-upper-middle-class.' Before the war, he wrote:

> you were either a gentleman or not a gentleman, and if you were a gentleman you struggled to behave as such, whatever your income might be... Probably the distinguishing mark of the upper-middle class was that its traditions were not to any extent commercial, but mainly military, official, and professional. People in this class owned no land, but they felt that they were landowners in the sight of God and kept up a semi-aristocratic outlook by going into the professions and the fighting services rather than into trade. Small boys used to count the plum stones on their plates and foretell their destiny by chanting 'Army, Navy, Church, Medicine, Law'.

The Turings were in this position. There was nothing grand about the life of their sons, except perhaps on the few Scottish holidays. Their luxuries were the cinema, the ice rink, and watching the stunt-man dive off the pier on a bicycle. But in the Ward establishment there was an incessant washing away of sins, washing away of smells, to distinguish them from the other children of the town. 'I was very young, not much more than six,' recalled Orwell, 'when I first became aware of class-distinctions. Before that age my chief heroes had generally been working-class people, because they always seemed to do such interesting things, such as being fishermen and blacksmiths and bricklayers. . . . But it was not long before I was forbidden to play with the plumber's children; they were "common" and I was told to keep away from them. This was snobbish, if you like, but it was also necessary, for middle-class people cannot afford to let their children grow up with vulgar accents.'

The Turings could afford very little, since even in the well paid ICS it was always necessary to save for the future. What they *had* to afford could be summed up in two words: public school. In this respect the war, the inflation, the talk of revolution changed nothing. The Turing boys had to go to public schools, and everything had to be subordinated to this demand. Never, indeed, would Mr Turing allow his sons to forget the debt they owed him for a public school education. Alan's duty was to go through the system without causing trouble, and in particular to learn Latin, which was required for entrance to a public school.

So as Germany collapsed, and the bitter armistice began, Alan was set to work on copy-books and Latin primers. He later told a joke against his own first exercise, in which he translated 'the table' as *omit mensa* because of the cryptic footnote 'omit' attached to the word 'the'. He was not interested in Latin, and for that matter he had great difficulty in writing. His brain seemed barely coordinated with his hand. A whole decade of fighting with scratchy nibs and leaking fountain-pens was to begin, in which nothing he wrote was free from crossings-out, blots, and irregular script which veered from stilted to depraved.

But at this stage he was still the bright, jolly little boy. On Christmas visits to the Trustram Eves in Earls Court, his uncle Bertie liked making Alan the butt of his practical jokes because of his innocent giggly humour. These occasions were more of a trial for John, who was now considered to be responsible for his younger brother's appearance and behaviour – a responsibility that no human being would ever lightly shoulder. To make matters worse, as John saw it,[6] he was dressed in sailor suits, according to the convention of the day (they suited him well); I know nothing in the whole range of the cussedness of inanimate objects to compete with a sailor suit. Out of the boxes there erupted collars and ties and neckerchiefs and cummerbunds and oblong pieces of flannel with lengthy tapes attached; but how one put these pieces together, and in what order, was beyond the wit of man. Not that my brother cared a button – an apt phrase, many seemed to be off – for it was all the same thing to him which shoe was on which foot or that it was only three minutes to the fatal breakfast gong. Somehow or another I managed by skimping such trumpery details as Alan's teeth, ears, etc., but I was exhausted by these nursery attentions and it was only when we were taken off to the pantomime that I was able to forget my fraternal cares. Even then Alan was quite a nuisance, complaining loudly of the scene of green dragons and other monsters in 'Where the Rainbow Ends'. ...

The Christmas pantomime was the high spot of the year, although Alan himself later recalled of Christmas 'that as a small child I was quite unable to predict when it would fall, I didn't even realise that it came at regular intervals.' Back at dreary Baston Lodge, his head was buried in maps. He asked for an atlas as a birthday present and pored over it by the hour. He also liked recipes and formulae, and wrote down the ingredients for a dockleaf concoction for the cure of nettle-stings. The only books he had were little nature-study notebooks, supplemented by his mother reading *The Pilgrim's Progress* aloud. Once she cheated by leaving out a long theological dissertation, but that made him very cross. 'You spoil the *whole thing*,' he shouted, and ran up to his bedroom. Perhaps he was responding to the uncompromising note of Bunyan's plain-speaking Englishman. But once the rules were agreed then they must be followed to the bitter end, without bending or cheating. His Nanny found the same when playing with him:[7]

> The thing that stands out most in my mind was his *integrity* and his *intelligence* for a child so young as he then was, also you couldn't camouflage anything from him. I remember one day Alan and I playing together. I played so that he should win, but he spotted it. There was commotion for a few minutes. ...

In February 1919, Mr Turing returned after three years'

separation. It was not easy for him to re-establish his authority with Alan, who had a good line in answering back. He told Alan once to untwist his boot-tongues. 'They should be flat as a pancake,' he said. 'Pancakes are generally rolled up,' piped back Alan. If Alan had an opinion, he said that he *knew*, or that he *always knew* ; he always knew that the forbidden fruit of the Garden of Eden was not an apple but a plum. In the summer, Mr Turing took them for a holiday at Ullapool, in the far north-west of Scotland, this time a distinctly posh holiday, complete with gillie. While Mr Turing and John lured the trout, and Mrs Turing sketched the loch, Alan gambolled in the heather. He had the bright idea of gathering the wild bees' honey for their picnic tea. As the bees buzzed past, he observed their flight-paths and by plotting the intersection point located the nest. The Turings were vividly impressed by this direction-finding, more than by the murky honey he retrieved.

But that December his parents steamed away and Alan was left again with the Wards, while John returned to Hazelhurst. Their father was transferred at last to the metropolis of Madras to serve in the Revenue Department, but Alan stagnated in the deathly ennui of St Leonards-on-Sea, concocting recipes. His development was so held back that he had not even learnt how to do long division by the time of his mother's return in 1921, when he was nearly nine.

His mother perceived him as changed from 'extremely vivacious – even mercurial – making friends with everyone' to being 'unsociable and dreamy'. There was a wistful, withdrawn expression in photographs of his ten-year-old face. She took him away from St Leonards and, after a summer holiday in Brittany, somewhat spoilt by the constant counting of francs, she taught him herself in London, where he alarmed her by looking for iron filings in the gutter with a magnet. Mr Turing, who in May 1921 had again been promoted to be Secretary to the Madras Government Development Department, responsible for agriculture and commerce throughout the Presidency, returned once more in December and they all went to St Moritz, where Alan learnt to ski.

Miss Taylor, the headmistress of St Michael's, had said that Alan 'had genius', but this diagnosis was not allowed to modify the programme. In the new year of 1922, Alan was launched on the next stage of the process and was sent off to Hazelhurst like his brother.

*

Hazelhurst was a small establishment of thirty-six boys of ages from nine to thirteen, run by the headmaster Mr Darlington, a Mr Blenkins who taught mathematics, Miss Gillett who taught drawing and music of a Moody and Sankey variety, and the Matron. John had loved his time there, and now in his last term was head boy. His younger brother proved to be a thorn in the flesh, for Alan found the Hazelhurst regime a distraction. It 'deprived him of his usual occupations,' as his mother saw it. Now that the whole day was organised into classes, games and meal-times, he had but odd minutes in which to indulge his interests. He arrived with a craze for paper-folding, and when he had shown the other boys what to do, John found himself confronted everywhere with paper frogs and paper boats. Another humiliation followed when Alan's passion for maps was discovered by Mr Darlington. This inspired him to set a geography test to the whole school, in which Alan came sixth, beating his brother, who found geography very boring. On another occasion Alan sat in the back row at a school concert, choking himself with laughter while John sang *Land of Hope and Glory* as a solo.

John left Hazelhurst at Easter for Marlborough, his public school. In the summer, Mr Turing again took the family to Scotland, this time to Lochinver. Alan exercised his knowledge of maps on the mountain paths, and they fished in the loch, Alan now competing with John. The brothers had a good line in non-violent rivalry, as for instance when they played a game to alleviate the awfulness of their grandfather Stoney's visits. This depended upon winning points by leading him on, or heading him away from one of his well-rehearsed club bore stories. And at Lochinver Alan defeated his family in what Mrs Turing considered the rather vulgar after-dinner sport of throwing discarded gooseberry skins as far as possible. Cleverly inflating them, he made them soar over the hedge.

Life when off duty, in this early afternoon of the Empire, could be very agreeable. But in September his parents saw Alan back to Hazelhurst, and as they drove away in their taxi, Alan rushed back along the school drive with arms flung wide in pursuit. They had to bite their lips and sail away to Madras. Alan continued to maintain his detached view of the Hazelhurst regime. He gained average marks in class, and in turn held an unflattering view of the instruction. Mr Blenkins initiated his class into elementary algebra, and Alan reported to John, 'He gave a *quite false impression* of what is meant by x.'

Although he enjoyed the feeble little plays and debates, he hated and feared the gym class and the afternoon games. The boys played hockey in winter, and Alan later claimed that it was the necessity of avoiding the ball that had taught him to run fast. He did enjoy being linesman, judging precisely where the ball had crossed the line. In an end-of-term sing-song, the following couplet described him:

> Turing's fond of the football field
> For geometric problems the touch-lines yield

Later another verse had him 'watching the daisies grow' during hockey, an image which inspired his mother to a whimsical pencil sketch. And although intended as a joke against his dreamy passivity, there might have been a truth in the observation. For something new had happened.

At the end of 1922, some unknown benefactor had given him a book, called *Natural Wonders Every Child Should Know*.[8] Alan told his mother later that this book had opened his eyes to science. Indeed, it must have been the first time that he became conscious that such a kind of knowledge as 'science' existed. But more than that, it opened the book of life. If anything at all can be said to have influenced him, it was this book which, like so many new things, came from the United States.

The book had first appeared in 1912 and its author, Edwin Tenney Brewster, had described it as

... the first attempt to set before young readers some knowledge of certain loosely related but very modern topics, commonly grouped together under the name, General Physiology. It is, in short, an attempt to lead children of eight or ten, first to ask and then to answer, the question: What have I in common with other living things, and how do I differ from them? Incidentally, in addition, I have attempted to provide a foundation on which a perplexed but serious-minded parent can himself base an answer to several puzzling questions which all children ask – most especially to that most difficult of them all: By what process of becoming did I myself finally appear in this world?

In other words, it was about sex and science, starting off with 'How the Chicken got inside the Egg', rambling through 'Some Other Sorts of Eggs' until arriving at 'What Little Boys and Girls are Made Of'. Brewster quoted 'the old nursery rhyme' and said that:

It has this much truth in it, that little boys and little girls are far from being alike, and it isn't worth while trying to make either one over into the other.

The precise nature of this difference was not revealed, and only after a skilful diversion on to the subject of the eggs of starfish and sea-urchins did Brewster eventually arrive back at the human body:

So we are not built like a cement or a wooden house, but like a brick one. We are made of little living bricks. When we grow it is because these living bricks divide into half bricks, and then grow into whole ones again. But how they find out when and where to grow fast, and when and where to grow slowly, and when and where not to grow at all, is precisely what nobody has yet made the smallest beginning at finding out.

The process of biological growth was the principal scientific theme of E. T. Brewster's book. Yet science had no explanations, only descriptions. In fact on 1 October 1911, when Alan Turing's 'living bricks' were first dividing and redividing, Professor D'Arcy Thompson was telling the British Association that 'the ultimate problems of biology are as inscrutable as of old.'

But equally inscrutable, *Natural Wonders* conspicuously failed to describe where the *first* cell in the human process came from, only dropping the elusive hint that 'the egg itself arose by the splitting of still another cell which, of course, was part of the parent's body.' The secret was left for the 'perplexed but serious-minded parent' to explain. Mrs Turing's way of dealing with the thorny topic was, in fact, highly consonant with Brewster's, for John at least was the recipient at Hazelhurst of a special letter starting with the birds and the bees, and ending with instructions 'not to go off the rails'. Presumably Alan was informed in the same way.

In other ways, however, *Natural Wonders* was indeed 'very modern', and certainly no little 'nature book'. It conveyed the idea that there had to be a reason for the way things were, and that the reason came not from God but from science. Long passages explained why little boys liked throwing things and little girls liked babies, and derived from the pattern of the living world the ideal of a Daddy to go out to work at the office and a Mama to stay at home. This picture of respectable American life was rather remote from the training of the sons of Indian civil servants, but more relevant to Alan was a picture of the brain:

> Do you see now why you have to go to school five hours a day, and sit on a hard seat studying still harder lessons, when you would much rather sneak off and go in swimming? It is so that you may build up these thinking spots in your brains. . . . We begin young, while the brain is still growing. With years and years of work and study, we slowly form the thinking spots over our left ears, which we are to use the rest of our days. When we are grown up, we can no more form new thinking places. . . .

So even school was justified by science. The old world of divine authority was reduced to a vague allusion in which Brewster, having described evolution, said that 'why it all happens or what it is all for' was precisely 'one of those things that no fellah can find out.' Brewster's living things were unequivocally *machines*:

> For, of course, the body is a machine. It is a vastly complex machine, many, many times more complicated than any machine ever made with hands; but still after all a machine. It has been likened to a steam engine. But that was before we knew as much about the way it works as we know now. It really is a gas engine; like the engine of an automobile, a motor boat, or a flying machine.

Human beings were 'more intelligent' than the other animals, but were not accorded a mention of 'soul'. The process of cellular division and differentiation was something no one had *yet* begun to understand – but there was no suggestion that it required the interference of angels. So if Alan was indeed 'watching the daisies grow', he could have been thinking that while it looked as though the daisy knew what to do, it really depended upon a system of cells working like a machine. And what about himself? How did *he* know what to do? There was plenty to dream about while the hockey ball whizzed past.

Besides watching the daisies, Alan liked inventing things. On 11 February 1923 he wrote:[9]

Dear Mother and Daddy,

I have got a lovely cinema kind of thing Micheal* sills gave it to me and you can draw new films for it and I am making a copy of it for you for an easter present I am sending it in another envelope if you want any more films for it write for them there are 16 pictures in each but I worked out that I could draw 'The boy stood at the tea table' you know the Rhyme made up from casabianca I was 2nd this week again. Matron sends her love GB said that as I wrote so thick I was to get some new nibs from T. Wells and I am writing with them now there is a lecture tomorrow Wainwright was next to bottom this week this is my patent ink

There was nothing about science, inventions, or the modern world in the Common Entrance examination – the public school admission test, which was the *raison d'être* of schools like Hazelhurst. *Casabianca* was nearer the mark. In the American *Natural Wonders* everything had to have a reason. But the British system was building different 'thinking spots' – the virtue of Casabianca, the boy on the burning deck, was that he carried out his instructions literally, losing his life in the process.

The masters did their best to discourage Alan's irrelevant interest in science, but could not stop his inventions – in particular, machines to help him in the writing problems that still plagued him:

April 1 (fool's day)

Guess what I am writing with. It is an invention of my own it is a fountain pen like this: – [*crude diagram*] you see to fill it scweeze E ['*squishy end of fountain pen filler*'] and let go and the ink is sucked up and it is full. I have arranged it so that when I press a little of the ink comes down but it keeps on getting clogged.

I wonder if John has seen Joan of Arc's Statue yet coz it is in Rouen. Last monday we had scouts v cubs it was rather exiting there was no weeks order this week I hope John likes Rouen I don't feel much like writing much today sorry. Matron says John sent something.

This provoked another couplet, about a fountain pen that 'leaked enough for four'. Another letter in July, written in green ink which

---

* Alan's spelling and punctuation, here and throughout, is faithfully reproduced.

was (predictably) forbidden, described an exceedingly crude idea for a typewriter .

John's stay in Rouen was part of a general alteration in the Turing family arrangements. Before going to Marlborough, he had told his father that he would like a change from the Wards, and this was agreed. The parents found a Hertfordshire vicarage to be their home as from the summer of 1923. Meanwhile, at Easter, John had parted from his brother for the first time, going to stay with a Mme Godier in Rouen. This went quite well, and in the summer Alan ('simply longing to go there') went with him to imbibe the culture and civilisation of France for a few weeks. Alan made a great impression on the *petit-bourgeoise* Mme Godier. It was *'comme il est charmant'* when he had been persuaded to wash behind the ears, and a telling-off for John if he had not. John loathed Mme Godier, and her fawning on Alan came as a relief, enabling him to slip off to the cinema. Both Turing boys, in fact, were singularly good-looking, with a subtle, vulnerable appeal; John rather the sharper, and Alan dreamier. The stay was not a great success. John had refused to take his bicycle this time because of the prospect of navigating wobbly Alan through the cobbled Rouen streets. So they were marooned listlessly in the *maison Godier*, or were obliged to take long walks. *'Il marche comme un escargot,'* declared Mme Godier of Alan, an observation which fitted Alan's snail-like progress along the gutter, but also the Turing family's picture of itself – that of the slow Turings, the gloomy Turings, always fighting on the losing side, and coming in last if not least.

Much happier was the new home in Hertfordshire, to which the boys went for the rest of the summer. It was the Georgian red-brick rectory at Watton-at-Stone, seat of the old Archdeacon Rollo Meyer, a charming and mellow man whose environment was that of the rose-bed and the tennis court, rather than the well-scrubbed, brisk discipline of the Wards. John and Alan both responded with joy, John to girls on the tennis court (he being fifteen and decidedly interested), Alan to being left alone, allowed to cycle in the woods and make his own mess as he pleased as long as he met minimum standards in the house. Alan's standing also went up in Mrs Meyer's eyes when a gypsy fortune teller at the church fête said that he would be a genius.

The Meyers' guardianship was shortlived, for Mr Turing

suddenly decided to resign from the Indian Civil Service. He was angry at his rival, a certain Campbell who had come out with him in 1896 and had obtained a lower grade in the entrance examination, being promoted to be Chief Secretary to the Madras Government. So he abandoned his own chances of further advancement, and Alan's parents never returned as Sir Julius and Lady Turing,* though they had the more tangible benefit of a £1000 per annum pension.

It was not a return to England, for Alan's father adopted a new role as tax exile. The Inland Revenue allowed him to escape the income tax if he spent only six weeks in the United Kingdom each year, so the Turings installed themselves in the French resort of Dinard, opposite St Malo on the Brittany coast. Henceforward the boys were to travel to France for Christmas and Easter vacations, while the parents would come to England for the summer.

Technically, Mr Turing did not resign until 12 July 1926, and in the meantime he was on leave, the development of Madras somehow continuing without him. But he lost no time in establishing a new sense of economy. Mrs Turing had to submit accounts detailing the housekeeping expenses to the centime. Holidays in St Moritz and Scotland were declared henceforth out of the question.

In many ways his premature retirement was a disaster. Both sons felt it was a mistake. Alan was to imitate in a particularly droll manner the huffy comments that his father would pass on 'XYZ Campbell', and his brother later wrote:[10]

> I doubt if I should have found my father an easy superior or subordinate for by all accounts he cared nothing for the hierarchy nor his own future in the Indian Civil Service and spoke his mind regardless of the consequences. One example will suffice. For a while he acted as principal private secretary to mild Lord Willingdon in the Madras Presidency and when a difference of opinion developed between them my father remarked 'After all you are not the Government of India'. Such thumping, suicidal indiscretion one can but admire from a safe distance.

This particular incident was always held against Mr Turing by his wife, the more so as she was particularly in awe of Lady Willingdon. The truth perhaps was that despite all the endless talk of duty, the

---

* Unlike Sir Archibald Campbell.

qualities required in a district officer were very different from those of rule-book-keeping and deference to rank. Governing millions of people spread over an area equal to that of Wales called for an independent judgment and force of personality which were less welcome in the more courtly circles of metropolitan Madras. They were certainly little needed in his retirement, in which the busy intrigue of India assumed a retrospective appeal. His remaining years were dogged by a sense of loss, disillusion, and an intense boredom which fishing and bridge parties could never alleviate. He was aggravated by the fact that his younger wife found the return to Europe an opportunity to emerge from the constricting mental atmosphere of Dublin and Coonoor. For he had little regard for her more intellectual ambitions, combined as they were with a rather nervous, fussy domesticity; while she suffered from his obsessive penny-pinching and sense of being betrayed. They were both emotionally demanding, but neither met the other's demands, and they came to communicate in little but planning the garden.

One result of the new arrangement was that Alan now saw some point to learning French, and it became Alan's favourite school subject. But he also liked it as a sort of code, in which he naively wrote a postcard to his mother about 'la revolution' at Hazelhurst that Mr Darlington was not supposed to be able to read. (The joke was on their Breton maid at Dinard, who often spoke of a socialist revolution being imminent.)

But it was science that entranced him, as his parents discovered when they arrived back to find him clutching Natural Wonders. Their reaction was not entirely negative. Mrs Turing's grandfather's second cousin, George Johnstone Stoney (1826–1911), had been a famous Irish scientist whom she had once met as a girl in Dublin. He was best known as the inventor of the word 'electron' which he coined in 1894 before the atomicity of electric charge had been established. Mrs Turing was very proud of having a Fellow of the Royal Society in her family, for ranks and titles made a great impression upon her. She would also show Alan the picture of Pasteur on the French postage stamps, which suggested the prospect of Alan as a benefactor of humanity. Perhaps she recalled that doctor missionary in Kashmir, all those years ago! – but there was also the simple fact that although she herself pressed her ideas into a suitably ladylike form, she still represented the Stoneys who had married applied science to the expanding empire. Alan's father, however, could well have pointed out that a scientist could expect no more than £500 a year, even in the Civil Service.

But he also helped Alan in his own way, for when back at school in May 1924 Alan wrote:

> . . . You (Daddy) were telling me about surveying in the train, I have found out or rather read how they find out the heights of trees, widths of rivers, valley's etc. by a combination of both I found out how they find heights of mountains without climbing them.

Alan had also read about how to draw a geographical section, and had added this accomplishment to 'family tree, chess, maps etc. (gennerally my own amusements)'. In the summer of 1924 the family stayed for a time at Oxford – a nostalgic exercise on Mr Turing's part – and then in September they holidayed at a boarding house in North Wales. The parents stayed on awhile when Alan went back by himself to Hazelhurst ('I tipped the porter all right and the taxi too . . . I did not tip the Frant chap but that was not expected of me. Was it?') where he made his own maps of the Snowdonia mountains. ('Will you compare my map with the Ordnance one and send it back.')

Maps were an old interest: family trees he also liked, and the particularly awkward Turing genealogy, with its leaps of the baronetcy from bough to bough and its enormous Victorian families, exercised his ingenuity. Chess was the most social of his activities:

> There was not going to be any Chess Tournament because Mr Darlington had not seen many people playing but he said that if I asked everyone who could play and made a list of everyone who had played this term we would have it. I managed to get enough people so we are having it.

He also found the work in class 1B to be 'much more interesting'. But all this paled before chemistry. Alan had always liked recipes, strange brews and patent inks, and had tried clay-firing in the wood when staying with the Meyers. The idea of chemical processes would not have been strange to him. And in the summer holidays at Oxford, his parents had allowed him to play with a box of chemicals for the first time.

Natural Wonders did not have much to say about chemistry, except in terms of poisons. A strong defence of Temperance, not to say Prohibition, flowed from Brewster's scientific pen:

> The life of any creature, man, animal or plant, is one long fight against being poisoned. The poisons get us in all sorts of ways . . . like alcohol, ether, chloroform, the various alkaloids, such as strychnin and atropin and cocain, which we use as medicines, and nicotin, which is the alkaloid of tobacco, the poisons of many toadstools, caffein which we get in tea and coffee. . . .

There was another section headed 'Of Sugar and Other Poisons', explaining the effect of carbon dioxide in the blood, causing fatigue, and the action of the brain:

> When the nerve center in the neck tastes a little carbon dioxid, it doesn't say anything. But the moment the taste begins to get strong (which is in less than a quarter minute after one starts running hard) it telephones over the nerves to the lungs:
>
> 'Here, here, here! What is the matter with you fellows. Get busy. Breathe hard. This blood is fairly sizzling with burnt up sugar!'

All this was grist to Alan's mill, although at this point what interested him was the more sober claim that:

> The carbon dioxid becomes in the blood ordinary cooking soda; the blood carries the soda to the lungs, and there it changes to carbon dioxid again, exactly as it does when, as cooking soda, or baking powder, you add it to flour and use it to raise cake.

There was nothing in *Natural Wonders* to explain chemical names or chemical change, but he must have picked up the ideas from somewhere else, for on arriving back at school on 21 September 1924 his letter reminded his parents 'Don't forget the science book I was to have instead of the Children's Encyclopedia,' and also:

> In Natural wonders every child should know it says that the Carbon dioxide is changed to cooking soda in the blood and back to carbon dioxide in the lungs. If you can will you send me the chemical name of cooking soda or the formula better still so that I can see how it does it.

Presumably he had seen the *Children's Encyclopedia*, if only to reject it as too childish and vague, and could well have learnt the basic ideas of chemistry from its multitude of little 'experiments' with household substances. The prophetic spark of enquiry lay in his trying to combine the ideas of chemical formulae on the one hand with the mechanistic description of the body on the other.

Chemistry was not the Turing parents' *forte*, but in November he found a more reliable source of information: 'I have come into great luck here: there is an Encyclopedia that is 1st form property.' And at Christmas 1924 he was given a set of chemicals, crucibles and test-tubes, and allowed to use them in the cellar of *Ker Sammy*, their villa in the Rue du Casino. He heaved great quantities of sea-weed back from the beach in order to extract a minute amount of iodine. This was much to the amazement of John, who with different eyes saw Dinard as an expatriate English colony of the bright 1920s, and spent his time on tennis, golf, dancing and flirting in the Casino. There was an English schoolmaster in the neighbourhood, whom Alan's parents employed to coach Alan for the Common Entrance examination, who found himself plied with questions about science. In March 1925, back at school, Alan wrote:

> I came out in the same place in Common Entrance* this term as last with 53% average. I got 69% in French.

But it was the chemistry that mattered:

> I wonder whether I could get an earthenware retort anywhere for some high-heat actions. I have been trying to learn some Organic Chemistry, when I began if I saw something like this

$$H(CH_2)_{17}CO_2H(CH_2)_2C$$

> I would try and work it out like this $C_{21}H_{40}O_2$ which might be all sorts of things it is a kind of oil. I find the Graphic formulae help too, thus Alcohol is

$$H(CH_2)_2OH \text{ or } C_2H_6O) \text{ is } H—\overset{\displaystyle H}{\underset{\displaystyle H}{\overset{|}{\underset{|}{C}}}}—\overset{\displaystyle H}{\underset{\displaystyle H}{\overset{|}{\underset{|}{C}}}}—O—H$$

> while Methyl ether $HCH_2.O.CH_2H$ or C_2H_6O

---

* These were practice papers

is   H—C—O—C—H

with H atoms above and below each C

you see they shew the molecular arrangement.

And then a week later:

> . . . The earthenware retort takes the place of a crucible when the essential
> product is a gas which is very common at high temperatures. I am making
> a collection of experiments in the order I mean to do them in. I always seem
> to want to make things from the thing that is commonest in nature and
> with the least waste in energy.

For Alan was now conscious of his own ruling passion. The longing
for the simple and ordinary which would later emerge in so many
ways was not for him a mere 'back to nature' hobby, a holiday from
the realities of civilisation. To him it was life itself, a civilisation from
which everything else came as a distraction.

To his parents the priorities were the other way round. Mr Turing
was not at all the man for airs and graces; a man who would insist on
walking rather than take a taxi, there was a touch of the desert island
mentality in his character. But nothing altered the fact that chemistry
was merely the amusement allowed to Alan on his holidays and
that what mattered was that at thirteen he had to go on to a public
school. In the autumn of 1925 Alan sat the Common Entrance for
Marlborough, and to the surprise of all did rather well. (He had not
been allowed to try for a scholarship.) But at this point John played a
decisive part in the life of his strange brother. 'For God's sake don't
send him here,' he said, 'it will crush the life out of him.'

Alan posed a difficult problem. It was not in question that he
must adapt to public school life. But what public school would cater
best for a boy whose principal concern was to do experiments with
muddy jam jars in the coal cellar? It was a contradiction in terms. As
Mrs Turing saw it,[11]

> Though he had been loved and understood in the narrower homely circle
> of his preparatory school, it was because I foresaw the possible difficulties
> for the staff and himself at a public school that I was at such pains to find
> the right one for him, lest if he failed in adaptation to public school life he
> might become a mere intellectual crank.

Her pains were not prolonged. She had a friend called Mrs Gervis, the wife of a science master at Sherborne School, a public school in Dorset. In spring 1926 Alan took the examination again, and was accepted by Sherborne.

Sherborne was one of the original English public schools, whose origins[12] lay in the abbey, which itself was one of the first sites of English Christianity, and in a charter of 1550 establishing the school for local education. In 1869, however, Sherborne had fallen into line as a boarding school on Dr Arnold's model. After a period of low repute, it had revived in 1909 when a Nowell Smith was appointed headmaster. By 1926 Nowell Smith had doubled the roll from two hundred to four hundred, and had established Sherborne as a moderately distinguished public school.

Mrs Turing paid a visit to Sherborne before Alan went there and was able to see the headmaster's wife. She gave Mrs Nowell Smith 'some hints about what to expect' and Mrs Nowell Smith 'contrasted her description with the more favourable accounts given by other parents of their boys.' It was probably at her suggestion that Alan was put down to board at Westcott House, whose housemaster was Geoffrey O'Hanlon.

The summer term was due to start on Monday 3 May 1926 which was, it so happened, the first day of the general strike. On the ferry from St Malo Alan heard that only the milk trains would be running. But he knew he could cycle the sixty miles west from Southampton to Sherborne:

> so I cycled as programme left luggage with baggage master started out of docks about 11 o'clock got map for 3/- including Southampton missing Sherborne by about 3 miles. Noted where Sherborne was just outside. With an awful strive, found General Post Office, sent wire O'Hanlon 1/-. Found cycle shop, had things done 6d. Left 12 o'clock had lunch 7 miles out 3/6 went on to Lyndhurst 3 miles got apple 2d. went on to Beerley 8 miles pedal a bit wrong had it done 6d. went on Ringwood 4 miles.
>
> The streets in Southampton were full of people who had struck. Had a lovely ride through the New Forest and then over a sort of moor into Ringwood and quite flat again to Wimborne.

Alan stayed overnight at the best hotel in Blandford Forum – an expedient that would hardly have been approved by his father. (Alan had to account for every penny that he spent: no mere figure of speech, for his letter ended 'Sending back £1-0-1 in £ note and penny stamp.') But the proprietors only charged a nominal amount and saw him off in the morning. Then:

> Just near Blandford some nice downs and suddenly merely undulating near all the way here but the last mile was all downhill.

From West Hill he could see his destination: the little Georgian town of Sherborne and the school itself by the abbey.

For a boy of his class to improvise a solution without a fuss was not at all the expected thing. The bicycle journey was regarded with astonishment, and was reported in the local newspaper.[13] While Winston Churchill called for the 'unconditional surrender' of the 'enemy' miners, Alan had made the most of the general strike for himself. He had enjoyed two days of freedom outside the usual system. But they were over very quickly. There was a book[14] about life at Sherborne, *The Loom of Youth* by Alec Waugh, and this described the sensations:

> The new boy's first week at a Public School is probably the most wretched he will ever pass in his life. It is not that he is bullied . . . it is merely that he is utterly lonely, is in constant fear of making mistakes, and so makes for himself troubles that do not exist.

When his hero wrote home at the end of his second day, 'it did not need a very clever mother to read between the lines and see her son was hopelessly miserable.' It was worse for Alan, for he could not even merge inconspicuously while all his belongings were stuck at Southampton by the strike. At the end of his first week:

> Its an awful nuisance here without any of my clothes or anything. . . . It's rather hard getting settled down. Do write soon. There was no work on Wednesday except for 'Hall' or prep. And then its a business finding my classrooms what books to get but I will be more or less settled after a week or so . . .

But a week later Alan was not much better off:

I am getting more and more settled down. But I won't be quite right until my things come. Fagging starts for us next Teusday. It is run on the same principle as the Gallic councils that tortured and killed the last man to arrive; here one fagmaster calls and all his fags run the last to arrive getting the job. You have to have cold showers in the morning here like cold baths at Marlborough. We have tea at 6.30 here on Mon., Wed., Frid. I manage to go without food from lunch to then. . . . The general strike had a part of it as a printer's strike the result of that is that 'Bennetts' booksellers had none of the books ordered and I am without a lot of them. As in most public schools new boys have to sing some song. The time has not yet come. I am not sure what to sing anyhow it won't be 'buttercup'. . . . The amount of work we are given for Hall here is sometimes absurdly small e.g. Read Acts chapters 3 and 4 and that is for ¾hr.

Yr loving son Alan

There was indeed a song-singing and another ceremony in which he was kicked up and down the day room in a waste paper basket. However, if *his* mother read between the lines, she subordinated sympathy to her sense of duty. Her comment on this letter was that it displayed Alan's 'whimsical sense of humour'.

He was now at last being taught science, and reported:

We do do Chemistry 2 hrs. a week. We have only got to about the stages of 'Properties of Matter', 'Physical and chemical change' etc. The master was quite amused by my Iodine making and I shewed him some samples. The headmaster is called 'Chief'. I seem to be doing Greek and not Hellenics. . . .

The master, Andrews, was no doubt 'amused' that Alan already knew so much. He had arrived 'delightfully ingenuous and unspoilt'. And the head boy of Westcott House, Arthur Harris, had rewarded Alan's cycling initiative by appointing him his 'fag', or servant. But neither scientific education nor initiative were exactly Sherborne priorities.

The headmaster used to expound the meaning of school life in his sermons.[15] Sherborne was not, he explained, entirely devoted to 'opening the mind', although 'historically . . . this was the primary meaning of school.' Indeed, said the headmaster, there was 'constantly a danger of forgetting the original object of school.' For the English public school had been consciously developed into what he called 'a nation in miniature'. With a savage realism, it dispensed with the lip service paid to such ideas as free speech, equal justice and parliamentary democracy, and concentrated upon the fact of precedence and power. As the headmaster put it:

> In form-room and hall and dormitory, on the field and on parade, in your relations with us masters and in the scale of seniority among yourselves, you have become familiar with the ideas of authority and obedience, of cooperation and loyalty, of putting the house and the school above your personal desires . . .

The great theme of the 'scale of seniority' was the balance of privilege and duty, itself reflecting the more worthy side of the British Empire. But this was a theme to which 'opening the mind' came as at best an irrelevance.

The Victorian reforms had made their mark, and competitive examinations played a part in public school life. Those who came as scholars had an opportunity to take on the role of an intelligentsia in the 'nation in miniature', tolerated provided they interfered with nothing that mattered. But Alan, who did not belong to this group, was quick to note the 'absurdly small' amount expected of him. And in fact it was the organised team games of rugby football ('footer') and cricket which for most boys would dominate the years at Sherborne and through which the emotional lessons were taught. Nor had the social changes of the Great War made any difference to the total, introverted, self-conscious system of house life, with its continuous public scrutiny and control of every individual boy. These were the true priorities.

In one respect only a token concession had been made to Victorian reform. There had been a science master at Sherborne since 1873, but this was primarily for the sake of the medical profession. It was not for the Workshop of the World, stigmatised as too mundanely utilitarian in spirit to occupy the time of a gentleman. The Stoneys might build the bridges of the Empire, but it was a higher caste which commanded them. Neither did science enjoy respect for its enquiry, irrespective of usefulness, into truth. Here again the public school had resisted the triumphant claims of nineteenth-century science. Nowell Smith divided the intellectual world into Classical, Modern and Science, in that order, and held that it is only the shallowest mind that can suppose that all the advance of discovery brings us appreciably nearer to the solution of the riddles of the universe which have haunted man from the beginning . . .

Such was the miniature, fossilised Britain, where masters and servants still knew their places, and where the miners were disloyal to their school. And while the boys were playing at being servants, loading the milk churns on to the trains until the strike was broken by the masters of their country, the shallow mind of Alan Turing had arrived in their midst. It was a mind that had no interest in the problems of would-be landowners, empire-builders, or administrators of the white man's burden; they belonged to a system that had no interest in him.

The word 'system', indeed, was one which was a constant refrain, and the system operated almost independently of individual personalities. Westcott House, which Alan joined, had taken its first boarders only in 1920, and yet already existed as though the traditional prefects and 'fags' and beatings in the washroom were laws of nature. This was true even though the housemaster, Geoffrey O'Hanlon, had a mind of his own. Then a bachelor in his forties, and nicknamed (rather snobbishly) Teacher, he had extended the original house building with his own private fortune derived from Lancashire cotton. He personally did not believe in moulding the boys to a common form, and failed to instil the religion of 'footer' with quite the enthusiasm of the other housemasters. His house enjoyed in consequence a dim reputation for 'slackness'. He encouraged music and art, disliked bullying, and stopped the song-singing initiation soon after Alan arrived. A Catholic classicist, he was the nearest thing to a liberal government in the 'nation in miniature'. Yet the system prevailed, in all but details. One could conform, rebel, or withdraw – and Alan withdrew.

'He appears self-contained and is apt to be solitary,' commented O'Hanlon.[16] 'This is not due to moroseness, but simply I think to a shy disposition.' Alan had no friend, and at least once in this year he was trapped underneath some loose floorboards in the house day-room by the other boys. He tried to continue chemistry experiments there, but this was doubly hated, as showing a swottish mentality, and producing nasty smells. 'Slightly less dirty and untidy in his habits,' wrote O'Hanlon at the end of 1926, 'rather more conscious of a duty to mend his ways. He has his own furrow to plough, and may not meet with general sympathy: he seems cheerful, though I'm not always certain he really is so.'

'His ways sometimes tempt persecution: though I don't think he is unhappy. Undeniably he is not a "normal" boy: not the worse for that, but probably less happy,' he wrote somewhat inconsistently at the end of the spring term of 1927. The headmaster commented more briskly:

> He should do very well when he finds his *métier*; but meanwhile he would do much better if he would try to do his best as a member of this school – he should have more *esprit de corps*.

Alan was not what Brewster called a 'proper boy', whose instincts, inherited from thousands of years of warfare, made him want to throw things at other people. In this respect he was more like his father, who had managed to escape games as a boy in Bedford. Mr Turing, who lacked his wife's excessive respect for schoolmasters, made a special request for Alan to be excused from cricket, and he was allowed by O'Hanlon to play golf instead. But he made himself 'a drip' by letting down his house contingent at the gym with his 'slackness'. He was also called *dirty*, thanks to his rather dark, greasy complexion, and a perpetual rash of ink stains. Fountain pens still seemed to spurt ink whenever his clumsy hands came near them. His hair, which naturally fell forward, refused to lie down in the required direction; his shirt moved out of his trousers, his tie out of his stiff collar. He still seemed unable to work out which coat button corresponded to which buttonhole. On the Officers Training Corps parade on Friday afternoons, he stood out with cap askew, hunched shoulders, ill-fitting uniform with puttees like lampshades winding up his legs. All his characteristics lent themselves to easy mockery, especially his shy, hesitant, high-pitched voice – not exactly stuttering, but hesitating, as if waiting for some laborious process to translate his thoughts into the form of human speech.

Mrs Turing saw the fulfilment of her worst fear, which was that Alan would not adapt to public school life. Nor was he the kind of boy who was unpopular in the house but pleased the masters in class. He failed there too. In his first term, he had been placed in a form called 'the Shell', with boys a year older than himself who were not good at the work. Then he was 'promoted', but only to the entrance form for those supposed of average ability. Alan took little notice. The masters streamed past – seventeen in those first four terms – and none understood the dreaming boy in a class of twenty-two. According to a classmate of the period:[17]

> he was the cruel butt of at least one master because he always managed to get ink on his collar so that the master could raise an easy laugh by saying 'Ink on your collar again, Turing!' A small and petty thing but it stuck in my mind as an example of how a sensitive and inoffensive boy . . . can have his life made hell at public school.

Reports were issued twice a term, and the unopened envelopes would lie accusingly on the breakfast table, while Mr Turing 'fortified himself with a couple of pipes and *The Times*.' Alan would say, unconvincingly, 'Daddy expects school reports to be like after-dinner speeches,' or 'Daddy should see some of the other boys' reports.' But Daddy was not paying for the other boys, and was seeing the hard-won fees disappear without detectable effect.

Daddy did not mind his divergences from conventional behaviour, or at least regarded them with an amused tolerance. In fact both John and Alan took after their father, all three believing in speaking their mind and applying their ideas with a determination punctuated by moments of recklessness. Within the family, the voice of public opinion was supplied by Mama, whose tastes and judgments were thought insipidly provincial by the others. It was she, not her husband nor John, who felt called upon to reform Alan. However, Mr Turing's tolerance did not extend to the waste of a precious public school education. His finances were particularly tight at this point. He had finally tired of exile, and had taken a small house on the edge of Guildford in Surrey, but besides paying income tax he now had to launch John on a career. He had dissuaded his son from the ICS, predicting that the 1919 reforms, introducing Indian representation into provincial government, spelt the beginning of the end. John had spiritedly thought of going into publishing instead, and his father's pet idea was that he should go to South America to make money out of guano, but in the end it was Mrs Turing's safer suggestion that he should be a solicitor which won. Mr Turing was obliged to pay £450 for his son to be articled and to support him for five years.

But Alan could not see the point of the schooling so dearly won for him. Even in French, once a favourite, the master wrote 'His lack of interest is very depressing except when something amuses him.' He developed a particularly annoying way of ignoring the teaching during the term and then coming top in the examination. Greek, however, which he was supposed to learn for the first time at Sherborne, he ignored completely. He was placed at the bottom of the bottom set for three terms, after which the point was conceded and he was grudgingly allowed to abandon it. 'Having secured one privileged exemption,' O'Hanlon wrote, 'he is mistaken in acting as though idleness and indifference will procure release from uncongenial subjects.'

In mathematics and science the masters wrote more approving reports, but there was always something to complain about. In the summer of 1927, Alan showed to his mathematics teacher, a certain Randolph, some work he had done for himself. He had found the infinite series for the 'inverse tangent function', starting from the trigonometrical formula for tan½x.* Randolph was appropriately amazed, and told Alan's form master that he was 'a genius'. But the news sank like a stone in the Sherborne pond. It merely saved Alan from a demotion, and even Randolph reported unfavourably:

> Not very good. He spends a good deal of time apparently in investigations in advanced mathematics to the neglect of elementary work. A sound groundwork is essential in any subject. His work is dirty.

The headmaster issued a warning:

> I hope he will not fall between two stools. If he is to stay at a Public School, he must aim at becoming *educated*. If he is to be solely a *Scientific Specialist*, he is wasting his time at a Public School.

The hint of expulsion thudded on to the breakfast table, endangering

---

* The series is: $\tan^{-1} x = x - \dfrac{x^3}{3} + \dfrac{x^5}{5} - \dfrac{x^7}{7} \cdots$

It was a standard result in sixth form work, but the point was that he discovered it without the use of the elementary calculus. Perhaps the most remarkable thing was his seeing that such a series should exist at all.

all that Mr and Mrs Turing had worked and prayed for respectively. But Alan discovered a way to beat the system that Nowell Smith called 'the essential glory and function of the English Public School'. He spent the second half of the term isolated in the sanatorium with mumps. Emerging to perform as well as usual in the examinations, he won a prize. The headmaster commented:

> He owes his place and prize entirely to mathematics and science, but he shewed improvement on the literary side. If he goes on as he is doing now, he should do very well.

In the summer holiday the Turings stayed again at a boarding house in Wales, this time at Ffestiniog. Alan and his mother strode up the peaks. Back at the boarding house was a Mr Neild who took great interest in Alan and gave him a book on mountain climbing in which he wrote a long inscription treating Alan's climbs as symbolic of his eventually attaining intellectual heights. One person, for a brief moment, had taken him seriously.

The human body, *Natural Wonders* explained, was a 'Living Apothecary Shop'. It was Brewster's way of describing the effects of the recently discovered hormones, whereby the 'different parts of the body' could 'signal to one another' with 'chemical messages' rather than through the nervous system. It was during 1927, when he became fifteen, that Alan gained his height, and presumably the more interesting and exciting changes took place at the same time.

It was also time for the puberty rite of the Church of England. Alan was confirmed on 7 November 1927. Like the Officers Training Corps, confirmation was one of those duties for which everyone had to volunteer. But he did believe in it, or at least in something, as he knelt before the bishop of Salisbury and renounced for himself the world, the flesh and the devil. Nowell Smith, however, took advantage of the occasion to remark:

> I hope he takes his confirmation seriously. If he does, he will not be content to neglect obvious duties in order to indulge his own tastes, however good in themselves.

To Alan, the 'duties' to translate silly sentences into Latin, polish the buttons on his Corps tunic, and suchlike, were far from 'obvious'.

He had his own kind of seriousness. The headmaster's words would more appropriately have been directed at the outward conformity that Alec Waugh had written about:

> As is the case with most boys, Confirmation had very little effect on Gordon. He was not an atheist; he accepted Christianity in much the same way as he accepted the Conservative party. All the best people believed in it, so it was bound to be right; but at the same time it had not the slightest influence over his actions. If he had any religion at this time it was House football . . . .

These were strong words for a book which had appeared in 1917 when Shirburnians were being sacrificed at the rate of one a week. It was because of such remarks that *The Loom of Youth* was forbidden at Sherborne, and any boy found with it was subject to an immediate beating.

Yet the renegade author had said little more, although in different language, than was revealed by the headmaster:

> Mind you, I am not attacking the Public School system. I believe in its enormous value, above all in the sense of duty and the loyalty and the law-abidingness which it inculcates. But it cannot escape the dangers attendant upon any system of discipline, the dangers of submitting to mere routine, of adopting ready-made sentiments at second-hand, of a slavish, or perhaps I should rather say a sheepish, want of independence of character.

'The system cannot escape these dangers,' he continued, 'but we individuals . . . can overcome them if we set the right way about it.' It was, however, very difficult for individuals to go against the grain of a total organisation. As Nowell Smith said, 'of all societies very few are so definite and easily understood as a school like this . . . we all here live under a common life, under a common discipline. Our life is organised with great thoroughness, and the organisation is directed to a definite aim. . . .' And the headmaster further observed that 'schoolboys, however much originality they may possess as individuals, are in their conduct to the highest degree conventional.' Nowell Smith was not a small-minded man and somehow managed to reconcile this system of education with his love of Wordsworth's poetry, of which he was an editor. Within the classicist there beat a romantic heart, and one which perhaps troubled him.

But the problem of inspiring 'independence of character' within a system of 'mere routine' arose principally in connection with what was called 'dirty talk', rather than with the more elevated questions treated by the romantic consciousness. The headmaster called upon individuals to show their true patriotism to Sherborne by avoiding it, and appealed to the boy of independent character, who

> brought up in a civilised home, has an instinctive dislike of swearing and coarse jokes and vulgar innuendoes, and yet from sheer cowardice will conceal his dislike, and perhaps force a laugh, and even begin to learn the vile jargon!

In an all-male school there was only one kind of 'vulgar innuendo' possible. Contact between the boys was fraught with sexual potential, a fact which was reflected in the effective ban on associations between boys of different houses, or of different ages. These bans, and the 'gossip' or 'scandal' associated with them, were not part of the *official* life of a public school, but were no less real for that. Nowell Smith might condemn the fact that there was 'one kind of language suitable for home or for a master's ears, and another kind for the study or the dormitory,' but it was a fact of school life. *Natural Wonders* explained that

> We say commonly that we think with our brains. That is true; but it is by no means the whole story. The brain has two halves, just alike, exactly as the body has. In fact, the two sides of the brain are even more precisely alike than the two hands. Nevertheless, we do all our thinking with one side only.

It was Alec Waugh's accusation that Sherborne provided a training in – metaphorically speaking – using two halves of the brain independently. 'Thinking', or rather official thinking, went on in one hemisphere, and ordinary life in the other. It was not hypocrisy: it was that no one in his senses would confuse the two worlds. It worked very well, and only went wrong when something happened to bridge the gap. Then, as Waugh said with some feeling, the real crime was to be found out.

In 1927 the school had changed somewhat in its unofficial conventions. When the boys read *The Loom of Youth* (as of course they did, because it was forbidden) they were rather surprised at the tolerance shown, or at least suggested, of sexual friendships. When the games teams met their counterparts from other public schools, they were amazed at the latitude allowed at the rival establishments. Sherborne boys were at this period asserting a more puritanical, less cynical orthodoxy than that of Alec Waugh's 1914. Nowell Smith was no longer appealing to the independent boys to stamp out what he called 'filth'. But he had not prevented the chemical messages from flowing in four hundred budding 'living apothecary shops', and not even the cold baths had put a stop to 'dirty talk'.

Alan Turing was a boy of independent character, but this subject presented him with a problem which was the opposite of the headmaster's. To most boys 'scandal' would be a quickly-forgotten bantering, alleviating the monotony of school. But to him, it touched the centre of life itself. For although he had surely learnt by now about the birds and the bees, his heart was to be elsewhere. The secret of how the babies were born was hidden well, but everyone knew there *was* a secret. He, however, had been made aware by Sherborne of a secret that in the outside world was not even supposed to exist. And it was *his* secret. For he was drawn by love and desire not only to 'the commonest in nature', but to his own sex.

He was a serious person, and not what Alec Waugh called 'the average boy'. He was not 'in the highest degree conventional', and he was suffering for it. For him there had to be a reason for everything; it had to make sense – and to make one sense, not two. But Sherborne was no help to him in this respect, except in making him more conscious of himself. To be independent he had to work his way through official and unofficial rules alike, and there were certainly no 'ready-made sentiments' for him. At Sherborne the two natural wonders of his life were 'Stinks' and 'Filth'.

If Nowell Smith sometimes had reservations about the public school system, no such doubts assailed Alan's form-master in the autumn of 1927, a certain A. H. Trelawny Ross. A man schooled at Sherborne, who had returned there immediately from Oxford in 1911, he learnt nothing and forgot nothing in his thirty years as a housemaster.[18] A stern foe of 'slackness', he shared none of the headmaster's qualms about slavishness. His style of English also contrasted with that of Nowell Smith, his 1928 'house letter' commencing thus:

> I have a bone to pick with my Captain of the Dayroom (height 4'11"). He
> has been telling people I am a woman-hater. This fib was started some years
> ago by a dame who did not find me gushing enough. My view actually is
> that a woman-hater has a mental kink, just as a female man-hater has, of
> whom there are plenty about. . . .

A narrow nationalist, who had not properly learnt the lesson of loyalty to school as well as to house, Ross was little interested in his form. However, he gave them the benefit of his knowledge and experience of life. He taught Latin translation for a week, Latin prose for a week, and English for a week, this consisting of spelling, 'how to start, write, and address a letter,' 'how to make a précis,' 'how a sonnet is built up, and by a typed summary of the main points, to show how to get good sensible, well-arranged written essays.'

In this respect Ross urged his sensible opinion that, 'As democracy advances, manners and morals recede', and urged the staff to read *The Rising Tide of Colour*. He held that the defeat of Germany had come about 'because she thought that Science and materialism were stronger than religious thought and observance.' He called the scientific subjects 'low cunning', and would sniff and say, 'This room smells of mathematics! Go out and fetch a disinfectant spray!'

Alan used the time on something he found more interesting. Ross caught him doing algebra during time devoted to 'religious instruction', and wrote at half-term:

> I can forgive his writing, though it is the worst I have ever seen, and I try
> to view tolerantly his unswerving inexactitude and slipshod, dirty, work,
> inconsistent though such inexactitude is in a utilitarian, but I cannot forgive
> the stupidity of his attitude towards sane discussion on the New Testament.
>
> He ought not to be in this form of course as far as form subjects go. He
> is ludicrously behind.

In December 1927, Ross placed him bottom in both English and Latin, attaching to the report an inky, blotted page which clearly indicated the negligible amount of energy conceded by Alan to the deeds of Marius and Sulla. Yet even Ross tempered his complaint with the comment 'I like him personally'. O'Hanlon wrote of his 'saving sense of humour'. At home, Alan's messy experiments might be tiresome, but he had a jolly way of coming out with scientific facts, and of telling jokes against his own clumsiness, naive and free from showing off, that it was hard not to like. He was certainly foolish in not making life easier for himself; lazy and perhaps arrogant in supposing he knew what was good for him; but he was not so much obstreperous as bewildered by demands which had nothing to do with his interests. Nor did he complain at home about Sherborne, for he seemed to regard it as the fact of life which indeed it was.

Anyone might like him personally, but as part of a system it was a very different story. At Christmas 1927 the headmaster wrote:

> He is the kind of boy who is bound to be rather a problem in any kind of school or community, being in some respects definitely anti-social. But I think in our community he has a good chance of developing his special gifts and at the same time learning some of the art of living.

With that judgment Nowell Smith suddenly retired, perhaps not sorry to relinquish the contradictions of his community, and the problem of Alan Turing's independent character.

The new year of 1928 marked a period of change for Sherborne. Nowell Smith's successor was a C. L. F. Boughey, who had been an assistant master at Marlborough. By chance, the headmaster's departure had coincided with the death of Carey, the school games master. Between them, as 'Chief' and 'The Bull', the two had divided the Sherborne world into *mens* and *corpus*, and ruled them respectively for twenty years. Carey was succeeded in his role by that bulldog figure Ross.

It also marked a change for Alan. His housemaster asked Blamey, an earnest and also rather isolated boy a year older than Alan, to share a two-boy study with him. Blamey was to try to make Alan more tidy, to 'help him conform, and try and show him there were other things in life besides mathematics.' In the first objective there was a lamentable failure; in the second he came up against the difficulty that Alan 'had wonderful concentration, and would become absorbed in some abstruse problem.' Blamey would consider it his duty to 'interrupt and say it is time to go to chapel, to games, or afternoon classes' as the case might be, he being a well-meaning person, who believed in making the system run as smoothly as possible.[19] O'Hanlon had written at Christmas of Alan that

> No doubt he is very aggravating, and he should know by now that I don't care to find him boiling heaven knows what witches' brew by the aid of two guttering candles on a naked window sill. However, he has borne his afflictions very cheerfully, and undoubtedly has taken more trouble, e.g. with physical training. I am far from hopeless.

Alan's only regret regarding the 'witches' brew' was that O'Hanlon 'had missed seeing at their height the very fine colours produced by the ignition of the vapour produced by super-heated candle-grease.' Alan was still fascinated by chemistry, but not interested in doing it in a way that pleased anyone else. Mathematics and science reports such as '. . . marred by inaccuracy, untidiness, and bad style . . . frightfully untidy both in written and experimental work . . .' continued to reflect his lack of ability to communicate effectively, while admitting that he was 'very promising'. 'His manner of presenting work is still disgusting,' wrote O'Hanlon, 'and takes away much of the pleasure it should give.' 'He doesn't understand what bad manners bad writing and messy figures are.' Ross had passed him on to another form, but he was still placed nearly at the bottom in the spring of 1928. 'His mind seems rather chaotic at present and he finds great difficulty in expressing himself. He should read more,' wrote the master, perhaps more enlightened than Ross.

It was in doubt whether he could take the School Certificate and go on to the sixth form. O'Hanlon and the science masters wanted him to try, and the rest opposed it. The decision had to be made by the new headmaster, who knew nothing of Alan. Boughey had proved himself a new broom, upsetting sacred traditions of the school. The head of the Classical Sixth was no longer automatically the head of school. The prefects had been alienated when he lectured the whole school on 'dirty talk'. (They felt he was judging Sherborne by Marlborough standards.) The staff were horrified when he issued a *fiat*, in front of the school, that there would be no memorial to Carey in the chapel. This incident sealed his doom. The official history[20] would record that

A natural shyness could give an impression of self-esteem and indifference to school affairs that had perhaps no great foundation in fact . . . he had to fight against an ill health that was largely the result of war service and he found it increasingly hard to make the public appearances or even to provide the constant private accessibility which a headmaster's position inevitably demands.

Whether as cause or effect, he was 'poisoned', as Brewster would have put it, by alcohol. The school settled down to a power struggle between Ross and Boughey, and it was the fight between old and new that settled Alan's future, for Boughey over-ruled Ross on principle and allowed him to be entered for the School Certificate.

During the holidays, Alan's father coached him in English. Mr Turing had a great love of literature, although he did not have a mind for abstract ideas. He could recite from memory pages from the Bible, Kipling and humorous Edwardian novels like *Three Men in a Boat*. All this was wasted on Alan, whose set work was *Hamlet*. For a brief moment he pleased his father by saying that at least there was one line he liked. The pleasure was dissipated when Alan explained it was the *last* line: 'Exeunt, bearing off the bodies. . . .'

For the summer term of 1928, Alan was moved to yet another form, that of the Reverend W. J. Bensly, to prepare for the examinations. He saw no reason to depart from his usual pattern, and continued to be placed at the bottom by Bensly, who rashly offered to donate a billion pounds to any charity named by Alan, should he pass in Latin. O'Hanlon, more perceptively, had predicted:

> He has as good brains as any boy that's been here. They are good enough for him to get through even in 'useless' subjects like Latin, French and English.

O'Hanlon saw some of the papers that Alan submitted. They were 'astonishingly legible and tidy'. He passed with credits in English, French, elementary mathematics, additional mathematics, physics, chemistry – and Latin. Bensly never paid up, authority having the privilege of being able to change the rules.

The School Certificate passed, the system allowed him a small part to play, that of the 'maths brain'. There was no mathematical sixth at Sherborne, as at some schools, notably Winchester. There was a science sixth for whom mathematics, Alan's best subject, was regarded as subsidiary. Nor was Alan promoted to the sixth form immediately; he was held back in the fifth for the autumn of 1928, but allowed to join the sixth form for their mathematics classes. These were taught by a young master, Eperson, just a year down from Oxford and a gentle, cultured person, the kind of master who would constantly be played up by the boys. Here was the chance for the system to redeem itself at last, the spirit breaking through the letter of the law. And in a negative way, Eperson did what Alan wanted, by leaving him alone:[21]

> All that I can claim is that my deliberate policy of leaving him largely to his own devices and standing by to assist when necessary, allowed his natural mathematical genius to progress uninhibited . . .

He found that Alan always preferred his own methods to those supplied by the text book, and indeed Alan had gone his own way all the time, making few concessions to the school system. During the machinations over the School Certificate, or even before, he had been studying the theory of relativity from Einstein's own popular account.[22] This employed only elementary mathematics, but gave full rein to ideas which went far beyond anything in the school syllabus. For if *Natural Wonders* had introduced him to the post-Darwinian world, Einstein took him into the twentieth-century revolution of physics. Alan produced a small red Memo Book of notes, which he gave to his mother.

'Einstein here throws doubt,' Alan commented, 'on whether Euclid's axioms, when applied to rigid bodies, hold . . . . He therefore sets out to test . . . the Galilei-Newtonian laws or axioms.' He had identified the crucial point, that Einstein *doubted the axioms*. Not for Alan the 'obvious duties', for nothing was obvious to him. His brother John, who by now regarded Alan with a rather patronising, but not hostile amusement, held that

> You could take a safe bet that if you ventured on some self-evident proposition, as for example that the earth was round, Alan would produce a great deal of incontrovertible evidence to prove that it was almost certainly flat, ovular, or much the same shape as a Siamese cat which had been boiled for fifteen minutes at a temperature of one thousand degrees Centigrade.

Cartesian doubt came as an incomprehensible intrusion into Alan's family and school environment, an intrusion that the English coped with more by laughter than by persecution. But doubt being a very difficult and rare state of mind, it had taken the whole intellectual world a very long time to ask whether the 'Galilei-Newtonian laws or axioms', apparently 'self-evident', were actually true. Only by the late nineteenth century was it recognised that they were inconsistent with the known laws of electricity and magnetism. The implications were frightening, and it had needed Einstein to take the step of saying that the assumed basis of mechanics was actually *incorrect*, thereby creating the Special Theory of Relativity in 1905. This then proved inconsistent with Newton's laws of gravity, and to remove these contradictions Einstein had gone even further, casting doubt even on Euclid's axioms of space to create the General Theory of Relativity in 1915. The point of what Einstein had done did not lie in this or that experiment. It lay, as Alan saw, in the ability to doubt, to take ideas seriously, and to follow them to a logical if upsetting conclusion. 'Now he has got his axioms,' wrote Alan, 'and is able to proceed with his logic, discarding the old ideas of time, space, etc.'

Alan also saw that Einstein avoided philosophical discussions of what space and time 'really were', and instead concentrated on something that could in principle be done. Einstein placed great emphasis on 'rods' and 'clocks' as part of an *operational* approach to physics, in which 'distance', for instance, only had meaning in terms of some well-defined measuring operation, not as an absolute ideal. Alan wrote:

> It is meaningless to ask whether the two p[oin]ts are always the same distance apart, as you stipulate that that distance is your unit and your ideas have to go by that definition. . . . These ways of measuring are really conventions. You modify your laws to suit your method of measurement.

No respecter of persons, he preferred a piece of working of his own to that supplied by Einstein 'because in this way I think it should seem less "magicky".' He reached the very end of the book, and gave a masterly derivation of the law* which in General Relativity would supplant Newton's axiom, that a body subject to no external force would move in a straight line with constant speed:

---

* Usually called 'the law of geodesic motion'.

He has now got to find the general law of motion for bodies. It will have, of course, to satisfy the general Principle of Relativity. He does not actually give the law, which I think is a pity, so I will. It is: 'The separation between any two events in the history of a particle shall be a maximum or minimum when measured along its world line.'

To prove it, he brings in the Principle of Equivalence, which says that: 'Any natural gravitational field is equivalent to some artificial one.' Suppose then that we substitute an artificial field for the natural one. Now as the field is artificial there is some system at that p[oin]t which is Galileian, and as it is Galileian the particle will be moving uniformly relative to it, i.e. it has a straight world line relatively to it. Straight lines in Euclidean space have always a maximum or minimum length between two p[oin]ts. Therefore the world line satisfies the conditions given above for one system, therefore it satisfies it for all.

As Alan explained, Einstein had not stated this law of motion in his popular account. Alan might just possibly have guessed it for himself. On the other hand, he could very well have found it in another work which was published in 1928, and which he was reading by 1929 – *The Nature of the Physical World* by Sir Arthur Eddington. Professor of Astronomy at Cambridge, Eddington had worked on the physics of the stars and the development of the mathematical theory of relativity. This influential book, however, was one of his popular works, in which he set out to convey the great change in the scientific world-picture that had taken place since 1900. Its rather impressionistic account of relativity did state the law of motion, although without proof, and might have supplied its form to Alan. Certainly, in one way or another, Alan had done more than study a book, for he had put several ideas together for himself.

This study arose out of his own initiative, and Eperson did not know about it. He was thinking quite independently of his environment, which offered him little but nagging and scolding. He had had to look to his totally bewildered mother for a little encouragement. But then something new happened to put him into contact with the world.

There was a boy in another house – Ross's house, in fact – whose name was Morcom. As yet he was nothing but 'Morcom' to Alan, although later[23] he became 'Christopher'. Alan had first noticed

Christopher Morcom early in 1927, and had been very struck by him, partly because he was surprisingly small for his form. (He was a year older than Alan and a year ahead in the school, but fair-haired and slight.) It was also, however, because he 'wanted to look again at his face, as he felt so attracted.' Later in 1927 Christopher had been away from school and then had returned looking, Alan noticed, very thin in the face. He shared with Alan a passion for science, but he was a very different person. The institutions that were for Alan such stumbling-blocks had been for Christopher Morcom the instruments of almost effortless advance, the source of scholarships, prizes and praise. He again returned late to school this term, but when he arrived Alan was waiting for him.

His utter loneliness was pierced at last. It was difficult to make friends with an older boy from another house. Nor was Alan good at conversation. But he found an *entrée* in mathematics. 'During the term Chris and I began setting one another our pet problems and discussing our pet methods.' It would be impossible to separate the different aspects of thought and feeling. This was first love, which Alan would himself come to regard as the first of many for others of his own sex. It had that sense of surrender ('worshipped the ground he trod on'), and a heightened awareness, as of brilliant colour bursting upon a black and white world. ('He made everyone else seem so ordinary.') At the same time, it was most important that Christopher Morcom was someone who took scientific ideas seriously. And gradually, though always with considerable reserve, he took Alan seriously. ('My most vivid recollections of Chris are almost entirely of the kind things he said to me sometimes.') So these elements were all present, and had the effect of giving Alan reason to communicate.

Before and after Eperson's classes Alan might talk to Christopher about relativity, or might show him other pieces of work. He had, for example, calculated π to thirty-six decimal places at about this time, perhaps making use of his own series for the inverse tangent function, and being much annoyed to find an error in the last decimal place. After a time, Alan found another opportunity to see Christopher. By accident he discovered that during a certain period on Wednesday afternoons set for private study, Chris went to the library and not to his house. (Ross did not allow boys to work unsupervised, fearing the sexual potential in unregulated associations.) 'I so enjoyed Chris' company there,' wrote Alan, 'that ever since I always used to go to the library instead of my study.'

Yet another chance arose through the gramophone society which the progressive Eperson had started. Christopher, a fine piano player, was a keen member. Alan had little interest in music, but sometimes on Sunday afternoons he would go to Eperson's lodgings with Blamey (who also had a gramophone and records in their shared study). There he could sit and steal glances at Christopher while the 78's played out their disjointed versions of the great symphonies. This was, incidentally, part of Blamey's noble effort to show Alan that there were other things in life besides mathematics. He also showed Alan how to make a crystal wireless receiver out of basic materials, having noticed that Alan had little pocket money for such things. Alan insisted on winding the coils for the variometer and was delighted to find that his clumsy hands had made something that actually worked, even if he could never aspire to rival Christopher's dexterity.

At Christmas, Eperson reported:

> This term has been spent, and the next two terms will have to be spent, in filling in the many gaps in his knowledge and *organising* it. He thinks very rapidly and is apt to be 'brilliant' but unsound in some of his work. He is seldom defeated by a problem, but his methods are often crude, cumbersome and untidy. But thoroughness and polish will no doubt come in time.

He would have found the Higher Certificate dull stuff, compared with the job of organising Einstein. But he cared more about his own failure to fit in with what was expected, now that Christopher had done 'hopelessly better' in the test at the end of term. In the new year of 1929 there was another shuffle, and Alan joined the sixth form proper, so that he did all his classes with Christopher. He made a point of sitting next to him in every class right from the start. Christopher, Alan wrote,

> made some of the remarks I was afraid of (I know better now) about the coincidence but seemed to welcome me in a passive way. It was not long before we began doing experiments together in Chemistry and we were continually changing our ideas on all sorts of subjects.

Unfortunately Christopher was away from classes with a cold for most of January and February, and Alan could work with him only for five weeks of the spring term.

Chris' work was always better than mine because I think he was very thorough. He was certainly very clever but he never neglected details, and for instance very seldom made arithmetical slips. He had a great power in practical work of finding out just what was the best way of doing anything. To give an example of his skill, he could estimate when a minute had passed to within half a second. He could sometimes see Venus in the day-time. Of course he was born with very good eyes, but still I think it is typical of him. His skill extended to all sorts of more everyday things, such as driving, fives and billiards.

One cannot help admiring such powers and I certainly wanted to be able to do that kind of thing myself. Chris always had a delightful pride in his performances and I think it was this that excited one's competitive instinct to do something which might fascinate him and which he might admire. This pride extended to a pride in his possessions. He used to demonstrate the virtues of his 'Research' fountain pen in a way which made my mouth water and then admitted he was trying to make me jealous.

Slightly inconsistently, Alan also wrote:

Chris always seemed to me very modest. He would never for instance tell Mr Andrews that his ideas weren't sound although the opportunity occurred again and again. More particularly he very much disliked to offend anyone in any way and often used to apologise (e.g. to masters) in cases where the average boy would not dream of doing so.

The average boy, as all school stories and magazines admitted, held the masters in contempt – especially in 'Stinks'. It was the most obvious contradiction of the system. But Christopher rose above it all:

A thing about Chris which I think is very unusual, is that he had a very definite code of morals. One day he was talking about an essay in an exam and how it had led to the subject of 'right and wrong'. 'I have some very definite ideas of "right and wrong",' he said. Somehow I never seemed to doubt that anything that Chris would do would be right, and I think there was a lot more in that than blind admiration.

Take dirty talk for instance. The idea of Chris having to do with such a thing seemed simply ludicrous, and of course I do not know anything at all about Chris at the house, but I should think in this respect he would prevent dirty talk by making people not want to do it rather than making them avoid shocking him. This of course tells you nothing but the way his personality impressed me. I remember an occasion when I made a remark to him on purpose, that would decidedly not pass in a drawing room, but which would not be thought anything of at school, just to see how he would take it. He made me feel sorry for saying it, without him in any way seeming silly or priggish.

Despite all these amazing virtues, Christopher Morcom was human. He had nearly got into trouble when he was dropping stones down train funnels from the railway bridge and struck a railwayman instead. Another exploit involved sending gas-filled balloons over the field to the Sherborne Girls' School. Nor was their time in the laboratories too solemn. Another boy, a tough athlete called Mermagen, joined them for physics, and the three of them had to work through the practical experiments in a little annex while Gervis taught his class. These classes were enlivened by Gervis's sausage-lamps, painted bulbs which he used as electrical resistances. 'Take another sausage-lamp, boy!' was his catchphrase, and the three of them worked out a comic sketch around the things, which Christopher was thinking of setting to music.

In the summer term of 1929 they were doing only the dull revision work for the Higher School Certificate, but even this was coloured by romance since 'As always it was my great ambition to do as well as Chris. I was always as well supplied with ideas as he, but have not the same thoroughness in carrying them out.' Alan had never before taken any notice of naggings to take care over details and style, since he had worked for himself, by himself. But now perhaps he recognised that what was good enough for Christopher Morcom was good enough for him, and that he should train himself to communicate in the way that the system required. He had not yet acquired the necessary skill. Andrews observed that he was 'at last trying to improve his style in written work,' but Eperson, writing that his work for the Higher Certificate showed 'distinct promise', re-emphasised the need to 'put a neat and tidy solution on paper.'

The examiner for the mathematics of the Higher Certificate[24] commented that:

> A. M. Turing showed an unusual aptitude for noticing the less obvious points to be discussed or avoided in certain questions and for discovering methods which would at once shorten or illumine the solutions. But he appeared to lack the patience necessary for careful computation of algebraic verification, and his handwriting was so bad that he lost marks frequently – sometimes because his work was definitely illegible, and sometimes because his misreading his own writing led him into mistakes. His mathematical ability was not of a standard to compensate entirely for the cumulative effects of these faults.

Alan obtained 1033 marks in the mathematics papers, compared with Christopher's 1436.

The Morcoms were a wealthy, vigorous scientific and artistic family, with a base in a Midlands engineering firm. They had developed a Jacobean dwelling near Bromsgrove in Worcestershire into a large country house, the Clock House, where they lived in some style. Christopher's grandfather had been an entrepreneur in stationary steam engines, and the Birmingham company of Belliss and Morcom, of which his father, Colonel Reginald Morcom, had recently become chairman, now also built steam turbines and air compressors. Christopher's mother was the daughter of Sir Joseph Swan, who starting from a very ordinary background had become in 1879 the inventor, independently of Edison, of the electric light. Colonel Morcom retained an active interest in scientific research, while Mrs Morcom matched his energy in her own pursuits. At the Clock House she ran a goat farm; she bought and renovated cottages in the neighbouring village of Catshill; she was out every day on some project or county duty. She had studied in London at the Slade School of Art, and in 1928 returned there, taking a flat and a studio near Victoria, and producing sculpture of force and style. It was typical of her flair and zest that when back at the Slade she still pretended to be 'Miss Swan', but then invited other art students back to the Clock House, involving herself in absurd disguises when she doubled as Mrs Morcom.

Rupert Morcom, the elder son, had entered Sherborne in 1920, and had won a scholarship in science to Trinity College, Cambridge;

he was now engaged in research at the Technische Hochschule in Zurich. Like Alan he was an avid experimenter, but one with the advantage that his parents had been able to construct a laboratory for him at home. His younger brother, who also had the use of the laboratory, now told Alan of all this, exciting great envy.

In particular, Christopher told Alan about an experiment that Rupert had taken up before going to Cambridge in 1925. It concerned a chemical effect which Andrews often used to draw the interest of the younger boys. By chance it involved Alan's old favourite, iodine. Solutions of iodates and sulphites, when mixed, would result in the precipitation of free iodine, but in a rather striking way. Alan later explained: 'It is a beautiful experiment. Two solutions are mixed in a beaker and after waiting for some very definite period of time, the whole suddenly becomes a deep blue. I have known it take a time, 30 secs., and then turn blue in 1/10 of a sec. or less.' Rupert had been investigating not the easy problem of working out the recombination of ions, but that of explaining this time delay. It required a knowledge of physical chemistry, and an understanding of differential equations, both beyond the school syllabus. Alan wrote:

> Chris and I wanted to find a relation between the time and the concentrations of the solutions and thereby verify Rupert's theories. Chris had already done some experiments upon it. We were looking forward very much to the experiment. The results unfortunately did not agree with theory and I made more experiments during the following holidays and invented a new theory. I sent the results to him and so we started to write to one another in the holidays.

He did more than write to Christopher – he invited him to Guildford. Ross, as housemaster, would have been horrified by this audacious step.[25] Christopher replied[26] (after some delay) on 19 August:

> . . . Before getting on to experiments I must thank you very much for your invitation to come and stay with you, but I am afraid I shall not be able to come as we are going away somewhere, probably abroad for about three weeks, just at that time . . . I am sorry not to be able to come; it is very kind of you to ask me.

As for the iodates, new ventures at the Clock House had rendered them definitely *passé*. There were experiments to measure air resistance, liquid friction, another problem in physical chemistry with Rupert ('I enclose the integral which you might like to try'), plans for a twenty foot long reflecting telescope, and

> . . . So far all I have done is to make an adding machine for pounds and ounces. It works surprisingly well. I think l have given up Maths for these holidays, having just read a very good book on Physics in general including relativity.

Alan laboriously copied the ingenious experiment on air resistance that Christopher had devised and wrote back with more ideas about chemistry and a mechanics problem, only for Christopher to pour cold water on both in a letter of 3 September:

> I haven't studied your conical pendulum carefully but I can't so far understand you're [*sic*] method. Incidentally I believe you're equations of motion have a mistake in them. . . .
> I am now helping my brother analyse American plasticine for an artist. . . . The procedure is to boil with organic solvents. . . . I made a quite good plasticine and very nearly like the stuff we want, by mixing this iron soap with flowers of sulphur . . . and adding a little mutton fat. Hope you are having good holidays; see you on 21st, Yrs, C. C. Morcom.

But chemistry had now given way to astronomy, to which Christopher had introduced Alan earlier in the year. Alan had been given Eddington's *Internal Constitution of the Stars* by his mother for his seventeenth birthday, and had also acquired a 1½-inch telescope. Christopher had a four-inch telescope ('He never tired of talking about his wonderful telescope if he thought one was interested') and had been given a star atlas for his eighteenth birthday. Besides astronomy, Alan was also reading deep into *The Nature of the Physical World*, for in his letter[27] of 20 November 1929 there was a paraphrase of part of its account:

> Schrödinger's quantum theory requires 3 dimensions for every electron he considers. Of course he does not believe that there are really about 10^{70} dimensions, but that this theory will explain the behaviour of an electron. He thinks of 6 dimensions, or 9, or whatever it may be without forming

any mental picture. If you like you can say that for every new electron you introduce these new variables analogous to the coordinates of space.

This came from Eddington's description of that other change in fundamental physical concepts, one much more mysterious than relativity. The quantum theory had done away with the billiard-ball corpuscles and the ethereal waves of the nineteenth century, and replaced both by entities which had characteristics both of particles and of waves; lumpy but nebulous.

Eddington had a lot to say, for the 1920s had been a decade of rapid advance in theoretical physics, following up the spate of discoveries at the turn of the century. In 1929 Schrödinger's formulation of the quantum theory of matter was only three years old. The two boys also read books by Sir James Jeans, the other Cambridge astronomer, and here too there were entirely new developments. It had only just been established that some nebulae were clouds of gas and stars on the margins of the Milky Way, and that others were completely separate galaxies. The mental picture of the universe had expanded a millionfold. Alan and Christopher discussed these ideas and 'usually didn't agree', wrote Alan, 'which made things much more interesting.' Alan kept 'some pieces of paper with Chris' ideas in pencil and mine in ink scrawled all over them. We even used to do this during French.'

The date 28/9/29 appeared on them, and so did the official work:

Monsieur . . . recevez monsieur mes salutations empresseés*
Cher monsieur . . . Veuillez agréer l'expression des mes sentiments distingués
Cher ami . . . Je vous serre cordialement la main . . . mes affectueux souvenirs
. . . votre affectioné

but also there were generalised noughts and crosses, a reaction involving iodine and phosphorus, and a diagram which suggested doubting Euclid's axiom that for every line there would be exactly one parallel line passing through a given point.

Alan kept these pages, as *souvenirs affectueux*, although he could never express his *sentiments distingués*. As for *serrer cordialement la main*, or more – that was probably pretty firmly repressed in his mind, although soon he would write: 'There were times when I felt

---

* This piece of work was marked 'Nine wrong genders. 5/25. Very poor.'

his personality particularly strongly. At present I am thinking of an evening when he was waiting outside the labs, and when I came too, he grasped me with his big hand and took me out to see the stars.'

Alan's father was delighted, if amazed, when the reports began to change their tone. His interest in mathematics was confined to the calculation of income tax, but he was proud of Alan, and so was John, who admired him for taking on the system and getting away with it. There had been a method in his madness all along. Unlike his wife, Mr Turing never claimed to have the faintest idea of what his son was doing, and this was the theme of a punning couplet that Alan once read out from his father's letter in his study:

> I don't know what the *'ell 'e meant*
> But that is what *'e said 'e meant* !

Alan seemed quite happy with this bluff and trusting ignorance. Mrs Turing, however, took the more accusing line of 'I told you so,' and made a good deal of the idea that her choice of school had been the right one. She had certainly paid a certain amount of attention to Alan, and it had not all been in the direction of moral improvement, for she liked to feel that she understood his love of science.

Alan was now in a position to think of winning a scholarship to university, a scholarship representing not only merit but a reasonable income, almost enough to live on as an undergraduate. An exhibition, awarded to second-class candidates, would mean significantly less. Christopher, now eighteen, was expected to win a Trinity College scholarship like his brother. It was ambitious of Alan to attempt the same at seventeen. In mathematics and science, Trinity held the highest reputation among the colleges of the university which was itself, after Göttingen in Germany, the scientific centre of the world.

The public schools were good at putting candidates through the daunting procedure for entrance scholarships to the ancient universities, and Sherborne also gave Alan a £30 per annum subsidy. But there was no automatic red carpet laid down. The scholarship examinations were distinguished by questions of an open-ended, imaginative kind, without a published syllabus. They gave a taste of future life. To Alan this was an excitement in itself, but there was more than this to stimulate his ambition. There was Christopher, who would so shortly be leaving Sherborne; there was some muddle over when this would be, but it would probably be at Easter 1930. To fail in the scholarship would be to lose Christopher for more than a year. Perhaps it was this uncertainty that provoked gloomy forebodings in November, when Alan had recurring thoughts that something would happen before Easter to prevent Christopher from going to Cambridge.

The Cambridge examinations opened up the prospect of a whole week in Christopher's company, unconstrained by the house system – 'I was looking forward as much to spending a week with Chris as to seeing Cambridge.' On Friday 6 December, Christopher's study-mate Victor Brookes was to be driving from London to Cambridge, and had offered to take Alan as well as Christopher. They went on the train together to London, where they stopped off to see Mrs Morcom. She took them to her studio, allowed them to play at chipping marble from a bust that she was working on, and then gave them lunch at her flat. Christopher used to tease Alan a good deal, and had a particular running joke about 'deadly stuff', the joke being to pretend that certain harmless substances were really poisonous. He joked about the vanadium in the special Morcom vanadium-steel cutlery being 'absolutely deadly'.

In Cambridge they could live the lives of young gentlemen for a week, with rooms of their own and no lights-out. There was dinner in the Hall of Trinity College, in evening dress, with the portrait of Newton looking down. It was an opportunity to meet and compare themselves with candidates from other schools. Alan made one new acquaintance, Maurice Pryce, with whom he established an easy *rapport* through almost identical interests in mathematics and physics. Pryce was taking the examination for the second time. A year before he had sat under Newton's portrait and had said to himself that now nothing else would suffice. And although Christopher was rather blasé about everything, that was what it was like for them all: nothing could be quite the same again.

It was, wrote Alan, 'a very good meal', and then they

went to play Bridge with some other Sherburnians in Trinity Hall. We were . . . to be back at our Colleges by 10 o'clock but at 4 minutes to 10 Chris wanted to play another hand. I wouldn't let him, and as it was, we were only back just in time. The next day, Saturday, we played cards again

'Rummy' this time. After ten o'clock Chris and I went on playing other games. I remember very clearly Chris' broad smile when we decided we didn't want to go to bed just yet. We played till 12-15. A few days later we tried to get into the Observatory. We had been invited by an astronomer friend of Chris' to go there if it was fine. Our idea of what was fine did not quite agree with his.

Christopher 'loved all games and was always finding out new ones (of the more trivial kind).' He used to 'try to make people believe things that were credible but just not true,' and at Cambridge persuaded Alan to advance his watch by twenty minutes. 'He was immensely pleased when I found out.' They also went to the cinema together, joined by Norman Heatley, who had been Christopher's friend at a preparatory school, and was now a Cambridge undergraduate. Christopher told him how Alan had a notation of his own for the calculus, and had to translate everything into standard formulae when he did examinations. This aspect of Alan's independence also worried Eperson, who found that 'on paper his solutions were often unorthodox, and required the writer's elucidation.' He doubted whether the Cambridge examiners would perceive the mind that struggled behind the hand.

On the way back from the cinema, Alan hung back and walked with Heatley, to test how much Christopher wanted his company. He was rewarded:

> Evidently I looked rather lonely as Chris beckoned to me (mostly I think with his eyes) to walk beside him. Chris knew I think so well how I liked him, but hated me shewing it.

Alan was conscious that he was a boy in another house, and that everything was open to comment. ('We never went on bicycle rides together. I think perhaps Chris was rather ragged about me at the house.') But this pleased him 'ever so much'.

After what Alan said had been the happiest week of his life, the boys went back to school on 13 December for the last few days of term. At the house supper, they sang about Alan:

> The maths brain lies often awake in his bed
> Doing logs to ten places and trig in his head

The results were published on 18 December in *The Times*, just after term ended. It was a Great Crash. Christopher had won a Trinity scholarship, and Alan had not. Writing in congratulation, Alan had a letter in return with a particularly friendly tone:

20/12/29

Dear Turing,

Thank you very much for your letter. I was as sorry you did not get a schol as I was pleased that I did. What Mr Gow says means that you would have certainly got an Exhibition if you had put it down . . .

. . . Have had two of the clearest nights I have known. I have never seen Jupiter better and I could see 5 or 6 belts and even some detail on one of the large central belts. Last night I saw no. I satellite come out from eclipse. It appeared quite suddenly (during a few seconds) at some distance from Jupiter and looked very attractive. It is the first time I have seen one. I also saw Andromeda Neb. very clearly but did not stay out long. Saw spectrum of Sirius, Pollux and Betelgeux and also bright line spectrum of Orion nebula. Am at moment making a spectrograph. Will write again later. Happy Christmas etc.   Yrs ever C. C. M.

Anything like 'making a spectrograph' was far beyond the resources Alan enjoyed at Guildford, but he got hold of an old spherical glass lampshade, filled it with plaster of Paris, covered it with paper (which made him think about the nature of curved surfaces) and set out to mark in the constellations of fixed stars. Typically, he insisted on doing it from his own observation of the night sky, although it would more easily and accurately have been done from an atlas. He trained himself to wake at four o'clock in the morning so that he could mark in some stars not visible in the December evening sky, thus waking up his mother, who thought she had heard a burglar. This done, he wrote to Christopher about it, also asking him whether he thought it would be advisable to try for a college other than Trinity next year. If this was a test of affection, he was again rewarded, for Christopher replied:

5/1/30

Dear Turing,

. . . I really can't give you any advice about exams because it is nothing to do with me and I feel it would not be quite write [*sic*]. John's is a very good College, but of course I should prefer personally that you came to Trinity where I should see more of you.

I should be very interested to see your star map when it is done but I suppose it is quite impracticable to bring it to school or anything. I have often wanted to make a star globe, but have never really bothered, especially now I have got the star atlas going down to 6th mag. . . .

Recently I have been trying to find Nebulae. We saw some quite good ones the other night, one very good planetary in Draco 7th mag. 10″. Also we have been trying to find a Comet 8th mag. in Delphinus . . . . I wonder if you will be able to get hold of a telescope to look for it with your 1½″ will be useless for such a small object. I tried to compute its orbit but failed miserably with 11 unsolved equations and 10 unknowns to be eliminated.

Have been getting on with plasticine. Rupert has been making horrid smelling soaps and fatty acids from . . . Rape Oil and Neal's Boot Oil. . . .

This letter was written from his mother's flat in London, where he was 'to see the dentist . . . and also to avoid a dance at home.' Next day he wrote again from the Clock House:

. . . I found the Comet at once in its assigned position. It was much more obvious and interesting than I had expected . . . I should say it is nearly 7th mag. It . . . should be obvious in your telescope. The best way is to learn the 4th & 5th mag. stars by heart, and move slowly to the right place, never losing sight of *all* the known stars. . . . In about half an hour I shall look again if it is clear (it has just clouded) and see if I can notice its motion among the stars and also see what it looks like with the powerful eyepiece (×250). The group of 5 4th mag. stars in Delphinus come into the field of the finder in pairs. Yrs.   C. C. Morcom.

But Alan had already seen the comet, though in a more haphazard manner

10/1/30

Dear Morcom,

Thank you very much for the map for finding the comet. On Sunday I think I must have seen it. I was looking at Delphinus and thinking it was Equuleus and saw something like this [*a tiny sketch*] rather hazy and about 3′ long. I am afraid I did not examine it very carefully. I then looked for the comet elsewhere in Vulpecula thinking it was Delphinus. I knew from the

Times that there was a comet in Delphinus that day.

. . . The weather really is annoying for this comet. Both on Wednesday and today I have had it quite clear until sunset and then a bank of cloud comes over the region of Aquila. On Wednesday it cleared away just after the comet had set. . . .

Yours A. M. Turing

Please don't always thank me for my letters so religeously. I'll let you thank me for writing them legibly (if I ever do) if you like.

Alan plotted the course of the comet, as it sped from Equuleus into Delphinus in the frosty heavens. He took the primitive star globe back to school to show to Christopher. Blamey had left at Christmas, and Alan now had to share another study, in which the inky sphere was poised. There were but few constellations marked in, but they amazed the younger boys with Alan's erudition.

Three weeks into the term, on 6 February, some visiting singers gave a concert of sentimental part-songs. Alan and Christopher were both present, and Alan was watching his friend, trying to tell himself, 'Well, this isn't the last time you'll see Morcom.' That night he woke up in the darkness. The abbey clock struck; it was a quarter to three. He got out of bed and looked out of the dormitory window to look at the stars. He often used to take his telescope to bed with him, to gaze at other worlds. The moon was setting behind Ross's house, and Alan thought it could be taken as a sign of 'goodbye to Morcom'.

Christopher was taken ill in the night, at just that time. He was taken by ambulance to London, where he underwent two operations. After six days of pain, at noon on Thursday 13 February 1930, he died.

2 The Spirit of Truth

I sing the body electric,
The armies of those I love engirth me and I engirth them,
They will not let me off till I go with them, respond to them,
And discorrupt them, and charge them full with the charge of the soul.
Was it doubted that those who corrupt their own bodies conceal
   themselves?
And if those who defile the living are as bad as they who defile the
   dead?
And if the body does not do fully as much as the soul?
And if the body were not the soul, what is the soul?

No one had told Alan that Christopher Morcom had contracted bovine tuberculosis from drinking infected cows' milk as a small boy; it had set up a pattern of internal damage, and his life had been constantly in danger. The Morcom family had gone to Yorkshire in 1927 to observe the total eclipse of the sun on 29 June, and Christopher had been taken terribly ill in the train coming back. He had undergone an operation, and that was why Alan had been struck by his thin features when he returned to school late that autumn.

'Poor old Turing is nearly knocked out by the shock,' a friend wrote from Sherborne to Matthew Blamey next day. 'They must have been awfully good friends.' It was both less and more than that. On his side, Christopher had at last been becoming friendly, rather than polite. But on Alan's side – he had surrendered half his mind, only to have it drop into a void. No one at Sherborne could have understood. But on the Thursday that Christopher died, 'Ben' Davis, the junior housemaster, did send to Alan a note telling him to prepare for the worst. Alan immediately wrote[1] to his mother, asking her to send flowers to the funeral, which was held on the Saturday, at dawn. Mrs Turing wrote back at once and suggested that Alan himself write to Mrs Morcom. This he did on the Saturday.

15/2/30

Dear Mrs Morcom,

I want to say how sorry I am about Chris. During the last year I worked with him continually and I am sure I could not have found anywhere another companion so brilliant and yet so charming and unconceited. I regarded my interest in my work, and in such things as astronomy (to which he introduced me) as something to be shared with him and I think he felt a little the same about me. Although that interest is partly gone, I know I must put as much energy if not as much interest into my work as if he were alive, because that is what he would like me to do. I feel sure that you could not possibly have had a greater loss.

Yours sincerely, Alan Turing

I should be extremely grateful if you could find me sometime a little snapshot of Chris, to remind me of his example and of his efforts to make me careful and neat. I shall miss his face so, and the way he used to smile at me sideways. Fortunately I have kept all his letters.

Alan had awoken at dawn, at the time of the funeral:

I am so glad the stars were shining on Saturday morning, to pay their tribute as it were to Chris. Mr O'Hanlon had told me when it was to take place so that I was able to follow him with my thoughts.

Next day, Sunday, he wrote again, perhaps in more composed form, to his mother:

16/2/30

Dear Mother,

I wrote to Mrs Morcom as you suggested and it has given me a certain relief. . . .

. . . I feel sure that I shall meet Morcom again somewhere and that there will be some work for us to do together, and as I believed there was for us to do here. Now that I am left to do it alone I must not let him down but put as much energy into it, if not as much interest, as if he were still here. If I succeed I shall be more fit to enjoy his company than I am now. I remember what G O'H said to me once 'Be not weary of well doing for in due season ye shall reap if ye faint not' and Bennett* who is very kind on these occasions

---

* John Bennett was a boy in the house, who himself died later in 1930 on a lone winter trek across the Rockies.

'Heaviness may endure for a night but joy cometh in the morning'. Rather Plymouth brotherish perhaps. I am sorry he is leaving. It never seems to have occurred to me to try and make any other friends besides Morcom, he made everyone seem so ordinary, so that I am afraid I did not really appreciate our 'worthy' Blamey and his efforts with me for instance. . . .

On receiving Alan's letter, Mrs Turing wrote to Mrs Morcom:

Feb. / 17 / 30

Dear Mrs Morcom,

Our boys were such great friends that I want to tell you how much I feel for you, as one mother for another. It must be terribly lonely for you, and so hard not to see here the fulfilment, that I am sure there will be, of all the promise of Christopher's exceptional brains and lovable character. Alan told me one couldn't help liking Morcom and he was himself so devoted to him that I too shared in his devotion and admiration: during exams he always reported Christopher's successes. He was feeling very desolate when he wrote asking me to send flowers on his behalf and in case he feels he cannot write to you himself I know he would wish me to send his sympathy with mine.

Yours sincerely, Ethel S. Turing

Mrs Morcom immediately invited Alan to stay at the Clock House in the Easter holiday. Her sister Mollie Swan sent him a photograph of Christopher. Sadly, the Morcoms had very few pictures of him, and this was a poor likeness, taken on an automatic machine with a reversed image. Alan replied:

20 / 2 / 30

Dear Mrs Morcom,

Thank you very much for your letter. I should enjoy coming to the Clockhouse immensely. Thank you so much. We actually break up on April 1, but I am going to Cornwall with Mr O'Hanlon my housemaster until the 11th – so that I could come any time that suits you between then and the beginning of May. I have heard so much about the Clockhouse – Rupert, the telescope, the goats, the Lab and everything.

Please thank Miss Swan very much for the photograph. He is on my table now, encouraging me to work hard.

Apart from the photograph, Alan had to keep his emotions to himself. He was allowed no mourning period, but had to go through

Corps and Chapel like everyone else. Alan's devotion to Christopher's memory had come as a surprise to the Morcoms. Christopher had always been reticent at home about his school friends, and had a way of referring to 'a person called' so-and-so as though he had never been mentioned before. 'A person called Turing' had featured in a few of his remarks about experiments, but no more than that, and the Morcom parents had only very briefly met Alan in December. They knew him only from his letters. At the beginning of March they changed their plans and decided to take the holiday in Spain which had been planned before Christopher died. So it was testimony to Alan's letters that on 6 March they invited him to take Christopher's place on the journey, instead of coming to their home. Alan wrote to his mother the next day:

> . . . I am half sorry it is not to be the Clockhouse as I should like very much to see it and everything that Morcom has told me about there – but I don't get invited to go to Gibraltar every day of the week.

On 21 March the Morcoms paid their farewell visit to Sherborne and Alan was allowed into Ross's house to see them in the evening. Term ended a week later and Alan went to Rock, on the north coast of Cornwall, with O'Hanlon, whose private income allowed him to treat groups of boys in this way. The party included the tough Ben Davis and three Westcott House boys, Hogg and Bennett and Carse. Alan wrote later to Blamey that he 'had a very good time there – plenty to eat and a pint of beer after lots of exercise.'

While he was away, Mrs Turing called on Mrs Morcom in her London flat. Mrs Morcom recorded in her diary (6 April):

> Mrs Turing came to see me at flat tonight. Had not met her before. We talked nearly all the time about Chris and she told me how much he had influenced Alan and how Alan thought he was still working with him and helping him. She stayed till nearly 11 and had to get back to Guildford. She had been to Bach Concert at Queens Hall.

After ten days in Cornwall, Alan made a quick stop-over at Guildford, where Mrs Turing hastily tried to put him in order (extracting the usual quota of old handkerchieves from the lining of his overcoat), and on 11 April he arrived at Tilbury, joining the Morcom party on the *Kaisar-i-Hind*. Besides Colonel and Mrs Morcom, and Rupert, this included a director of Lloyds Bank and a Mr Evan Williams, chairman of Powell Dyffryn, the Welsh mining company. Mrs Morcom wrote in her diary:

> . . . Sailed about noon. Wonderful day with bright sunshine until 3.30 when we began to come into mist and slowed down. Before tea we dropped anchor and remained just outside the mouth of the Thames until midnight. Ships all around us blowing fog-horns and sounding bells. . . . Rupert and Alan very excited about the fog and it really is rather alarming.

Alan shared a cabin with Rupert, who did his best to draw him out on Jeans and Eddington, but found Alan very shy and hesitant. Each night before going to sleep, Alan spent a long time looking at the photograph. On the first morning of the voyage, Alan began to talk to Mrs Morcom about Christopher, releasing his feelings in speech for the first time. The next day, after deck tennis with Rupert, was spent the same way, telling her how he had felt attracted to Christopher before getting to know him, about his presentiments of catastrophe and the moon setting. ('It is not difficult to explain these things away – but, I wonder!') On Monday, as they rounded Cape Vincent, Alan showed her the last letters he had received from Christopher.

They only spent four days on the Peninsula, driving over the hairpin bends to Granada where, it being Holy Week, they saw a religious procession in the starlight. On Good Friday they were back in Gibraltar and embarked on a homebound liner the next day. Alan and Rupert took early Communion on board ship on Easter Sunday.

Rupert was by now impressed with Alan's originality of thought, but he did not think of Alan as in a different class from the Trinity mathematicians and scientists he had known. Alan's future seemed unsure. Should he read science or mathematics at Cambridge? Was he sure of a scholarship at all? Somewhat in terms of a last resort, he spoke to Evan Williams about scientific careers in industry. Williams explained the problems of the coal industry, for instance the analysis of coal-dust for toxicity, but Alan was suspicious of this and remarked to Rupert that it might be used to cheat the miners by flourishing a scientific certificate at them.

They had done the trip in style, staying at the best hotels, but what Alan wanted most was to visit the Clock House. Mrs Morcom sensed this and gracefully asked him to 'help' her look through Christopher's papers and sort them. So on the Wednesday, Alan went to her studio in London, and then after a visit to the British Museum joined her on the Bromsgrove train. For two days he saw the laboratory, the uncompleted telescope, the goats (they had replaced the guilty cow) and everything Christopher had told him about. He had to go home on Friday, 25 April, but surprised Mrs Morcom by coming up to London the next day, presenting her with a parcel of Christopher's letters. On the Monday he wrote:

28/4/30

Dear Mrs Morcom,

I am only just writing to thank you for having me on your trip and to tell you how much I enjoyed it. I really don't think I have had such a jolly time before, except that wonderful week at Cambridge with Chris. I must thank you too for all the little things belonging to Chris that you have let me have. It means a great deal to me to have them . . . .

Yours affectionately, Alan

I was so glad you let me come on to the Clockhouse. I was very much impressed with the house and everything connected with it, and was very pleased to be able to help putting Chris' things in order.

Mrs Turing had also written:

27/4/30

Dear Mrs Morcom,

Alan got home last night looking so well and happy – He loved his time with you but specially precious to him was the visit to the Clockhouse: he went off to Town today to see someone but he said he would tell me of that part another day – and I knew he meant that it was an experience quite apart. We've had no real talk yet but I am sure it has helped him to exchange memories with you and he is treasuring with the tenderness of a woman the pencils and the beautiful star map and other souvenirs you gave him. . . .

I hope you won't think it an impertinence – but after our talk and your telling me how true to his name Chris was – and I believe is – in helping the weak – I thought how beautiful it would be to have a panel in his memory of S. Christopher in the School Chapel – a panel of your doing, and what an inspiration it would be for the boys who are so reminded that there are the followers of S. Christopher today and that genius and humble service can go hand in hand as in Chris. . . .

Mrs Morcom had already put into effect a similar idea. She had commissioned a stained-glass window of St Christopher – not however for Sherborne, but for their parish church at Catshill. Nor was it the 'humble service' of Mrs Turing that it was to express, but the life that went on. Back at school, Alan wrote to Mrs Morcom:

3/5/30

... I am hoping to do as well as Chris in the Higher Certificate this term. I often think about how like I am to Chris in a few ways through which we became real friends, and wonder if I am left to do something that he has been called away from.

Mrs Morcom had also called upon Alan to help choose books for the school prizes that Christopher was posthumously to receive:

I think Chris would almost certainly have got *The Nature of the Physical World* (Eddington) and *The Universe around us* (Jeans) for the Digby prize and possibly one of *The Internal Constitution of the Stars* (Eddington), *Astronomy and Cosmogony* (Jeans). I think you would like *The Nature of the Physical World*.

The Morcom family endowed a new prize at Sherborne, a science prize to be awarded for work which included an element of originality. Alan had plodded on with the iodate experiment, and now he undertook to write it up for the prize. Christopher it was, even from the grave, who induced him to communicate and to compete. He wrote to his mother:

18 May 1930

... I have just written to a Mellor the author of a Chemistry book to see if he can give me a reference about the experiment I was doing in the summer last year. Rupert said he would look it up in Zurich if I could get him a reference. It's annoying I couldn't get hold of anything before.

Alan was also interested in perspective drawing:

This week's efforts in drawing are not on any better paper ... I don't think much really of Miss Gillet's efforts. I remember she did once or twice say something in a vague sort of way about parallel lines being drawn concurrent, but she usually had the slogan 'vertical lines remain vertical' on the tip of her tongue. I wonder how she managed drawing things below

her. I have not been doing much by way of drawing bluebells and things like
that but mostly perspective.

Mrs Turing wrote to Mrs Morcom:

May 21 1930

. . . Alan has taken up drawing which I was anxious for him to do long ago:
I think this is quite likely an inspiration from you. He is quite devoted to
you and I think he was just wishing for an excuse to pay you a call when he
went up to Town the day after saying 'Goodbye' to you! You were all most
awfully good to him, and in many ways opened up a new world to him . . . .
Whenever we were alone he wanted to talk just of Chris and you and Col.
Morcom and Rupert.

Alan hoped this summer to gain an improved mark in the
Higher Certificate. His name was put down for Pembroke College,
Cambridge, which awarded a number of scholarships on Higher
Certificate marks alone, although he half-hoped to fail, so that he
would have a chance of trying for Trinity. He did fail for he found
the mathematics paper much more difficult than in the previous
year, and his marks showed no improvement. But Eperson reported:

. . . I think he has succeeded in improving his style of written work, which
is more convincing and less sketchy than last year . . .

and Gervis:

He is doing much better work than this time last year partly because he
knows more but chiefly because he is getting a more mature style.

Andrews was presented with Alan's submission for the new
Morcom science prize, and later said:[2]

I first realised what an unusual brain Alan had when he presented me with
a paper on the reaction between iodic acid and sulphur dioxide. I had used
the experiment as a 'pretty' demonstration – but he had worked out the
mathematics of it in a way that astonished me . . .

The iodates won him the prize. 'Mrs Morcom is extraordinarily
nice and the whole family is extremely interesting,' Alan wrote to
Blamey. 'They have founded a prize in Chris' memory which I very
appropriately won this year.' He also wrote:

> I have started learning German. It is possible that I may be made to go
> to Germany sometime during next year but I don't much want to. I am
> afraid I would much rather stay and hibernate in Sherborne. The worst of
> it is that most of the people left in Group III nauseate me rather. The only
> respectable person in it since February has been Mermagen and he doesn't
> do Physics seriously or Chemistry at all.

The master who taught him German wrote: 'He does not seem to
have any aptitude for languages.' It was not what he wanted to think
about in his hibernation.

One Sunday that summer, the boys of Westcott House arrived
back from their afternoon walks to find Alan, who was by now
accorded a certain awed respect, engaged upon an experiment. He
had set up a long pendulum in the stairwell, and was checking that,
as the day went on, the plane of its motion would remain fixed while
the Earth rotated beneath it. It was only the elementary Foucault
pendulum experiment, such as he might have seen in the Science
Museum in London. But it caused great astonishment at Sherborne,
and made an impression second only to his arrival by bicycle in
1926. Alan also told Peter Hogg that it had to do with the theory
of relativity, which ultimately it did: one problem that concerned
Einstein was how the pendulum kept its place fixed relative to the
distant stars. How did the pendulum know about the stars? Why
should there be an absolute standard of rotation, and why should it
agree with the disposition of the heavens?

But if the stars still exerted their attraction, Alan also had to
work out his thoughts about Christopher. Mrs Morcom had asked
him in April to write his recollections of her son for an anthology.
Alan found this task very hard to fulfil:

> My impressions of Chris that I have been writing for you seem to have
> become more a description of our friendship than anything else so I thought
> I would write it as such for you and write something less to do with me that
> you could print with the others.

In the end he would make three attempts but every one of them
strayed from manly detachment, too honest to disguise his feelings.
The first pages were sent off on 18 June, and explained:

> My most vivid recollections of Chris are almost entirely of the kind things
> he said to me sometimes. Of course I simply worshipped the ground he

trod on – a thing which I did not make much attempt to disguise, I am sorry to say.

Mrs Morcom asked for more, and Alan promised to try again when he was on holiday:

20/6/30

. . . I think I know what you mean about those little points of which you want a record. I shall have a lot of quiet time in Ireland to think them out for you. I couldn't do it before that because there is not much longer this term and camp is not a very suitable atmosphere. A lot of the things I cut out were things which were to me typical of Chris but when I read them through later I realized that to anyone who did not know both Chris and myself a little bit at least, they would not mean much. I tried to get over that just to shew a little bit what Chris was to me. Of course *you* know . . . .

The OTC camp, in the first week of the summer holiday, also obstructed the invitation to stay at the Clock House which Mrs Morcom had extended to both Alan and his mother. Fortunately there was an outbreak of infectious illness at Sherborne and camp was cancelled.

Alan arrived at the Clock House on Monday 4 August. Mrs Morcom recorded '. . . Have just been along to tuck him up. He has my room but is sleeping in sleeping pack where Chris slept last autumn . . .' Next day Mrs Turing joined them. Colonel Morcom gave Alan permission to work in the laboratory on an experiment that he and Chris together had begun. There was a day out to the county show and a visit to Christopher's grave. On the Sunday evening, Mrs Morcom wrote:

. . . I went with Mrs Turing and Alan in Lanchester. They were leaving soon after 7 pm for Ireland. Stayed until 7 talking to them . . . Alan came in to talk to me this morning and said how he loves being here. He says he feels Chris' blessing here.

The Turings crossed over to Ireland and holidayed in Donegal. Alan fished with John and his father, climbed the hills with his mother, and kept his thoughts to himself.

*

At the end of the summer term O'Hanlon had conferred the accolade: 'A good term. With some obvious minor failings, he has character.' Alan had become more prepared to go along with the system. It was not that he had ever rebelled, for he had only withdrawn; nor was it now a reconciliation, for he was still withdrawn. But he would take the 'obvious duties' as conventions rather than impositions, as long as they interfered with nothing important. In the autumn term of 1930 his contemporary Peter Hogg became head of house and, as the other third-year sixth former, Alan was made a prefect. O'Hanlon wrote to Mrs Turing: 'That he will be loyal I am well assured: and he has brains: also a sense of humour. These should carry him through . . .' He did his share of disciplining the younger boys of the house. One new boy was David Harris, brother of the Arthur Harris who had been head of house four years before. As duty prefect, Alan caught him having left his football clothes off the peg for the second time. Alan said, 'I'm afraid I shall have to beat you,' and so he did, rendering Harris a hero among his peers for being the first of the new boys thus to suffer. Harris held on to the gas ring and Alan launched the strokes. However, without the right shoes on he slid all over the shiny washroom floor and the strokes landed at random, one on Harris's spine, one on his leg. It was not the way to win respect. Alan Turing was a kindly but 'weak' prefect, one whom the younger boys could chafe, blowing out his candle in the dormitory or putting sodium bicarbonate in his chamberpot. (There were no lavatories attached to the house dormitories.) Old Turog, he was called, after the Turog loaf, and was always good for having his leg pulled. A similar incident, which took place in 'Hall' was witnessed[3] by Knoop, one of the older boys who saw Alan as 'brain where I was brawn':

> During this period of 1½ hours punishment was normally carried out by pupils. Our studies at Westcott House were down a long corridor with studies on either side shared by from 2 to 4 boys. On this particular evening during this silent period, we heard footsteps come up the corridor, a knock on a door, a mumble of voices and then two lots of footsteps come up the corridor to the locker/washroom, then we heard the swish of a cane, a crash of crockery and a loaf as cane connected with bottom, this was stroke one, exactly the same happened on the second stroke, by that time me and

my companions were splitting our sides with laughter. What happened was Turing on his back stroke had knocked down some tea making crockery belonging to prefects, he did this on two consecutive strokes and from the noise we could all tell what was going on in the washroom, the third and final stroke did not connect with crockery as by that time it was lying shattered on the floor.

Much more upsetting, his diary,[4] which he kept under lock, was taken and damaged by another boy. There was, however, a limit to what Alan would take:[5]

> Turing . . . was quite a lovable creature but rather sloppy in appearance. He was a year or more older than me, but we were quite good friends.
>
> One day I saw him shaving in the washroom, with his sleeves loose and his general appearance rather execrable. I said, in a friendly way, 'Turing, you look a disgusting sight.' He seemed to take it not amiss, but I tactlessly said it a second time. He took offence and told me to stay there and wait for him. I was a bit surprised, but (as the house washroom was the place for beatings) I knew what to expect. He duly re-appeared with a cane, told me to bend over and gave me four. After that he put the cane back and resumed his shaving. Nothing more was said; but I realised that it was my fault and we remained good friends. That subject was never mentioned again.

But apart from the important matters of 'Discipline, self-control, the sense of duty and responsibility', there was Cambridge to think about:

2/11/30

Dear Mrs Morcom,

I have been waiting to hear from Pembroke to write to you. I heard indirectly a few days ago that they will not be able to give me a scholarship. I was rather afraid so; my marks were spread too evenly amongst the three subjects. . . . I am full of hope for the December exam. I like the papers they give us there so much better than the Higher Certificate ones. I don't seem though to be looking forward to it like I was last year. If only Chris were there and we were to be up there for another week together.

Two of my books for the 'Christopher Morcom' prize have come. I had great fun yesterday evening learning some of the string figures out of 'Mathematical Recreations' . . . I have been made a school prefect this term, to my great surprise as I wasn't even a house-prefect last term. Last term they started having at least two in each house which rather accounts for it.

I have just joined a Society here called the Duffers. We go (if we feel inclined) every other Sunday to the house of some master or other and after tea someone reads a paper he has written on some subject. They are always very interesting. I have agreed to read a paper on 'Other Worlds'. It is now about half written. It is great fun. I don't know why Chris never joined.

Mother has been out to Oberammergau. I think she enjoyed it very much but she has not told me much about it yet . . .

Yours affectionately, A. M. Turing

Alan's elevation to School Prefect was a great comfort to his mother. But much more significant was a new friendship in his life.

There was a boy three years younger than Alan in the house, Victor Beuttell, who was also one who neither conformed, nor rebelled, but dodged the system. He also, like Alan, was labouring under a grief that no one knew about, for his mother was dying of bovine tuberculosis. Alan saw her when she came to visit Victor, himself in great peril with double pneumonia, and asked what was wrong. It struck a terrible chord. Alan also learnt something else that few knew, which was that Victor had been caned so severely by a prefect in another house that his spine had been damaged. This turned him against the beating system, and he never caned Victor (who was frequently in trouble), but passed him on to another prefect. The link between them was one of compassion, but it developed into friendship. Though at odds with the axioms of the public school, which normally would forbid boys of different ages from spending time together, a special dispensation from O'Hanlon, who kept a card index on the boys' activities and watched closely over them, allowed it to continue.

They spent a good deal of time playing with codes and ciphers. One source of ideas might have been the *Mathematical Recreations and Essays*,[6] which Alan had chosen as Christopher Morcom Prize, and which indeed had served a generation of school prize-winners since it appeared in 1892. The last chapter dealt with simple forms of cryptography. The scheme that Alan liked was not a very mathematical one. He would punch holes in a strip of paper, and supply Victor with a book. Poor Victor had to plod through the pages until he found one where through the holes in the strip appeared letters that spelt out a message such as HAS ORION GOT A BELT. By this time, Alan had passed on his enthusiasm for astronomy to Victor, and had explained the constellations to him. Alan also showed him a way to construct Magic Squares (also from *Mathematical Recreations*), and they played a lot of chess.

As it happened, Victor's family was also linked with the Swan electric light industry, for his father, Alfred Beuttell, had made a small fortune by inventing and patenting the Linolite electric strip reflector lamp in 1901. The lamp was manufactured by Swan and Edison, while Mr Beuttell, who had broken away from his own father's business in carpet wholesaling, acquired further experience as an electrical engineer. He had also enjoyed a fine life until the First World War, flying, motor racing, sailing, and gambling successfully at Monte Carlo.[7]

A very tall, patriarchal figure, Alfred Beuttell dominated his two sons, of whom Victor was the elder. In his character Victor took more after his mother, who in 1926 had published a curious pacifist, spiritualist book. He combined her bright-eyed, rather magical charm, with his father's strong good looks. In the 1920s Alfred Beuttell had gone back into research into lighting, and in 1927 had taken out patents on a new invention, the 'K-ray Lighting System'. It was designed to allow uniform illumination of pictures or posters. The idea was to frame a poster in a glass box, whose front surface would be curved in such a way that it reflected light from a strip light at the top exactly evenly over the poster. (Without such a reflection, the poster would be much brighter at the top than at the bottom.) The problem was to find the right formula for the curvature of the glass. Alan was introduced to the problem by Victor, and suddenly produced the formula, without being able to explain it, which agreed with Alfred Beuttell's calculation. But Alan went further, and pointed out the complication which arose through the thickness of the glass, which would cause a second reflection at the front surface. This made necessary a change in the curve of the K-ray System, which was soon put into application for exterior hanging signs, the first contract being with J. Lyons and Co. Ltd, the catering chain.

It was characteristic. As with the iodate and sulphite calculation, it always delighted Alan that a mathematical formula could actually work in the physical world. He had always liked practical demonstrations, even though he was not good at them, and although pushed into the corner as the intellectual 'maths brain', did not make the error of considering thought as sullied or lowered by having a concrete manifestation.

There was a parallel development, in that he did not permit the Sherborne 'games' religion to instil in him a contempt for the body. He would have liked to have been as successful with *corpus* as with *mens*, and found the same difficulties with both: a lack of coordination and ease of expression. But he had discovered by now that he could run rather well. He would come in first place on the house runs, when rainy weather obliged the cancellation of all-important Footer. Victor would go out with him for runs, but after two miles or so would say 'It's no good, Turing, I shall have to go back', only to find Alan overtaking him on the return from a much longer course.

Running suited him, for it was a self-sufficient exercise, without equipment or social connotations. It was not that he had sprinting speed, nor indeed much grace, for he was rather flat-footed, but he developed great staying power by forcing himself on. It was not important to Sherborne, where what mattered was that (to Peter Hogg's surprise) he became a 'useful forward' in the house team. But it was noticed with admiration by Knoop, and it was certainly important to Alan himself: He was not the first intellectual to impose this kind of physical training upon himself, and to derive lasting satisfaction from proving his stamina in running, walking, cycling, climbing, and enduring the elements. It was part of his 'back to nature' yearnings. But necessarily there were other elements involved; he perceived tiring himself out by running as an alternative to masturbation. It would probably be hard to overestimate the importance to his life of the conflicts surrounding his sexuality from this time onwards – both in controlling the demands of his body, and in a growing consciousness of emotional identity.

In December it was the same arrival at Waterloo, on the way to Cambridge, but no trip to Mrs Morcom's studio. Instead his mother and John (now an articled clerk in the City) were there to meet him, and Alan said he would go and see Howard Hughes' aerial film *Hell's Angels*. At Cambridge he failed again to win a Trinity scholarship. But his greater confidence was not entirely misplaced, for he was elected to a scholarship at the college of his second choice, King's. He was placed eighth in order of the Major Scholars, with £80 per annum.*

Everyone congratulated him. But he had set himself to *do something*, something that Christopher had been 'called away from'. For a person with a mathematical mind, an ability to deal with very abstract relations and symbols as though with tangible everyday objects, a King's scholarship was a demonstration like sight-reading a sonata or repairing a car – clever and satisfying, but no more. Many had won better scholarships, and at an earlier age. More to the point than the word 'brilliant' which now came to schoolmasters' lips was the couplet that Peter Hogg sang at the house supper:

> Our Mathematician comes next in our lines
> With his mind deep in Einstein – and study light fines.

For he had thought deeply about Einstein and had broken the rules to do so.

Alan hibernated for two more terms – it was the usual thing. There was not much in the way of temporary employment in the conditions of 1931. By now he had settled upon mathematics rather than science as his future course at Cambridge. In February 1931 he acquired G. H. Hardy's *Pure Mathematics*, the classic work with which university mathematics began. He took the Higher Certificate for a third time, this time with mathematics as major subject, and at last gained a distinction. He also entered again for the Morcom prize and won it. This time it came with a Prize Record Book, which Alan wrote 'was most fascinatingly done and bears such a spirit of Chris in the clear bright illumination.' The Morcoms had commissioned it in a contemporary neo-mediaeval style, which stood out sharply from the fusty Sherborne background.

In the Easter holiday, on 25 March, he went on a walking and hitch-hiking trip with Peter Hogg (a keen ornithologist) and an older boy, George Maclure. On their way from Guildford to Norfolk they spent one night in a working men's hostel, which suited Alan, indifferent to anything more fancy (though it shocked his mother). One day, rather typically, he walked on by himself while the other

---

* For comparison: a skilled worker earned about £160 per annum; unemployment benefit ran at £40 per annum for a single man.

two accepted a lift. He also spent five days on the OTC course at Knightsbridge barracks, qualifying in drill and tactics. This rather amazed John, who detected an unwonted enthusiasm in Alan for dressing up as a soldier. Perhaps he found this rare contact with men from outside the upper-middle-class cocoon to be strangely exciting.

David Harris became his fag, and found him well-meaning but absent-minded as a master. One of Boughey's revolutionary innovations was that prefects were allowed to have prefects from other houses to tea on Sunday afternoons, and occasionally Harris had to cook baked beans on toast when Alan availed himself of the concession. Alan had reached the summit of privilege. He continued with perspective drawing, stimulated by Victor's interest and considerable artistic talent. They had many discussions on perspective and geometry. Alan entered a line drawing of the Abbey for a school art competition in July, and gave it to Peter Hogg. (Victor won a prize for his water-colour painting.) And then *Valete*, A. M. Turing, School Prefect, Sergeant in the OTC, Member of Duffers! Alan collected a number of prizes and a £50 per annum Cambridge subsidy from the Sherborne endowments. He was also awarded a King Edward VI gold medal for mathematics. At the Commemoration, he received the faint praise which was to be his only mention while at school in the Sherborne magazine[8], marking out his proper place in the scheme of things. The scholarship winners were:

> G. C. Laws, who had been extraordinarily helpful to him (the Headmaster), a real mainstay to the tone of the place and a perpetually genial and cheerful and thoroughly best type of Shirburnian. (Applause.) The other open scholarship, mathematics, was gained by A. M. Turing who, in his sphere, was one of the most distinguished boys they had had recently.

O'Hanlon described this as 'a very successful close' to 'an interesting career, with varied experiences', expressing gratitude for Alan's 'essentially loyal help'.

Mrs Morcom had invited Alan and Mrs Turing to stay again in the summer. A letter from Alan of 14 August, answering some more of Mrs Morcom's questions, and enclosing all of Christopher's letters, said that his mother should have written to make the arrangements. But, for some reason, no visit was made. Instead, for the first two weeks of September, Alan went with O'Hanlon to Sark. Peter Hogg, Arthur Harris, and two old friends of O'Hanlon made up the party. They stayed at an eighteenth-century farmhouse, and spent the days on the rocky shores of the island, where Alan bathed naked. Arthur Harris was sketching in water colours, when Alan came up behind him, pointing to a heap of horse manure that lay on the road ahead. 'I hope you're going to put *that* in,' he said.

Few new students crossed the threshold of King's College without some trepidation induced by its grandeur. Yet the translation to Cambridge was by no means a plunge into an entirely new environment, for in many ways the university resembled a very large public school – without its violence, but inheriting many of its attitudes. Anyone familiar with the subtle relationship of loyalties to house and school would find nothing perplexing in the system of college and university. The 11 pm curfew, the obligation to wear a gown after sunset, the prohibition on unchaperoned visits by the other sex, were lightly borne by the great majority of those *in statu pupillari*. They felt newly free, to drink and smoke and spend the day as they chose.

Cambridge was positively feudal in its arrangements. The majority of the undergraduates came from public schools, and the minority who came from a lower-middle-class background, having won scholarships from grammar schools, had to adapt to the peculiar relationship between 'gentlemen' and 'servants'. As for *ladies*, they were supposed to be content with their two colleges.

As with public schools, there was a great deal about the ancient universities which had less to do with learning than with social status, with courses in geography and estate management for those of a less academic turn of mind. But the jolly raggings, debaggings and destruction of earnest students' rooms had ended with the Twenties. With the depression, the Thirties had begun, stringent and serious. And nothing could interfere with that precious freedom – a room of one's own. Cambridge rooms had double doors, and the convention was that the occupant who 'sported his oak' by locking the outer door was not at home. At last Alan could work, or think, or just be miserable – for he was far from happy – however and whenever he chose. His room could be as muddled and as untidy as he liked, so long as he made his peace with the college servants. He might be disturbed by Mrs Turing, who would scold him for the dangerous way he cooked breakfast on the gas ring. But these interruptions were very occasional, and after this first year Alan saw his parents only on fleeting visits to Guildford. He had gained his independence, and was at last left alone.

But there were also the university lectures, which on the whole were of a high standard; the Cambridge tradition was to cover the entire mathematics course with lectures which were in effect definitive textbooks, by lecturers who were themselves world authorities. One of these was G. H. Hardy, the most distinguished British mathematician of his time, who returned in 1931 from Oxford to take up the Sadleirian Chair at Cambridge.

Alan was now at the centre of scientific life, where there were people such as Hardy and Eddington who at school had been only names. Besides himself, there were eighty-five students who thus embarked upon the mathematics degree course, or 'Tripos' as Cambridge had it, in 1931. But these fell into two distinct groups: those who would offer Schedule A, and those who would sit for Schedule B as well. The former was the standard honours degree, taken like all Cambridge degrees in two Parts, Part I after one year, and Part II two years later. The Schedule B candidates would do the same, but in the final year they would also offer for examination an additional number – up to five or six – of more advanced courses. It was a cumbersome system, which was changed the following year, the Schedule B becoming 'Part III'. But for Alan Turing's year it meant neglecting study for Part I, which was something of a historical relic, hard questions on school mathematics, and instead beginning immediately on the Part II courses, leaving the third year free to study for the advanced Schedule B papers.

The scholars and exhibitioners would be expected to offer Schedule B, and Alan *par excellence* was among them, one of those who could feel themselves entering another country, in which social rank, money and politics were insignificant, and in which the greatest figures, Gauss and Newton, had both been born farm boys. David Hilbert, the towering mathematical intellect of the previous thirty years, had put it thus:[9] 'Mathematics knows no races . . . for mathematics, the whole cultural world is a single country', by which he meant no idle platitude, for he spoke as the leader of the German delegation at the 1928 international congress. The Germans had been excluded in 1924 and many refused to attend in 1928.

Alan responded with joy to the absolute quality of mathematics, its apparent independence of human affairs, which G. H. Hardy expressed another way:[10]

> 317 is a prime, not because we think so, or because our minds are shaped in one way rather than another, but *because it is so*, because mathematical reality is built that way.

Hardy was himself a 'pure' mathematician, meaning that he worked in those branches of the subject independent not only of human life, but of the physical world itself. The prime numbers, in particular, had this immaterial character. The emphasis of pure mathematics also lay upon absolutely logical deduction.

On the other hand, Cambridge also laid emphasis on what it called 'applied' mathematics. This did not mean the application of mathematics to industry, economics or the useful arts, there being in English universities no tradition of combining high academic status with practical benefits. It referred instead to the interface of mathematics and physics, generally physics of the most fundamental and theoretical kind. Newton had developed the calculus and the theory of gravitation together, and in the 1920s there had been a similar fertile period, when it was discovered that the quantum theory demanded techniques which were miraculously to be found in some of the newer developments of pure mathematics. In this area the work of Eddington, and of others such as P. A. M. Dirac, placed Cambridge second only to Göttingen, where much of the new theory of quantum mechanics had been forged.

Alan was no foreigner to an interest in the physical world. But at this point, what he needed most was a grip on rigour, on intellectual toughness, on something that was absolutely right. While the Cambridge Tripos – half 'pure' and half 'applied' – kept him in touch with science, it was to pure mathematics that he turned as to a friend, to stand against the disappointments of the world.

Alan did not have many other friends – particularly in this first year, in which he still mentally belonged to Sherborne. The King's scholars mostly formed a self-consciously élite group, but he was one of the exceptions. He was a shy boy of nineteen, who had had an education more to do with learning silly poems by rote, or writing formal letters, than with ideas or self-expression. His first friend, and link with the others of the group, was David Champernowne, one of the other two mathematical scholars. He came from the mathematical sixth form of Winchester College, where he had been a scholar, and was more confident socially than Alan. But the two shared a similar 'sense of humour', being alike unimpressed by institutions or traditions. They also shared a hesitancy in speech, although David Champernowne's was more slight than Alan's. It was and remained a rather detached, public school kind of friendship, but important to Alan was that 'Champ' was not shocked by unconventionality. Alan told him about Christopher, showing him a diary that he had kept of his feelings since the death.

They would go to college tutorials together. To begin with, it was a case of Alan catching up, for David Champernowne had been much better taught, and Alan's work was still poorly expressed and muddled. Indeed, his friend 'Champ' had the distinction of publishing a paper[11] while still an undergraduate, which was more than Alan did. The two supervisors of mathematics at King's were A. E. Ingham, serious but with a wry humour, the embodiment of mathematical rigour, and Philip Hall, only recently elected a Fellow, under whose shyness lay a particular friendly disposition. Philip Hall liked taking Alan, and found him full of ideas, talking excitedly in his own strange way, in which his voice went up and down in pitch rather than in stress. By January 1932 Alan was able to write in an impressively off-hand way:

> I pleased one of my lecturers rather the other day by producing a theorem, which he found had previously only been proved by one Sierpinski, using a rather difficult method. My proof is quite simple so Sierpinski* is scored off.

But it was not all work, because Alan joined the college Boat Club. This was unusual for a scholar, for the university was stuck with the polarising effect of the public schools, and students were

---

* W. Sierpinski, a prominent twentieth-century Polish pure mathematician.

supposed to be either 'athletes' or 'aesthetes'. Alan fitted into neither category. There was also the other problem of mental and physical balance, for he fell in love again, this time with Kenneth Harrison, who was another King's scholar of his year, studying the Natural Sciences Tripos. Alan talked to him a good deal about Christopher, and it became clear that Kenneth, who also had fair hair and blue eyes, and who also was a scientist, had become a sort of reincarnation of his first great flame. One difference, however, was that Alan did speak up for his own feelings, as he would never have dared with Christopher. They did not meet with reciprocation, but Kenneth admired the straightforwardness of his approach, and did not let it stop them from talking about science.

At the end of January 1932, Mrs Morcom sent back to Alan all the letters between him and Christopher which he had surrendered to her in 1931. She had copied them out – quite literally – in facsimile. It was the second anniversary of his death. Mrs Morcom sent a card asking him to dinner on 19 February at Cambridge, and he in turn made the arrangements for her stay. It was not the most convenient weekend, he being engaged with the Lent boat races and obliged to be abstemious at dinner. But Alan found time to show her round: Mrs Morcom noted that his rooms were 'very untidy', and they went on to see where Alan and Christopher had stayed in Trinity for the scholarship examination, and where Mrs Morcom imagined Christopher would have sat in Trinity chapel.

In the first week of April, Alan went to stay at the Clock House again, this time with his father. Alan slept in Christopher's sleeping bag. They all went together to see the window of St Christopher, now installed in Catshill parish church, and Alan said that he could not have imagined anything more beautiful of its kind. Christopher's face had been incorporated into the window – not as the sturdy St Christopher fording the stream, but as the secret Christ. On Sunday he went to communion there, and at the house they held an evening gramophone concert. Mr Turing read and played billiards with Colonel Morcom, while Alan played parlour games with Mrs Morcom. Alan went out one day for a long walk with his father, and they had another day at Stratford-upon-Avon. On the last evening, Alan asked Mrs Morcom to come and say goodnight to him, as he lay in Christopher's place in bed.

The Clock House still held the spirit of Christopher Morcom. But how could this be? Could the atoms of Alan's brain be excited by a non-material 'spirit', like a wireless set resonating to a signal from the unseen world? It was probably on this visit[12] that Alan wrote out for Mrs Morcom the following explanation:

NATURE OF SPIRIT

It used to be supposed in Science that if everything was known about the Universe at any particular moment then we can predict what it will be through all the future. This idea was really due to the great success of astronomical prediction. More modern science however has come to the conclusion that when we are dealing with atoms and electrons we are quite unable to know the exact state of them; our instruments being made of atoms and electrons themselves. The conception then of being able to know the exact state of the universe then really must break down on the small scale. This means then that the theory which held that as eclipses etc. are predestined so were all our actions breaks down too. We have a will which is able to determine the action of the atoms probably in a small portion of the brain, or possibly all over it. The rest of the body acts so as to amplify this. There is now the question which must be answered as to how the action of the other atoms of the universe are regulated. Probably by the same law and simply by the remote effects of spirit but since they have no amplifying apparatus they seem to be regulated by pure chance. The apparent non-predestination of physics is almost a combination of chances.

As McTaggart shews matter is meaningless in the absence of spirit (throughout I do not mean by matter that which can be a solid a liquid or a gas so much as that which is dealt with by physics e.g. light and gravitation as well, i.e. that which forms the universe). Personally I think that spirit is really eternally connected with matter but certainly not always by the same kind of body. I did believe it possible for a spirit at death to go to a universe entirely separate from our own, but I now consider that matter and spirit are so connected that this would be a contradiction in terms. It is possible however but unlikely that such universes may exist.

Then as regards the actual connection between spirit and body I consider that the body by reason of being a living body can 'attract' and hold on to a 'spirit', whilst the body is alive and awake the two are firmly connected. When the body is asleep I cannot guess what happens but when the body dies the 'mechanism' of the body, holding the spirit is gone and the spirit finds a new body sooner or later perhaps immediately.

As regards the question of why we have bodies at all; why we do not or cannot live free as spirits and communicate as such, we probably could do so but there would be nothing whatever to do. The body provides something for the spirit to look after and use.

Alan could have found many of these ideas in his reading of Eddington while still at school. He had told Mrs Morcom that she would like *The Nature of the Physical World*, and this would have been because of the olive branch that Eddington held out from the throne of science towards the claims of religion. He had found a resolution of the classical problem of determinism and free will, of mind and matter, in the new quantum mechanics.

The idea that Alan said 'used to be supposed in Science' was familiar to anyone who studied applied mathematics. In any school or university problem, there would always be just sufficient information supplied about some physical system to determine its entire future. In practice, predictions could not be performed except in the most simple of cases, but in principle there was no dividing line between these and systems of any complexity. It was also true that some sciences, thermodynamics and chemistry for instance, considered only averaged-out quantities, and in those theories information could appear and disappear. When the sugar has dissolved in the tea, there remains no evidence, on the level of averages, that it was ever in the form of a cube. But in principle, at a sufficiently detailed level of description, the evidence would remain in the motion of the atoms. That was the view as summed up by Laplace[13] in 1795:

> Given for one instant an intelligence which could comprehend all the forces by which nature is animated and the respective situations of the beings who compose it – an intelligence sufficiently vast to submit these data to analysis – it would embrace in the same formula the movements of the greatest bodies and those of the lightest atom; for it, nothing would be uncertain and the future, as the past, would be present to its eyes.

From this point of view, whatever might be said about the world on *other* levels of description (whether of chemistry, or biology, or psychology, or anything else), nevertheless, there was *one* level of description, that of microscopic physical detail, in which every event was completely determined by the past. In the Laplacian view, there was no possibility of any undetermined events. They might *appear* undetermined, but that would only be because one could not in practice perform the necessary measurements and predictions.

The difficulty was that there was one kind of description of the world to which people were strongly attached, namely that of ordinary language, with deciding and choosing, justice and responsibility. The problem lay in the lack of any connection between the two kinds of description. The physical 'must' had no connection with the psychological 'must', for no one would feel like a puppet pulled by strings because of physical law. As Eddington declared:

> I have an intuition much more immediate than any relating to the objects of the physical world; this tells me that nowhere in the world as yet is there any trace of a deciding factor as to whether I am going to lift my right hand or my left. It depends on the unfettered act of volition not yet made or foreshadowed. My intuition is that the future is able to bring forth deciding factors which are not secretly hidden from the past.

But he was not content to keep 'science and religion in watertight compartments', as he put it. For there was no obvious way in which the body was excused obedience to the laws of matter. There had to be some connection between the descriptions – some unity, some integrity of vision. Eddington was not a dogmatic Christian, but a Quaker who wished to preserve some idea of free consciousness, and an ability to perceive a 'spiritual' or 'mystical' truth directly. He had to reconcile this with the scientific view of physical law. And how, he asked, could 'this collection of ordinary atoms be a thinking machine?' Alan's problem was the same, only with the intensity of youth. For he believed that Christopher was still helping him – perhaps by 'an intuition much more immediate than any relating to the objects of the physical world.' But if there was no immaterial mind, independent of the physics of the brain, then there was nothing to survive, nor any way in which a surviving spirit might act upon his brain.

The new quantum physics offered such a reconciliation, because it seemed that certain phenomena were absolutely undetermined. If a beam of electrons was directed at a plate in which there were two holes, then the electrons would divide between the two, but there seemed no way of predicting the path that any particular electron would follow, not even in principle. Einstein, who in 1905 had made a very important contribution to the early quantum theory with a description of the related photo-electric effect, was never convinced that this was really so. But Eddington was more readily persuaded, and was not shy of turning his expressive pen to explain to a general audience that determinism was no more. The Schrödinger theory, with its waves of probability, and the Heisenberg Uncertainty Principle (which, formulated independently, turned out to be equivalent to Schrödinger's ideas) gave him the idea that mind could act upon matter *without* in any way breaking physical laws. Perhaps it could select the outcome of otherwise undetermined events.

It was not as simple as that. Having painted the picture of mind controlling the matter of the brain in this way, Eddington admitted that he found it impossible to believe that manipulating the wave-function of just *one* atom could possibly give rise to a mental act of decision. 'It seems that we must attribute to the mind power not only to decide the behaviour of atoms individually but to affect systematically large groups – in fact to tamper with the odds on atomic behaviour .' But there was nothing in quantum mechanics to explain how *that* was to be done. At this point his argument became suggestive in character, rather than precise – and Eddington did tend to revel in the obscurity of the new theories. As he went on, the concepts of physics became more and more nebulous, until he compared the quantum-mechanical description of the electron with the 'Jabberwocky' in *Through the Looking Glass*:

> *Something unknown is doing we don't know what* – that is what our theory amounts to. It does not sound a particularly illuminating theory. I have read something like it elsewhere:–
>
> The slithy toves
> Did gyre and gimble in the wabe.

Eddington was careful to say that in some sense the theory actually worked, for it produced numbers which agreed with the outcome of experiments. Alan had grasped this point back in 1929: 'Of course he does not believe that there are really about 10^{70} dimensions, but that this theory will explain the behaviour of an electron. He thinks of 6 dimensions, or 9, or whatever it may be without forming any mental picture.' But it seemed no longer possible to ask what waves or particles really were, for their hard nineteenth-century billiard-ball concreteness had evaporated. Physics had become a symbolic representation of the world, and nothing more, Eddington argued, edging towards a philosophical idealism (in the technical sense) in which everything was in the mind.

This was the background of Alan's assertion that 'We have a will which is able to determine the action of the atoms probably in a small portion of the brain, or possibly all over it.' Eddington's ideas had bridged the gap between the 'mechanism' of the body, which Alan had learnt from *Natural Wonders*, and the 'spirit' in which he wanted to believe. He had found another source of support in the Idealist philosopher McTaggart, and added ideas about reincarnation. But he had in no way advanced upon or even clarified Eddington's view, having ignored the difficulties which Eddington had pointed out in discussing the action of the 'will'. Instead, he had taken a slightly different direction, one fascinated with the idea of the body amplifying the action of the will, and more generally concerned with the nature of the connection between mind and body in life and death.

These ideas did, in fact, show the shape of things to come, though in 1932 there was little outward evidence of future development. In June he had found himself placed in the second class in the Part I of the Tripos. 'I can hardly look anyone in the face after it. I won't try to offer an explanation, I shall just have to get a 1st in Mays* to shew I'm not really so bad as that,' he wrote to Mrs Morcom. But more significant, in reality, was the fact that he had ordered as his last prize from Sherborne a book that promised a serious account of the interpretation of quantum mechanics. It was an ambitious choice of study, a book only published in 1932. It was the *Mathematische Grundlagen der Quantenmechanik*, the Mathematical Foundations of Quantum Mechanics, by the young Hungarian mathematician John von Neumann.

On 23 June it was his twentieth birthday, and then on 13 July what would have been Christopher's twenty-first. Mrs Morcom sent Alan a 'Research' fountain pen, such as Christopher had shown off, as a present. Alan wrote from Cambridge, where he spent the 'Long Vacation Term':

---

* 'Mays' were the semi-official second-year examinations.

14/7/32

Dear Mrs Morcom,

. . . I remembered Chris' birthday and would have written to you but for the fact that I found myself quite unable to express what I wanted to say. Yesterday should I suppose have been one of the happiest days of your life.

How very kind of you it was to think of sending me a 'Research' pen. I don't think anything else (of that kind) could remind me better of Chris; his scientific appreciation and dexterous manipulation of it. I can so well remember him using it.

But if he was twenty, and preparing to confront the work of European mathematicians, he was still a boy away from home, away from Sherborne. The summer holidays were spent much as those of the previous year:

Daddy and I have just been to Germany, for just over a fortnight. We spent most of the time walking in the Schwarzwald, though Daddy of course was not up to much more than 10 miles a day. My knowledge of the language wasn't altogether of the kind that [was] most wanted. I have learnt nearly all my German by reading half a German mathematical book.* I got home somehow or other . . .

Yours affectionately, Alan M. Turing

Alan had another holiday camping with John in Ireland, where he amazed his family by turning up at Cork in a pig-boat. Then for the first two weeks of September he joined O'Hanlon for a second and last time on Sark. Alan was 'a lively companion even to the extent of mixed bathing at midnight,' wrote[14] O'Hanlon, who had struck a modern note by allowing two girls on the party. Alan had taken some fruit-flies with him, as he was studying genetics in a rather haphazard way. Back at Guildford the *Drosophilae* escaped and infested the Turing home for weeks, not at all to Mrs Turing's pleasure. O'Hanlon was sufficiently detached from the 'nation in miniature' to write[15] of Alan as 'human and lovable', saying:

I look back on holidays in Cornwall and Sark among the great enjoyments of my life: in all his companionship and whimsical humour, and the diffident shake of the head and rather high pitched voice as he propounded some

---

* Not the von Neumann book, however, which he only received in October 1932.

question or objection or revealed that he had proved Euclid's postulates or was studying decadent flies – you never knew what was coming.

The all-encompassing system still allowed some moments of freedom. And Sherborne had also left Alan with one friendship that lasted – with Victor. Alan's younger friend had been obliged to leave school at the same time, his father suffering from financial loss at what was the worst of the Depression. He had failed his School Certificate (telling Alan that it was because of too much time spent on chess and codes) but quickly caught up by passing it at a London crammers, and began what Alan called 'his grim life as a chartered accountant'. At Christmas 1932 Alan stayed with the Beuttells for two weeks and worked in Alfred Beuttell's office near Victoria. The visit was overshadowed by the fact that Victor's mother had died on 5 November. The deep shadow was a link, for both boys were having to deal with the fact of early death. The link was close enough to break Alan's usual reserve as to his beliefs – just as Mrs Morcom had broken it – and rather grudgingly to discuss his ideas about religion and survival. Victor believed very strongly, not only in the essential Christian ideas, but in extra-sensory perception and in reincarnation. To him, Alan appeared as one who wanted so much to believe, but whose scientific mind made him an unwilling agnostic, and who was under great tension as a result. Victor saw himself as a 'crusader', trying to keep Alan on the straight and narrow, and they had fierce arguments, the more so as Alan did not like being challenged by a boy of seventeen. They talked about who had rolled the stone away, and how the five thousand had really been fed. What was myth and what was fact? They argued about the after-life, and the pre-life too. Victor would say to Alan, 'Look, no one has ever been able to *teach* you any mathematics – perhaps you have remembered it from a previous life.' But, as Victor saw it, Alan could not believe in such a thing 'without a mathematical formula'.

Victor's father, meanwhile, had thrown himself into research and work to overcome his bereavement. Alan's work in his office was concerned with calculations required for his commission as lighting consultant to the Freemasons' new headquarters in Great Queen Street. Alfred Beuttell was a pioneer in the scientific measurement of illumination, and the development of a lighting code[16] based on 'first principles' as part of the 'reduction of the physiology of vision to a scientific and mathematical basis'. His work for the Masons involved elaborate calculations to estimate the illumination at the floor level, in terms of the candle power of lights installed and the reflecting properties of the walls. Alan, who was not allowed into the Masonic building, had to work from imagination to check Mr Beuttell's figures.

Alan became friendly with Mr Beuttell, who told him about his success in Monte Carlo as a young man, when he had managed to live for a month on his winnings. He showed Alan his gambling system, which Alan took back to Cambridge and studied. On 2 February 1933 he wrote back with the result of his analysis, which was that the system yielded an expected gain of exactly zero, and that accordingly Mr Beuttell's winnings had been entirely due to luck and not to skill. He also sent a formula he had worked out for the illumination of the floor of a hemispheric room lit from its centre – not, admittedly, an immediately useful result, but a very neat one.

Standing up to Mr Beuttell's ideas about his gambling system took some courage, as he was a forceful man, whose heart of gold was buried deep, with strong opinions on many subjects. An eclectic Christian tending to Theosophy, he was a great believer in the unseen world, and told Alan that his invention of the Linolite electric lamp had been sent to him from beyond. This Alan found too much to swallow. But he also had ideas about the brain, which he had formed since the early 1900s, according to which it worked on electric principles, with differences of potential determining moods. An electric brain! – there lay a more scientific idea. They had long discussions on these lines.

Alan and Victor also went down to Sherborne together for the house supper, and after Christmas Alan wrote to Blamey, saying:

> I still haven't quite decided what I am going to do when I grow up. My ambition is to become a don at King's. I am afraid it may be more ambition than profession though. I mean it is not very likely I shall ever become one.
>
> Glad you had a good beano for your coming of age. Personally when my time comes I shall retire into a corner of England far from home and sulk. In other words I don't want to come of age (Happiest days of my life at school etc.)

Sherborne was part of him; and, essentially loyal to his past, he did not make the mistake of trying to cast it out. Although, indeed, the official speeches about training, leadership and the future of the Empire had left him almost untouched, there were aspects of the distinctive English public-school culture in which he genuinely shared. Its dowdy, Spartan amateurism, in which possessions and consumption played a small role, were his. So was its combination of conventionality and weird eccentricity; so too, to some degree, was its anti-intellectualism. For Alan Turing did not think of himself as placed in a superior category by virtue of his brains, and only insisted upon playing what happened to be his own special part. And if the public school was founded upon deprivation and suppression, this was of a kind which gave its products the privilege of knowing that their thoughts and actions were considered significant. In setting out to *do something* in life, Alan exhibited in a pure form the sense of moral mission that headmasterly sermons sought so laboriously to inculcate.

But he could not stay with one foot in the nineteenth century; Cambridge had introduced him to the twentieth. There had been a moment in 1932 when, after a college Feast, Alan wandered quite drunk into David Champernowne's rooms, only to be told to 'get a grip on himself'. 'I must get a grip on myself, I must get a grip on myself,' Alan repeated in a very droll fashion, so that Champ always liked to think that this had marked a turning point. Be this as it may, it was indeed the year 1933 which brought Alan closer to the problems of the modern world, and in which he began to interact with it.

On 12 February 1933, Alan marked the third anniversary of Christopher's death:

Dear Mrs Morcom,

I expect you will be thinking of Chris when this reaches you. I shall too, and this letter is just to tell you that I shall [be] thinking of Chris and of you tomorrow. I am sure that he is as happy now as he was when he was here.

Your affectionate Alan.

Others were to remember that week for another reason: on 9 February the Oxford Union resolved that under no circumstances would it fight for King and Country. There were parallel sentiments at Cambridge, not necessarily of complete pacifism, but of a kind which rejected any war fought for that slogan. Patriotism was not enough, after the First World War; there might legitimately be a defence of 'collective security' but not a 'national war'. Newspapers and politicians reacted as though the Enlightenment had never happened, but enlightened scepticism was particularly alive at King's, and Alan began to find that it was more than a rather grand and frightening house in a giant public school.

King's enjoyed special privileges within the university system, and was distinguished by its opulence, thanks to a fortune amassed by John Maynard Keynes. But it also prized a *moral* autonomy that had been at its most pure and intense in the early 1900s, as Keynes described:[17]

> ... We entirely repudiated a personal liability on us to obey general rules. We claimed the right to judge every individual case on its merits, and the wisdom, experience, and self-control to do so successfully. This was a very important part of our faith, violently and aggressively held, and for the outer world it was our most obvious and dangerous characteristic. We repudiated entirely customary morals, conventional wisdom. We were, that is to say, in the strict sense of the term, immoralists. The consequences of being found out had, of course, to be considered for what they were worth. But we recognised no moral obligation on us, no inner sanction, to conform or to obey. ...

E. M. Forster had more gently, but more widely, portrayed an insistence on the priority of individual relationships over every kind of institution. In 1927 Lowes Dickinson, the King's historian and first advocate of a 'League of Nations', wrote[18] in his autobiography:

> I have seen nothing lovelier than Cambridge at this time of year. But Cambridge is a lovely backwater. The main stream is Jix* and Churchill and Communists and Fascists and hideous hot alleys in towns, and politics, and that terrible thing called the 'Empire', for which everyone seems to be willing to sacrifice all life, all beauty, all that is worthwhile, and has it any worth at all? It's a mere power engine.

They spoke of *mere* power, that was the point. Even Keynes, involved in state affairs and devoted to economics, did so in the belief that

---

* Joynson Hicks, the reactionary Home Secretary.

with such tawdry problems solved, people could start to think about something important. It was an attitude very different from the cult of duty, which made a virtue out of playing the expected part in the power structure. King's College was very different from Sherborne School.

It was also part of the King's attitude to life that it regarded games, parties and gossip to be natural pleasures, and assumed that clever people would still enjoy ordinary things. Although King's had only gradually moved away from its original role as a sister foundation to Eton, there were among its dons those who made a positive effort to encourage candidates who did not come from public schools and tried to make them feel at home. There was great emphasis on the mixing between dons and undergraduates in what was a small college, with less than sixty students in each year. No other college was like this, and so Alan Turing gradually woke up to the fact that by chance he had arrived in a unique environment, which was as much his element as any institution could be. It corroborated what he always knew, which was that his duty was to think for himself. The match was not perfect, for a number of reasons, but it was still a great stroke of fortune. At Trinity he would have been a lonelier figure; Trinity also inherited the moral autonomy, but without the personal intimacy that King's encouraged.

The year 1933 only brought to the surface ideas which in King's had a long history. Alan shared in the climate of dissent:

26/5/33

Dear Mother,

Thank you for socks etc. . . . Am thinking of going to Russia some time in vac but have not yet quite made up my mind.

I have joined an organisation called the 'Anti-War Council'. Politically rather communist. Its programme is principally to organize strikes amongst munitions and chemical workers when government intends to go to war. It gets up a guarantee fund to support the workers who strike.

. . . There has been a very good play on here by Bernard Shaw called 'Back to Methuselah'.

Yours, Alan

For a short time, Anti-War Councils sprang up across Britain and united pacifists, communists and internationalists against a 'national' war. Selective strikes. had, in fact, prevented the British government from intervening on the Polish side against the Soviet Union in 1920. But for Alan the real point lay not in political commitments, but in the resolve to question authority. Since 1917 Britain had been deluged by propaganda to the effect that Bolshevik Russia was the kingdom of the devil, but in 1933 anyone could see that something had gone completely wrong with the western trading and business system. With two million people unemployed, there was no precedent for what was above all a *baffling* situation, in which no one knew what should be done. Soviet Russia, after its second revolution of 1929, offered the solution of state planning and control, and there was great interest among intellectual circles in how it was working. It was the testing-ground of the Modern. Alan probably enjoyed riling his mother with a nonchalant 'rather communist': the point lay not in this or that label, but in the fact that his generation were going to think for themselves, take a wider view of the world than their parents had done, and not be frightened by bogey words.

Alan did not in fact go to see Russia for himself. But even if he had, he would have found himself ill-disposed to become an enthusiast for the Soviet system. Nor did he become a 'political' person in the Cambridge of the 1930s. He was not sufficiently interested in 'mere power'. Buried in the *Communist Manifesto* was the declaration that the ultimate aim was to make society 'an association, in which the free development of each is the condition for the free development of all.' But in the 1930s, to be a communist meant identifying with the Soviet regime, which was a very different matter. Those at Cambridge who perceived themselves as members of a responsible British prefect class might well identify with the Russian rulers as with a sort of better British India, collectivising and rationalising the peasants for their own good. For products of the English public school, apt to despise trade, it was but a small step to reject capitalism, and place faith in greater state control. In many ways the Red was a mirror image of the White. Alan Turing, however, was not interested in organising anyone, and did not wish to be organised by anyone else. He had escaped from one totalitarian system, and had no yearning for another.

Marxism claimed to be scientific, and it spoke to the modern need for a rationale of historical change that could be justified by science. As the Red Queen told Alice, 'You may call it "nonsense" if you like, but *I've* heard nonsense, compared with which that would be as sensible as a dictionary.' But Alan was not interested in the problems of history, while the marxist attempts to explain the exact sciences in terms of 'prevailing modes of production' were very remote from his ideas and experience. The Soviet Union judged relativity and quantum mechanics by political criteria, while the English theorist Lancelot Hogben sustained an economic explanation of the development of mathematics only by restricting attention to its most elementary applications. Beauty and truth, which motivated Alan Turing as they had always inspired mathematicians and scientists, were lacking. The Cambridge communists took upon themselves something of the character of a fundamentalist sect, with the air of being saved, and the element of 'conversion' met in Alan Turing the same scepticism as he had already turned upon Christian beliefs. With his fellow sceptic Kenneth Harrison he would mock the communist line.

On economic questions, indeed, Alan came to think highly of Arthur Pigou, the King's economist who had played a slightly earlier part than Keynes in patching up nineteenth-century liberal capitalism. Pigou held that more equal distribution of income was likely to increase economic welfare, and was an early advocate of the welfare state. Broadly similar in their outlook, both Pigou and Keynes were calling for increased state spending during the 1930s. Alan also began to read the *New Statesman*, and could broadly be identified with the middle-class progressive opinion to which it was addressed, concerned both for individual liberty and for a more rationally organised social system. There was much talk about the benefits of scientific planning (so that Aldous Huxley's 1932 satire *Brave New World* could treat it as the intellectuals' already dated orthodoxy), and Alan went to talks on progressive ventures such as the Leeds Housing Scheme.* But he would not have seen *himself* as one of the scientific organisers and planners.

In fact his idea of society was that of an aggregate of individuals, much closer to the views of democratic individualism held by J. S. Mill than that of socialists. And to keep his individual self intact,

---

* This gave him a tenuous link with his mother, who had shares in a Bethnal Green housing association. Alan's reaction was approval that they planned the flats for the families who needed them rather than *vice versa*.

self-contained, self-sufficient, uncontaminated by compromise or hypocrisy,* was his ideal. It was an ideal far more concerned with the moral than with the economic or political; and closer to the traditional values of King's than to the developing currents of the 1930s.

Like many people (E. M. Forster among them) he found a special pleasure in discovering Samuel Butler's *Erewhon*. Here was a Victorian writer who had doubted the moral axioms, playing with them in Looking-Glass fashion by attaching the taboos on sex to the eating of meat, describing Anglican religion in terms of transactions in ornamental money, and exchanging the associations of 'sin' with those of 'sickness'. Alan also much admired Butler's successor Bernard Shaw, enjoying his light play with serious ideas. For the well-read sophisticate of the 1930s, Butler and Shaw were already out-worn classics, but for one from Sherborne School they still held a liberating magic. Shaw had taken up what Ibsen† called 'the revolution of the Spirit', and wanted to show true individuals on the stage, those who lived not by 'customary morals' but by inner conviction. But Shaw also asked hard questions about what kind of society could contain such true individuals: questions highly pertinent to a young Alan Turing. *Back to Methuselah*, which Alan thought 'a very good play' in May 1933, was an attempt at what Shaw called 'politics *sub specie aeternitatis*'. With its science-fiction view of Fabian ideas, treating with contempt the sordid realities of Asquith and Lloyd George, it suited Alan's idealist frame of mind.

One subject, however, did *not* feature in Bernard Shaw's plays, and only very rarely in the *New Statesman*.[19] In 1933 its drama critic reviewed *The Green Bay Tree*, which was about 'a boy . . . adopted for immoral purposes by a wealthy degenerate,' and said it was 'well worth seeing for anyone who finds a pervert a less boring subject for the drama than a man with a diseased liver.' In this respect, King's College was unique. Here it was possible to doubt an axiom which Shaw left unquestioned and Butler skated over nervously.

It was only possible because no one breached the line that separated the official from the unofficial worlds. The consequences

---

* 'Regarding Aunt J's funeral', Alan wrote in January 1934 to his mother, 'I am not v. keen on going, and I think it would be consummate hypocrisy if I did. But if you think anyone will be the better for my attending I will see whether it can be managed.'

† Alan also considered Ibsen's plays 'remarkably good'.

of being found out were the same in King's as anywhere else, and the same double life was imposed by the outside world. It was a ghetto of sexual dissent, with the advantages and disadvantages of ghetto life. The internal freedom to express heretical thoughts and feelings was certainly of benefit to Alan. He was, for instance, helped by the fact that Kenneth Harrison derived from his father, himself a graduate of King's, a liberal understanding of other people's homosexual feelings. But the world of Keynes and Forster, the parties and comings and goings of Bloomsbury people, lay far above Alan's head. There was a glossiness about King's, whose greatest strength lay in the arts, and drama in particular, in which he had no share. He would have been too easily deterred and frightened by the more theatrical elements in expressing his homosexuality. If at Sherborne his sexuality was described in terms of 'filth' and 'scandal', he now also had to come to terms with that other kind of labelling that the world found so important: that of the *pansy*, an affront and traitor to masculine supremacy. He did not find a place in this compartment; nor did the King's aesthete set, flourishing in its protected corner, reach out to a shy mathematician. As in so many ways, Alan was the prisoner of his own self-sufficiency. King's could only protect him while he worked out the problems for himself.

It was the same with regard to religious belief, for while agnosticism was all but *de rigueur* in King's, he was not the person to follow a trend, only to be stimulated and liberated by the freedom to ask hitherto forbidden questions. In developing his intellectual life, he did not form the social connections that a less shy person could have made. Unlike most of his close acquaintants, he was a member neither of the 'Ten Club' nor of the Massinger Society – two King's undergraduate societies of which the first read plays and the other discussed far into the night, over mugs of cocoa, papers on culture and moral philosophy. He was too awkward, even uncouth, to fit into these comfortable gatherings. Nor was he elected to the exclusive university society, the Apostles, which drew much of its membership from King's and Trinity. In many ways, he was too ordinary for King's.

In this respect he had something in common with one of his new friends, James Atkins, who was the third mathematical scholar of Alan's year. James and Alan got on well together, in an amiable manner that lacked any deep conversations about Christopher or science, and it was James whom Alan asked to come with him for a few days walking in the Lake District.

They were away from 21 to 30 June, so that Alan did achieve his objective of being away from home on 23 June, his 'coming of age'. In fact they were walking that day from the youth hostel at Mardale over High Street to Patterdale. The weather was unusually hot and sunny, leading Alan at one point to sunbathe naked, and perhaps encouraging him in the gentle sexual approach that he made a few days later, as they rested on the hillside. This almost accidental but electric moment was perhaps less important to Alan than to James, who had been particularly repressed at his public school and was catching up years of self-knowledge, mentally and physically. There was no repetition during the holiday, while he thought it over. In the following two weeks, he found himself roused to feelings of affection and desire for Alan, and expected to see him when he returned to Cambridge on 12 July for the long vacation term. This was not so much to study mathematics as to take part in concerts during the International Congress of Musical Research, for James found in music the absoluteness that Alan found in pure mathematics.

James did not know that the same day Alan had gone to the Clock House to remember Christopher. At Easter, he had stayed there again, taken communion at his shrine, and had written:

20/4/33

My dear Mrs Morcom,

I was so pleased to be at the Clockhouse for Easter. I always like to think of it specially in connection with Chris. It reminds us that Chris is in some way alive *now*. One is perhaps too inclined to think only of him alive at some future time when we shall meet him again; but it is really so much more helpful to think of him as just separated from us for the present.

His July visit coincided with the dedication of the memorial window on 13 July, which would have been Christopher's twenty-second birthday. The local children had the day off school, and laid flowers beneath the stained-glass window. A family friend preached on 'Kindness' in Christopher's memory. They all sang Christopher's favourite hymn:

> Gracious Spirit, Holy Ghost
> Taught by Thee we covet most
> Of thy gifts at Pentecost
> Holy heavenly Love

In a marquee at the Clock House, a conjuror amused the children over their buns and lemonade; Rupert demonstrated Christopher's experiment with the iodates and sulphites, and his uncle explained it to them. They blew bubbles and sent up balloons.

Alan returned to Cambridge two or three weeks after this bitter-sweet ceremony, and so it was not long before James indicated that he would like to continue the sexual contact that Alan had sparked off. But there was always a sense that Alan never again showed the initiative which the summer sun had elicited, and there was a complexity which James could never penetrate. The associations of Christopher, which Alan did not share with James, might have been part of the reason. The visit would have refreshed the memory of pure, intense romantic love, of a kind which did not exist within his relationship with James. Instead, they were satisfied with an easy-going sexual friendship in which there was no pretence of being in love. But at least Alan knew that he was not alone.

Sometimes he seemed ruffled. At the Founder's Feast in December 1933, there was an incident when an undergraduate from James's old school said to Alan in an obnoxious manner, 'Don't look at me like that, I'm not a homosexual.' Alan, upset by this attack, said to James, 'If you want to go to bed, it'll be one-sided.' But this was the exceptional moment in a relationship which continued – to a lessening degree – for several years.

No one else knew of this, although in more general terms, as the Feast incident illustrated, Alan was not particularly secretive about his sexuality. There was another undergraduate for whom Alan (as he told James) had longings, and their names were linked by scurrilous clues like 'See under 2 down' for a crossword puzzle in an abortive King's rag magazine. In the autumn of 1933 Alan also made another friend, with whom the main link was discussion of sex. This was Fred Clayton, who was a very different character. While both Alan and James were reserved, but got on with it without making a fuss, with Fred it was the reverse. His father was headmaster of a small village school near Liverpool, and he had not been through the public-school training. A rather small, rather young classics scholar, he had been cox of the boat in which Alan rowed, but their acquaintanceship developed as Fred became aware of Alan as someone whose sexuality seemed to be made no secret, either by himself or by others.

Fred was very interested in an exchange of views and emotional experiences, feeling himself very puzzled by sex, and confronted by ex-public schoolboys much more conscious of homosexual attraction. He had taken advantage of the King's freedom of discussion, and had been told by a Fellow that he 'seemed a pretty normal bisexual male'. But it was not as simple as that, and nothing was ever simple for Fred Clayton.

Alan told his friend about how much he resented having been circumcised, and also of his earliest memories of playing with the gardener's boy (presumably at the Wards' house), which he thought had perhaps decided his sexual pattern. Rightly or wrongly, he gave Fred, and others too, the impression that public schools could be relied upon for sexual experiences – although more important perhaps was that schooldays continued to loom large in his consciousness of sexuality. Fred read Havelock Ellis and Freud, and also made discoveries in the classics which he would convey to his mathematical friend, not noted otherwise for an interest in Latin and Greek.

Puzzlement was an entirely reasonable reaction, in the conditions of 1933, when even in King's there was so little to go on for those outside the most chromium-plated circles. These conversations were whispers in a crushing, deafening silence. It was not the effect of the law, whose prohibition of all male homosexual activity played but a tiny part in the Britain of the 1930s, in direct terms. It was more as J. S. Mill had written[20] of heresy:

> ... the chief mischief of the legal penalties is that they strengthen the social stigma. It is that stigma which is really effective, and so effective is it, that the profession of opinions which are under the ban of society is much less common in England than is, in many other countries, the avowal of those which incur risk of judicial punishment.

Modern psychology had made a twentieth-century difference; the 1920s had given to the *avant-garde* the name of Freud to conjure with. But his ideas were used in practice to discuss what had 'gone wrong' with homosexual people, and such intellectual openings were outweighed by the continual efforts of the official world to render homosexuality invisible – a process in which the academic world played its part along with prosecutions and censorship. As for respectable middle class opinion, it was represented by the *Sunday Express* in 1928, greeting *The Well of Loneliness* with the words, 'I had rather give a healthy boy or a healthy girl a phial of prussic acid than this novel.' The general rule remained that of unmentionability above all else, leaving even the well-educated homosexual person with nothing more encouraging than the faint signals from the ancient world, the debris of the Wilde trials, and the rare exceptions to the rules supplied by the writings of Havelock Ellis and Edward Carpenter.

In a peculiar environment such as Cambridge, it might be a positive advantage to enjoy homosexual experience, simply in terms of the opportunity for physical release. The deprivation was not one of laws but of the spirit – a denial of identity. Heterosexual love, desire and marriage were hardly free from problems and anguish, but had all the novels and songs ever written to express them. The homosexual equivalents were relegated – if mentioned at all – to the comic, the criminal, the pathological, or the disgusting. To protect the self from these descriptions was hard enough, when they were embedded in the very words, the only words, that language offered. To keep the self a complete and consistent whole, rather than split into a facade of conformity, and a secret inner truth, was a miracle. To be able to *develop* the self, to increase its inner connections and to communicate with others – that was next to impossible.

Alan was at the one place that could support that development. Here, after all, was the circle round which Forster passed the manuscript of his novel *Maurice* which conveyed so much about being 'an unmentionable of the Oscar Wilde sort'. How to *complete* the work, that was one problem. It had to have its own integrity of feeling, yet be credible as a story of the real world. There was a fundamental contradiction, which was not resolved by having his hero escape into the 'greenwood' of a happy ending.

There was another contradiction, in that this attempt at communication remained secret for fifty years.[21] But here at least was the place where these contradictions were understood, and although Alan's self-contained nature placed him on the edge of King's society, he was protected from the harshness of the outside world.

If Alan enjoyed *Back to Methuselah*, it would also have been because Shaw dramatised his theory of the Life Force, which raised the same questions as the 'spirit'. One of Shaw's characters said 'Unless this withered thing religion, and this dry thing science, come alive in our hands, alive and intensely interesting, we may just as well go out and dig the garden until it is time to dig our graves.' This was Alan's problem in 1933, although he could not accept Shaw's easy solution. Bernard Shaw had no compunction about rewriting science if it did not agree with his ideas; determinism had to go, if it conflicted with a Life Force. Shaw fixed on Darwin's theory of evolution, which he discussed as if it were an account of every kind of change, social and psychological change included, and rejected it as a 'creed': he wrote[22] that:

> What damns Darwinian Natural Selection as a creed is that it takes hope out of evolution, and substitutes a paralysing fatalism which is utterly discouraging. As Butler put it, it 'banishes Mind from the universe.' The generation that felt nothing but exultant relief when it was delivered from the tyranny of an Almighty Busybody by a soulless Determinism has nearly passed away, leaving a vacuum which Nature abhors.

Science, for Shaw, existed to provide a hopeful 'creed', which revealed religion no longer did. There *had* to be a Life Force, of which the super-intelligent Oracle of A.D. 3000 could say 'Our physicists deal with it. Our mathematicians express its measurements in algebraic equations.'

But for Alan, science had to be true, rather than comforting. Nor did that mathematician and physicist John von Neumann have anything to say that lent credence to a Life Force. His *Mathematische Grundlagen der Quantenmechanik* had arrived in October 1932, but perhaps Alan had put off tackling it until the summer, when he also obtained books on quantum mechanics by Schrödinger and Heisenberg. On 16 October 1933 he wrote:

> My prize book from Sherborne is turning out very interesting, and not at
> all difficult reading, although the applied mathematicians seem to find it
> rather strong.

Von Neumann's account was very different from Eddington's. In his formulation, the *state* of a physical system evolved perfectly deterministically; it was the *observation* of it that introduced an element of absolute randomness. But if this process of observation were itself observed from outside, it could be regarded as deterministic. There was no way of saying where the indeterminacy was; it was not localised in any particular place. Von Neumann was able to show that this strange logic of observations – quite unlike anything encountered with everyday objects – was consistent in itself, and agreed with known experiments. It left Alan sceptical about the interpretation of quantum mechanics, but certainly gave no support to the idea of the mind manipulating wave-functions in the brain.

Alan would not only have found von Neumann's book 'very interesting' because it was tackling a subject of such philosophical importance to himself. It would also have been because of the way in which von Neumann approached his scientific subject as much as possible by logical thought. For science, to Alan Turing, was thinking for himself, and seeing for himself, and not a collection of facts. Science was doubting the axioms. He had the pure mathematician's approach to the subject, allowing a free rein to thought, and seeing afterwards whether or not it had application to the physical world. He would often argue on these lines with Kenneth Harrison, who took the more traditional scientific view of experiments and theories and verification.

The 'applied mathematicians' would have found von Neumann's study of quantum mechanics to be 'rather strong' because it required a considerable knowledge of recent pure-mathematical developments. He had taken the apparently different quantum theories of Schrödinger and Heisenberg, and by expressing their essential ideas in a much more abstract mathematical form, shown their equivalence. It was the logical consistency of the theory, not the experimental results, that von Neumann's work treated. This suited Alan, who sought that kind of toughness, and it made a beautiful example of how the expansion of pure mathematics for its own sake had borne unexpected fruit in physics.

Before the war, Hilbert had developed a certain generalisation of Euclidean geometry, which involved considering a space with infinitely many dimensions. This 'space' had nothing to do with physical space. It was more like an imaginary graph on which could be plotted all musical sounds, by thinking of a flute, or violin, or piano tone as made up of so much of the fundamental, so much of the first harmonic, so much of the second harmonic, and so on – each kind of sound requiring (in principle) the specification of infinitely many ingredients.* A 'point' in such a 'space', a 'Hilbert space', would correspond to such a sound; then two points could be added (like adding sounds), and a point could be multiplied by a factor (like amplifying a sound).

Von Neumann had noticed that 'Hilbert space' was exactly what was needed to make precise the idea of the 'state' of a quantum-mechanical system, such as that of an electron in a hydrogen atom. One characteristic of such 'states' was that they could be added like sounds, and another was that there would generally be infinitely many possible states, rather like the infinite series of harmonics above a ground. Hilbert space could be used to define a rigorous theory of quantum mechanics, proceeding logically from clear-cut axioms.

The unforeseen application of 'Hilbert space' was just the kind of thing that Alan would produce to support his claim for pure mathematics. He had seen another vindication in 1932, when the positron was discovered. For Dirac had predicted it on the basis of an abstract mathematical theory, which depended upon combining the axioms of quantum mechanics with those of special relativity. But in arguing about the relationship between mathematics and science, Alan Turing found himself tackling a perplexing, subtle, and to him personally important aspect of modern thought.

The distinction between science and mathematics had only been clarified in the late nineteenth century. Until then it might be supposed that mathematics necessarily represented the relations of

---

* The analogy is not intended to be exact; Hilbert space and quantum mechanical 'states' differ in an essential way from anything in ordinary experience.

numbers and quantities appearing in the physical world, although this point of view had really been doomed as soon as such concepts as the 'negative numbers' were developed. The nineteenth century, however, had seen developments in many branches of mathematics towards an *abstract* point of view. Mathematical symbols became less and less obliged to correspond directly with physical entities.

In school algebra – eighteenth-century algebra, in effect – letters would be used as symbols for numerical quantities. The rules for adding and multiplying them would follow from the assumption that they were 'really' numbers. But by the twentieth century, this view had been abandoned. A rule such as 'x + y = y + x' could be regarded as a rule for a game, as in chess, stating how the symbols could be moved around and combined legitimately. The rule might possibly be *interpreted* in terms of numbers, but it would not be necessary nor indeed always appropriate to do so.

The point of such abstraction was that it liberated algebra, and indeed all mathematics, from the traditional sphere of counting and measurement. In modern mathematics, symbols might be used according to any rules whatever, and might be interpreted in ways far more general than in terms of numerical quantities, if indeed they bore any interpretation at all. Quantum mechanics presented a fine example of where the expansion and liberation of mathematics for its own sake had paid off in physics. It had proved necessary to create a theory not of numbers and quantities, but of 'states' – and 'Hilbert space' offered exactly the right symbolism for these. Another related development in pure mathematics, which quantum physicists were now busy exploiting, was that of the 'abstract group'. It had come about through mathematicians putting the idea of 'operation' into a symbolic form, and treating the result as an abstract exercise.* The

---

* The word 'group', as used in mathematics, has a technical meaning quite distinct from its use in ordinary language. It refers to the idea of a set of operations, but only when that set of operations meets certain precise conditions. These may be illustrated by considering the rotations of a sphere. If A, B and C are three different rotations, then one can see that:

(i)   there exists a rotation which exactly reverses the effect of A.

(ii) there exists a rotation which has exactly the same effect as performing A, and then B.

Let this rotation be called 'AB'. Then

(iii) AB, followed by C, has the same effect as A, followed by BC.

effect of abstraction had been to generalise, to unify, and to draw new analogies. It had been a creative and constructive movement, for by changing the rules of these abstract systems, new kinds of algebra with unforeseen applications had been invented.

On the other hand, the movement towards abstraction had created something of a crisis within pure mathematics. If it was to be thought of as a game, following arbitrary rules to govern the play of symbols, what had happened to the sense of absolute truth? In March 1933 Alan acquired Bertrand Russell's *Introduction to Mathematical Philosophy*, which addressed itself to this central question.

The crisis[23] had first appeared in the study of geometry. In the eighteenth century, it was possible to believe that geometry was a branch of science, being a system of truths about the world, which Euclid's axioms boiled down into an essential kernel. But the nineteenth century saw the development of geometrical systems different from Euclid's. It was also doubted whether the real universe was actually Euclidean. In the modern separation of mathematics from science, it became necessary to ask whether Euclidean geometry was, regarded as an abstract exercise, a complete and consistent whole.

It was not clear that Euclid's axioms really did define a complete theory of geometry. It might be that some extra assumption was being smuggled into proofs, because of intuitive, implicit ideas about points and lines. From the modern point of view it was necessary to *abstract* the logical relationships of points and lines, to formulate them in terms of purely symbolic rules, to forget about their 'meaning' in terms of physical space, and to show that the resulting abstract game made sense in itself. Hilbert, who was

---

These are essentially the conditions required for the rotations to form a 'group'. Abstract group theory then arose by taking these conditions, representing them appropriately with symbols, and then abandoning the original concrete embodiment. The resulting theory might profitably be *applied* to rotations, as indeed it was, in quantum mechanics. It could also apply to the apparently unrelated field of ciphering. (Ciphers enjoy the 'group' properties: a cipher must have a well-defined decipherment operation which reverses it, and if two ciphering operations are performed in succession, the result is another cipher.) But by the 1930s it was accepted that 'groups' could be explored in the abstract, without any concrete representation or application in mind.

always down-to-earth, liked to say: 'One must always be able to say "tables, chairs, beer-mugs", instead of "points, lines, planes".'

In 1899, Hilbert succeeded in finding a system of axioms which he could prove would lead to all the theorems of Euclidean geometry, without any appeal to the nature of the physical world. However, his proof required the assumption that the theory of 'real numbers'* was satisfactory. 'Real numbers' were what to the Greek mathematicians were the measurements of lengths, infinitely subdivisible, and for most purposes it could be assumed that the use of 'real numbers' was solidly grounded in the nature of physical space. But from Hilbert's point of view this was not good enough.

Fortunately it was possible to describe 'real numbers' in an essentially different way. By the nineteenth century it was well understood that 'real numbers' could be represented as infinite decimals, writing the number π for instance as 3.14159265358979.... A precise meaning had been given to the idea that a 'real number' could be represented as accurately as desired by such a decimal – an infinite sequence of integers. But it was only in 1872 that the German mathematician Dedekind had shown exactly how to define 'real numbers' in terms of the integers, in such a way that no appeal was made to the concept of measurement. This step both unified the concepts of number and length, and had the effect of pushing Hilbert's questions about geometry into the domain of the integers, or 'arithmetic', in its technical mathematical sense. As Hilbert said, all he had done was 'to reduce everything to the question of consistency for the arithmetical axioms, which is left unanswered.'

At this point, different mathematicians adopted different attitudes. There was a point of view that it was absurd to speak of the axioms of arithmetic. Nothing could be more primitive than the integers. On the other hand, it could certainly be asked whether there existed a kernel of fundamental properties of the integers, from which all the others could be derived. Dedekind also tackled this question, and showed in 1888 that all arithmetic could be derived from three ideas: that there is a number 1, that every number has a successor, and that

---

* There is nothing 'real' about 'real numbers'. The term is a historical accident, arising from the equally misleading terms 'complex numbers' and 'imaginary numbers'. The reader not familiar with these expressions could think of 'real numbers' as 'lengths defined with a hypothetical infinite precision.'

a principle of induction allows the formulation of statements about *all* numbers. These could be written out as abstract axioms in the spirit of the 'tables, chairs and beer-mugs' if one so chose, and the whole theory of numbers could be constructed from them without asking what the symbols such as '1' and '+' were supposed to mean. A year later, in 1889, the Italian mathematician G. Peano gave the axioms in what became the standard form.

In 1900 Hilbert greeted the new century by posing seventeen unsolved problems to the mathematical world. Of these, the second was that of proving the consistency of the 'Peano axioms' on which, as he had shown, the rigour of mathematics depended. 'Consistency' was the crucial word. There were, for instance, theorems in arithmetic which took thousands of steps to prove – such as Gauss's theorem that every integer could be expressed as the sum of four squares. How could anyone know for sure that there was not some equally long sequence of deductions which led to a contradictory result? What was the basis for credence in such propositions about all numbers, which could never be tested out? What was it about those abstract rules of Peano's game, which treated '+' and '1' as meaningless symbols, that guaranteed this freedom from contradictions? Einstein doubted the laws of motion. Hilbert doubted even that two and two made four – or at least said that there had to be a reason.

One attack on this question had already been made in the work of G. Frege, starting with his 1884 *Grundlagen der Arithmetik*. This was the *logistic* view of mathematics, in which arithmetic was derived from the logical relationships of the entities in the world, and its consistency guaranteed by a basis in reality. For Frege, the number '1' clearly meant something, namely the property held in common by 'one table', 'one chair', 'one beer-mug'. The statement '2 + 2 = 4' had to correspond to the fact that if any two things were put together with any other two things, there would be four things. Frege's task was to abstract the ideas of 'any', 'thing', 'other', and so forth, and to construct a theory that would derive arithmetic from the simplest possible ideas about existence.

Frege's work was, however, overtaken by Bertrand Russell, whose theory was on the same lines. Russell had made Frege's ideas more concrete by introducing the idea of the 'set'. His proposal was that a set which contained just one thing could be characterised by the feature that if an object were picked out of that set, it would always be the same object. This idea enabled one-ness to be defined in terms of same-ness, or equality. But then equality could be defined in terms of satisfying the same range of predicates. In this way the concept of number and the axioms of arithmetic could, it appeared, be rigorously derived from the most primitive notions of entities, predicates and propositions.

Unfortunately it was not so simple. Russell wanted to define a set-with-one-element, without appealing to a concept of counting, by the idea of equality. Then he would define the number 'one' *to be* 'the set of all sets-with-one-element'. But in 1901 Russell noticed that logical contradictions arose as soon as one tried to use 'sets of all sets'.

The difficulty arose through the possibility of self-referring, self-contradictory assertions, such as 'this statement is a lie.' One problem of this kind had emerged in the theory of the infinite developed by the German mathematician G. Cantor. Russell noticed that Cantor's paradox had an analogy in the theory of sets. He divided the sets into two kinds, those that contained themselves, and those that did not. 'Normally', wrote Russell, 'a class is not a member of itself. Mankind, for example, is not a man.' But the set of abstract concepts, or the set of all sets, would contain itself. Russell then explained the resulting paradox in this way:

> Form now the assemblage of classes which are not members of themselves. This is a class: is it a member of itself or not? If it is, it is one of those classes that are not members of themselves, *i.e.* it is not a member of itself. If it is not, it is not one of those classes that are not members of themselves, *i.e.* it is a member of itself. Thus of the two hypotheses – that it is, and that it is not, a member of itself – each implies its contradictory. This is a contradiction.

This paradox could not be resolved by asking what, if anything, it *really meant*. Philosophers could argue about that as long as they liked, but it was irrelevant to what Frege and Russell were trying to do. The whole point of this theory was to derive arithmetic from the most primitive logical ideas in an automatic, watertight, depersonalised way, without any arguments *en route*. Regardless of what Russell's paradox *meant*, it was a string of symbols which, according to the rules of the game, would lead inexorably to its own contradiction. And that spelt disaster. In any purely logical system there was no room for a single inconsistency. If one could ever arrive at '2 + 2 = 5' then it would follow that '4 = 5', and '0 = 1', so that any number was equal to 0, and so that every proposition whatever was equivalent to '0 = 0' and therefore true. Mathematics, regarded in this game-like way, had to be totally consistent or it was nothing.

For ten years Russell and A. N. Whitehead laboured to remedy the defect. The essential difficulty was that it had proved self-contradictory to assume that any kind of lumping together of objects could be called 'a set'. Some more refined definition was required. The Russell paradox was by no means the only problem with the theory of sets, but it alone consumed a large part of *Principia Mathematica*, the weighty volumes which in 1910 set out their derivation of mathematics from primitive logic. The solution that Russell and Whitehead found was to set up a hierarchy of different kinds of sets, called 'types'. There were to be primitive objects, then sets of objects, then sets of sets, then sets of sets of sets, and so on. By segregating the different 'types' of set, it was made impossible for a set to contain itself. But this made the theory very complicated, much more difficult than the number system it was supposed to justify. It was not clear that this was the only possible way in which to think about sets and numbers, and by 1930 various alternative schemes had been developed, one of them by von Neumann.

The innocuous-sounding demand that there should be some demonstration that mathematics formed a complete and consistent whole had opened a Pandora's box of problems. In one sense, mathematical propositions still seemed as true as anything could possibly be true; in another, they appeared as no more than marks on paper, which led to mind-stretching paradoxes when one tried to elucidate what they meant.

As in the Looking-Glass garden, an approach towards the heart of mathematics was liable to lead away into a forest of tangled technicalities. This lack of any simple connection between mathematical symbols and the world of actual objects fascinated Alan. Russell had ended his book saying, 'As the above hasty survey must have made evident, there are innumerable unsolved problems in the subject, and much work needs to be done. If any student is led into a serious study of mathematical logic by this little book, it will have served the chief purpose for which it has been written.' So the *Introduction to Mathematical Philosophy* did serve its purpose, for Alan thought seriously about the problem of 'types' – and more generally, faced Pilate's question: *What is truth?*

Kenneth Harrison was also acquainted with some of Russell's ideas, and he and Alan would spend hours discussing them. Rather to Alan's annoyance, however, he would ask 'but what use is it?' Alan would say quite happily that of course it was completely useless. But he must also have talked to more enthusiastic listeners, for in the autumn of 1933 he was invited to read a paper to the Moral Science Club. This was a rare honour for any undergraduate, especially one from outside the faculty of Moral Sciences, as philosophy and its allied disciplines were called at Cambridge. It would have been a quite unnerving experience, speaking in front of professional philosophers, but he wrote with customary *sang froid* to his mother:

26/11/33

. . . I am reading a paper to the Moral Science Club on Friday. Something by way of being Mathematical Philosophy. I hope they don't know it all allready.

The minutes[24] of the Moral Science Club recorded that on Friday 1 December 1933:

The sixth meeting of the Michaelmas term was held in Mr Turing's rooms in King's College. A. M. Turing read a paper on 'Mathematics and logic'. He suggested that a purely logistic view of mathematics was inadequate; and that mathematical propositions possessed a variety of interpretations, of which the logistic was merely one. A discussion followed.

R. B. Braithwaite (signed).

Richard Braithwaite, the philosopher of science, was a young Fellow of King's; and it might well have been through him that the invitation was made. Certainly, by the end of 1933, Alan Turing had his teeth into two parallel problems of great depth. Both in quantum physics and in pure mathematics, the task was to relate the abstract and the physical, the symbolic and the real.

*

German mathematicians had been at the centre of this enquiry, as in all mathematics and science. But as 1933 closed, that centre was a gaping, jagged hole, with Hilbert's Göttingen ruined. John von Neumann had left for America, never to return, and others had arrived in Cambridge. 'There are several distinguished German Jews coming to Cambridge this year,' wrote Alan on 16 October. 'Two at least to the mathematical faculty, viz. Born and Courant.' He might well have attended the lectures on quantum mechanics that Born gave that term, or those of Courant* on differential equations the next term. Born went on to Edinburgh, and Schrödinger to Oxford, but most exiled scientists found America more accommodating than Britain. The Institute for Advanced Study, at Princeton University, grew particularly quickly. When Einstein took up residence there in 1933, the physicist Langevin commented, 'It is as important an event as would be the transfer of the Vatican from Rome to the New World. The Pope of physics has moved and the United States will become the centre of the natural sciences.'

It was not Jewish ancestry alone that attracted the interference of Nazi officialdom, but scientific ideas themselves, even in the philosophy of mathematics:[25]

> A number of mathematicians met recently at Berlin University to consider the place of their science in the Third Reich. It was stated that German mathematics would remain those of the 'Faustian man', that logic alone was no sufficient basis for them, and that the Germanic intuition which had produced the concepts of infinity was superior to the logical equipment which the French and Italians had brought to bear on the subject. Mathematics was a heroic science which reduced chaos to order. National Socialism had the same task and demanded the same qualities. So the 'spiritual connexion' between them and the New Order was established – by a mixture of logic and intuition . . . .

To English minds, the wonder was that any state or party could interest itself in abstract ideas.

Meanwhile to the *New Statesman*, Hitler's rancour at the Treaty of Versailles only vindicated what Keynes and Lowes Dickinson

---

* Alan acquired a copy, soon heavily annotated, of Hilbert and Courant's *Methoden der Mathematischen Physik* in July 1933.

had always said. The difficulty was that being fair to Germany now meant making concessions to a barbarous regime. Conservative opinion, however, perceived the new Germany in terms of a balance of nation states, in which it was a renewed potential threat to Britain, but also a strong 'bulwark' against the Soviet Union. It was in this context that the Cambridge Anti-War movement revived in November 1933. Alan wrote:

12/11/33

There has been a lot happening this week. The Tivoli Cinema had arranged to shew a film called 'Our Fighting Navy' which was blatant militarist propaganda. The Anti-War movement organized a protest. The organization wasn't very good and we only got 400 signatures of wh[ich] 60 or more were from King's. The film was eventually withdrawn, but this was on account of the shindy that the militarists made outside the cinema when they had heard of our protest and had got it into their heads that we were going to break up the Cinema.

A further comment, that 'There was a very successful A[nti]-W[ar] demonstration yesterday', referred to the Armistice Day wreath-laying ceremony, which this year had more the flavour of a political statement. This was not wholly pacifist in spirit. Alan's friend James Atkins had decided that he was a pacifist, and Alan himself that he was not. But very influential was the suggestion that the First World War had been whipped up by the self-interest of the armament manufacturers. There was great feeling, in which probably Alan shared, that glorification of weapons should not be allowed to make a second great war more likely.

It was Eddington, who as a Quaker was a pacifist and inter-nationalist, who stimulated the next outward and visible step in Alan's career. This time it was not in connection with the 'Jabberwocky' of quantum mechanics, but through his course of lectures on the methodology of science[26] which Alan attended in the autumn of 1933. Eddington touched upon the tendency of scientific measurements to be distributed, when plotted on a graph, on what was technically called a 'normal' curve. Whether it was the wingspans of *Drosophilae*, or Alfred Beuttell's winnings at Monte Carlo, the readings would tend to bunch around a central value, and die away on either side, in a specific way. To explain why this should be so was a problem of fundamental importance in the theory of probability and statistics. Eddington offered an outline of why it was to be expected, but this did not satisfy Alan who, sceptical as ever, wanted to prove an exact result by rigorous pure-mathematical standards.

By the end of February 1934 he had succeeded. It did not require a conceptual advance, but still this was the first substantial result of his own. Typically, for him, it was one that connected pure mathematics with the physical world. But when he showed his work to someone else, he was told that the Central Limit Theorem, as the result was called, had already been proved in 1922 by a certain Lindeberg.[27] Working in his self-contained way, he had not thought to find out first whether his objective had already been attained. But he was also advised that, provided due explanation was given, it might still be acceptable as original work for a King's fellowship dissertation.

From 16 March to 3 April 1934, Alan joined a Cambridge party to go skiing in the Austrian Alps. It had a vaguely Quaker, internationalist link with Frankfurt University, whose ski-hut near Lech on the Austro-German border they used. The flavour of cooperation was soured by the fact that the German ski coach was an ardent Nazi. On his return, Alan wrote:

29/4/ 34

. . . We had a very amusing letter from Micha, the German leader of the skiing party . . . He said '. . . but in thoughts I am in your middle' . . .

I am sending some research I did last year to Czüber* in Vienna, not having found anyone in Cambridge who is interested in it. I am afraid however that he may be dead, as he was writing books in 1891.

But first the final Tripos examination had to be got out of the way; Part II from 28 to 30 May and then the Schedule B papers[28] from 4 to 6 June. In between the examinations he had to rush down to Guildford to see his father. Mr Turing, who was now sixty, underwent a prostate operation after which he was never again in the good health he had so far enjoyed.

He passed with distinction, making him what was called a 'B-star Wrangler' along with eight others. It was only an examination,

---

* The author of one of the books which described the Central Limit Theorem.

and Alan deprecated the fuss that his mother made over sending telegrams, and tried to persuade her not to come to the Degree Day formalities on 19 June. But it did mean the award by King's of a research studentship at £200 per annum, and this enabled him to stay on to try for a fellowship – a serious ambition of which he could now feel more confident than he had in 1932. Several others of his year stayed, including Fred Clayton and Kenneth Harrison. David Champernowne had switched to economics and had not yet taken his degree. James had found himself disoriented by the abstract nature of Part II, and gained a Second. He was not sure how to begin his career, and for the next few months, during which he came to visit Alan several times, did some private tuition work.

By the end of Alan's undergraduate period, his depression was lifting and new industry was arising, just as in the world outside. He had begun to put down firm Cambridge roots, and to cut a figure as one less subdued and more ready with wit and good humour. It was still true that he belonged neither to an 'aesthete' nor to an 'athlete' compartment. He had continued to row in the boat club, and got on amiably with the other members, once downing a pint of beer in one go. He played bridge with others of his year, though with the usual defect of serious mathematicians he could not be trusted to add up the scores. The visitor to his room would find a disarray of books and notes and unanswered letters about socks and underpants from Mrs Turing. Round the walls were stuck various mementoes – Christopher's picture, for one – but also, for those with eyes to see, magazine pictures with male sex-appeal. He also liked to root around in sales and street markets, and picked up a violin in London, on Farringdon Road, for which he took some lessons. This did not produce very aesthetic results, but there was a little of the 'aesthete' side in him, inasmuch as it debunked the pompous and stiff-upper-lip models of behaviour. It was all somewhat mystifying to Mrs Turing, when at Christmas 1934 Alan asked for a teddy bear, saying he had never had one as a little boy. The Turings usually dutifully exchanged more useful and improving presents. But he had his way, and Porgy the bear was installed.

Graduation meant little change in his general way of life, except that he gave up rowing and resumed running. After the degree day he took a cycling trip to Germany, asking an acquaintance, Denis

Williams, to come with him. A first-year student of the Moral Sciences Tripos, Denis knew Alan from the Moral Science Club, the King's boat club and the skiing trip. They took their bicycles on the train as far as Cologne, and then did thirty miles or so a day. One purpose of the trip was to visit Göttingen, where Alan consulted some authority, presumably in connection with the Central Limit Theorem.

A peculiar gangster regime there might be in Berlin, but Germany was still best for student travel, with cheap fares and youth hostels. They could hardly avoid seeing the swastika flags draped everywhere, but to English eyes they seemed less sinister than ridiculous. Once they stopped in a mining village, where they heard the miners singing on their way to work – a welcome contrast to the contrived Nazi displays. In the youth hostel Denis chatted with a German traveller, bidding goodbye amiably with a 'Heil Hitler', as foreign students generally did simply as a matter of polite conformity to local custom. (There had also been cases of them being assaulted when they failed to do so.) Alan came in and happened to see this. He said to Denis, 'You shouldn't have said that, he was a Socialist.' He must have spoken with the German earlier, and Denis was struck by the fact that someone had identified himself to Alan as an opponent of the regime. But it was not that Alan reacted as a signed-up anti-fascist, it was that he could not go through with a ritual with which he did not agree. To Denis it was more like another incident on their trip, when two working-class boys from England happened to catch up with them and Denis said that it would be polite to invite them over to have a drink. 'Noblesse oblige', said Alan, which made Denis feel very small and insincere.

They happened to be in Hanover a day or two after 30 June 1934, when the SA was overthrown. Alan's knowledge of German, although it was culled from mathematical textbooks, was better than Denis's, and he translated from the newspaper an account of how Roehm first had been given the chance to commit suicide and had then been shot. They were rather surprised by the attention given to his demise by the English press. But then, this was a symbolic event with resonances going beyond the plain fact that Hitler thereby gained supreme power. It removed a major contradiction within the Nazi party, trumpeting its intention to turn Germany into a giant stud farm. To grateful conservatives it was the end of 'decadent' Germany. Later, when Hitler was thoroughly unpopular, the opposite connection could be drawn, and Nazidom painted as itself 'decadent' and 'perverse'. Behind it lay the powerful *leitmotiv* that Hitler so skilfully orchestrated: that of the homosexual traitor.

For some Cambridge students a sight of the new Germany, and a brush with its crudities, might engender a powerful anti-fascist commitment. That step was not for Alan Turing. He was always friendly to the anti-fascist cause, but nothing would make him a 'political' person. His was the other road to freedom, that of dedication to his craft. Let others do what they could; he would achieve something right, something true. He would continue the civilisation that the anti-fascists defended.

In the summer and autumn of 1934, he continued to work on his dissertation.[29] The deadline for its submission was 6 December, but Alan handed it in a month early, and was ready for a next step. Eddington, who had played so important a part in his early development, had suggested his dissertation problem to him. The next suggestion came from Hilbert, although not so directly. In the spring of 1935, while his dissertation went the rounds of the King's Fellows, Alan went to a Part III course on Foundations of Mathematics. It was given by M. H. A. Newman.

Newman, then nearly forty, was with J. H. C. Whitehead the foremost British exponent of topology. This branch of mathematics could be described as the result of abstracting from geometry such concepts as 'connected', 'edge' and 'neighbouring' which did not depend upon measurement.* In the 1930s it was unifying and generalising much of pure mathematics. Newman was a progressive figure in a Cambridge where classical geometry was more strongly represented.

The basis of topology was the theory of sets, and so Newman had been drawn into the foundations of set theory. He had

---

* A simple example of a topological problem is that of the 'four colour theorem'. This states that a map such as one of the English counties can always be coloured with just four colours, in such a way that no two adjoining counties share the same colour. Alan himself took some interest in this problem, but it was to remain an unproved assertion until 1976.

also attended the 1928 international congress at which Hilbert represented the Germany excluded in 1924. Hilbert had revived his call for an investigation into the foundations of mathematics. And it was in Hilbert's spirit, rather than as a continuation of Russell's 'logistic' programme, that Newman lectured. Indeed, the Russell tradition had petered out, for Russell himself had left Cambridge in 1916 when first convicted and deprived of his Trinity College lectureship; and of his contemporaries, Wittgenstein had turned in a different direction, Harry Norton had gone mad, and Frank Ramsey had died in 1930. This left Newman as the only person in Cambridge with a deep knowledge of modern mathematical logic, although there were others, Braithwaite and Hardy amongst them, who were interested in the various approaches and programmes.

The Hilbert programme was essentially an extension of the work on which he had started in the 1890s. It did not attempt to answer the question which Frege and Russell had tackled, that of what mathematics *really was*. In that respect it was less philosophical, less ambitious. On the other hand, it was more far-reaching in that it asked profound and difficult questions *about* the systems such as Russell produced. In fact Hilbert posed the question as to what were, in principle, the limitations of a scheme such as that of *Principia Mathematica*. Was there a way of finding out what could, and what could not, be proved within such a theory? Hilbert's approach was called the *formalist* approach, because it treated mathematics as if a game, a matter of form. The allowable steps of proof were to be considered like the allowable moves in a game of chess, with the axioms as the starting position of the game. In this analogy, 'playing chess' corresponded to 'doing mathematics', but statements *about* chess (such as 'two knights cannot force checkmate') would correspond to statements *about* the scope of mathematics. And it was with such statements that the Hilbert programme was concerned.

At that 1928 congress, Hilbert made his questions quite precise. First, was mathematics *complete*, in the technical sense that every statement (such as 'every integer is the sum of four squares') could either be proved, or disproved. Second, was mathematics *consistent*, in the sense that the statement '$2 + 2 = 5$' could never be arrived at by a sequence of valid steps of proof. And thirdly, was mathematics *decidable*? By this he meant, did there exist a definite method which could, in principle, be applied to any assertion, and which was guaranteed to produce a correct decision as to whether that assertion was true.

In 1928, none of these questions was answered. But it was Hilbert's opinion that the answer would be 'yes' in each case. In 1900 Hilbert had declared 'that every definite mathematical problem must necessarily be susceptible of an exact settlement . . . in mathematics there is no *ignorabimus*'; and when he retired in 1930 he went further:[30]

> In an effort to give an example of an unsolvable problem, the philosopher Comte once said that science would never succeed in ascertaining the secret of the chemical composition of the bodies of the universe. A few years later this problem was solved. . . . The true reason, according to my thinking, why Comte could not find an unsolvable problem lies in the fact that there is no such thing as an unsolvable problem.

It was a view more positive than the Positivists. But at the very same meeting, a young Czech mathematician, Kurt Gödel, announced results which dealt it a serious blow.

Gödel was able to show[31] that arithmetic must be *incomplete:* that there existed assertions which could neither be proved nor disproved. He started with Peano's axioms for the integers, but enlarged through a simple theory of types, so that the system was able to represent sets of integers, sets of sets of integers, and so on. However, his argument would apply to any formal mathematical system rich enough to include the theory of numbers, and the details of the axioms were not crucial.

He then showed that all the operations of 'proof', these 'chess-like' rules of logical deduction, were themselves arithmetical in nature. That is, they would only employ such operations as counting and comparing, in order to test whether one expression had been correctly substituted for another – just as to see whether a chess move was legal or not would only be a matter of counting and comparing. In fact, Gödel showed that the formulae of his system could be encoded as integers, so that he had integers representing statements *about* integers. This was the key idea.

Gödel continued to show how to encode proofs as integers, so that he had a whole theory of arithmetic, encoded *within* arithmetic.

It was an exploitation of the fact that if mathematics were regarded purely as a game with symbols, then it might as well employ numerical symbols as any other. He was able to show that the property of 'being a proof' or of 'being provable' was no more and no less arithmetical than the property of 'being square' or 'being prime'.

The effect of this encoding process was that it became possible to write down arithmetical statements which referred *to themselves*, like the person saying 'I am lying.' Indeed Gödel constructed one particular assertion which had just such a property, for in effect it said 'This statement is unprovable.' It followed that this assertion could not be proved *true*, for that would lead to a contradiction. Nor could it be proved *false*, for the same reason. It was an assertion which could not be proved or disproved by logical deduction from the axioms, and so Gödel had proved that arithmetic was *incomplete*, in Hilbert's technical sense.

There was more to it than this, for one remarkable thing about Gödel's special assertion was that since it was not provable, it was, in a sense, *true*. But to say it was 'true' required an observer who could, as it were, look at the system from outside. It could not be shown by working *within* the axiomatic system.

Another point was that the argument assumed that arithmetic was *consistent*. If, in fact, arithmetic were inconsistent, then *every* assertion would be provable. So, more precisely, Gödel had shown that formalised arithmetic must either be inconsistent, or incomplete. He was also able to show that arithmetic could not be proved consistent within its own axiomatic system. To do so, all that would be required would be a proof that there was a single proposition (say, $2 + 2 = 5$) which could not be proved true. But Gödel was able to show that such a statement of existence had the same character as the sentence that asserted its own unprovability. And in this way, he had polished off the first two of Hilbert's questions. Arithmetic could *not* be proved consistent, and it was certainly *not* consistent *and* complete. This was an amazing new turn in the enquiry, for Hilbert had thought of his programme as one of tidying up loose ends. It was upsetting for those who wanted to find in mathematics something that was absolutely perfect and unassailable; and it meant that new questions came into view.

Newman's lectures finished with the proof of Gödel's theorem, and thus brought Alan up to the frontiers of knowledge. The *third* of Hilbert's questions still remained open, although it now had to be posed in terms of 'provability' rather than 'truth'. Gödel's results did not rule out the possibility that there was some way of distinguishing the provable from the non-provable statements. Perhaps the rather peculiar Gödelian assertions could somehow be separated off. Was there a definite method, or as Newman put it, a *mechanical process* which could be applied to a mathematical statement, and which would come up with the answer as to whether it was provable?

From one point of view this was a very tall order, going to the heart of everything known about creative mathematics. Hardy, for instance, had said[32] rather indignantly in 1928 that:

> There is of course no such theorem, and this is very fortunate, since if there were we should have a mechanical set of rules for the solution of all mathematical problems, and our activities as mathematicians would come to an end.

There were plenty of statements about numbers which the efforts of centuries had failed either to prove or disprove. There was Fermat's so-called Last Theorem, which conjectured that there was no cube which could be expressed as the sum of two cubes, no fourth power as sum of two fourth powers, and so on. Another was Goldbach's conjecture, that every even number was the sum of two primes. It was hard to believe that assertions which had resisted attack so long could in fact be decided automatically by some set of rules. Furthermore, the difficult problems which had been solved, such as Gauss's Four Square Theorem, had rarely been proved by anything like a 'mechanical set of rules', but by the exercise of creative imagination, constructing new abstract algebraic concepts. As Hardy said, 'It is only the very unsophisticated outsider who imagines that mathematicians make discoveries by turning the handle of some miraculous machine.'

On the other hand, the progress of mathematics had certainly brought more and more problems within the range of a 'mechanical' approach. Hardy might say that 'of course' this advance could never encompass the whole of mathematics, but after Gödel's theorem, nothing was 'of course' any more. The question deserved a more penetrating analysis.

Newman's pregnant phrase 'by a mechanical process' revolved

in Alan's mind. Meanwhile, the spring of 1935 saw two other steps forward. The fellowship election was held on 16 March. Philip Hall had just become an elector, and argued for Alan, saying that his full strength had not been shown in his rediscovery of the Central Limit Theorem. But his advocacy was not needed. Keynes, Pigou and the Provost, John Sheppard, all had an assessment of him for themselves. He was elected, the first of his year, as one of the forty-six Fellows. The boys of Sherborne School enjoyed a half-holiday, and there was a clerihew that circulated:

> Turing
> Must have been alluring
> To get made a don
> So early on.

He was still only twenty-two. The fellowship carried with it £300 a year for three years, which would normally be extended to six, and no explicit duties. He was entitled to room and board when he chose to reside at Cambridge, and to dine at High Table. On his first evening in the senior common room, he played Rummy and won a few shillings from the Provost. But he tended to prefer the company at dinner of his friends David Champernowne, Fred Clayton and Kenneth Harrison. It did not change his style of life, but did make him free for three years to pursue thought in any way he chose – as free as anyone could be without a private income. He supplemented his fellowship by supervising undergraduates in next-door Trinity Hall. If they came to his rooms hoping for a glimpse of King's eccentricity, they were sometimes rewarded, as when Alan sat Porgy the teddy bear by the fire, in front of a book supported by a ruler, and greeted them with 'Porgy is very *studious* this morning.'

The election coincided with what Alan called a 'small-scale discovery' which constituted a first publishable paper. It was a neat result in group theory, which he announced to Philip Hall (whose own research lay in that field) on 4 April, saying he was 'thinking of doing this sort of thing seriously.' It was submitted and published[33] by the London Mathematical Society later in the month.

The result was a small improvement to a paper by von Neumann,[34] which developed the theory of 'almost periodic functions'* by defining

---

* A recent development in pure mathematics, extending and generalising the idea of 'periodicity'.

them with reference to 'groups'. As it happened, von Neumann arrived at Cambridge later that month. He was spending a summer away from Princeton, and gave a lecture course at Cambridge on the subject of 'almost periodic functions'. Alan certainly met him this term, and most likely through attending this course.

They were very different men. When Alan Turing was born, János von Neumann was the eight-year-old son of a rich Hungarian banker.[35] There was for him no public-school training, and by 1922, before Alan was floating his paper boats at Hazelhurst, the eighteen-year-old von Neumann had published his first paper. Budapest János became Göttingen Johann, one of Hilbert's disciples, and then in 1933 became Princeton Johnny, adopting English as his fourth language. The paper on 'almost periodic functions' was his fifty-second, part of an immense output which had moved from the axioms of set theory and quantum mechanics, to the topological groups which were the pure-mathematical underpinning of quantum theory, but taking in numerous other topics on the side.

John von Neumann was one of the most important figures in twentieth-century mathematics, but he was a man who added worldly to intellectual success. He enjoyed a commanding manner, a sophisticated, racy humour, a training in engineering, a wide knowledge of history – and a salary of $10,000 over and above his substantial private income. He cut a figure very different from that of the twenty-two-year old in the shabby sports jacket, sharp but shy with a hesitant voice that had trouble with one language, let alone four. But mathematics did not see these things, and it might well have been the result of a meeting of minds when on 24 May Alan wrote home: '. . . I have applied for a visiting Fellowship at Princeton for next year.'*

An additional reason would be, however, that Alan's friend Maurice Pryce, whom he had met at the scholarship examinations in 1929, and with whom he had kept in touch, was ready to go to Princeton in September, having secured a fellowship there. In any case, it was becoming more and more clear that Princeton was the new Göttingen; there was a flow of first-rate mathematicians and physicists to and fro across the Atlantic. It was an aspect of the continuing transfer of power from Europe, and from Germany in

---

* It is not clear from the context whether 'next year' means 1935–6 or 1936–7.

particular, to America. No one who wanted to *do something*, as Alan did, could any longer ignore the United States.

Alan continued work in group theory during 1935.[36] He also thought of working in quantum mechanics, and approached R. H. Fowler, Professor of Mathematical Physics, for a suitable problem to work on. Fowler suggested trying to explain the dielectric constant of water, one of his favourite research topics. But Alan made no progress. And this problem, as indeed the whole field of mathematical physics, which offered so much to attract the ambitious young mathematicians of the 1930s, was put aside. For he had seen something new, something at the centre of mathematics, something at *his* centre. It owed almost nothing to the Tripos; it used only the commonest in Nature. It was profoundly ordinary, and yet led to a spectacular idea.

It had become his habit to run long distances in the afternoons, along the river and elsewhere, even as far as Ely. It was at Grantchester, so he said later, lying in the meadow, that he saw how to answer Hilbert's third question. It must have been in the early summer of 1935. 'By a mechanical process', Newman had said. So Alan Turing dreamed of machines.

'For, of course, the body is a machine. It is a vastly complex machine, many, many times more complicated than any machine ever made by hands; but still after all a machine.' Such was Brewster's paradoxical assertion. At one level, the body was living, not a machine. But at another, more detailed level of description, that of the 'living bricks', it was all *determined*. It was not the power of the machine that was the point of the remark; it was its lack of will.

It was not the determinism of physics, or chemistry, or of biological cells, that was involved in Hilbert's question about decidability. It was something more abstract. It was the quality of being fixed in advance, in such a way that nothing new could arise. And the operations were to be operations on symbols, rather than on things of any particular mass or chemical composition.

Alan had to *abstract* this quality of being determined, and apply it to the manipulation of symbols. People had spoken, as Hardy did, of 'mechanical rules' for mathematics, of 'turning the handle' of a miraculous machine, but no one had actually sat down to design one. This was

what he set out to do. For although he was not really 'the very unsophisticated outsider' of whom Hardy spoke, he attacked the problem in
a peculiarly naive way, undaunted by the immensity and complexity of
mathematics. He started from nothing, and tried to envisage a machine
that could tackle Hilbert's problem, that of deciding the provability of
any mathematical assertion presented to it.

There were, of course, machines in existence which manipulated
symbols. The typewriter was one such. Alan had dreamt of inventing typewriters as a boy; Mrs Turing had a typewriter; and he could
well have begun by asking himself what was meant by calling a
typewriter 'mechanical'. It would mean that its response, to any
particular action of the operator, was perfectly certain. One could
describe in advance exactly how the machine would behave in any
contingency. But there was more to be said even about a humble
typewriter than that. The response would depend upon the current
condition, or what Alan called the current *configuration*, of the
machine. In particular, a typewriter would have an 'upper case' configuration and a 'lower case' configuration. This was an idea which
Alan put in a more general and abstract form. He would consider
machines which at any time would be in one of a finite number of
possible 'configurations'. Then if, as with the typewriter keyboard,
there were only a finite number of things that could be done to the
machine, a complete account of the behaviour of the machine could
be given, once for all, in finite form.

However, the typewriter had a further feature which was
essential to its function. Its typing point could move, relative to the
page. Its typing action would be independent of the position of this
point on the page. Alan incorporated this idea too into his picture of
the more general machine. It was to have internal 'configurations',
and a variable position on a printing line. The action of the machine
would not depend upon its position.

Neglecting details as to margins, line control, and so forth, these
ideas would suffice to give a complete description of the nature of
a typewriter. An exact account of the configurations and positions
allowed, and of how the character keys determined the symbols
printed, the shift key the change of configuration from 'lower'
to 'upper', and the space bar and backspace the printing position,
would bring out the features most relevant to its function. If an

engineer took this account, and created a physical machine which met its specifications, the result would be a typewriter, regardless of its colour, weight, or other attributes.

But a typewriter was too limited to serve as a model. It dealt with symbols, but it could only *write* them, and it required a human operator to choose the symbols and changes of configuration and position, one at a time. What, Alan Turing asked, would be the most *general* kind of machine that dealt with symbols? To be a 'machine' it would have to retain the typewriter's quality of having a finite number of configurations, and an exactly determined behaviour in each. But it would have to be capable of much more. And so he imagined machines which were, in effect, super-typewriters.

To simplify the description he imagined his machines working with just one line of writing. This was only a technicality, which allowed margins and line control to be forgotten. But it was important that the supply of paper was to be assumed unlimited. In his picture, the typing point of his super-typewriter could progress indefinitely to left or right. For the sake of definiteness, he imagined the paper as being in the form of a *tape*, marked off into unit squares, such that just one symbol could be written on any one square. Thus his machines were to be finitely defined, but they would be allowed an unlimited amount of space on which to work.

Next, the machine would be able to *read*, or using his word, to 'scan' the square of tape on which it rested. It would still of course be able to write symbols, but now also to *erase* them. But it would only be able to move one place to the left or the right at a time. What role remained to the human operator of the typewriter? He did mention the possibility of what he called 'choice machines', in which an external operator would have the job of making decisions at certain points. But the thrust of his argument was directed at what he called *automatic* machines, in which human intervention would play no part. For the goal of his development was the discussion of what Hardy had called 'a miraculous machine' – a mechanical process which could work on Hilbert's decision problem, reading a mathematical assertion presented to it, and eventually writing a verdict as to whether it was provable or not. The whole point was that it should do so without the interference of human judgment, imagination or intelligence.

Any 'automatic machine' would work away by itself, reading and writing, moving to and fro, all in accordance with the way in which it was constructed. At every step its behaviour would be completely determined by the configuration it was in and the symbol it had read. To be precise, the construction of the machine would determine, for each combination of configuration and symbol scanned:

> whether to write a new (specified) symbol in a blank square, to leave the existing one unchanged, or to erase it and leave a blank square
>
> whether to remain in the same configuration, or to change to some other (specified) configuration
>
> whether to move to the square on the left, or to the right, or to stay in the same position.

If all this information, defining an automatic machine, were written out, it would form a 'table of behaviour' of a finite size. It would completely define the machine in the sense that whether physically constructed or not, the table would hold all the relevant information about it. From this abstract point of view, the table *was* the machine.

Every different possible table would define a machine with a different kind of behaviour. There would be infinitely many possible tables, corresponding to infinitely many possible machines. Alan had rendered the vague idea of a 'definite method' or a 'mechanical process' into something very precise: a 'table of behaviour'. And so now he had a very precise question to answer: was there or was there not one of these machines, one of these tables, that could produce the decision that Hilbert asked for?

An example machine: The following 'table of behaviour' completely defines a machine with the character of an adding machine. Started with the 'scanner' somewhere to the left of two groups of 1's, separated by a single blank space, it will add the two groups, and stop. Thus, it will transform

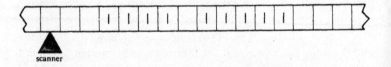

scanner

into

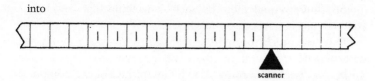

scanner

The task of the machine is to fill in the blank space, and to erase the last '1'. It will therefore suffice to provide the machine with four configurations. In the first it moves along the blank tape looking for the first group of '1's. When it moves into the first group, it goes into the second configuration. The blank separator sends it into the third configuration, in which it moves along the second group until it encounters another blank, which acts as the signal to turn back, and to enter the fourth and final configuration in which it erases the last '1' and marks time for ever.

The complete table is:

Symbol scanned

	blank	1
Config. 1	move right; config. 1	move right; config. 2
Config. 2	write '1' move right; config. 3	move right; config. 2
Config. 3	move left; config. 4	move right; config. 3
Config. 4	no move; config. 4	erase; no move; config. 4

Even a very simple machine of this kind, as shown in the example, would be doing more than sums. The machine would effect acts of *recognition*, such as 'finding the first symbol to the right'. A rather more complicated machine could perform multiplication, by repeated acts of copying out one group of 1s, while erasing one at a time of another group of 1s, and recognising when it had finished. Such a machine could also effect acts of *decision*, as for instance in deciding whether one number was divisible by another, or whether a given number was prime or composite. Clearly there was scope for exploiting this principle to mechanise a vast range of 'definite methods'. But could there be such a machine that could decide Hilbert's question about provability?

This was much too hard a problem to approach by trying to write a 'table' to solve it. But there was an approach which led to the answer by a back-door route. Alan hit on the idea of the 'computable numbers'. The crucial notion was that any 'real number' which was defined by some definite rule could be calculated by one of his machines. For instance, there would be a machine to calculate the decimal expansion of π, rather as he had done at school. For it would require no more than a set of rules for adding, multiplying, copying, and so forth. π being an infinite decimal, the work of the machine would never end, and it would need an unlimited amount of working space on its 'tape'. But it would arrive at every decimal place in *some* finite time, having used only a finite quantity of tape. And everything about the process could be defined by a finite table, left alone to work on a blank tape.

This meant that he had a way of representing a number like π, an infinite decimal, by a *finite* table. The same would be true of the square root of three, or the logarithm of seven – or any other number defined by some rule. Such numbers he called the 'computable numbers'.

More precisely, the machine itself would know nothing about decimals or decimal places. It would simply produce a sequence of digits. A sequence that could be produced by one of his machines, starting on a blank tape, he called a 'computable sequence'. Then an infinite computable sequence, prefaced by a decimal point, would define a 'computable number' between 0 and 1. It was in this more strict sense that any computable number between 0 and I could be defined by a finite table. It was important to his argument that the computable numbers would always be expressed as infinite sequences of digits, even if these were all 0 after a certain point.

But these finite tables could now be put into something like alphabetical order, beginning at the most simple, and working through larger and larger ones. They could be put in a list, or counted; and this meant that all the computable numbers could be put in a list. It was not a practical proposition actually to do it, but in principle the idea was perfectly definite, and would result in the square root of three being say 678th in order, or the logarithm of π being 9369th. It was a staggering thought, since this list would include every number that could be arrived at through arithmetical operations, finding roots of equations, and using mathematical functions like sines and logarithms – every number that could possibly arise in computational mathematics. And once he had seen this, he knew the answer to Hilbert's question. Probably it was this that he suddenly saw on the Grantchester meadow. He would have seen the answer because there was a beautiful mathematical device, ready to be taken off the shelf.

Fifty years earlier, Cantor had realised that he could put all the fractions – all the ratios or *rational* numbers – into a list. Naively it might be thought that there were many more fractions than integers. But Cantor showed that, in a precise sense, this was not so, for they could be counted off, and put into a sort of alphabetical order. Omitting fractions with cancelling factors, this list of all the rational numbers between 0 and 1 would begin:

1/2 1/3 1/4 2/3 1/5 1/6 2/5 3/4 1/7 3/5 1/8 2/7 4/5 1/9
3/7 1/10. . .

Cantor went on to invent a certain trick, called the Cantor diagonal argument, which could be used as a proof that there existed *irrational* numbers. For this, the rational numbers would be expressed as infinite decimals, and the list of all such numbers between 0 and 1 would then begin:

1	.5000000000000000000. . . .
2	.3333333333333333333. . . .
3	.2500000000000000000. . . .
4	.6666666666666666666. . . .
5	.2000000000000000000. . . .
6	.1666666666666666666. . . .
7	.4000000000000000000. . . .
8	.7500000000000000000. . . .
9	.1428571428571428571. . . .
10	.6000000000000000000. . . .
11	.1250000000000000000. . . .
12	.2857142857142857142. . . .
13	.8000000000000000000. . . .
14	.1111111111111111111. . . .
15	.4285714285714285714. . . .
16	.1000000000000000000. . . .
.	. . . .
.	. . . .

The trick was to consider the diagonal number, beginning

.5306060020040180. . . .

and then to change each digit, as for instance by increasing each by 1 except by changing a 9 to a 0. This would give an infinite decimal beginning

.6417171131151291 . . . .

a number which could not possibly be rational, since it would differ from the first listed rational number in the first decimal place, from the 694th rational number in the 694th decimal place, and so forth. Therefore it could not be in the list; but the list held all the rational numbers, so the diagonal number could not be rational.

It was already well-known – it was known to Pythagoras – that there were irrational numbers. The point of Cantor's construction was actually rather different from this. It was to show that no list could possibly contain all the 'real numbers', that is, all infinite decimals. For any proposed list would serve to define another infinite decimal which had been left out. Cantor's argument showed that in a quite precise sense there were *more* real numbers than integers. It opened up a precise theory of what was meant by 'infinite'.

However, the point relevant to Alan Turing's problem was that it showed how the rational could give rise to the irrational. In exactly the same way, therefore, the computable could give rise to the *uncomputable*, by means of a diagonal argument. As soon as he had made that observation, Alan could see that the answer to Hilbert's question was 'no'. There could exist no 'definite method' for solving all mathematical questions. For an uncomputable number would be an example of an unsolvable problem.

There was still much work to do before his result was clear. For one thing, there was something paradoxical about the argument. The Cantor trick itself would seem to be a 'definite method'. The diagonal number was *defined* clearly enough, it appeared – so why could it not be computed? How could something that was constructed in this mechanical way be uncomputable? What would go wrong, if it were attempted?

Suppose one tried to design a 'Cantor machine' to produce this diagonal uncomputable number. Roughly speaking, it would start with a blank tape, and write the number 1. It would then have to produce the *first* table, and then execute it, stopping at the *first* digit that it wrote, and adding on one. Then it would start again, with the number 2, produce the *second* table, executing it as far as the *second* digit, and writing this down, adding on one. It would have to continue doing this for ever, so that when its counter read '1000', it would produce the *thousandth* table, execute it as far as the *thousandth* digit, add on one to this and write it down.

One part of this process could certainly be done by one of his machines. For the process of 'looking up the entries' in a given table, and working out what the corresponding machine would do, was itself a 'mechanical process.' A machine could do it. There was a difficulty in that the tables were naturally thought of in two-dimensional form, but then it was only a technical matter to encode them in a form in which they could be put on a 'tape'. In fact, they could be encoded as integers, rather as Gödel had represented formulae and proofs as integers. Alan called them 'description numbers', so that there was a description number corresponding to each table. In one way this was just a technicality, a means of putting tables on to the tape, and arranging them in an 'alphabetical order'. But underneath there lay the same powerful idea that Gödel had used, that there was no essential distinction between 'numbers' and operations *on* numbers. From a modern mathematical point of view, they were all alike symbols.

With this done, it followed that one particular machine could simulate the work done by *any* machine. He called it the *universal* machine. It would be designed to read description numbers, decode them into tables, and execute them. It could do what any other machine would have done, if it were provided with the description number of that machine on its tape. It would be a machine to do everything, which was enough to give anyone pause for thought. It was, furthermore, a machine of perfectly definite form. Alan worked out an exact table for the universal machine.

This was not the trouble with mechanising the Cantor process. The difficulty lay in the other requirement, that of producing the tables, in their 'alphabetical order', for the computable numbers.

Suppose that the tables were encoded as description numbers. In practice, they would not use up all the integers; in fact, the system Alan devised would encode even the simplest tables into enormous numbers. But that would not matter. It would be essentially a 'mechanical' matter to work through the integers in turn, and to pass over those which did not correspond to proper tables. That was a technicality, almost a matter of notation. The real problem was more subtle. The question was this: given (say) the 4589th properly defined table, how could one tell that it would produce a 4589th digit? Or indeed, that it would produce any digits at all? It might trundle back and forth in a repeated cycle of operations for ever, without producing more figures. If this were the case, the Cantor machine would be stuck, and could never finish its job.

The answer was that one could *not* tell. There was no way of checking in advance that a table would produce an infinite sequence. There might be a method for some particular table. But there was no mechanical process – no machine – that could work on *all* instruction tables. There was nothing better than the prescription: 'take the table and try it out'. But this procedure would take an infinite time to find out whether infinitely many digits emerged. There was no rule that could be applied to any table, and be guaranteed to produce the answer in a finite time, as was required for the printing of the diagonal number. The Cantor process, therefore, could not be mechanised, and the uncomputable diagonal number could not be computed. There was no paradox after all.

Alan called the description numbers which gave rise to infinite decimals the 'satisfactory numbers'. So he had shown that there was no definite method of identifying an 'unsatisfactory number'. He had pinned down a clearly specified example of something Hilbert said did not exist – an *unsolvable problem*.

There were other ways of demonstrating that no 'mechanical process' could eliminate the unsatisfactory numbers. The one he himself favoured was one which brought out the connection with *self-reference* in the question. For supposing that such a 'checking' machine did exist, able to locate the unsatisfactory numbers, it could be applied to *itself.* But this, he showed, led to a flat contradiction. So no such checking machine could exist.

Either way, he had found an unsolvable problem, and it required only a technical step to show that this settled Hilbert's question about mathematics, in the exact form in which it had been posed. Alan Turing had dealt the death-blow to the Hilbert programme. He had shown that mathematics could never be exhausted by any finite set of procedures. He had gone to the heart of the problem, and settled it with one simple, elegant observation.

But there was more to what he had done than a mathematical trick, or logical ingenuity. He had created something new – the idea of his machines. And correspondingly, there remained a question as to whether his definition of the machine really did include everything that could possibly be counted as a 'definite method'. Was this repertoire of reading, writing, erasing, moving and stopping enough? It was crucial that it was so, for otherwise the suspicion would always lurk that some extension of the machine's faculties would allow it to solve a greater range of problems. One approach to this question led him to demonstrate that his machines could certainly compute any number normally encountered in mathematics. He also showed that a machine could be set up to churn out every provable assertion within Hilbert's formulation of mathematics. But he also gave some pages of discussion[37] that were among the most unusual ever offered in a mathematical paper, in which he justified the definition by considering what *people* could possibly be doing when they 'computed' a number by thinking and writing down notes on paper:

> Computing is normally done by writing certain symbols on paper. We may suppose this paper is divided into squares like a child's arithmetic book. In elementary arithmetic the two-dimensional character of the paper is sometimes used. But such a use is always avoidable, and I think that it will be agreed that the two-dimensional character of paper is no essential of computation. I assume then that the computation is carried out on one-dimensional paper, *i.e.* on a tape divided into squares. I shall also suppose that the number of symbols which may be printed is finite. If we were to allow an infinity of symbols, then there would be symbols differing to an arbitrarily small extent.

An 'infinity of symbols', he wished to argue, did not correspond to anything in reality. It might be argued that there was an infinity of symbols, in that

> an Arabic numeral such as 17 or 999999999999999 is normally treated as a single symbol. Similarly in any European language words are treated as single symbols (Chinese, however, attempts to have an enumerable infinity of symbols).

But he disposed of this objection with the observation that

> The differences from our point of view between the single and compound symbols is that the compound symbols, if they are too lengthy, cannot be observed at one glance. This is in accordance with experience. We cannot tell at a glance whether 9999999999999999 and 999999999999999 are the same.

Accordingly, he felt justified in restricting a machine to a finite repertoire of symbols. Next came a most important idea:

> The behaviour of the computer at any moment is determined by the symbols which he is observing, and his 'state of mind' at that moment. We may suppose that there is a bound B to the number of symbols or squares which the computer can observe at one moment. If he wishes to observe more, he must use successive observations. We will also suppose that the number of states of mind which need to be taken into account is finite. The reasons for this are the same character as those which restrict the number of symbols. If we admitted an infinity of states of mind, some of them will be 'arbitrarily close' and will be confused. Again, the restriction is not one which seriously affects computation, since the use of more complicated states of mind can be avoided by writing more symbols on the tape.

The word 'computer' here meant only what that word meant in 1936: a person doing calculations. Elsewhere in the paper he appealed to the idea that 'the human memory is necessarily limited,' but this was as far as he went in a discussion of the nature of the human mind. It was a bold act of imagination, a brave suggestion that 'states of mind' could be counted, on which to base his argument. It was especially noteworthy because in quantum mechanics, physical states *could* be 'arbitrarily close'. He continued with his discussion of the human computer:

Let us imagine the operations performed by the computer to be split up into 'simple operations' which are so elementary that it is not easy to imagine them further divided. Every such operation consists of some change in the physical system consisting of the computer and his tape. We know the state of the system if we know the sequence of symbols on the tape, which of these are observed by the computer (possibly with a special order) and the state of mind of the computer. We may suppose that in a simple operation not more than one symbol is altered. Any other changes can be split up into simple changes of this kind. The situation in regard to the squares whose symbols may be altered in this way is the same as in regard to the observed squares. We may, therefore, without loss of generality, assume that the squares whose symbols are changed are always 'observed' squares.

Besides these changes of symbols, the simple operations must include changes of distribution of observed squares. The new observed squares must be immediately recognisable by the computer. I think it is reasonable to suppose that they can only be squares whose distance from the closest of the immediately previously observed squares does not exceed a certain fixed amount. Let us say that each of the new observed squares is within L squares of an immediately previously observed square.

In connection with 'immediate recognisability', it may be thought that there are other kinds of square which are immediately recognisable. In particular, squares marked by special symbols might be taken as immediately recognisable. Now if these squares are marked only by single symbols there can be only a finite number of them, and we should not upset our theory by adjoining these marked squares to the observed squares. If, on the other hand, they are marked by a sequence of symbols, we cannot regard the process of recognition as a simple process. This is a fundamental point and should be illustrated. In most mathematical papers the equations and theorems are numbered. Normally the numbers do not go beyond (say) 1000. It is, therefore, possible to recognise a theorem at a glance by its number. But if the paper was very long, we might reach Theorem 157767733443477; then, further on in the paper, we might find '. . . hence (applying Theorem 157767734443477) we have . . .'. In order to make sure which was the relevant theorem we should have to compare the two numbers figure by figure, possibly ticking the figures off in pencil to make sure of their not being counted twice. If in spite of this it is still thought that there are other 'immediately recognisable' squares, it does not upset my contention so long as these squares can be found by some process of which my type of machine is capable . . . .

The simple operations must therefore include:

(a) Changes of the symbol on one of the observed squares

(b) Changes of one of the squares observed to another square within L squares of one of the previously observed squares.

It may be that some of these changes necessarily involve a change of state of mind. The most general single operation must therefore be taken to be one of the following:

(A) A possible change (a) of symbol together with a possible change of state of mind;

(B) A possible change (b) of observed squares, together with a possible change of state of mind.

The operation actually performed is determined, as has been suggested [above] by the state of mind of the computer and the observed symbols. In particular, they determine the state of mind of the computer after the operation is carried out.

'We may now construct a machine,' Alan wrote, 'to do the work of this computer.' The drift of his argument, indeed, was obvious, with each 'state of mind' of the human computer being represented by a configuration of the corresponding machine.

These 'states of mind' being a weak point of the argument, he added an alternative justification of the idea that his machines could perform any 'definite method' which did not need them:

We suppose [still] that the computation is carried out on a tape; but we avoid introducing the 'state of mind' by considering a more physical and definite counterpart of it. It is always possible for the computer to break off from his work, to go away and forget all about it, and later to come back and go on with it. If he does this he must leave a note of instructions (written in some standard form) explaining how the work is to be continued. This note is the counterpart of the 'state of mind'. We will suppose that the computer works in such a desultory manner that he never does more than one step at a sitting. The note of instructions must enable him to carry out one step and write the next note. Thus the state of progress of the computation at any stage is completely determined by the note of instructions and the symbols on the tape . . .

But these arguments were quite different. Indeed, they were complementary. The first put the spotlight upon the range of thought within the individual – the number of 'states of mind'. The second conceived of the individual as a mindless executor of given directives. Both approached the contradiction of free will and determinism, but one from the side of internal will, the other from that of external constraints. These approaches were not explored in the paper, but left as seeds for future growth.*

Alan had been stimulated by Hilbert's decision problem, or the *Entscheidungsproblem* as it was in German. He had not only answered it, but had done much more. Indeed, he would entitle his paper 'On Computable Numbers, *with an application* to the Entscheidungsproblem.' It was as though Newman's lectures had tapped some stream of enquiry which had been flowing all the time and which found in this question an opportunity to emerge. He had *done something*, for he had resolved a central question about mathematics, and had done so by crashing in as an unknown and unsophisticated outsider. But it was not only a matter of abstract mathematics, not only a play of symbols, for it involved thinking about what people did in the physical world. It was not exactly science, in the sense of making observations and predictions. All he had done was to set up a new model, a new framework. It was a play of imagination like that of Einstein or von Neumann, doubting the axioms rather than measuring effects. Even his model was not really new, for there were plenty of ideas, even in *Natural Wonders*, about the brain as a machine, a telephone exchange or an office system. What he had done was to combine such a naive mechanistic

---

* The arguments also implied two rather different interpretations of the machine 'configuration'. From the first point of view, it was natural to think of the configuration as the machine's *internal state* – something to be inferred from its different responses to different stimuli, rather as in behaviourist psychology. From the second point of view, however, it was natural to think of the configuration as a written *instruction*, and the table as a list of instructions, telling the machine what to do. The machine could be thought of as obeying one instruction, and then moving to another instruction. The universal machine could then be pictured as reading and decoding the instructions placed upon the tape. Alan Turing himself did not stick to his original abstract term 'configuration', but later described machines quite freely in terms of 'states' and 'instructions', according to the interpretation he had in mind. This free usage will accordingly be employed in what follows.

picture of the mind with the precise logic of pure mathematics. His machines – soon to be called *Turing machines* – offered a bridge, a connection between abstract symbols and the physical world. Indeed, his imagery was, for Cambridge, almost shockingly industrial.

Obviously there was a connection between the Turing machine and his earlier concern with the problem of Laplacian determinism. The relationship was indirect. For one thing, it might be argued that the 'spirit' he had thought about was not the 'mind' that performed intellectual tasks. For another, the description of the Turing machines had nothing to do with physics. Nevertheless, he had gone out of his way to set down a thesis of 'finitely many mental states', a thesis implying a material basis to the mind, rather than stick to the safer 'instruction note' argument. And it would appear that by 1936 he had indeed ceased to believe in the ideas that he had described to Mrs Morcom as 'helpful' as late as 1933 – ideas of spiritual survival and spiritual communication. He would soon emerge as a forceful exponent of the materialist view and identify himself as an atheist. Christopher Morcom had died a second death, and *Computable Numbers* marked his passing.

Beneath the change there lay a deeper consistency and constancy. He had worried about how to reconcile ideas of will and spirit with the scientific description of matter, precisely because he felt so keenly the power of the materialist view, and yet also the miracle of individual mind. The puzzle remained the same, but now he was approaching it from the other side. Instead of trying to defeat determinism, he would try to account for the appearances of freedom. There had to be a reason for it. Christopher had diverted him from the outlook of *Natural Wonders*, but now he had returned.

There was another point of constancy, in that he was still looking for some definite, down-to-earth resolution of the paradox of determinism and free will, not a wordy philosophical one. Earlier, in this search, he had favoured Eddington's idea about the atoms in the brain. He would remain very interested in quantum mechanics and its interpretation, a problem that von Neumann had by no means resolved, but the Jabberwocky would not be *his* problem. For now he had found his own *métier*, by formulating a new way of thinking about the world. In principle, quantum physics might include everything, but in practice to say anything about the world would require many different levels of description. The Darwinian 'determinism' of natural selection depended upon the 'random' mutation of individual genes; the determinism of chemistry was expressed in a framework where the motion of individual molecules was 'random'. The Central Limit Theorem was an example of how order could arise out of the most general kind of disorder. A cipher system would be an example of how disorder could arise by means of a determinate system. Science, as Eddington took care to observe, recognised many different determinisms, many different freedoms. The point was that in the Turing machine, Alan had created his own determinism of the automatic machine, operating within the *logical* framework he held to be appropriate to the discussion of the mind.

He had worked entirely on his own, not once discussing the construction of his 'machines' with Newman. He had a few words with Richard Braithwaite at the High Table one day on the subject of Gödel's theorem. Another time he put a question about the Cantor method to Alister Watson, a young King's Fellow (a communist, as it happened) who had turned from mathematics to philosophy. He described his ideas to David Champernowne, who got the gist of the universal machine, and said rather mockingly that it would require the Albert Hall to house its construction. This was fair comment on Alan's design in *Computable Numbers*, for if he had any thoughts of making it a practical proposition they did not show in the paper.[38] Just south of the Albert Hall, in the Science Museum, were lurking the remains of Babbage's 'Analytical Engine', a projected universal machine of a hundred years before. Quite probably Alan had seen them, and yet if so, they had no detectable influence upon his ideas or language. His 'machine' had no obvious model in anything that existed in 1936, except in general terms of the new electrical industries, with their teleprinters, television 'scanning', and automatic telephone exchange connections. It was his own invention.

A long paper, full of ideas, with a great deal of technical work and evidence of more left unpublished, *Computable Numbers* must have dominated Alan's life from spring 1935 through the following year. In the middle of April 1936, returning from Easter at Guildford, he called on Newman and gave him the draft typescript.

There were many questions to be asked about the discoveries that Gödel and he had made, and what they meant for the description of mind. There was a profound ambiguity to this final settlement of Hilbert's programme, though it certainly ended the hope of a too naive rationalism, that of solving every problem by a given calculus. To some, including Gödel himself, the failure to prove consistency and completeness would indicate a new demonstration of the superiority of mind to mechanism. But on the other hand, the Turing machine opened the door to a new branch of deterministic science. It was a model in which the most complex procedures could be built out of the elementary bricks of states and positions, reading and writing. It suggested a wonderful mathematical game, that of expressing any 'definite method' whatever in a standard form.

Alan had proved that there was no 'miraculous machine' that could solve all mathematical problems, but in the process he had discovered something almost equally miraculous, the idea of a universal machine that could take over the work of *any* machine. And he had argued that anything performed by a human computer could be done by a machine. So there could be a single machine which, by reading the descriptions of other machines placed upon its 'tape', could perform the equivalent of human mental activity. A single machine, to replace the human computer! An electric brain!

The death of George V, meanwhile, marked a transition from protest at the old order to fear of what the new might hold in store. Germany had already defeated the new Enlightenment; had already injected iron into the idealist soul. March 1936 saw the re-occupation of the Rhineland, it meant that the future lay with militarism. Who then could have seen the connection with the fate of an obscure Cambridge mathematician? Yet connection there was. For one day Hitler was to lose the Rhineland; and it would be then, and only then, that the universal machine could emerge into the world of practical action. The idea had come out of Alan Turing's private loss. But between the idea and its embodiment had to come the sacrifice of millions. Nor would the sacrifices end with Hitler; there was no solution to the world's *Entscheidungsproblem*.

3 New Men

I hear it was charged against me that I sought to destroy institutions,
But really I am neither for nor against institutions,
(What indeed have I in common with them? or what with the
　　destruction of them?)
Only I will establish in the Mannahatta and in every city of these
　　States inland and seaboard,
And in the fields and woods, and above every keel little or large that
　　dents the water,
Without edifices or rules or trustees or any argument,
The institution of the dear love of comrades.

Almost on the same day that Alan announced his discovery to Newman, someone else completed a demonstration that the Hilbert *Entscheidungsproblem* was unsolvable. It was at Princeton, where the American logician Alonzo Church had finished his argument for publication[1] on 15 April 1936. Church's essential idea, showing the existence of an 'unsolvable problem', had been announced a year earlier, but only at this point did he put it exactly in the form of answering Hilbert's question.

A new idea had found its way into two human minds simultaneously and independently. At first, this was not known at Cambridge, and Alan wrote to his mother on 4 May:

> I saw Mr Newman four or five days after I came up. He is very busy with other things just at present and says he will not be able to give his whole attention to my theory for some week or so yet. However he examined my note for C.R.* and approved it after some alterations. I also got it vetted by a French expert, and sent it off. I have had no acknowledgement of it, which is rather annoying. I don't think the full text will be ready for a fortnight or

---

* An abstract in French for the scientific journal *Comptes Rendus*. Mrs Turing helped with the French and the typing.

more yet. It will probably be about fifty pages. It is rather difficult to decide
what to put into the paper now and what to leave over till a later occasion.

When Newman did read it in mid-May, he could hardly believe
that so simple and direct an idea as the Turing machine would
answer the Hilbert problem over which many had been labouring
for the five years since Gödel had disposed of the other Hilbert
questions. His first impression was that it must be wrong, for some
more sophisticated machine would be able to solve the 'unsolvable
problem', and that one would then continue on and on. But finally
he satisfied himself that no finitely defined machine could possibly
do more than was allowed by the Turing construction.

Then Church's paper arrived from across the Atlantic. It pre-
empted the result, and threw into jeopardy the publication of Alan's
work, scientific papers not being allowed to repeat or copy one
another. But what Church had done was something rather different,
and in a certain sense weaker. He had developed a formalism called
the 'lambda-calculus'* and, with the logician Stephen Kleene, had
discovered that this formalism could be used to translate all the
formulae of arithmetic into a standard form. In this form, proving
theorems was a matter of converting one string of symbols of the
lambda-calculus into another string, according to certain rather
simple rules. Church had then been able to show that the problem
of deciding whether one string could be converted into another
string was unsolvable, in the sense that there existed no formula
of the lambda-calculus which could do it. Having found one such
unsolvable problem, it had become possible to show that the exact
question that Hilbert had posed must also be unsolvable. But it was
not obvious that 'a formula of the lambda-calculus' corresponded
to the notion of a 'definite method'. Church gave verbal arguments
for the assertion that any 'effective' method of calculation could be
represented by a formula of the lambda-calculus. But the Turing
construction was more direct, and provided an argument from first
principles, closing the gap in Church's demonstration.

So Alan was able to submit his paper on 28 May 1936 to the
London Mathematical Society for publication in its *Proceedings*, and

---

* The lambda-calculus represented an elegant and powerful symbolism for mathematical
processes of abstraction and generalisation.

Newman wrote to Church:

31 May 1936

Dear Professor Church,

An offprint which you kindly sent me recently of your paper in which you define 'calculable numbers', and shew that the Entscheidungsproblem for Hilbert logic is insoluble, had a rather painful interest for a young man, A. M. Turing, here, who was just about to send in for publication a paper in which he had used a definition of 'Computable numbers' for the same purpose. His treatment – which consists in describing a machine which will grind out any computable sequence – is rather different from yours, but seems to be of great merit, and I think it of great importance that he should come and work with you next year if that is at all possible. He is sending you the typescript of his paper for your criticisms.

If you find that it is right, and of merit, I should be greatly obliged if you could help Turing to get to Princeton next year, by writing to the Vice-Chancellor, Clare College, Cambridge, in support of Turing's application for the Procter Fellowship. If he fails to win this he can still just manage to come, I think, since he is a Fellow of King's College, but it would be a very tight fit. Is there any possibility of a small supplementary grant at the Princeton end? . . . I should mention that Turing's work is entirely independent: he has been working without any supervision or criticism from anyone. This makes it all the more important that he should come into contact as soon as possible with the leading workers on this line, so that he should not develop into a confirmed solitary.

There was no one in England who could referee the paper for publication in the London Mathematical Society *Proceedings*, and in fact Church himself was the only person who could reasonably do so. Newman wrote to the Secretary of the London Mathematical Society, F. P. White, explaining the position:

31 May 1936

Dear White,

I think you know the history of Turing's paper on Computable numbers. Just as it was reaching its final state an offprint arrived, from Alonzo Church of Princeton, of a paper anticipating Turing's results to a large extent.

I hope it will nevertheless be possible to publish the paper. The methods are to a large extent different, and the result is so important that different treatments of it should be of interest. The main result of both Turing and Church is that the Entscheidungsproblem on which Hilbert's disciples have been working for a good many years – i.e. the problem of finding a mechanical way of deciding whether a given row of symbols is the enunciation of a theorem provable from the Hilbert axioms – is insoluble in its general form. . . .

Alan reported to his mother on 29 May:

> I have just got my main paper ready and sent in. I imagine it will appear in October or November. The situation with regard to the note for Comptes Rendus was not so good. It appears that the man I wrote to, and whom I asked to communicate the paper for me had gone to China, and moreover the letter seems to have been lost in the post, for a second letter reached his daughter.
>
> Meanwhile a paper has appeared in America, written by Alonzo Church, doing the same things in a different way. Mr Newman and I have decided however that the method is sufficiently different to warrant the publication of my paper too. Alonzo Church lives at Princeton so I have decided quite definitely about going there.

He had applied for a Procter Fellowship. Princeton offered three of these, one in the gift of Cambridge, one of Oxford, one of the Collège de France. He was not to be successful, for the Cambridge one went that year to R. A. Lyttleton, the mathematician and astronomer. But he must have found that his King's fellowship would provide just enough funds.

Meanwhile, it was now necessary for the publication of the paper that he should include a demonstration that its definition of 'computable' – that is, as anything that could be computed by a Turing machine – was exactly equivalent to what Church had called 'effectively calculable', meaning that it could be described by a formula in the lambda-calculus. So he studied Church's work from the papers which he and S. C. Kleene had produced in 1933 and 1935, and sketched out the required demonstration in an appendix to the paper which was finished on 28 August. The correspondence of ideas was quite straightforward, since Church had used a definition (that of a formula being 'in normal form') which corresponded to the Turing definition of 'satisfactory' machines, and had then used a Cantor diagonal argument to produce an unsolvable problem.

If he had been a more conventional worker, he would not have attacked the Hilbert problem without having read up all of the available literature, including Church's work. He then might not have been pre-empted – but then, he might never have created the new idea of the logical machine, with its simulation of 'states of mind', which not only closed the Hilbert problem but opened up quite new questions. It was the advantage and the disadvantage of working as what Newman called 'a confirmed solitary'. Both with the Central Limit Theorem and with the *Entscheidungsproblem*, he had been the Captain Scott of mathematics, coming in a splendid second place. And while he was not the person to think of mathematics and science as a sort of competitive game, it was obviously a disappointment. It meant months of delay, and obscured the originality of his own attack. It confused his moment of coming out into the world.

As for the Central Limit Theorem, his fellowship dissertation was entered for the Cambridge mathematical essay competition, the Smith's Prize, that summer. This caused a flurry down at Guildford, where Mrs Turing and John spent a frantic half-hour on hands and knees doing up the parcel, which Alan had left until the last minute before sending off. John had married in August 1934 and Alan had by now become an uncle. Neither his brother, nor his parents, had the faintest inkling of the philosophical problems which underlay his work, or which underlay his life. News of Alan's successes came as glowing reports from a higher and higher Sixth Form. Mrs Turing, with her interest in the spiritual world, would have been the most sensitive to Alan's concern with free will, but even she never saw this fundamental connection. For Alan never expatiated on his inner problems, and only occasionally did rather cryptic hints of them emerge.

The university, like King's, took a charitable view of Alan's rediscovery of the theorem, and it won him the prize and hence £31. By now he had taken up sailing as a holiday pastime, and thought of putting the money towards buying a boat. But he decided against it, perhaps needing it for his year in America.

Victor Beuttell came to stay with Alan at Cambridge in the early summer. Alan was returning the hospitality that the Beuttells had offered him but another reason for Victor's visit was that he had now joined the family firm and had been set to work on developing the K-ray system. The geometry that he had discussed with Alan at school had helped him, but he was hoping to have Alan's advice on the new problem which was to make a double-sided system so that both sides of a poster could be illuminated evenly by a single light source. (It was required by a brewery chain.) Alan, however, said he was too preoccupied with his own work, and instead they went off to watch the May Bumps boat races.

Once they were talking about art and sculpture, and it was in this connection that Alan suddenly amazed Victor by saying that he found the male form beautiful, and the female unattractive. Victor now found himself a double crusader, and tried to convince Alan that Jesus had indicated the right course by befriending Mary Magdalene. Alan had no answer to this, but then this was not a problem of reason. All he could do was to express the sensation of being in a Looking-Glass world, in which from his point of view the conventional ideas were the wrong way about. This was probably the first time that he opened the subject outside the King's ambience.

It was difficult for Victor, who was a not particularly mature twenty-one, to know how to react. An element of trust now came into his staying in Alan's room, though Alan remained 'a perfect gentleman'. But Victor did not reject Alan's friendship. Instead, they continued to agree to disagree on this subject as they did on religion. They talked of what hereditary or environmental influences might determine it one way or the other. But whatever these were, it was clear that here was part of Alan that *was so*; that part of his reality was shaped that way. For him, without a God, there was nothing to appeal to but some inner consistency. As in mathematics, that consistency could not be proved by a rule-book, and there was no *deus ex machina* to hand down decisions as to right and wrong. The axioms of his life were becoming clear by now, although how to live them out was quite another question. He had wanted the commonest in nature; he liked ordinary things. But he found himself to be an ordinary English homosexual atheist mathematician. It would not be easy.

Alan also paid a visit to the Clock House before going out west, the first for three years. Mrs Morcom was now semi-invalid, but still mentally as vigorous as ever. She noted in her diary:

> September 9 (Wednesday) . . . Alan Turing came . . . He has come for a farewell visit before going out to America for 9 months (Princetown) to study under 2 great authorities on his subject: Godel (Warsaw) Alonso Church and Kleene. We had talk before dinner and again later to bring us up to date with our news. . . . He and Edwin played billiards.

> September 10: . . . Alan and Veronica to farms and Dingleside . . . . V and Alan tea up here with me. Had long talk with Alan about his work and whether in his subject (some abstruse branch of logic) one would come to 'dead end' etc.

> September 11: Alan went down to church alone to see Chris' window and the little garden which he hadn't seen before since it was finished – only the day he came to the dedication of the window . . . Alan taught me game called 'Go' – rather like Peggity.

> September 12: . . . Rupert and Alan had tea in my room and then I took them all by surprise by coming down to dinner. There were 10 of us – a jolly party. Gramophone concert . . . Men billiards.

> September 13 . . . Alan did problems with R[eginald] . . . Alan Rup[ert] and 2 girls bathed at Cadbury's pool . . . Rup[ert] and Alan tea with me . . . Alan tried to explain what he is working at . . . they went off to catch 7.45 New Street.

Alan lost Rupert when it came to the satisfactory and the unsatis-factory description numbers. It would have been hard for Mrs Morcom to feel that this 'abstruse branch of logic' had anything to do with the scientific imagination of her lost son, so that Alan had done something that Christopher had been called away from.

Mrs Turing saw Alan off at Southampton on 23 September, when he embarked on the Cunard liner, the *Berengaria*. He had picked up a sextant in the Farringdon Road market to amuse himself on the voyage. He also went equipped with all the standard upper-middle-class British prejudices about America and Americans, and the five days on the Atlantic did little to disabuse him. From '41°20′ N, 62°W', he complained:[2]

> It strikes me that Americans can be the most insufferable and insensitive
> creatures you could wish. One of them has just been talking to me and
> telling me of all the worst aspects of America with evident pride. However
> they may not all be like that.

The towers of the Manhattan skyline swam into view next morning,
on 29 September, and Alan entered the New World:

> We were practically in New York at 11.00 a.m. on Tuesday but what with
> going through quarantine and passing the immigration officers we were not
> off the boat until 5.30 p.m. Passing the immigration officers involved waiting
> in a queue for over two hours with screaming children round me. Then, after
> getting through the customs I had to go through the ceremony of initiation
> to the U.S.A., consisting of being swindled by a taxi-driver. I considered his
> charge perfectly preposterous, but as I had already been charged more than
> double English prices for sending my luggage, I thought it was possibly right.

Alan inherited his father's belief that to take a taxi was the height
of extravagance. But America, with its infinite variety, was not
all 'like that', and Princeton, where he arrived late that evening
on the train had little in common with the 'mass of canaille' of
the cheapest Tourist Class. For if Cambridge embodied *class*,
then Princeton spoke *wealth*. Perhaps of all the elite American
universities, Princeton was the most self-contained, insulated from
the squalor of the depression. One could look out and never know
that America had a problem. In fact, it hardly looked like America
at all, for with its mock Gothic architecture, its restriction to male
students, its rowing on the artificial Carnegie Lake, Princeton tried
to outdo the detachment of Oxford and Cambridge. It was the
Emerald City in the land of Oz. And as if the isolation from ordinary
America were not enough, the Graduate College was separated off
from the undergraduate life, to stand upon its gentle prominence,
overlooking a clean sweep of fields and woods. The tower of the
Graduate College imitated that of Magdalen College Oxford, and it
was popularly called the Ivory Tower, because of that benefactor of
Princeton, the Procter who manufactured Ivory Soap.

Mathematics at Princeton had been greatly augmented by the
endowment of five million dollars for the foundation, in 1932, of
the Institute for Advanced Study. Until 1940 the Institute had no

separate building of its own. Those whom it funded, almost all mathematicians and theoretical physicists, shared the space of Fine Hall, home of the regular Princeton mathematical faculty. Although for technical purposes the distinction had to be drawn, in practice no one knew nor cared who was Princeton University and who was IAS. The doubled department had attracted some of the greatest names in world mathematics, and especially the exiles from Germany. It was in some ways an all-American foundation, in others like some immigrant ship still traversing the Atlantic. The richly funded Princeton fellowships also attracted research students of a world class, although more from England than from any other country. There were none from King's, but Alan's friend Maurice Pryce from Trinity was in residence for a second year. Here, amidst the huddled élite of the exiled European intelligentsia, lay the opportunity for Alan Turing to follow up his major result. His first report home, on 6 October, betrayed no lack of self-confidence.

> The mathematics department here comes fully up to expectations. There is a great number of the most distinguished mathematicians here. J. v. Neumann, Weyl, Courant, Hardy, Einstein, Lefschetz, as well as hosts of smaller fry. Unfortunately there are not nearly so many logic people here as last year. Church is here of course, but Gödel, Kleene, Rosser and Bernays who were here last year have left. I don't think I mind very much missing any of these except Gödel. Kleene and Rosser are, I imagine, just disciples of Church and have not much to offer that I could not get from Church. Bernays is I think getting rather 'vieux jeu' that is the impression I get from his writing, but if I were to meet him I might get a different impression.

Of these, Hardy was only visiting from Cambridge for a term.

> At first he was very standoffish or possibly shy. I met him in Maurice Pryce's rooms the day I arrived, and he didn't say a word to me. But he is getting much more friendly now.

Hardy was something of a Turing of an earlier generation; he was another ordinary English homosexual atheist, who just happened to be one of the best mathematicians in the world. He was more fortunate than Alan in that his chief interest, the theory of numbers, fell cleanly within the classical framework of pure mathematics. He did not have Alan's problem, of having to create his own subject.

And his work was much more regular, more professional, than ever Alan's was. But both were refugees from the system, for whom Keynesian Cambridge was the only possible home, although neither belonged to the more glamorous circles. Both were passive resisters, though Hardy was slightly less passive; he had been president of the Association of Scientific Workers out of principle, and had Lenin's picture in his rooms. As the older man, his views were that much more firmly cast. Bertrand Russell once wittily distinguished catholic from protestant sceptics, according to the tradition they had rejected, and on this model Alan was, at this stage, more of a Church of England atheist. Hardy, however, played upon the English refusal to take ideas seriously, by becoming an atheist evangelical. At the same time, he found the pleasures of ritual in his devotion to the game of cricket. There was no one who knew more about it, although when in America he transferred his allegiance to baseball. He would organise cricket matches at Trinity, with Disbelief playing against Belief and the Almighty challenged to rain out the unbelievers. Hardy delighted in making a game out of anything, especially atheism.

Alan would have attended his advanced lectures and classes at Cambridge, and therefore felt aggrieved at being ignored. Although 'friendly', the relationship was not one that overcame a generation and multiple layers of reserve. And if this was true of his acquaintanceship with Hardy, who saw the world through such very similar eyes, it was all the more so of Alan's other professional contacts with elders. Although he was emerging as a figure of the serious academic world, he found it hard to shed the outlook and manners of an undergraduate.

The list of names in Alan's letter in itself meant little except that he might attend their lectures and seminars. Einstein would be seen occasionally in the corridors, but was almost incommunicado. S. Lefschetz was a pioneer in topology, which was at the centre of Princeton mathematics, and indeed a principal growth point of modern mathematics, but Alan's personal contact with him was probably characterised by an occasion when Lefschetz questioned whether he would understand L. P. Eisenhart's lecture course on Riemannian geometry, a question Alan considered insulting. Courant and Weyl, with von Neumann, covered the whole mainstream of pure and applied mathematics, bringing something of Hilbert's

Göttingen to life again on the western shore. But of them it was probably only von Neumann who had contact with Alan, through shared interests in group theory.

As for the logicians, Gödel had returned to Czechoslovakia. Kleene and Rosser had made more substantial contributions to logic than Alan's letter suggested, but had taken up positions elsewhere, and he would never meet either of them. The Swiss logician P. Bernays, a close associate of Hilbert, and another exile from Göttingen, had returned to Zürich. Thus the impression that Alan had given to Mrs Morcom, of working with two or three major authorities, was incorrect. It was a matter of working with Church alone, except inasmuch as there were graduates studying logic on a lower level. And Church was a retiring man himself, not given to a great deal of discussion. In short, Princeton did not cure Alan of being a 'confirmed solitary'. He wrote:

> I have seen Church two or three times and I get on with him very well. He seems quite pleased with my paper and thinks it will help him to carry out a programme of work he has in mind. I don't know how much I shall have to do with this programme of his, as I am developing [sic] the thing in a slightly different direction, and shall probably start writing a paper on it in a month or two. After that I may write a book.

Whatever these plans were, they did not come to fruition; there was no paper which fell into this description, nor a book.

He conscientiously attended Church's lectures, which were rather on the ponderous and laborious side. In particular, he took notes of Church's theory of types, reflecting his continued interest in that aspect of mathematical logic. There were something like ten students present, including a younger American, Venable Martin, whom Alan befriended and helped with understanding the course. Alan remarked:

> The graduate students include a very large number who are working in mathematics and none of them mind talking shop. It is very different from Cambridge in that way.

At Cambridge it was thought in very bad taste at High Table, or anywhere, for a person to speak only of his speciality. But this was not a feature of the English university that Princeton had imported along with the architecture. The English students, all from Oxford or Cambridge, would be amused at such American greetings as 'Hi, pleased to meet you, what courses are you taking?' English work was hidden under a decent show of well-bred amateurishness. This pretended negligence astonished the earnest devotees of the work ethic. But for Alan, who was excluded from the smarter circles of Cambridge society for his lack of sophistication, the more straightforward approach was an attraction. In that way America suited him – but not in other respects. To his mother, he wrote on 14 October:

> Church had me out to dinner the other night. Considering that the guests were all university people I found the conversation rather disappointing. They seem, from what I can remember of it, to have discussed nothing but the different states they came from. Description of travel and places bores me intensely.

He enjoyed the play of *ideas*, and in the same letter he let slip a hint of ideas in which Bernard Shaw himself might have found a plot:

> You have often asked me about possible applications of various branches of mathematics. I have just discovered a possible application of the kind of thing I am working on at present. It answers the question 'What is the most general kind of code or cipher possible', and at the same time (rather naturally) enables me to construct a lot of particular and interesting codes. One of them is pretty well impossible to decode without the key, and very quick to encode. I expect I could sell them to H.M. Government for quite a substantial sum, but am rather doubtful about the morality of such things. What do you think?

Ciphering would be a very good example of a 'definite method' applied to symbols, something that could be done by a Turing machine. It would be essential to the nature of a cipher that the encipherer behave like a machine, in accordance with whatever rules had been fixed in advance with the receiver of the message.

As for a 'most general code or cipher possible', in a sense any Turing machine could be regarded as encoding what it read on its tape, into what it wrote on the tape. However, to be useful there would have to be an *inverse* machine, which could reconstruct the original tape. His result, whatever it was, must have started on these lines. But as for the 'particular and interesting codes' he offered no further clue.

Nor did he touch again on the conflict indicated by the word 'morality': what was he to do? Mrs Turing, of course, was a Stoney; she assumed that science existed for the sake of useful applications, and she was not the person to doubt the moral authority of His Majesty's Government. But the intellectual tradition to which Alan belonged was quite different. It was not only for the detachment of Cambridge, but for a very significant section of modern mathematical opinion that G. H. Hardy spoke when he wrote:[3]

> The 'real' mathematics of the 'real' mathematicians, the mathematics of Fermat and Euler and Gauss and Abel and Riemann, is almost wholly 'useless' (and this is true of 'applied' as of 'pure' mathematics). It is not possible to justify the life of any genuine professional mathematician on the ground of the 'utility' of his work. . . . The great modern achievements of applied mathematics have been in relativity and quantum mechanics, and these subjects are, at present at any rate, almost as 'useless' as the theory of numbers. It is the dull and elementary parts of applied mathematics, as it is the dull and elementary parts of pure mathematics, that work for good or ill.

In making explicit his response to the growing separation of mathematics from applied science, Hardy attacked the shallowness of the current 'left-wing' Lancelot Hogben interpretation of mathematics in terms of social and economic utility, an interpretation based on the 'dull and elementary' aspects of the subject. Hardy spoke more for himself, however, in holding that 'useful' mathematics had in any case worked more for ill than for good, being preponderantly military in application. He held the total uselessness of his own work in the theory of numbers to be a positive virtue, rather than a matter for apology:

> No-one has yet discovered any warlike purpose to be served by the theory of numbers or relativity, and it seems very unlikely that anyone will do so for many years.

Hardy's own near-pacifist convictions stemmed from before the Great War, but no one touched by the Anti-War movements of the

1930s could fail to be unaware of a view that military applications were to be shunned. If Alan had now discovered something like a 'warlike purpose' in the play of symbols, he was faced, at least in embryo, with a mathematician's dilemma. Behind the off-hand, teasing words to his mother, there lay a serious question.

Meanwhile the English students were brightening the Graduate College life with amusements of their own:

> One of the Commonwealth Fellows, Francis Price (not to be confused with Maurice Pryce . . .) arranged a hockey match the other day between the Graduate College and Vassar, a women's college (amer.)/university (engl.) some 130 miles away. He got up a team of which only half had ever played before. We had a couple of practice games and went to Vassar in cars on Sunday. It was raining slightly when we arrived, and what was our horror when we were told the ground was not fit for play. However, we persuaded them to let us play a pseudo-hockey match in their gymn. at wh[ich] we defeated them 11-3. Francis is trying to arrange a return match, which will certainly take place on a field.

The amateurism was deceptive, since Shaun Wylie, the topologist, and Francis Price, the physicist, both from New College, Oxford, were players of a national standard. Alan was hardly in the same class (even if he was not now 'watching the daisies grow'), but enjoyed the games. Soon they were playing three times a week amongst themselves, and sometimes against local girls' schools.

The effete English playing a women's game might well have amazed the native Princeton students, but within the establishment there was a somewhat embarrassing anglophilia, in that all the most stuffy and mannered aspects of the English system were admired. In the summer of 1936, the Princeton chapel had been packed for a memorial service for George V. There was a professor in the Graduate College who harped upon his admiration for the royalty in a way that to educated English ears seemed only vulgar. As for George V's successor, the revelations of Edward VIII's Mediterranean cruise and Mrs Simpson created a particular sensation at Princeton. Alan wrote to his mother on 22 November:

I am sending you some cuttings about Mrs Simpson as representative sample of what we get over here on this subject. I don't suppose you have even heard of her, but some days it has been 'front page stuff' here.

Indeed, the British newspapers maintained their silence until 1 December, when the bishop of Bradford remarked that the King stood in need of God's grace, and Baldwin showed his hand. On 3 December, Alan wrote:

I am horrified at the way people are trying to interfere with the King's marriage. It may be that the King should not marry Mrs Simpson, but it is his private concern. I should tolerate no interference by bishops myself, and I don't see that the King need either.

But the King's marriage was not a private matter, but one that reflected upon the British state. It was a prophetic episode for Alan, 'horrified' at government interference with an individual life. For his class, the horror was rather that the King himself had betrayed King and Country, a logical paradox more upsetting than any that Russell or Gödel had found.

On 11 December the Windsors went into their butterfly life of exile, and the reign of George VI began. Alan wrote that day:

I suppose this business of the King's abdication has come as rather a shock to you. I gather practically nothing was known of Mrs Simpson in England till about ten days ago. I am rather divided in my opinion of the whole matter. At first I was wholly in favour of the King retaining the throne and marrying Mrs Simpson, and if this were the only issue it would still be my opinion. However I have heard tales recently which seem to alter it rather. It appears that the King was extremely lax about state documents, leaving them about and letting Mrs Simpson and friends see them. There had been distressing leakages. Also one or two other things of same character, but this is the one I mind about most. However, I respect David Windsor for his attitude.

Alan's respect extended to the acquisition of a gramophone record of the abdication speech. He further wrote on 1 January:

I am sorry that Edward VIII has been bounced into abdicating. I believe the Government wanted to get rid of him and found Mrs Simpson a good opportunity. Whether they were wise to try to get rid of him is another matter. I respect Edward for his courage. As for the Archbishop of Canterbury I consider his behaviour disgraceful. He waited until Edward was safely out of the way and then unloaded a whole lot of quite uncalled-for abuse. He didn't dare do it whilst Edward was King. Further he had no objections to the King having Mrs Simpson as a mistress, but marry her, that wouldn't do at all. I don't see how you can say that Edward was guilty of wasting his ministers' time and wits at a critical moment. It was Baldwin who opened the subject.

The archbishop's broadcast, of 13 December, had denounced the King for abandoning his duties for a mere 'craving for private happiness'; the pursuit of happiness had never been accorded a high priority by the British rulers. Alan's views on marriage and morals were those of a modernist; in a discussion at King's with his theological contemporary Christopher Stead he had said that people should let their natural feelings take their course – and as for bishops, a class of person particularly dear to Mrs Turing, they epitomised for him the *ancien régime*. He talked to Venable Martin, his American friend from Church's logic class, about the 'very shabby way' in which the King had been treated.

As for work, on 22 November he wrote to Philip Hall:

I have not made any very startling discoveries over here, but I shall probably be publishing two or three small papers: just bits and pieces. One of them will be a proof of Hilbert's inequality if it really turns out to be new, and another on groups which I did about a year ago and Baer thinks is worth publishing. I shall write these things up and then have another go at the Math[ematical] logic.

I find that 'Go' is only played very little here now, but I have had two or three games.

Princeton is suiting me very well. Beyond the way they speak there is only one – no two! – feature[s] of American life which I find really tiresome, the impossibility of getting a bath in the ordinary sense, and their ideas on room temperature.

By 'the way they speak', Alan meant such complaints as:[4]

These Americans have various peculiarities in conversation which catch the ear somehow. Whenever you thank them for anything, they say 'You're welcome'. I rather liked it at first, thinking I was welcome, but now I find it comes back like a ball thrown against a wall, and become positively apprehensive. Another habit they have is to make the sound described by authors as 'Aha'. They use it when they have no suitable reply to a remark, but think that silence could be rude.

The proofs of *Computable Numbers* had been sent to him at Princeton just after he had arrived, so that publication of the paper was imminent. Meanwhile, Alonzo Church had suggested that Alan might be able to give one of the regular seminars, to launch his discovery into the mainstream of Princeton mathematics. On 3 November he had written home:

Church has just suggested to me that I should give a lecture to the Mathematical Club here on my Computable Numbers. I hope I shall be able to get an opportunity to do this, as it will bring the thing to people's attention a bit. I don't expect the lecture will come off for some time yet.

In fact he only had to wait a month, but then there was a disappointment:

There was rather bad attendance at the Maths Club for my lecture on Dec. 2. One should have a reputation if one hopes to be listened to. The week following my lecture G. D. Birkhoff came down. He has a very good reputation and the room was packed. But his lecture wasn't up to the standard at all. In fact everyone was just laughing about it afterwards.

It was also disappointing that when at the end of 1936 *Computable Numbers* at last appeared in print, there was so little reaction. Church reviewed it for the *Journal of Symbolic Logic*, and thereby put the words 'Turing machine' into published form. But only two people asked for offprints: Richard Braithwaite back at King's and Heinrich Scholz,[5] the almost lone representative of logic left in Germany, who wrote back saying that he had given a seminar on it at Münster, and begged almost plaintively for two copies of any future papers, explaining how difficult it was for him to keep abreast of developments otherwise. The world was rather less of a single country for mathematics now. Alan wrote home on 22 February:

I have had two letters asking for reprints. . . . They seemed very much interested in the paper. I think possibly it is making a certain amount of impression. I was disappointed by its reception here. I expected Weyl who had done some work connected quite closely with it some years ago, at least to have made a few remarks about it.

He might also have expected John von Neumann to have made a few remarks about it. Here was a truly powerful Wizard playing against Alan's version of the innocent Dorothy. Like Weyl, he had been very interested in the Hilbert programme and had once hoped to fulfil it, although his active interest in mathematical logic had ended with Gödel's theorem. He once claimed[6] that after 1931 he never read another paper in logic, but this was at most a half-truth, for he was a prodigious reader, working long before anyone else got up in the morning, and covering the whole gamut of mathematical literature. Yet there was not a word about him at this point in Alan's letters to his mother or Philip Hall.

As for the general readership of the LMS *Proceedings*, there were a number of reasons why Alan's paper was unlikely to make an impression upon them. Mathematical logic seemed to be a marginal area of research, which many mathematicians would consider either as tidying up what was obvious anyway, or as creating difficulties where none really existed. The paper started attractively, but soon plunged (in typical Turing manner) into a thicket of obscure German Gothic type in order to develop his instruction tables for the universal machine. The last people to give it a glance would be the applied mathematicians who had to resort to practical computation in some field such as astrophysics or fluid dynamics, where the equations did not allow an explicit solution. There was little encouragement offered to them to do so. *Computable Numbers* made no concession to practical design, not even for the limited range of logical problems to which the machines were applied in the paper itself. For instance, he had made a convention that the machines should print out the 'computable numbers' on alternate squares of the 'tape', and use the intervening squares as working space. But it would have been much easier if he had made a more generous allowance of working space. So there was little about his work to attract anyone from outside the narrow circle of mathematical logic – with the possible exception of pure mathematicians who would be interested in the distinction between the computable numbers and the real numbers. It had nothing obvious to do with what Lancelot Hogben called 'the world's work'.

There was one person, one of those few who were professionally interested in mathematical logic, who read the paper with a very considerable personal interest. This was Emil Post, a Polish-American mathematician teaching at the City College of New York, who since the early 1920s had anticipated some of Gödel and Turing's ideas in unpublished form.[7] In October 1936 he had submitted to Church's *Journal of Symbolic Logic* a paper[8] which proposed a way of making precise what was meant by 'solving a general problem'. It referred specifically to Church's paper, the one which solved the Hilbert decision problem but required an assertion that any definite method could be expressed as a formula in his lambda-calculus. Post proposed that a definite method would be one which could be written in the form of instructions to a mindless 'worker' operating on an infinite line of 'boxes', who would be capable only of reading the instructions and

(a) Marking the box he is in (assumed empty),
(b) Erasing the mark in the box he is in (assumed marked),
(c) Moving to the box on his right,
(d) Moving to the box on his left,
(e) Determining whether the box he is in, is or is not marked.

It was a very remarkable fact that Post's 'worker' was to perform exactly the same range of tasks as those of the Turing 'machine'. And the language coincided with the 'instruction note' interpretation that Alan had given. The imagery was perhaps that much more obviously based upon the assembly line. Post's paper was much less ambitious than *Computable Numbers*; he did not develop a 'universal worker' nor himself deal with the Hilbert decision problem. Nor was there any argument about 'states of mind'. But he guessed correctly that his formulation would close the conceptual gap that Church had left. In this it was only by a few months that *he* had been pre-empted by the Turing machine, and Church had to certify that the work had been completely independent. So even if Alan Turing had never been, his idea would soon have come to light in one form or another. It had to. It was the necessary bridge between the world of logic and the world in which people did things.

*

In another sense, it was that very bridge between the world of logic, and the world of human action, that Alan Turing found so difficult. It was one thing to have ideas, but quite another to impress them upon the world. The processes involved were entirely different. Whether Alan liked it or not, his brain was embodied in a specific academic system, which like any human organisation, responded best to those who pulled the strings and made connections. But as his contemporaries observed him, he was in this respect the least 'political' person. He rather expected truth to prevail by magic, and found the business of advancing himself, by putting his goods in the shop window, too sordid and trivial to bother with. One of his favourite words was 'phoney', which he applied to anyone who had gained some position or rank on what Alan considered an inadequate basis of intellectual authority. It was a word that he applied to the referee of one of the group-theory papers he submitted in the spring, who had made a mistaken criticism of it.

He knew that he ought to make more effort on his own behalf, and he could not help noticing that his friend Maurice Pryce was someone who both had the intellectual ability, and made sure that it was used to its best advantage.

Both of them had come a long way since that week in Trinity in December 1929. Alan had been the first to be elected a Fellow (thanks to King's looking generously at his dissertation subject). But Maurice had just now been elected a Fellow of Trinity, which was that bit more impressive. And everyone could see that it was he who was the rising star. Their interests had developed in a complementary way, for Maurice had taken up quantum electrodynamics, while keeping up an interest in pure mathematics. But both alike were interested in fundamental problems. They had quite often met at Cambridge lectures, sometimes exchanging notes over tea; it had transpired that the Pryces also lived at Guildford, and Maurice had once been invited to tea at 8 Ennismore Avenue, where Mrs Turing had welcomed him as one of the deserving poor from the grammar school. Alan had visited and admired the laboratory that Maurice had fixed up in the Pryce garage.

At Princeton, Maurice had been supervised in his first year by Pauli, the Austrian quantum physicist, but this year was loosely under the wing of von Neumann. And Maurice knew everyone; everyone knew him. He would be seen at the von Neumanns' luxuriant parties, spectacles 'like eighteenth-century operas', although there were less of these this year, because the von Neumann marriage was in difficulties. And if any of the English graduate students came to know John von Neumann, to find him sociable, exuberant, and a pretended playboy with an encyclopaedic knowledge, then it was Maurice Pryce – and certainly not Alan Turing. But at the other end of the scale, it was also Maurice who knew how to engage the reclusive Hardy in conversation. He could get on with everyone, and indeed it was he who made Alan himself feel welcome in the New World.

King's had sheltered Alan from the more pushy aspects of academic life, which in America were more noticeable. He fitted no better into the American Dream, of winning through the competition, than into the conservative British idea of life, of playing a programmed part in the system.

But King's also sheltered him from hard realities in another way. At Cambridge he could joke about it. When Victor visited in May 1936 there had been a small scandal, a certain old Shirburnian being caught with 'a lady' in his room and sent down. Alan commented wryly that it was not a sin of which *he* was guilty. Alan was not a moaner, and tried always to show a sense of humour, but there was nothing particularly funny about the problem that he faced in coming out into the world.

In *Back to Methuselah*, Bernard Shaw imagined super-intelligent beings of 31,920 A.D. growing out of the concerns of art, science and sex ('these childish games – this dancing and singing and mating') and turning away to think about mathematics. ('They are fascinating, just fascinating. I want to get away from our eternal dancing and music, and just sit down by myself and think about numbers.') This was all very well for Shaw, for whom mathematics could symbolise intellectual enquiry beyond his reach. But Alan had to think about mathematics at twenty-four, when he had by no means tired of the 'childish games'. He did not divide his mind into rigidly separate compartments, once saying that he derived a sexual pleasure from mathematics. Likewise with his new friend Venable Martin, he went to H. P. Robertson's lectures on relativity in the new year of 1937, and also went canoeing, perhaps in the stream that fed Carnegie Lake. At one point he[9] 'indirectly indicated' an 'interest in having a homosexual relation', but his friend made it clear that he was not interested. Alan never broached the matter again and it did not affect their relationship in other ways.

The New Jersey poet would have understood. But Alan did not see the America of Walt Whitman, only the land of sexual prohibition. The country of Daddy and Mama had adopted homosexuality as a deeply Un-American activity, especially since the twentieth-century clean-up had got under way. At Princeton there was no one talking about a 'pretty normal bisexual male'. Alan was lucky to be rebuffed by so tolerant a person as Venable Martin.

He faced the difficulty that confronted any homosexual person who had successfully resolved the internal psychological conflicts attendant on waking up in a Looking-Glass world. The individual mind was not the whole story; there was also a social reality which was not at all the mirror image of heterosexual institutions. The late 1930s offered him no help in coping with it. Except for those with eyes to see through the stylised heterosexuality of Fred Astaire and Busby Berkeley, the times favoured ever more rigid models of 'masculine' and 'feminine'. There was, all the time, quite another America of cruising blocks and steam baths and late-night bars, but to an Alan Turing this might as well have been on another planet. He was not ready to make the social adaptation that his sexuality, at least outside Cambridge, demanded.

He could reasonably have felt that there was *no* acceptable adaptation; that this particular mind-body problem had no solution. For the time being his shyness kept him from confronting the harshness of this social reality, and he continued to try to cope at an individual level, making gentle approaches to some of those he met through his work. It was not a great success.

Alan did spend some time in New York at Thanksgiving, but this was because duty called him to accept the invitation of a right-wing cleric who was a friend of Father Underhill,* Mrs Turing's favourite priest. ('He is a kind of American Anglo-Catholic. I liked him but found him a bit diehard. He didn't seem to have much

---

* He became bishop of Bath and Wells in 1937.

use for President Roosevelt.') Alan spent his time 'pottering about
Manhattan getting used to their traffic and subways (underground)'
and went to the Planetarium. More relevant to Alan's emotional
state, perhaps, was the holiday at Christmas when Maurice Pryce
took him on a skiing trip in New Hampshire for two weeks:

> He suggested going on the 16th and on the 18th we left. A man called
> Wannier attached himself to the party at the last moment. Probably just
> as well; I always quarrel if I go on holiday with one companion. It was
> charming of Maurice to ask me to join him. He has been very kind to me
> whilst I have been here. We spent the first few days at a cottage where we
> were the only guests. Afterwards we moved to a place where there were
> several Commonwealth Fellows and others of various nationalities. Why
> we moved I don't know, but I imagine Maurice wanted more company.

Perhaps Alan wanted Maurice more to himself, for his friend was
something of a grown-up Christopher Morcom. They drove back
through Boston, where the car broke down, and on their return,

> Maurice and Francis Price arranged a party with a Treasure Hunt last
> Sunday. There were 13 clues of various kinds, cryptograms, anagrams, and
> others completely obscure to me. It was all very ingenious, but I am not
> much use at them.

One clue was 'Role of wily Franciscan', which wittily attracted the
party into the bathroom that Francis Price and Shaun Wylie shared,
to locate the next clue in the toilet paper. Shaun Wylie himself
was amazingly good at anagrams. The treasure hunt bemused the
more earnest Americans with its 'undergraduate humour', and
'typical English whimsy'. There were charades and play-readings,
in which Alan joined. At lunchtimes they would play chess and Go
and another game called Psychology. Tennis began as the snow
melted, and the hockey was energetically continued. '*Virago Delenda
Est*', wrote Francis Price on the notice board as they set off for an
away match, and some bolder spirit crossed out the first 'a'. On the
playing-fields of Princeton, from which in May 1937 they watched
the flames of the *Hindenburg* illuminate the horizon, the new men
rehearsed an Anglo-American alliance.

Alan enjoyed all this, but his social life was a charade. Like any
homosexual man, he was living an imitation game, not in the sense

of conscious play-acting, but by being accepted as a person that he was not. The others thought they knew him well, as in conventional terms they did; but they did not perceive the difficulty that faced him as an individualist jarring with the reality of the world. He had to find himself as a homosexual in a society doing its best to crush homosexuality out of existence; and less acute, though equally persistent as a problem in his life, he had to fit into an academic system that did not suit his particular line of thought. In both cases, his autonomous self-hood had to be compromised and infringed. These were not problems that could be solved by reason alone, for they arose by virtue of his physical embodiment in the social world. Indeed, there were no solutions, only muddles and accidents.

At the beginning of February 1937 the offprints of *Computable Numbers* arrived and some he sent out to personal friends. One went to Eperson (who had now left Sherborne for the more suitable Church of England), and one to James Atkins, who had now taken up a career as a schoolmaster, and was teaching mathematics at Walsall Grammar School. James also had a letter[10] from Alan which described, in a rather detached way, that he had been feeling depressed and mentioned that he had even thought of a scheme for ending his life. It involved an apple and electrical wiring.

Perhaps this was a case of depression after the triumph; the writing of *Computable Numbers* would have been like a love affair, now over but for mopping-up operations. Now he had the problem of *continuing*. Had he killed off the spirit? Was his work a 'dead end'? He had done something, but what was it for? It was all very well for Bernard Shaw's Ancients to live on truth alone, but it was asking a great deal of him. Indeed, it was not his ideal. 'As regards the question of why we have bodies at all, why we do not or cannot live free as spirits and communicate as such, we probably could do so but there would be nothing whatever to do. The body provides something for the spirit to look after and use.' But what was his body to *do*, without the loss of innocence, or the compromise of truth?

The months from January to April 1937 were absorbed in writing up a paper on the lambda-calculus, and two on group theory.[11] Of these, the logic paper represented a small development of Kleene's ideas. The first group theory paper was work related to that of Reinhold Baer, the German algebraist now attached to the IAS, which had mostly been done in 1935. But the second was a new departure, which arose through contact with von Neumann. It was a problem suggested by the emigré Polish mathematician S. Ulam, that of asking whether continuous groups could be approximated by finite groups, rather like approximating a sphere by polyhedra. Von Neumann had passed the problem on to Alan, who successfully dealt with it by April, when it was submitted. This was fast work, although as he had shown that continuous groups could *not* in general be approximated in this way, it was a rather negative result. Nor, he wrote, was he 'taking these things so seriously as the logic.'

Meanwhile the possibility had arisen of staying at Princeton for a second year. Alan wrote home on 22 February:

> I went to the Eisenharts regular Sunday tea yesterday and they took me in relays to try and persuade me to stay another year. Mrs. Eisenhart mostly put forward social or semi-moral semi-sociological reasons why it would be a good thing to have a second year. The Dean weighed in with hints that the Procter Fellowship was mine for the asking (this is worth $2000 p.a.) I said I thought King's would probably prefer that I return, but gave some vague promise that I would sound them on the matter. The people I know here will all be leaving, and I don't much care about the idea of spending a long summer in this country. I should like to know if you have any opinions on the subject. I think it is most likely I shall come back to England.

Dean Eisenhart was an old-fashioned figure, who in his lectures would apologise for using the modern abstract group, but very kind. He and his wife made noble efforts to entertain the students at their tea-parties. Whatever his parents thought, Philip Hall had sent Alan the notice of vacancies for Cambridge lectureships, and this Alan would much have preferred if he could gain one. A lectureship would in effect mean a permanent home at Cambridge, which was the only possible resolution of his problems in life, as well as being due recognition of his achievements. Alan wrote back to him on 4 April:

> I am putting in for it, but offering fairly heavy odds against getting it.

He also wrote to his mother, who was just setting off on a pilgrimage to Palestine:

> Maurice and I are both putting in for it, though I don't suppose either of us will get it: I think it is a good thing to start putting in for these things early, so as to get one's existence recognised. It's a thing I am rather liable to neglect. Maurice is much more conscious of what are the right things to do to help his career. He makes great social efforts with the mathematical big-wigs.

As he forecast, he failed to gain a Cambridge appointment. Ingham wrote from King's, encouraging him to stay for another year, and this made up his mind. He wrote on 19 May:

> I have just made up my mind to spend another year here, but I shall be going back to England for most of the summer in accordance with previous programme. Thank you very much for your offer of help with this: I shall not need it, for if I have this Procter as the Dean suggests I shall be a rich man, and otherwise I shall go back to Cambridge. Another year here on the same terms would be rather an extravagance . . .
>
> My boat sails June 23. I might possibly do a little travelling here before the boat goes, as there will be very little doing here during the next month and it's not a fearfully good time of year for work. More likely I shall not as I don't usually travel for the sake of travelling.
>
> I am sorry Maurice won't be here next year. He has been very good company.
>
> I am glad the Royal Family are resisting the Cabinet in their attempt to keep Edward VIII's marriage quiet.

Since he was staying another year, he decided he should take a PhD degree, as Maurice had done. For his thesis, Church had suggested a topic that had come up in his lecture course, relating to the implications of Gödel's theorem. Alan had written in March that he was 'working out some new ideas in logic. Not so good as the computable numbers, but quite hopeful.' These ideas would see him through.

As for the Procter Fellowship, it did indeed fall into his lap. It was for the Vice-Chancellor of Cambridge University to nominate the Fellow, so there were letters of recommendation sent to him. One of these was from the Wizard himself, who wrote:[12]

1 June 1937

Sir,

Mr A. M. Turing has informed me that he is applying for a Proctor [*sic*] Visiting Fellowship to Princeton University from Cambridge for the academic year 1937–1938. I should like to support his application and to inform you that I know Mr Turing very well from previous years: during the last term of 1935, when I was a visiting professor in Cambridge, and during 1936–1937, which year Mr Turing has spent in Princeton, I had opportunity to observe his scientific work. He has done good work in branches of mathematics in which I am interested, namely: theory of almost periodic functions, and theory of continuous groups.

I think that he is a most deserving candidate for the Proctor Fellowship, and I should be very glad if you should find it possible to award one to him.

I am, Respectfully, John von Neumann

Von Neumann would have been asked to write the letter, because his name carried such weight. But why did he make no mention of *Computable Numbers*, a far more substantial piece of work than the papers to which he referred? Had Alan failed to make him aware of it, even after the paper had been printed, and reprints sent round? If Alan had an *entrée* with von Neumann, the first thing he should have done was to exploit it to help bring *Computable Numbers* to attention. It would be typical of what was perceived as his lack of worldly sense, if he had been too shy to push his work upon the 'mathematical big-wig'.

Against Alan's prediction, and perhaps to his mild chagrin, the redoubtable Maurice Pryce had been appointed to a Cambridge lectureship, as had Ray Lyttleton, the current Procter Fellow. And Alan did after all spend some time in travel, for Maurice Pryce sold him his car, a 1931 V8 Ford, which had taken him all over the continent on the tour that as a Commonwealth Fellow he had been obliged to make in summer 1936. Maurice taught him to drive, which was not an easy task, for Alan was ham-handed and not good with machines. Once he nearly reversed into the Carnegie Lake and drowned them both. Then on about 10 June they took off together for a Turing family visit, which no doubt Mrs Turing had long been urging upon Alan. It was to a cousin on her mother's side who had emigrated from Ireland. Jack Crawford, now nearly seventy, was the retired Rector of Wakefield, Rhode Island.

The visit was not quite the expected grim chore of conventional politeness, for Alan approved of Jack Crawford, who in his youth had studied at the then Royal College of Science in Dublin:

> I enjoyed the time I spent at Cousin Jack's. He is an energetic old bird. He has a little observatory with a telescope that he made for himself. He told me all about the grinding of mirrors . . . . I think he comes into competition with Aunt Sybil for the Relations Merit Diploma. Cousin Mary is a little bit of a thing you could pick up and put in your pocket. She is very hospitable and rather timid: she worships Cousin Jack.

They were ordinary people, who made Alan feel more at home than did the grand Princeton figures. In their old-fashioned country way, they put Alan and Maurice in the same double bed.

The compartments of lfe were fractured. Maurice was amazed – he had not had the slightest suspicion. Alan apologised and withdrew at once. Then he blazed out, not with a trace of shame, but with *anger*, with a story of how his parents had been away in India so long, and of his years in boarding schools. It had all been said before, in *The Loom of Youth:*

> Then Jeffries' wild anger, the anger that had made him so brilliant an athlete, burst out: 'Unfair? Yes, that's the right word; it is unfair. Who made me what I am but Fernhurst? . . . And now Fernhurst, that has made me what I am, turns round and says, "You are not fit to be a member of this great school!" and I have to go. . . .'

The deeply embarrassing moment brought to light a vein of self-pity that he otherwise never showed, as well as an analysis which he must have known to be facile. It would not do. It was time to look forward, not backward – but to what? How was he to continue? Maurice accepted the explanation, and they never spoke of it again. Alan boarded the *Queen Mary* on his twenty-fifth birthday, and on 28 June disembarked at Southampton. He missed the Fourth of July softball match at the Graduate College, between the British Empire and the Revolting Colonies.

★

Back for three months in the mild Cambridge summer of 1937, there were three major projects on hand. First there was some tidying up of *Computable Numbers*. Bernays, in Zürich, had perhaps rather annoyingly found some errors[13] in his proof that the Hilbert decision problem, in its precise form, was unsolvable, and these had to be put right by a correction note in the LMS *Proceedings*. He also completed a formal demonstration[14] that his own 'computability' coincided exactly with Church's 'effective calculability'. By now there existed yet a third definition of the same sort of idea. This was the 'recursive function', which was a way of making absolutely precise the notion of defining a mathematical function in terms of other more elementary functions; Gödel had suggested it, and it had been taken up by Kleene. It was implicit in Gödel's proof of the incompleteness of arithmetic. For when Gödel showed that the concept of proof by chess-like rules was an 'arithmetical' concept, one as 'arithmetical' as finding a highest common factor or something of the kind, he was really saying it could be done by a 'definite method'. This idea, when formalised and extended somewhat, led to the definition of the 'recursive function'. And now it had turned out that the general recursive function was exactly equivalent to the computable function. So Church's lambda-calculus, and Gödel's way of defining arithmetical functions, both turned out to be equivalent to the Turing machine. Gödel himself later acknowledged[15] the Turing machine construction as the most satisfactory definition of a 'mechanical procedure'. At the time, it was a very striking and surprising fact, that three independent approaches to the idea of doing something in a definite way, had converged upon equivalent concepts.

The second project concerned the 'new ideas in logic' for a doctoral thesis. The basic idea was to see if there was any way in which to escape the force of Gödel's result that there would always be true but unprovable assertions within arithmetic. This was not a new question, for Rosser, now at Cornell, had produced a paper[16] in March 1937 which took it up. But Alan planned a rather more general attack on the question.

His third project was a very ambitious one, for he had decided to try his strength on the central problem of the theory of numbers. It was not a new interest, for he had possessed Ingham's book on the subject since 1933. But in 1937 Ingham sent him some recent papers,[17] on learning that he wished to make an attack himself. It was ambitious because the question he chose was one that had long absorbed and defeated the greatest pure mathematicians.

Although the prime numbers were such ordinary things, it was easy to pose quite baffling questions about them in a few words. One question had been solved very early on. Euclid had been able to show that there were infinitely many prime numbers, so that although in 1937 the number $2^{127} - 1 = 170141183460469231731687303715884$ 105727 was the largest known prime, it was also known that they continued for ever. But another property that was easy to guess, but very hard to prove, was that the primes would always thin out, at first almost every other number being prime, but near 100 only one in four, near 1000 only one in seven, and near 10,000,000,000 only one in 23. There had to be a reason for it.[18]

In about 1793, the fifteen-year-old Gauss noticed that there was a regular pattern to the thinning-out. The spacing of the primes near a number n was proportional to the number of digits in the number n; more precisely, it increased as the natural logarithm of n. Throughout his life Gauss, who apparently liked doing this sort of thing, gave idle leisure hours to identifying all the primes less than three million, verifying his observation as far as he could go.

Little advance was made until 1859, when Riemann developed a new theoretical framework in which to consider the question. It was his discovery that the calculus of the complex numbers*

---

* The 'complex' number calculus exemplified the progress of mathematical abstraction. Originally, complex numbers had been introduced to combine 'real' numbers with the 'imaginary' square root of minus one, and mathematicians had agonised over the question of whether such things really 'existed'. From the modern point of view, however, complex numbers were simply defined abstractly as *pairs* of numbers, and pictured as points in a plane. A simple rule for the definition of the 'multiplication' of two such pairs was then sufficient to generate an enormous theory. Riemann's work in the nineteenth century had played a large part in its 'pure' development; but it was also found to be of great usefulness in the development of physical theory. Fourier analysis, treating the theory of vibrations, was an example of this. The quantum theory developed since the 1920s went even further in according complex numbers a place in fundamental physical concepts. None of these mathematical ideas are essential to what follows, although such connections between 'pure' and 'applied' were certainly relevant to a number of aspects of Alan Turing's later work.

could be used as a bridge between the fixed, discrete, determinate prime numbers on the one hand, and smooth functions like the logarithm – continuous, averaged-out quantities – on the other. He thereby arrived at a certain formula for the density of the primes, a refinement of the logarithm law that Gauss had noticed. Even so, his formula was not exact, and nor was it proved.

Riemann's formula ignored certain terms which he was unable to estimate. These error terms were only in 1896 proved to be small enough not to interfere with the main result, which now became the Prime Number Theorem, and which stated in a precise way that the primes thinned out like the logarithm – not just as a matter of observation, but proved to be so for ever and ever. But the story did not end here. As far as the tables went it could be seen that the primes followed this logarithmic law quite amazingly closely. The error terms were not only small compared with the general logarithmic pattern; they were *very* small. But was this also true for the whole infinite range of numbers, beyond the reach of computation? – and if so what was the reason for it?

Riemann's work had put this question into a quite different form. He had defined a certain function of the complex numbers, the 'zeta-function'. It could be shown that the assertion that the error terms remained so very small was essentially equivalent to the assertion that this Riemann zeta-function took the value zero only at points which all lay on a certain line in the plane. This assertion had become known as the Riemann Hypothesis. Riemann had thought it 'very likely' to be true, and so had many others, but no proof had been discovered. In 1900 Hilbert had made it his Fourth Problem for twentieth-century mathematics, and at other times called it 'the most important in mathematics, absolutely the most important'. Hardy had bitten on it unsuccessfully for thirty years.

This was the central problem of the theory of numbers, but there was a constellation of related questions, one of which Alan picked for his own investigation. The simple assumption that the primes thinned out like the logarithm, without Riemann's refinements to the formula, seemed always to overestimate the actual number of primes by a certain amount. Common sense, or 'scientific induction', based on millions of examples, would suggest that this would always be so, for larger and larger numbers. But in

1914 Hardy's collaborator J. E. Littlewood had shown that this was not so, for there existed some point where the simple assumption would underestimate the cumulative total of primes. Then in 1933 a Cambridge mathematician, S. Skewes, had shown[19] that if the Riemann Hypothesis were true, a crossover point would occur before

$$10^{10^{10^{34}}}$$

which, Hardy commented, was probably the largest number ever to serve any definite purpose in mathematics.* It could be asked whether this enormous bound might be reduced, or whether one could be found that did not depend upon the truth of the Riemann Hypothesis, and these were the problems that Alan now undertook.

One new departure at Cambridge was his acquaintance with Ludwig Wittgenstein, the philosopher. He would have seen Wittgenstein before at the Moral Science Club, and Wittgenstein (like Bertrand Russell) had received a copy of *Computable Numbers*. But it was in this summer of 1937 that Alister Watson, the King's Fellow, introduced them and they met sometimes in the botanical gardens. Watson had written a paper[20] on the foundations of mathematics for the Moral Science Club, in which he made use of the Turing machine. Wittgenstein, whose first work had been as an engineer, always liked practical, down-to-earth constructions and would have approved of the way that Alan had made a vague idea

---

* 10^{34} is 10,000,000,000,000,000,000,000,000,000,000,000 – a number comparable with the number of elementary particles in a large building. But $10^{10^{34}}$ is far bigger: as 1 followed by 10^{34} zeroes it would require books with the mass of Jupiter to print it in decimal notation. It could be thought of as the number of possible man-made objects. Skewes' number was much bigger again, as 1 followed by $10^{10^{34}}$ zeroes! In actual fact mathematicians had certainly thought about numbers far larger than these, here we have only gone through three stages of growth, but it is not difficult to make up a new notation to express the idea of going through ten such stages, or 10^{10}, or $10^{10^{10}}$; or of regarding even these as just the first step in a process of super-growth, and then defining super-super-growth, and then . . . . Such definitions, indeed, had already played a role in the theory of 'recursive functions', one of the other approaches to the idea of 'definite method' which had been found equivalent to that of the Turing machine. But Skewes' number was certainly remarkably large for a problem which could be expressed in such elementary terms.

so definite. Curiously, the failure of the Hilbert programme had also meant the end of the point of view advanced by Wittgenstein in his first phase, in the *Tractatus Logico-Philosophicus*, that every well-posed problem could be solved.

Alan probably had a boating holiday – either on the Norfolk Broads, or at Bosham on Chichester Harbour. He also stayed for a while with the Beuttells in London. Although Mr Beuttell in principle espoused liberal causes of feminism and profit-sharing, his own firm was run on strictly autocratic lines, and so was his family. Victor's younger brother Gerard was studying physics at Imperial College, but his father was extremely annoyed that he spent his time flying model aeroplanes to investigate wind currents, and put a stop to his studies. Alan was furious to hear this, saying that Gerard had a contribution to make to science,[21] and was doubly upset because he respected his father. He also roared with approval when he heard that Gerard had told his father, in connection with his infringing some petty rule of the family firm, that he would only obey 'sensible rules'.

It was also in London that Alan met James again for a weekend. They stayed at a rather sordid bed-and-breakfast place near Russell Square. They went to see a film or two and Elmer Rice's play *Judgment Day* about the Reichstag-fire trial. Alan must have found it a relief to be with someone who did not reject his sexual advances, although it was always clear that James aroused in him neither deep feelings, nor a special physical attraction. The relationship was not able to develop beyond this point. After this weekend, James had almost no further experience for about twelve years. Although Alan was more exploratory, this would be his story too. His life would not change until much water had flowed under the bridge.

On 22 September, Alan met up at Southampton with an American friend from Graduate College, Will Jones. They had arranged to travel back together, and boarded the German liner, the *Europa*. Will Jones had spent the summer at Oxford, and it was he who chose the German ship, simply because it was faster. A more dutiful anti-fascist than Alan would not have used it, but on the other hand a more conventional person would not have spent the voyage in learning Russian, enjoying the shocked German expressions as he wielded a textbook emblazoned with hammer and sickle.

On the boat, wrote Alan on arrival, he

Was very glad to have Will Jones as travelling companion. There didn't seem to be anyone very interesting on board so Will and I whiled away the time with philosophical discussions, and spent half of one afternoon in trying to find the speed of the boat.

Back at Princeton, Alan and Will Jones spent much time talking together. Will Jones came from the old white South of deepest Mississippi, and had studied philosophy at Oxford. So this was not the stereotyped meeting of Yankee brashness and old world elegance, far from it. Will Jones came from quite another America, just as Alan represented a plain-speaking, pragmatic, liberal England. As a philosopher with a serious interest in science, Will Jones also rose above the usual boundaries of arts and sciences. He was currently writing a dissertation on the claim of Kant that moral categories could be justified even if human actions were as determined as the motions of the planets. He canvassed Alan's opinion as to whether quantum mechanics had affected the argument – so much the problem to which Alan had addressed himself five or so years before! But now he gave the impression that he had long been happy with the Russellian view, that at some level the world must evolve in a mechanistic way. He was not now very interested in philosophical, as opposed to scientific, discussions of the problem of free will. Perhaps the trace of his former conflict lay in his very vehemence in the materialist direction. 'I think of people as *pink*-coloured collections of *sense*-data,' he once joked. If only it were so easy! Symbolically, the Research fountain pen that Mrs Morcom had given him in 1932 was lost on the voyage.

Will Jones also had Alan explain to him some of the theory of numbers, and enjoyed the way that Alan did it, showing how from the most simple axioms, all the properties could be precisely derived – an approach quite different from the rote-learning of school mathematics. Alan never talked to Will about his emotional problems, but it might well be that he derived moral support in a much more general way, for Will appreciated in him the embodiment of the moral philosophy of G. E. Moore and Keynes.

Alan and Will had become aquainted through being members of the same circle of friends the previous year, and another of the circle had also returned to Princeton. This was Malcolm MacPhail, a Canadian physicist, who became involved in a sideline that Alan took up.[22]

> It was probably in the fall of 1937 that Turing first became alarmed about a possible war with Germany. He was at that time supposedly working hard on his famous thesis but nevertheless found time to take up the subject of cryptanalysis with characteristic vigour. . . . on this topic we had *many* discussions. He assumed that words would be replaced by numbers taken from an official code book and messages would be transmitted as numbers in the binary scale. But, to prevent the enemy from deciphering captured messages even if they had the code book, he would multiply the number corresponding to a specific message by a horrendously long but secret number and transmit the product. The length of the secret number was to be determined by the requirement that it should take 100 Germans working eight hours a day on desk calculators 100 years to discover the secret factor by routine search!
>
> Turing actually designed an electric multiplier and built the first three or four stages to see if it could be made to work. For the purpose he needed relay-operated switches which, not being commercially available at that time, he built himself. The Physics Department at Princeton had a small but well equipped machine shop for its graduate students to use, and my small contribution to the project was to lend Turing my key to the shop, which was probably against all the regulations, and show him how to use the lathe, drill, press etc. without chopping off his fingers. And so, he machined and wound the relays; and to our surprise and delight the calculator worked.

Mathematically, this project was not advanced, for it used only multiplication. But although it used no advanced theory, it involved applications of 'dull and elementary' mathematics which were by no means well known in 1937.

For one thing, the binary representation of numbers would have seemed a novelty to anyone then engaged in practical computations. Alan had used binary numbers in *Computable Numbers*. There they brought in no point of principle, but made it possible to represent all the computable numbers as infinite sequences of 0s and 1s alone. In a practical multiplier, however, the advantage of binary numbers was more concrete: it was that the multiplication table then reduced to:

×	0	1
0	0	0
1	0	1

The binary multiplication table being so trivial, the work of a multiplier would be reduced to carrying and adding operations.

A second aspect of his project was its connection with elementary logic. The arithmetical operations with 0s and 1s could be thought of in terms of the logic of propositions. Thus the trivial multiplication table, for instance, could be considered as equivalent to the function of the word AND in logic. For if p and q were propositions, then the following 'truth-table' would show in what circumstances 'p AND q' was true:

p

AND	false	true
false	false	false
true	false	true

q

It was the same game, with a different interpretation. All of this would have been entirely familiar to Alan, the calculus of propositions appearing on the first page of any text on logic. It was sometimes called 'Boolean Algebra' after George Boole, who had formalised what he optimistically called 'the laws of thought' in 1854. Binary arithmetic could all be expressed in terms of Boolean algebra, using AND, OR, and NOT. His problem in designing the multiplier would be to use Boolean algebra to minimise the number of these elementary operations required.

This, as a paper exercise, would be very similar to that of designing a 'Turing machine' for the same problem. But in order to embody it in working machinery, it required some means of arranging for different physical 'configurations'. This was achieved by the switches, for the whole point of a switch would be that it could be in one of two states, 'on' or 'off', '0' or '1', 'true' or 'false'. The switches that he used were operated by relays, and in this way electricity played its first direct part in his urge to connect logical ideas with something that physically worked. There was nothing new about the electromagnetic relay, which had been invented by the American physicist Henry a hundred years earlier. Its physical principle was the same as that of the electric motor, an electric current, passing through a coil, causing a magnetic head to move. However, the point of a relay was that the magnetic head would either open or close another electrical circuit. It would act as a switch. The name 'relay' derived from its use in early telegraph systems, to allow an enfeebled electric signal to set off a fresh, clean click. It was this all-or-nothing *logical* function of relays that made them necessary by the million in the automatic telephone exchanges proliferating in the United States and Britain alike.

It was not well known in 1937 that the logical properties of combinations of switches could be represented by Boolean algebra, or by binary arithmetic, but this was not hard for a logician to see. Alan's task was to embody the logical design of a Turing machine in a network of relay-operated switches. The idea would be that when a number was presented to the machine, presumably by setting up currents at a series of input terminals, the relays would click open and closed, currents would pass through, and emerge at output terminals, thus in effect 'writing' the enciphered number. It would not actually use 'tapes' but from a logical point of view it came to the same thing. Turing machines were coming to life, for the first stages of his relay multiplier actually *worked*. Alan's rather surreptitious access to the physics workshop was, however, symbolic of the problem that he faced in creating that life, by overcoming the line drawn between mathematics and engineering, the logical and the physical.

As a cipher the idea was surprisingly feeble, the more so when set against his claim of a year earlier. Did he not credit the Germans with being able to find the highest common factor of two or more numbers in order to find the 'secret number' used as key? Even if some added sophistication removed this loophole, it would still suffer from the crippling practical disadvantage that a single incorrectly transmitted digit would render the entire message indecipherable.

It might be that it was never intended very seriously, and that he had gone off at a tangent in meeting the challenge of designing a binary multiplier. But as a reader of the *New Statesman*,* sent to him from England, he had no particular reason to be frivolous in speaking of Germany. Every week there were frightening articles about German policy inside and outside the new Reich. Even if the prospect of war work was more the excuse to take up a 'dull and elementary' (but fascinating) sideline than anything like the call of duty, he would not have been alone if he found that Nazi Germany had resolved his qualms about 'morality'.

There was also another machine he had in mind, but this had nothing to do with Germany, except in the very different sense that it came out of the work of Riemann. Its purpose was to calculate the Riemann zeta-function. Apparently he had decided that the Riemann hypothesis was probably false, if only because such great efforts had failed to prove it. Its falsity would mean that the zeta-function *did* take the value zero at some point which was off the special line, in which case this point could be located by brute force, just by calculating enough values of the zeta-function.

This programme had already been started; indeed Riemann himself had located the first few zeroes and checked that they all lay on the special line. In 1935-6, the Oxford mathematician E. C. Titchmarsh had used the punched-card equipment which was then used for the calculation of astronomical predictions to show that (in a certain precise sense) the first 104 zeroes of the zeta-function did all lie on the line. Alan's idea was essentially to examine the next few thousand or so in the hope of finding one off the line.

There were two aspects to the problem. Riemann's zeta-function was defined as the sum of an infinite number of terms, and although this sum could be re-expressed in many different ways, any attempt

---

* Certainly one attraction to Alan of the *New Statesman* would have been its exceptionally demanding puzzle column. In January 1937 he was delighted when his friend David Champernowne defeated such runners-up as M. H. A. Newman and J. D. Bernal in giving a witty solution, phrased in Carrollian language, to a problem set by Eddington called 'Looking Glass Zoo'. (It depended upon a knowledge of the matrices used by Dirac in his theory of the electron.) But Alan's comments on the Abdication, naive in idealism perhaps but certainly not ill-informed, indicate very clearly that his interest in the journal would not have been confined to this feature.

to evaluate it would in some way involve making an approximation. It was for the mathematician to find a good approximation, and to prove that it was good: that the error involved was sufficiently small. Such work did not involve computation with numbers, but required highly technical work with the calculus of complex numbers. Titchmarsh had employed a certain approximation which – rather romantically – had been exhumed from Riemann's own papers at Göttingen where it had lain for seventy years. But for extending the calculation to thousands of new zeroes a fresh approximation was required; and this Alan set out to find and to justify.

The second problem, quite different, was the 'dull and elementary' one of actually doing the computation, with numbers substituted into the approximate formula, and worked out for thousands of different entries. It so happened that the formula was rather like those which occurred in plotting the positions of the planets, because it was of the form of a sum of circular functions with different frequencies. It was for this reason that Titchmarsh had contrived to have the dull repetitive work of addition, multiplication, and of looking up of entries in cosine tables done by the same punched-card methods that were used in planetary astronomy. But it occurred to Alan that the problem was very similar to another kind of computation which was also done on a large practical scale – that of tide prediction. Tides could also be regarded as the sum of a number of waves of different periods: daily, monthly, yearly oscillations of rise and fall. At Liverpool there was a machine[23] which performed the summation automatically, by generating circular motions of the right frequencies and adding them up. It was a simple *analogue* machine; that is, it created a physical analogue of the mathematical function that had to be calculated. This was a quite different idea from that of the Turing machine, which would work away on a finite, discrete, set of symbols. This tide-predicting machine, like a slide rule, depended not on symbols, but on the measurement of lengths. Such a machine, Alan had realised, could be used on the zeta-function calculation, to save the dreary work of adding, multiplying, and looking up of cosines.

Alan must have described this idea to Titchmarsh, for a letter[24] from him dated 1 December 1937 approved of this programme of extending the calculation, and mentioned: 'I have seen the tide-predicting machine at Liverpool, but it did not occur to me to use it in this way.'

There were some diversions. The hockey playing continued, although without Francis Price and Shaun Wylie the team had lost its sparkle. Alan found himself involved in making the arrangements. He also played a good deal of squash. At Thanksgiving he drove north to visit Jack and Mary Crawford for a second time. ('I am getting more competent with the car.') Before Christmas, Alan took up an invitation from his friend Venable Martin to go and stay with him. He came from a small town in South Carolina.

> We drove down from here in two days and then I stayed there for two or three days before I came back to Virginia to stay with Mrs Welbourne. It was quite as far south as I had ever been – about 34°. The people seem to be all very poor down there still, even though it is so long since the civil war.

Mrs Welbourne was 'a mysterious woman in Virginia' who had a habit of inviting English students from the Graduate College for Christmas. 'I didn't make much conversational progress with any of them,' Alan had to confess of her family. Alan and Will Jones organised another treasure hunt, although it lacked the *élan* of the previous year; one of the clues was in his collected Shaw. And in April he and Will made a trip to visit St John's College, Annapolis, and Washington. 'We also went and listened to the Senate for a time. They seemed very informal. There were only six or eight of them present and few of them seemed to be attending.' They looked down from the gallery and saw Jim Farley, Roosevelt's party boss. It was another world.

The main business of the year was the completion of his PhD thesis,[25] investigating whether there was any way of escaping the force of Gödel's theorem. The fundamental idea was to add further axioms to the system, in such a way that the 'true but unprovable' statements could be proved. But arithmetic, looked at in this way, had a distinctly hydra-headed nature. It was easy enough to add an axiom so that one of Gödel's peculiar statements could be proved. But then Gödel's theorem could be applied to the enlarged set of axioms, producing yet another 'true but unprovable' assertion. It could not be enough to add a *finite* number of axioms; it was necessary to discuss adding infinitely many.

This was just the beginning, for as mathematicians well knew, there were many possible ways of doing 'infinitely many' things in order. Cantor had seen this when investigating the notion of ordering the integers. Suppose, for example, that the integers were ordered in the following way: first *all* the even numbers, in ascending order, and then *all* the odd numbers. In a precise sense, this listing of the integers would be 'twice as long' as the usual one. It could be made three times as long, or indeed infinitely many times as long, by taking first the even numbers, then remaining multiples of 3, then remaining multiples of 5, then remaining multiples of 7, and so on. Indeed, there was no limit to the 'length' of such lists. In the same way, extending the axioms of arithmetic could be done by *one* infinite list of axioms, or by two, or by infinitely many infinite lists – there was again no limit. The question was whether any of this would overcome the Gödel effect.

Cantor had described his different orderings of the integers by 'ordinal numbers', and Alan described his different extensions of the axioms of arithmetic as 'ordinal logics'. In one sense it was clear that no 'ordinal logic' could be 'complete', in Hilbert's technical sense. For if there were infinitely many axioms, they could not all be written out. There would have to be some finite rule for generating them. But in that case, the whole system would still be based on a finite number of rules, so Gödel's theorem would still apply to show that there were still unprovable assertions.

However, there was a more subtle question. In his 'ordinal logics', the rule for generating the axioms was given in terms of substituting an 'ordinal formula' into a certain expression. This was itself a 'mechanical process'. But it was *not* a 'mechanical process' to decide whether a given formula was an ordinal formula. What he asked was whether all the incompleteness of arithmetic could be concentrated in one place, namely into the unsolvable problem of deciding which formulae were 'ordinal formulae'. If this could be done, then there would be a sense in which arithmetic was complete; everything could be proved from the axioms, although there would be no mechanical way of saying what the axioms were.

He likened the job of deciding whether a formula was an ordinal formula to 'intuition'. In a 'complete ordinal logic', any theorem in arithmetic could be proved by a mixture of mechanical reasoning, and steps of 'intuition'. In this way, he hoped to bring the Gödel incompleteness under some kind of control. But he regarded his results as disappointingly negative. 'Complete logics' did exist, but they suffered from the defect that one could not count the number of 'intuitive' steps that were necessary to prove any particular theorem. There was no way of measuring how 'deep' a theorem was, in his sense; no way of pinning down exactly what was going on.

One nice touch on the side was his idea of an 'oracle' Turing machine, one which would have the property of being able to answer one particular unsolvable problem (like recognising an ordinal formula). This introduced the idea of relative computability, or relative unsolvability, which opened up a new field of enquiry in mathematical logic. Alan might have been thinking of the 'oracle' in *Back to Methuselah*, through whose mouth Bernard Shaw answered the unsolvable problems of the politicians with 'Go home, poor fool'!

Less clear from his remarks in the paper was to what extent he regarded such 'intuition', the ability to recognise a true but unprovable statement, as corresponding to anything in the human mind. He wrote that:

> Mathematical reasoning may be regarded rather schematically as the exercise of a combination of two faculties, which we may call *intuition* and *ingenuity*. (We are leaving out of account that most important faculty which distinguishes topics of interest from others; in fact, we are regarding the function of the mathematician as simply to determine the truth or falsity of propositions.) The activity of the intuition consists in making spontaneous judgments which are not the result of conscious trains of reasoning. . . .

and he claimed that his ideas on 'ordinal logics' represented one way of formalising this distinction. But it was not established that 'intuition' had anything to do with the incompleteness of finitely defined formal systems. After all, no one knew of this incompleteness until 1931, while intuition was a good deal older. It was the same ambiguity as in *Computable Numbers*, which mechanised mind, yet pointed out something beyond mechanisation. Did this have a significance for human minds? His views were not clear at this stage.

As for the future, his intention was to return to King's, provided that, as expected, they renewed his Fellowship which was, in March 1938, at the end of its first three years. On the other hand, his father wrote advising him (not very patriotically, perhaps) to find an appointment in the United States. For some reason King's College was slow in notifying him that the extension of his fellowship had been made. Alan wrote to Philip Hall on 30 March:

> I am writing a thesis for a Ph.D, which is proving rather intractable, and I am always rewriting parts of it. . . .
>
> I am rather worried about the fact that I have heard nothing about re-election to my Fellowship. The most plausible explanation is simply that there has been no re-election, but [I] prefer to think there is some other reason. If you would make some cautious enquiries and send me a postcard I should be very grateful.
>
> I hope Hitler will not have invaded England before I come back.

After the union with Austria on 13 March everyone was beginning to take Germany seriously. Meanwhile, Alan did dutifully go to Eisenhart and ask him 'about possible jobs over here; mostly for Daddy's information, as I think it is unlikely I shall take one unless you are actually at war before July. He didn't know of one at present, but said he would bear it all in mind.' But then a job materialised. Von Neumann himself offered a research assistantship at the IAS.

This might have meant a certain priority being given to von Neumann's research areas – at that time in the mathematics associated with quantum mechanics and other areas of theoretical physics, and not including logic or the theory of numbers. On the other hand, a position with von Neumann would be the ideal start to the American academic career, which, presumably, Alan's father thought wise. Competition was intense, and the market, already in depression, was flooded by European exiles. The stamp of approval from von Neumann would carry great weight.

In professional terms, this was a big decision. But all Alan wrote of the opportunity to Philip Hall on 26 April was: 'Eventually a possibility of a job here turned up', and to Mrs Turing on 17 May 'I had an offer of a job here as von Neumann's assistant at $1500 a year but decided not to take it.' For he had cabled King's to check that the fellowship had been renewed, and since it had, the decision was clear.

Despite himself, he had made his name in the Emerald City. It was not entirely necessary to have a reputation in order to be listened to.

By this time, von Neumann was aware of *Computable Numbers*, even if he had not been a year earlier. For when he travelled with Ulam to Europe in this summer of 1938, he proposed[26] a game of 'writing down on a piece of paper as big a number as we could, defining it by a method which indeed has something to do with some schemata of Turing's.'* But whatever the attractions, remunerations and compliments, the real issue was much simpler. He wanted to go home to King's.

The thesis, which in October he had hoped to finish by Christmas, was delayed. 'Church made a number of suggestions which resulted in the thesis being expanded to an appalling length.' A clumsy typist himself, he engaged a professional, who in turn made a mess of it. It was eventually submitted on 17 May. There was an oral examination on 31 May, conducted by Church, Lefschetz and H. F. Bohnenblust. 'The candidate passed an excellent examination, not only in the special field of mathematical logic, but also in other fields.' There was a quick test in scientific French and German as well. It was vaguely absurd, to be examined in this way while at the same time he was refereeing the PhD thesis of a Cambridge candidate. This, as it happened, he had to reject. (To Philip Hall on 26 April: 'I hope my remarks don't encourage the man to go and rewrite the thing. The difficulty with these people is to find out a really good way of being blunt. However I think I've given him something to keep him quiet a long time if he really is going to rewrite it.') The PhD was granted on 21 June. He made little use of the title, which had no application at Cambridge, and which elsewhere was liable to prompt people to retail their ailments.

His departure from the land of Oz was rather different from that in the fable. The Wizard was not a phoney, and had asked him to stay. While Dorothy had disposed of the Wicked Witch of the West, in his case it was the other way round. Though Princeton was fairly secluded from the orthodox, Teutonic side of America, it shared in a kind of conformity that made him ill at ease. And his problems remained unresolved. He was inwardly confident – but as in the

---

* Ulam writes further that 'von Neumann had great admiration for him and mentioned his name and "brilliant ideas" to me already, I believe, in early 1939. . . . At any rate von Neumann mentioned to me Turing's name several times in 1939 in conversations, concerning mechanical ways to develop formal mathematical systems.'

Murder in the Cathedral which he saw performed in March ('very much impressed') he was living and partly living.

In one way, however, he resembled Dorothy. For all the time, there was something that he could do, and which was just waiting for the opportunity to emerge. On 18 July Alan disembarked at Southampton from the *Normandie*, with the electric multiplier mounted on its breadboard and wrapped up securely with brown paper. 'Will be seeing you in the middle of July', he had written to Philip Hall, 'I also expect to find the back lawn criss-crossed with 8ft. trenches.' It had not come to that, but there were more discreet preparations, in which he could take part himself.

Alan was right in thinking that HM Government was concerned with codes and ciphers.* It maintained a department to do the technical work. In 1938 its structure was still a legacy of the Great War, a continuation of the organisation discreetly known as *Room 40* that the Admiralty had set up.

After the initial break of a captured German code book, passed to the Admiralty by Russia in 1914, a great variety of wireless and cable signals had been deciphered by a mostly civilian staff, recruited from universities and schools. The arrangement had the peculiar feature that the director, Admiral Hall, enjoyed control over diplomatic messages (for instance the famous Zimmermann telegram). Hall was no stranger to the exercise of power.[27] It was he who showed Casement's diary to the press, and there were more important instances of his[28] 'acting on intelligence independently of other departments in matters of policy that lay beyond the concerns of the Admiralty.' The organisation survived the armistice, but in 1922 the Foreign Office succeeded in detaching it from the Admiralty. By then it had been renamed as the 'Government Code and Cypher School', and was supposed

---

* In what follows, *code* refers to any conventional system of communicating text, whether secret or not. *Cipher* is used for communications designed to be incomprehensible to third parties. *Cryptography* is the art of writing in cipher; *cryptanalysis* that of deciphering what has been concealed in cipher. *Cryptology* covers both the devising and breaking of ciphers. At the time, these distinctions were not made, and Alan Turing himself referred to cryptanalysis as 'cryptography'.

to study[29] 'the methods of cypher communication used by foreign powers' and to 'advise on the security of British codes and cyphers.' It now came technically under the control of the head of the secret service,* himself nominally responsible to the Foreign Secretary.

The director of GC and CS, Commander Alastair Denniston, was allowed by the Treasury to employ thirty civilian Assistants,[30] as the high-level staff were called, and about fifty clerks and typists. For technical civil-service reasons, there were fifteen Senior and fifteen Junior Assistants. The Senior Assistants had all served in Room 40, except perhaps Feterlain, an exile from Russia who became head of the Russian section. There was Oliver Strachey, who was brother of Lytton Strachey and husband of Ray Strachey, the well-known feminist, and there was Dillwyn Knox, the classical scholar and Fellow of King's until the Great War. Strachey and Knox had both been members of the Keynesian circle at its Edwardian peak. The Junior Assistants had been recruited as the department expanded a little in the 1920s; the most recently appointed of them, A. M. Kendrick, had joined in 1932.

The work of GC and CS had played an important part in the politics of the 1920s. Russian intercepts leaked to the press helped to bring down the Labour government in 1924. But in protecting the British Empire from a revived Germany, the Code and Cypher School was less vigorous. There was a good deal of sucess in reading the communications of Italy and Japan, but the official history[31] was to describe it as 'unfortunate' that 'despite the growing effort applied at GC and CS to military work after 1936, so little attention was devoted to the German problem.'

One underlying reason for this was economic. Denniston had to plead for an increase in staff to match the military activity in the Mediterranean. In the autumn of 1935, the Treasury allowed an increase of thirteen clerks, although only on a temporary basis of six months at a time. It was a typical communication[32] from Denniston to the Treasury in January 1937 that read:

---

* The British spying organisation, variously entitled SIS, MI6. Apart from this top-level administrative overlap it was and remained essentially distinct from the cryptanalytic department.

The situation in Spain . . . remains so uncertain that there is an actual increase in traffic to be handled since the height of the Ethiopian crisis, the figures for cables handled during the last three months of 1934, 1935 and 1936 being

1934	10,638
1935	12,696
1936	13,990

During the past month the existing staff has only been able to cope with the increase in traffic by working overtime.

During 1937, the Treasury agreed to an increase in the permanent staff. But this measure did not meet a situation in which:[33]

The volume of German wireless transmissions . . . was increasing; it was steadily becoming less difficult to intercept them at British stations; yet even in 1939, for lack of sets and operators, by no means all German Service communications were being intercepted. Nor was all intercepted traffic being studied. Until 1937-38 no addition was made to the civilian staff as opposed to the service personnel at GC and CS; and because of the continuing shortage of German intercepts, the eight graduates then recruited were largely absorbed by the same growing burden of Japanese and Italian work that had led to expansion of the Service sections.

It was not simply a question of numbers and budgets, however. This elderly department was failing to rise to the mechanical challenge of the late 1930s. The years after the First World War had been 'the golden age of modern diplomatic codebreaking'.[34] But now the German communications presented GC and CS with a problem beyond their powers – the Enigma machine:[35]

By 1937 it was established that, unlike their Japanese and Italian counterparts, the German Army, the German Navy and probably the Air Force, together with other state organisations like the railways and the SS used, for all except their tactical communications, different versions of the same cypher system – the Enigma machine which had been put on the market in the 1920s but which the Germans had rendered more secure by progressive modifications. In 1937 GC and CS broke into the less modified and less secure model of this machine that was being used by the Germans, the Italians and the Spanish nationalist forces. But apart from this the Enigma still resisted attack, and it seemed likely that it would continue to do so.

The Enigma machine was the central problem that confronted British Intelligence in 1938. But they believed it was unsolvable. Within the existing system, perhaps it was. In particular, this department of classicists, a sort of secret shadow of King's down in Broadway Buildings, did not include a mathematician.

No addition was made to permanent staff in 1938 to meet this striking deficiency. But[36] 'plans were made to take on some 60 more cryptanalysts in the event of war.' And this was where Alan Turing came into the story, for he was one of the recruits. He might possibly have been in touch with the government since 1936. Or he might have stepped off the *Normandie* with the intention of demonstrating his multiplier. But more likely he was suggested to Denniston through one of the elder dons who had worked in Room 40 in the First World War. One of these was Professor Adcock, a Fellow of King's since 1911. Had Alan ever spoken of codes and ciphers on the King's High Table, his enthusiasm could quickly have been communicated to GC and CS. One way or another, he was a natural recruit. On his return in the summer of 1938, he was taken on to a course at the GC and CS headquarters.

Alan and his friends could see that war was likely, despite all the hopes of 1933, and found it important to see that they were used in some sensible way, rather than in leading cannon-fodder over the top. It was hard to separate this feeling from that of wanting to avoid injury, and the government's policy for reserving intellectual talent came as some relief, releasing them from guilt. In this way, Alan Turing made his fateful decision, and chose to begin his long association with the British government. For all his suspicion of 'HM Government', it must have been exciting to be allowed to see the back of the shop. But it meant that he had for the first time surrendered a part of his mind, with a promise to keep the government's secrets.

Though stern and demanding, the government that he joined, like the White Queen who took Alice on her journey, was in a muddled state, struggling with safety pins and string. The failure to make a serious effort at the Enigma was but one aspect of an incoherent strategy, which all the world could see in September 1938. Until that month, British people could still convince themselves that there were logical 'solutions' to German 'grievances' within the existing framework. After that month, moral debates about fairness and self-determination finally ceased to cloak the essential reality of power. The Cambridge population re-assembled for what was to be 'the year under the terror' in the words of Frank Lucas, a King's don. The White Queen had squealed before the prick of the needle actually came. London children had been evacuated to Newnham College, and the male undergraduates had felt themselves on the brink of enlistment. Nothing was clear, but that something dreadful was in the offing. Radical agitation emphasised the devastating power expected of the modern air raid, while the government seemed to have nothing in mind but the building of bombers to execute a counter-attack.

The old world might be nearing its end, but there was a little escape into fantasy on offer from the new. *Snow White and the Seven Dwarfs* arrived at Cambridge in October, and Alan played exactly the part expected by Cambridge of King's dons by going to see it with David Champernowne. He was very taken with the scene where the Wicked Witch dangled an apple on a string into a boiling brew of poison, muttering

> Dip the apple in the brew
> Let the Sleeping Death seep through

He liked to chant the prophetic couplet over and over again.

Alan also invited Shaun Wylie over from Oxford as guest at the college feast. Shaun Wylie and David Champernowne had been fellow scholars at Winchester. Alan had mentioned the multiplying cipher idea to Champ, but he told Shaun about the summer course, saying that he had put his name forward to the authorities as a possible recruit. The Princeton treasure hunts therefore had a serious consequence. He also said that he had been studying probability theory, and would like to experiment with tossing coins, but would feel silly if someone came in, although in King's he need hardly have worried about appearing eccentric. They also played war games. David Champernowne had 'Denis Wheatley's exciting new war game – Invasion', for which they invented new rules to make it a better game. Maurice Pryce, then in his second year as a university lecturer, had a conversation with Alan about the new idea of uranium fission, and Maurice found an equation for the conditions required for a chain reaction to start.*

Presumably Alan had again applied for a lectureship, but if so he had again been disappointed. However, he had offered to the faculty a course for the spring term on Foundations of Mathematics. (Newman was not giving one this year.) This they accepted,[37] awarding the rather nominal £10 fee, as was the custom for mathematically respectable, but not officially commissioned Part III lectures. He was also asked to assess the claims of Friedrich Waismann, the philosopher from the Vienna Circle, exiled in Britain and expelled for misbehaviour from Wittgenstein's retinue, who wanted to lecture on Foundations of Arithmetic. So Alan had carved out a small niche for himself.

On 13 November 1938, Neville Chamberlain attended the Armistice Day service in the University Church, and a bishop gratifyingly referred to the 'courage, insight and perseverance of the Prime Minister in his interviews with Herr Hitler that saved the peace of Europe six weeks ago.' But some Cambridge opinion was more in touch with reality. In King's, Professor Clapham chaired a committee for the reception of Jewish refugees allowed in by the government after the November wave of violence in Germany. These were events with a particular meaning for Alan's friend Fred Clayton, who between 1935 and 1937 had spent time studying first in Vienna and then in Dresden, with experiences very different from the jolly hockey-sticks of Princeton.

They meant two very difficult and hurtful things. On the one hand, he was highly conscious of the implications of the Nazi regime. On the other, there were two boys, one the younger son of a Jewish widow living in the same house in Vienna, one at the school where he had taught in Dresden. The November 1938 events had put the Vienna family in great peril, and he received appeals for help from Frau S——. He tried to help her get her sons to England,

---

* David Champernowne also discussed the principle of the chain reaction with Alan after reading an article about it by J. B. S. Haldane in the *Daily Worker*.

and this was achieved just before Christmas by the Quakers' Relief Action. They found themselves in a refugee camp on the coast at Harwich, and wrote to Fred, who soon made a visit. In the dank, freezing, slave-market atmosphere some other young refugees rendered some German and English songs, and the passage from Schiller's *Don Karlos* about Elizabeth receiving those fleeing from the Netherlands. Fred was already very fond of Karl, an affection which fatherless Karl returned, and went away to help find someone to foster him.

On hearing this story, Alan's reaction was wholehearted. One wet Sunday in February 1939 he cycled with Fred to the camp at Harwich. He had conceived the idea of sponsoring a boy who wanted to go to school and university. Most of the boys were only too glad to be free of school for good. Of the very few exceptions, one was Robert Augenfeld – 'Bob' from the moment of his arrival in England – who had decided when he was ten that he wanted to be a chemist. He came from a Viennese family of considerable distinction and his father, who had been an aide-de-camp in the First World War, had instructed him to insist he should continue with his education. He had no means of support in England, and Alan agreed to sponsor him. It was impractical, for Alan's fellowship stipend would run to nothing of the kind, although he had probably saved some of Procter's money. His father wrote asking 'Is it wise, people will misunderstand?' which annoyed Alan, although David Champernowne thought his father had a good point.

But the immediate practical problems were soon solved. Rossall, a public school on the Lancashire coast, had offered to take in a number of refugee boys without fee. Fred's protégé Karl was going to take a place there, and this was arranged for Bob as well. Bob had to travel up north to be interviewed, where Rossall accepted him with the proviso that he should first improve his English at a preparatory school. On the way he had been looked after by the Friends in Manchester, and they in turn approached a rich, Methodist, mill-owning family to take him in. (Karl was fostered in just the same way.) This settled his future, and although Alan was ultimately responsible for him, and Bob always felt a great debt, he did not have to pay for more than some presents and school kit to help the boy get started. His recklessness had been justified, although it certainly helped that Bob was mentally as tough as Alan, having survived the loss of everything he knew, and being determined to fight for his own future education.

Meanwhile Alan was becoming more closely involved with the problems of GC and CS. At Christmas there was another training session at the headquarters in Broadway. Alan went down and stayed at a hotel in St James's Square with Patrick Wilkinson, the slightly senior classics don at King's, who had also been drawn in. Thereafter, every two or three weeks, he would make visits to help with the work. He found himself attached to Dillwyn Knox, the Senior Assistant, and to young Peter Twinn, a physics postgraduate from Oxford, who had joined as a new permanent Junior when a vacancy was advertised in February. Alan would be allowed to take back to King's some of the work they were doing on the Enigma. He said he 'sported his oak' when he studied it, as well he might. It was wise of Denniston not to wait until hostilities opened before letting his reserve force see the problems. But they were getting nowhere. A general knowledge of the Enigma machine was not enough upon which to base an attack.

It would have amazed Mrs Turing, if she had known that her younger son was being entrusted with state secrets. Alan had by this time developed a skilful technique for dealing with his family, and his mother in particular. They all thought of him as devoid of common sense, and he in turn would rise to the role of absent-minded professor. 'Brilliant but unsound', that was Alan to his mother, who undertook to keep him in touch with all those important matters of appearance and manners, such as buying a new suit every year (which he never wore), Christmas presents, aunts' birthdays, and getting his hair cut. She was particularly quick to note and comment on anything that smacked of lower-middle-class manners. Alan tolerated this at home, using his *persona* as the boy genius to advantage. He avoided confrontation – in the case of religious observance by singing Christmas hymns while he worked over Easter and *vice versa*, or by referring in conversation to 'Our Lord' with a perfectly straight face. He was not exactly telling lies, but successfully avoiding hurt by deception. This was not something he would do for anyone else, but for him, as for most people, the family was the last bastion of deceit.

There was, however, another side to the relationship: Mrs Turing did sense that he had done something incomprehensibly important, and was most impressed by the interest aroused in his work abroad. Once a letter came from Japan! For some reason she was particularly struck by the fact that Scholz was going to mention Alan's work in the 1939 revision of the German *Encyklopädie der mathematischen Wissenschaften*.[38] It needed such official-sounding reverberations for her to feel that anything had happened. Alan in turn was not above using his mother as a secretary; she sent out some of the reprints of *Computable Numbers* while he was in America. He also made an effort to explain mathematical logic and complex numbers to her – but with a complete lack of success.

It was in the spring of 1939 that he gave his first Cambridge lecture course. He started with fourteen Part III students, but 'no doubt the attendance will drop off as the term advances,' he wrote home. He must have kept at least one, for he had to set questions on his course for the examination in June. One of these asked for a proof of the result of *Computable Numbers*. It must have been very pleasing to be able to set as an examination problem in 1939, the question that Newman had posed as unanswered only four years before.

But at the same time, Alan joined Wittgenstein's class on Foundations of Mathematics. Although this had the same title as Alan's course, it was altogether different. The Turing course was one on the chess game of mathematical logic; extracting the neatest and tightest set of axioms from which to begin, making them flower according to the exact system of rules into the structures of mathematics, and discovering the technical limitations of that procedure. But Wittgenstein's course was on the *philosophy* of mathematics; what mathematics *really was*.

Wittgenstein's classes were unlike any others; for one thing, the members had to pledge themselves to attend every session. Alan broke the rules and received a verbal rap on the knuckles as a result: he missed the seventh class, very possibly because of his journey to the Clock House where, on 13 February, an entire side chapel of the parish church was dedicated to Christopher, on the ninth anniversary of his death. This particular course extended over thirty-one hours, twice a week for two terms. There were about fifteen in the class,

Alister Watson among them, and each had to go first for a private interview with Wittgenstein in his austere Trinity room. These interviews were renowned for their long and impressive silences, for Wittgenstein despised the making of polite conversation to a far more thoroughgoing degree than did Alan. At Princeton, Alan had told Venable Martin of how Wittgenstein was 'a *very* peculiar man', for after they had talked about some logic, Wittgenstein had said that he would have to go into a nearby room to think over what had been said.

Sharing a brusque, outdoor, spartan, tie-less appearance (though Alan remained faithful to his sports jacket, in contrast to the leather jacket worn by the philosopher), they were rather alike in this intensity and seriousness. Neither one could be defined by official positions (Wittgenstein, then fifty, had just been appointed Professor of Philosophy in succession to G. E. Moore), for they were unique individuals, creating their own mental worlds. They were both interested only in fundamental questions, although they went in different directions. But Wittgenstein was much the more dramatic figure. Born into the Austrian equivalent of the Carnegies, he had given away a family fortune, spent years in village school-teaching, and lived alone for a year in a Norwegian hut. And even if Alan was a son of Empire, the Turing household had precious little in common with the Palais Wittgenstein.

Wittgenstein[39] wanted to ask about the relationship of mathematics to 'words of ordinary everyday language'. What, for instance, did the chess-like 'proofs' of pure mathematics have to do with 'proof' as in 'The proof of Lewy's guilt is that he was at the scene of his crime with a pistol in his hand'? As Wittgenstein kept saying, the *connection* was never clear. *Principia Mathematica* only pushed the problem to another place: it still required people to agree on what it meant to have 'a proof'; it required people to agree what counting and recognising and symbols meant. When Hardy said that 317 was a prime *because it was so*, what did this mean? Did it only mean that people would always agree if they did their sums right? How did they know what were the 'right' rules? Wittgenstein's technique was to ask questions which brought words like *proof, infinite, number, rule*, into sentences about real life, and to show that they might make nonsense. As the only working mathematician in the class, Alan tended to be treated as responsible for everything that mathematicians ever said or did, and he rather nobly did his best to defend the abstract constructions of pure mathematics against Wittgenstein's attack.

In particular, there was an extended argument between them about the whole structure of mathematical logic. Wittgenstein wanted to argue that the business of creating a watertight, automatic logical system had nothing to do with what was ordinarily meant by truth. He fastened upon the feature of any completely logical system, that a single contradiction, and a self-contradiction in particular, would allow the proof of *any* proposition:

WITTGENSTEIN: . . . Think of the case of the Liar. It is very queer in a way that this should have puzzled anyone – much more extraordinary than you might think. . . . Because the thing works like this: if a man says 'I am lying' we say that it follows that he is not lying, from which it follows that he is lying and so on. Well, so what? You can go on like that until you are black in the face. Why not? It doesn't matter . . . it is just a useless language-game, and why should anybody be excited?

TURING: What puzzles one is that one usually uses a contradiction as a criterion for having done something wrong. But in this case one cannot find anything done wrong.

WITTGENSTEIN: Yes – and more: nothing has been done wrong. . . . where will the harm come?

TURING: The real harm will not come in unless there is an application, in which a bridge may fall down or something of that sort.

WITTGENSTEIN: . . . The question is: Why are people afraid of contradictions? It is easy to understand why they should be afraid of contradictions in orders, descriptions, etc., *outside* mathematics. The question is: Why should they be afraid of contradictions inside mathematics? Turing says, 'Because something may go wrong with the application.' But nothing need go wrong. And if something does go wrong – if the bridge breaks down – then your mistake was of the kind of using a wrong natural law. . . .

TURING: You cannot be confident about applying your calculus until you know that there is no hidden contradiction in it.

WITTGENSTEIN: There seems to me to be an enormous mistake there . . . . Suppose I convince Rhees of the paradox of the Liar, and he says, 'I lie, therefore I do not lie, therefore I lie and I do not lie, therefore we have a contradiction, therefore $2 \times 2 = 369$.' Well, we should not call this 'multiplication', that is all. . . .

TURING: Although you do not know that the bridge will fall if there are no contradictions, yet it is almost certain that if there are contradictions it will go wrong somewhere.

WITTGENSTEIN: But nothing has ever gone wrong that way yet . . . .

But Alan would not be convinced. For any pure mathematician, it would remain the beauty of the subject, that argue as one might about its meaning, the system stood serene, self-consistent, self-contained. Dear love of mathematics! Safe, secure world in which nothing could go wrong, no trouble arise, no bridges collapse! So different from the world of 1939.

He did not complete his research into the Skewes problem, which was left as an error-strewn manuscript[40] and never taken up by him again. But he continued to pursue the more central problem, that of examining the zeroes of the Riemann zeta-function. The theoretical part, that of finding and justifying a new method of calculating the zeta-function, was finished at the beginning of March, and was submitted for publication.[41] This left the computational part to be done. In this respect there had been a development. Malcolm MacPhail had written[42] in connection with the electric multiplier:

How is your University fixed with storage batteries and lathes and so on which you can use for your machine? It's such a pity that you will have to alter it. Hope you do not find it too much bunched together to be hard to work with. By the way if you are going to have time to work on it this fall and want some help don't hesitate to ask my brother. I told him about the machine and how it worked. He's very enthusiastic about your method of drawing wiring diagrams which rather surprised me. You know how conservative and old-fashioned engineers tend to be.

It so happened that his brother, Donald MacPhail, was a research student attached to King's, studying mechanical engineering. The multiplier made no progress, but Donald MacPhail did now join in the zeta-function machine project.

Alan was not the only person to be thinking about mechanical computation in 1939. There were a number of ideas and initiatives, reflecting the growth of new electrical industries. Several projects were on hand in the United States. One of these was the 'differential analyser' that the American engineer Vannevar Bush had designed at the Massachusetts Institute of Technology in 1930. This could set up physical analogues of certain differential equations – the class of problem of most interest in physics and engineering. A similar machine had then been built by the British physicist D. R. Hartree out of Meccano components at Manchester University. This in turn had been followed by the commissioning of another differential analyser at Cambridge, where in 1937 the mathematical faculty had sanctioned a new Mathematical Laboratory to house it. One of Alan's fellow 'B-stars' of 1934, the applied mathematician M. V. Wilkes, had been appointed as its junior member of staff.

Such a machine would have been useless for the zeta-function problem. Differential analysers could simulate only one special kind of mathematical system, and that only to a limited and very approximate extent. Similarly the Turing zeta-function machine would be entirely specific to the even more special problem on hand. It had no connection whatever with the Universal Turing Machine. It could hardly have been less universal. On 24 March he applied[43] to the Royal Society for a grant to cover the cost of constructing it, and on their questionnaire wrote,

> Apparatus would be of little permanent value. It could be added to for the purpose of carrying out similar calculations for a wider range of $t^\star$ and might be used for some other investigations connected with the zeta-function. I cannot, think of any applications that would not be connected with the zeta-function.

Hardy and Titchmarsh were quoted as referees for the application, which won the requested £40. The idea was that although the machine could not perform the required calculation exactly, it could locate the places where the zeta-function took a value near zero, which could then be tackled by a more exact hand computation. Alan reckoned it would reduce the amount of work by a factor of fifty. Perhaps as important, it would be a good deal more fun.

The Liverpool tide-predicting machine made use of a system of strings and pulleys to create an analogue of the mathematical

---

$\star$ That is, for looking at even more zeroes of the zeta-function.

problem of adding a series of waves. The length of the string, as it wrapped itself round the pulleys, would be measured to obtain the total sum required. They started with the same idea for the zeta-function summation, but then changed to a different design. In this, a system of meshing gear wheels would rotate to simulate the circular functions required. The addition was to be done not by measuring length, but weight. There would in fact be thirty wave-like terms to be added, each simulated by the rotation of one gear wheel. Thirty weights were to be attached to the corresponding wheels, at a distance from their centres, and then the moment of the weights would vary in a wave-like way as the wheels rotated. The summation would be performed by balancing the combined effect of the weights by a single counterweight.

The frequencies of the thirty waves required ran through the logarithms of the integers up to 30. To represent these irrational quantities by gear wheels they had to be approximated by fractions. Thus for instance the frequency determined by the logarithm of 3 was represented in the machine by gears giving a ratio* of $34 \times 31 / 57 \times 35$. This required four gear wheels, with 34, 31, 57 and 35 teeth respectively, to move each other so that one of them could act as the generator of the 'wave'. Some of the wheels could be used two or three times over, so that about 80, rather than 120 gear wheels were required in all. These wheels were ingeniously arranged in meshing groups, and mounted on a central axis in such a way that the turning of a large handle would set them in simultaneous motion. The construction of the machine demanded a great deal of highly accurate gear-cutting to make this possible.

Donald MacPhail drew up a blueprint of the design,[44] dated 17 July 1939. But Alan did not leave the engineering work to him. In fact his room, in the summer of 1939, was liable to be found with a sort of jigsaw puzzle of gear wheels across the floor. Kenneth Harrison, now a Fellow, was invited in for a drink and found it in this state. Alan tried and lamentably failed to explain what it was all for. It was certainly far from obvious that the motion of these wheels would say anything about the regularity with which the prime numbers thinned out, in their billions of billions out to infinity. Alan made a

---

* He used logarithms to base 8, so this ratio approximated $\log_8 3$.

start on doing the actual gear-cutting, humping the blanks along to the engineering department in a rucksack, and spurning an offer of help from a research student. Champ lent a hand in grinding some of the wheels, which were kept in a suitcase in Alan's room, much impressing Bob when he came down from his school at Hale in August.

Kenneth Harrison had been much amazed, for he well knew from conversations with Alan that a pure mathematician worked in a symbolic world and not with things. The machine seemed to be a contradiction. It was particularly remarkable in England, where there existed no tradition of high status academic engineering, as there was in France and Germany and (as with Vannevar Bush) in the United States. Such a foray into the practical world was liable to be met with patronising jokes within the academic world. For Alan Turing personally, the machine was a symptom of something that could not be answered by mathematics alone. He was working within the central problems of classical number theory, and making a contribution to it, but this was not enough. The Turing machine, and the ordinal logics, formalising the workings of the mind; Wittgenstein's enquiries; the electric multiplier and now this concatenation of gear wheels – they all spoke of making some connection between the abstract and the physical. It was not science, not 'applied mathematics', but a sort of applied logic, something that had no name.

By now he had edged a little further up the Cambridge structure, for in July the faculty asked that he should give his lectures on Foundations of Mathematics again in spring 1940, this time for the full fee of £50. In the normal course of events he could have expected fairly soon to be appointed to a university lectureship, and most likely to stay at Cambridge forever, as one of its creative workers in logic, number theory and other branches of pure mathematics. But this was not the direction in which his spirit moved.

Nor was it the direction of history. For there was to be no normal course of events. In March, the remains of Czecho-Slovakia slid into German control. On 31 March, the British government gave its guarantee to Poland, and committed itself to defending east European frontiers, while alienating the Soviet Union, already the world's second industrial power. It was a device to deter Germany, not to aid Poland, there being no way in which Britain could render assistance to its new ally.

It might have seemed that there was, equally, no way in which Poland could help the United Kingdom. Yet there was. In 1938, the Polish secret service had dropped a hint that they held information on the Enigma. Dillwyn Knox had gone to negotiate for it, but returned empty-handed, complaining that the Poles were stupid and knew nothing. The alliance with Britain and France had changed the position. On 24 July, British and French representatives attended a conference in Warsaw and this time came away with what they wanted.

A month later everything changed again, the Anglo-Polish alliance becoming even more impractical than before. As far as Intelligence was concerned, the year had gained little for Britain. There was now a new wireless interception station at St Albans, replacing the old arrangement whereby the Metropolitan Police did the work at Grove Park. But there was still[45] 'a desperate shortage of receivers for wireless interception', despite the pleas of GC and CS since 1932. The great exception was the fluke, handed over on a silver platter by the Poles.

The news-stands were announcing the Ribbentrop-Molotov pact as Alan set off from Cambridge for a week's sailing holiday, together with Fred Clayton and the refugee boys. They went to Bosham, his usual holiday haunt, where he had hired a boat. Several anxieties lay beneath the quiet surface. The boys, who had not been sailing before, thought the two men incompetent, and altered their watches so that they would turn back in good time. 'The lame leading the blind,' was what Bob thought of it. Fred, however, was more worried about the emotional undertones of the holiday. Alan teased him a good deal, mocking the idea that after a couple of terms at Rossall a boy would be innocent of sexual experience.*

One day they sailed across to Hayling Island, and went ashore to look at the RAF planes lined up on the airfield. The boys were not very impressed with what they saw. The sun went down and the tide went out, and the boat was stuck in the mud. They had to leave it and wade across to the island to get back by bus, their legs

---

* Alan was wrong.

encrusted with thick black mud. Karl said they looked like soldiers in long black boots.

It was at Bosham that King Cnut had shown his advisers that his powers did not extend to stemming the tide. The thin line of aircraft, charged with turning back the bombers, did not that August evening inspire much greater confidence. And who could have guessed that this shambling, graceless yachtsman, squelching bare-legged in the mud and grinning awkwardly at embarrassed Austrian boys, was to help Britannia rule the waves?

For now he would give no 1940 lectures. Nor indeed would he ever return to the safe world of pure mathematics. Donald MacPhail's design would never be realised, and the brass gear wheels would lie packed away in their case. For other, more powerful wheels were turning: not only Enigma wheels, but tank wheels. The bluff was called, so the deterrent had failed to work. Yet Hitler had miscalculated, for this time British duty would be done. Parliament kept the government to its word, and there would be war with honour.

It was much as *Back to Methuselah* had prophesied in 1920:

> And now we are waiting, with monster cannons trained on every city and seaport, and huge aeroplanes ready to spring into the air and drop bombs every one of which will obliterate a whole street, until one of you gentlemen rises in his helplessness to tell us, who are as helpless as himself, that we are at war again.

Yet they were not quite as helpless as they seemed. At eleven o'clock on 3 September, Alan was back at Cambridge, sitting in his room with Bob, when Chamberlain's voice came over the wireless. His friend Maurice Pryce would soon be giving serious thought to the practical physics of chain reactions. But Alan had committed himself to the other, logical, secret. It would do nothing for Poland. But it would connect him with the world, to a degree surpassing the wildest dream.

4 The Relay Race

Gliding o'er all, through all,
Through Nature, Time, and Space,
As a ship on the waters advancing,
The voyage of the soul – not life alone,
Death, many deaths I'll sing.

Alan reported next day, 4 September, to the Government Code and Cypher School, which had been evacuated in August to a Victorian country mansion, Bletchley Park. Bletchley itself was a small town of ordinary dullness, a brick-built Urban District in the brickfields of Buckinghamshire. But it lay at the geometric centre of intellectual England, where the main railway from London to the north bisected the branch line from Oxford to Cambridge. Just to the north-west of the railway junction, on a slight hill graced by an ancient church, and overlooking the clay pits of the valley, stood Bletchley Park.

The trains were busy with the evacuation of 17,000 London children into Buckinghamshire, swelling Bletchley's population by twenty-five per cent. 'The few who returned,' said an urban district councillor, 'no one on earth would have billeted, and they did the wisest thing eventually to return to the hovels from whence they came.' In these circumstances, the arrival of a few select gentlemen for the Government Code and Cypher School would have caused little stir, although it was said that when Professor Adcock first arrived at the station, a little boy shouted 'I'll read your secret writing, mister!' in a most disconcerting way. Later on there were complaints by local residents about the do-nothings at Bletchley Park, and it was said that the MP had to be prevented from asking a question in Parliament. They had the pick of accommodation: the few hostelries of mid-Buckinghamshire. Alan was billeted at the Crown Inn at Shenley Brook End, a tiny hamlet three miles north of Bletchley Park, whither he cycled each day. His landlady, Mrs Ramshaw, was one of those who lamented that an able-bodied young man was not doing his bit. Sometimes he helped out in the bar.

The early days at Bletchley resembled the arrangements of a displaced senior common room, obliged through domestic catastrophe to dine with another college, but nobly doing its best not to complain. In particular there was a strong King's flavour, with old-timers Knox, Adcock and Birch, and the younger Frank Lucas and Patrick Wilkinson as well as Alan. The shared background in Keynesian Cambridge was probably helpful for Alan. In particular it offered a link with Dillwyn Knox, a figure not generally noted for geniality or acessibility by Alan's contemporaries. GC and CS was by no means a vast establishment. On 3 September, Denniston wrote[1] to the Treasury:

> Dear Wilson,
>   For some days now we have been obliged to recruit from our emergency list men of the Professor type who the Treasury agreed to pay at the rate of £600 a year. I attach herewith a list of these gentlemen already called up together with the dates of their joining.

Alan was not quite the first, for according to Denniston's list there were nine of these 'men of the Professor type' at Bletchley by the time that he arrived with seven others the next day. Over the following year, about sixty more outsiders were brought in.

The 'emergency in-take quadrupled the cryptanalytic staff of the Service sections and nearly doubled the total cryptanalytic staff.' But only three of these first recruits came from the science side. Besides Alan, there were only W. G. Welchman and John Jeffries.[*] Gordon Welchman was the senior figure, lecturer in mathematics at Cambridge since 1929 and six years older than Alan. His field was algebraic geometry, a branch of mathematics then strongly represented at Cambridge, but one which never attracted Alan; their paths had not crossed before.

Welchman had not been involved with GC and CS before the outbreak of war as Alan had, and thus found himself, as a newcomer, relegated by Knox to the task of analysing the pattern of German

---

[*] J. R. F. Jeffries, Research Fellow in mathematics at Downing College, Cambridge, contracted tuberculosis in early 1941 and died.

call-signs, frequencies, and so forth. As it transpired, this was a job of immense significance, and his work rapidly brought such 'traffic analysis' to a quite new standard. It made possible the identification of the different Enigma key-systems, as was soon to prove so important, and opened GC and CS eyes to a much wider vision of what could be done. But no one could decipher the messages themselves. There was just a 'small group which, headed by civilians and working on behalf of all three Services, struggled with the Enigma.' This group consisted first of Knox, Jeffries, Peter Twinn and Alan. They established themselves in the stables building of the mansion, soon dubbed 'the Cottage', and developed the ideas that the Poles had supplied at the eleventh hour.

There was no glamour about ciphers. In 1939 the job of any cipher clerk, although not without skill, was dull and monotonous. But ciphering was the necessary consequence of radio* communication. Radio had to be used for aerial, naval, and mobile land warfare, and a radio message to one was a message to all. So messages had to be disguised, and not just this or that 'secret message', as with spies and smugglers, but the whole communication system. It meant mistakes, restrictions, and hours of laborious work on each message, but there was no choice.

The ciphers used in the 1930s did not depend on any great mathematical sophistication, but on the simple ideas of *adding on* and *substituting*. The 'adding on' idea was hardly new; Julius Caesar had concealed his communications from the Gauls by a process of adding on three to each letter, so that an A became D, a B became an E, and so on. More precisely, this kind of adding was what mathematicians called 'modular' addition, or addition without carrying, because it meant Y becoming B, Z becoming C, as though the letters were arranged around a circle.

Two thousand years later, the idea of modular addition by a *fixed* number would hardly be adequate, but there was nothing out-of-date about the general idea. One important type of cipher used the

---

* For consistency the word 'radio' is used henceforth, although at the time this was the American term, English people calling it 'wireless', or more formally 'wireless telegraph'. At the time of Roosevelt's re-election in 1936, Alan wrote from Princeton 'all the results are coming out over the wireless ("radio" they say in the native language). My method of getting the results is to go to bed and read them in the paper next morning.'

idea of 'modular addition', but instead of a fixed number, it would be a varying sequence of numbers, forming a *key*, that would be added to the message.

In practice, the words of the message would first be encoded into numerals by means of a standard code-book. The job of the cipher clerk would then be to take this 'plain-text', say

6728 5630 8923,	and to take the 'key' say
9620 6745 2397,	and form the cipher-text
5348 1375 0210	by modular addition.

For this to be of any use, however, the legitimate receiver had to know what the key was, so that it could be subtracted and the 'plain-text' retrieved. There had to be some *system*, by which the 'key' was agreed in advance between sender and receiver.

One way of doing this was by means of the *one-time* principle. This was one of the few sound ideas of 1930s cryptography, as well as the simplest. It required the key to be written out explicitly, twice over, and one copy given to the sender, one to the receiver of the transmission. The argument for the security of this system was that provided the key were constructed by some genuinely completely random process, such as shuffling cards or throwing dice, there could be nothing for the enemy cryptanalyst to go on. Given cipher text '5673', for instance, the analyst might guess that the plain-text was in fact '6743' and the key therefore '9930', or might guess that the plaintext was '8442' and the key '7231', but there would be no way of verifying such a guess, nor reason to prefer one guess to another. The argument depended upon the key being absolutely patternless, and spread evenly over the possible digits, for otherwise the analyst *would* have reason to prefer one guess to another. Indeed, discerning a pattern in the apparently patternless was essentially the work of the cryptanalyst, as of the scientist.

In the British system, one-time pads were produced, to be used up one at a time. Provided the key was random, no page was used twice, and the pads were never compromised, the system was foolproof. But it would involve the manufacture of a colossal quantity of key, equal in volume to the maximum that the particular communication link might require. Presumably this thankless task was undertaken by the ladies of the Construction Section of GC and

CS, which on the outbreak of war was evacuated not to Bletchley but to Mansfield College, Oxford. As for the system in use, that was no joy either. Malcolm Muggeridge, who was employed in the secret service, found it[2]

> a laborious business, and the kind of thing I have always been bad at. First, one had to subtract from the groups of numbers in the telegram corresponding groups from a so-called one-time pad; then to look up what the resultant groups signified in the code book. Any mistake in the subtraction, or, even worse, in the groups subtracted, threw the whole thing out. I toiled away at it, getting into terrible muddles and having to begin again . . .

Alternatively, a cipher system might be based upon the 'substitution' idea. In its simplest form, this was used for puzzle-page cryptograms, such as they had solved in Princeton treasure hunts. It meant that one letter of the alphabet would be substituted for another, according to some fixed rule like:

A B C D E F G H I J K L M N O P Q R S T U V W X Y Z
K S G J T D A Y O B X H E P W M I Q C V N R F Z U L

so that TURING would become VNQOPA. Such a simple, or 'mono-alphabetic' cipher could easily be solved by looking at letter frequencies, common words, and so forth, and in fact the only point of puzzle-page problems was that the setter would include peculiar words like XERXES to make this difficult. Such a system would be too simple-minded for military application. But there were systems in use in 1939 which were not much more advanced. One sophistication lay in the use of several alphabetic substitutions, used in rotation or according to some other simple scheme. The few manuals and text-books[3] of cryptology in existence devoted themselves mainly to such 'poly-alphabetic' ciphers.

Slightly more complex was the use of a system which substituted not for single letters, but for the 676 possible letter pairs. One British cipher system of this period was of this nature, combining the idea with the use of a code book. It was used by the Merchant Navy.[4]

The cipher clerk would first have to render the message into Merchant Navy Code, thus:

Text	Coded
Text	*Coded*
Expect to arrive at	V Q U W
14 C	F U D
40 U	Q G L

The next step required an even number of rows, so the clerk would have to add a nonsense word to make it up:

Balloon   Z J V Y

Then the encipherment would be done. The clerk would take the first vertical pair of letters, here VC, and look it up in a table of letter pairs. The table would specify some other pair, say XX. The clerk would continue to go through the message, substituting for each letter pair in this way.

There was little more to it, except that as with the 'adding on' kind of cipher, the process was futile unless the legitimate receiver knew which substitution table was being used. To preface the transmission with 'Table number 8', say, would allow the enemy analyst to collect and collate the transmissions enciphered with the same table, and make an attack. There had to be some element of disguise involved. So printed with the table there was another list of eight-letter sequences such as 'BMTVKZMD'. The clerk would choose one of these sequences and add it to the beginning of the message proper. The receiver, equipped with the same list, could then see which table was being used.

This simple rule illustrated a very general idea. In practical cryptography, as opposed to the setting of isolated puzzles, there would usually be some part of the message transmitted which did not convey the text itself, but which conveyed instructions on how to decipher it. Such elements of the transmission, which would be disguised and buried within it, were called *indicators*. Even a one-time pad system might employ indicators, to give a check on which page of the pad was being used. In fact, unless everything were spelt out in advance and in complete, rigid, detail, without any chance of ambiguity or error, there would have to be some form of indicator.

It must surely have struck Alan, who had been thinking at least since 1936 about 'the most general kind of code or cipher', that this mixing of instructions and data within a transmission was reminiscent of his 'universal machine', which would first decipher the 'description number' into an instruction, and then apply that instruction to the contents of its tape. Indeed *any* cipher system might be regarded as a complicated 'mechanical process' or Turing machine, involving not just the rules for adding or substituting, but rules for how to find, apply and communicate the method of encipherment itself. Good cryptography lay in the creation of an entire body of rules, not in this or that message. And serious cryptanalysis consisted of the work of recovering them; reconstructing the entire mechanical process which the cipher clerks performed, through an analysis of the entire mass of signals.

Maybe the Merchant Navy cipher system was not the last word in baffling complexity, but for operational use in ordinary ships, it was near the limit of practicality for hand methods. Anyone might dream up more secure systems, but if a ciphering operation became too long and complicated, it would only result in more delays and mistakes. However, if cipher machines were used, to take over part of the clerk's 'mechanical process', the situation could be very different.

In this respect Britain and Germany were running a symmetrical war, using very similar machines. Virtually every German official radio communication was enciphered on the Enigma machine. The British state relied, less totally, on the Typex. This machine was used throughout the army and in most of the RAF; the Foreign Office and the Admiralty retained their own hand systems depending on books. Enigma and Typex alike mechanised the basic operations of substitution and adding on, in such a way that a more complex system came within practical grasp. They did nothing that could not have been done by the looking up of tables in books, but enabled the work to be done more quickly and accurately.

There was no secret about the existence of such machines. Everyone knew of it – everyone, at least, who had a 1938 edition of Rouse Ball's *Mathematical Recreations and Essays* as a school prize. A revised chapter written by the U.S. Army cryptanalyst, Abraham Sinkov, wheeled out all the antiquated grilles, Playfair ciphers, and so forth, but also mentioned that

Quite recently there has been considerable research carried on in an attempt to invent cipher machines for the automatic encipherment and decipherment of messages. Most of them employ periodic polyalphabetic systems.

A 'periodic' polyalphabetic cipher would run through some sequence of alphabetic substitutions, and then repeat that sequence.

The most recent machines are electrical in operation, and in many cases the period is a tremendously large number. . . . These machine systems are much more rapid and much more accurate than hand methods. They can even be combined with printing and transmitting apparatus so that, in enciphering, a record of the cipher message is kept and the message transmitted; in deciphering, the secret message is received and translated, all automatically. So far as present cryptanalytic methods are concerned, the cipher systems derived from some of these machines are very close to practical unsolvability.

Nor was there anything secret about the basic Enigma machine. It had been exhibited in 1923, soon after its invention, at the congress of the International Postal Union. It was sold commercially and used by banks. In 1935 the British had created Typex by adding certain attachments to it, while a few years earlier the German cryptographic authorities had modified it in a different way to create the machine which, though bearing the original name of Enigma, was much more effective than the commercially available device.

This did not mean that the German Enigma with which Alan Turing now had to grapple, was something ahead of its time, or even the best that the technology of the late 1930s could have produced. The only feature of the Enigma that brought it into the twentieth century, or at least the late nineteenth, was that it was indeed 'electrical in operation'. It used electrical wirings to perform automatically a series of alphabetical substitutions, as shown in the first figure. But an Enigma would be used in a fixed state only for enciphering *one* letter, and then the outermost rotor would move round by one place, creating a new set of connections between input and output, as shown in the second figure.

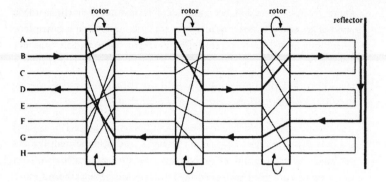

The Basic Enigma

For the sake of simplicity, the diagram has been drawn for an alphabet of only eight letters, although in fact the Enigma worked on the ordinary 26-letter alphabet. It shows the state of the machine at some particular moment in its use. The lines marked correspond to current-carrying wires. A simple switch system at the input has the effect that if a key (say the B key) is depressed, a current flows (as shown in the diagram by bold lines) and lights up a bulb in the output display panel (in this case, under the letter D). For the hypothetical 8-letter Enigma, the next state of the machine would be:

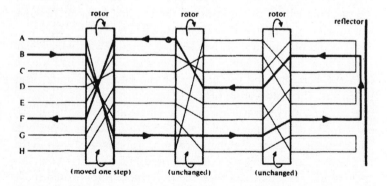

For the 26-letter Enigma, there would be $26 \times 26 \times 26 = 17576$ possible states of the rotors. They were geared essentially* like

---

* There were complications, but not affecting the account that follows.

any kind of adding machine or comptometer, so that the middle rotor would move on one step when the first had made a complete revolution, and the innermost move a step when the middle one had made a complete turn. The 'reflector', however, would not move, it being a fixed set of wires connecting the outputs of the innermost rotor.

So the Enigma was polyalphabetic, with a period of 17576. But this was not a 'tremendously large number'. Indeed, it would require a book only the size of a ready reckoner for all the alphabets to be written out. This mechanism was not, in itself, a leap into a new degree of sophistication. There was also a warning sounded by Rouse Ball in the old 1922 edition of his book that Alan had studied at school:

> The use of instruments giving a cipher, which is or can be varied constantly and automatically, has often been recommended . . . but the risk of some instrument . . . falling into unauthorized hands must be taken into account. Since equally good ciphers can be constructed without the use of mechanical devices I do not think their employment can be recommended.

For what was done by a machine might all the more easily be undone by a machine. The inner complexity of the Enigma, however clever it might look, would be worthless unless it created a cipher system which could not be broken even by an enemy in possession of a copy of the machine. It might only serve to give a false sense of security.

Nor was the technical construction of the Enigma as advanced as that suggested by Sinkov's description of contemporary developments. The cipher clerk using it still had the tedious and time-consuming task of noting which letter had been illuminated, and writing it down. There was no automatic printing or transmission, which had to be done laboriously in Morse code. Far from being a weapon of the modern Blitzkrieg, this plodding device drew on nothing more technologically advanced than the electric light bulb.

From the cryptanalyst's point of view, however, the physical labours of the cipher clerk, and the physical construction of the machine, were irrelevant. What mattered was the *logical* description – just like a Turing machine. Everything relevant to the Enigma was contained in its 'table', a list of its states and what it would do in each state. And from a logical point of view, the action of the

Enigma, in any given, fixed, state, enjoyed a very special property. It was a symmetrical property inherent in the 'reflecting' nature of the machine. For any Enigma, in any state, it would be true that if A were enciphered into E, then in that same state, E would be enciphered as A. The substitution alphabets resulting from an Enigma state would always be *swappings*.

For the hypothetical 8-letter machine in the state shown in the first diagram, the substitution would be:

    plain   A B C D E F G H
    cipher   E D G B A H C F

For the machine in the state shown in the second diagram, it would be:

    plain   A B C D E F G H
    cipher   E F G H A B C D

These could be written as swappings: (A E) (B D) (C G) (F H) in the first case and (A E) (B F) (C G) (D H) in the second.

There was a practical advantage to this Enigma property. It meant that the deciphering operation was identical with the enciphering operation. (In group-theory terms, the cipher was self-inverse). The receiver of the message had only to set up the machine in exactly the same way as the sender, and feed in the cipher-text, to recover the plain-text. There was no need to incorporate 'encipher' and 'decipher' modes into the Enigma machine, which made its operation that much less liable to mistakes and confusion. But it was associated with a grave weakness, in that the substitutions thus performed were always of this very special kind, with the particular feature that no letter could ever be enciphered into itself.

This was the basic structure of the Enigma. But there was much more to the machine actually in military use. For one thing, the three rotors were not fixed in place, but could be removed and replaced in any order. Until late 1938 there was a stock of just three rotors, which therefore allowed a total of six arrangements. In this way, the machine offered $6 \times 17576 = 105456$ different alphabetic substitutions.

Obviously, the rotors had to be marked in some way on the outside so that the different positions could be identified. However, here entered yet another element of complexity. Each rotor was encircled by a ring bearing the 26 letters, so that with the ring fixed in position, each letter would label a rotor position.* (In fact, the letter would show through a window at the top of the machine.) However, the position of the ring, relative to the wirings, would be changed each day. The wirings might be thought of as labelled by numbers from 1 to 26, and the position of the ring by the letters A to Z appearing in the window. So a ring-setting would determine where the ring was to sit on the rotor, with perhaps the letter G on position 1, H on position 2, and so forth.

It would be part of the task of the cipher clerk to make the ring-settings, and thereafter he would use the letters on the ring to define the rotor-settings. From the cryptanalyst's point of view, this meant that even if it were openly announced that rotor-setting 'K' was being used, this would not give away what at Bletchley they would call the *core-position* – the actual physical position of the wiring. This could only be deduced if the ring-setting were also known. However, the analyst might know the *relative* core-positions; thus settings K and M would necessarily correspond to core-positions two places apart. So it was known that if K were at position 9, then M would be at position 11.

The more important complicating feature, however, was the attachment of a *plugboard*. It was this that distinguished the military from the commercial Enigma, and made it something that had unnerved the British analysts. It had the effect of performing automatically an extra swapping of letters, both before entering the rotors, and after emerging from them. Technically, this was achieved by attaching wires, with plugs at each end, into a plugboard with 26 holes – rather like making connections on a telephone switchboard. It required ingenious electrical connections, and the use of double wires, to have the required effect. Until late 1938, it was usual in the German use of the machine to have only six or seven pairs of letters connected in this way.

Thus with the rotors and reflector of the basic machine in such a state as to effect the substitution

---

* The rather tiresome complication introduced by the ring-setting is, unfortunately, required to make sense of what the Poles achieved. It will play no part thereafter.

A B C D E F G H I J K L M N O P Q R S T U V W X Y Z

C O A I G Z E V D S W X U P B N Y T J R M H K L Q F

and with the plugboard wires set up to connect the pairs

(A P) (K O) (M Z) (I J) (C G) (W Y) (N Q),

the result of pressing the A key would be to send a current through the plugboard wire to P, then through the rotors and out again to N, then through the plugboard wire to Q.

Because of the symmetrical use of the plugboard both before and after the passage of the current through the rotors, it would preserve the self-inverse character of the basic Enigma, and the feature that no letter could be enciphered into itself. If A were enciphered into Q then, in the same state of the machine, Q would be enciphered into A.

So the plugboard left unaffected this useful – but dangerous – aspect of the basic Enigma. But it enormously increased the sheer number of states of the Enigma machine. There would be 1,305,093,289,500 ways* of connecting seven pairs of letters on the plugboard, for each of the 6×17576 rotor states.

Presumably the German authorities believed that these modifications to the commercial Enigma had brought it 'very close to practical unsolvability'. And yet, when Alan joined up at Bletchley on 4 September, he found it humming with the disclosures made by the Polish cryptanalysts.[5] It was all still fresh and new, for only on 16 August had the technical material reached London. And this revealed the methods by which, for seven years, the Poles had been deciphering Enigma messages.

The first thing, the *sine qua non*, was that the Poles had been able to discover the wirings of the three rotors. It was one thing to know that an Enigma machine was being used; quite another thing – but

---

* i.e. $\left(\dfrac{26!}{7! \, 12! \, 2^7} \right)$

absolutely essential – to know the specific wirings employed. To do this, in the peacetime conditions of 1932, was itself an impressive feat. It had been made possible by the French secret service who had obtained, through spying, a copy of the instructions for using the machine in September and October 1932. They had passed it to the Poles. They had also passed it to the British. The difference was that the Polish department employed three energetic mathematicians, who were able to use the papers to deduce the wirings.

Highly ingenious observations, good guessing, and the use of elementary group theory, produced the rotor wirings, and the structure of the reflector. The guessing, as it happened, was necessary to ascertain how the letters on the keyboard were connected to the enciphering mechanism. They might have been connected in some jumbled order to introduce another element of complexity into the machine. But they guessed and verified that the Engima design made no use of this potential freedom. The letters were joined to the rotor in alphabetical order. The result was that logically, if not physically, they had captured a copy of the machine, and could proceed to exploit that fact.

They were only able to make these observations, on account of the very particular way in which the machine was used. And they were only able to progress towards a regular decipherment of Enigma material by exploiting that method of use. They had not broken the *machine*; they had beaten the *system*.

The basic principle of using an Enigma machine was that its rotors and rings and plugboard would be set up in some particular way, and then the message would be encrypted, the rotors automatically stepping round as this was done. But for this to be of any use in a practical communication system, the receiver of the message also had to know the initial state of the machine. It was the fundamental problem of any cipher system. The machine was not enough; there had to be also an agreed, fixed, 'definite method' of using it. According to the actual method employed by the Germans, the initial state of the machine was partly decided at the time of use by the cipher clerk. Inevitably, therefore, it made use of indicators, and it was through the indicator system that the Poles enjoyed their success.

To be explicit, the order of the three rotors was laid down in written instructions, and so was the plugboard and ring-setting.

The task of the cipher clerk was to choose the remaining element, the initial settings for the three rotors. This amounted to choosing some triplet of letters, say 'WHJ'. The most naive indicator system would have been simply to transmit 'WHJ', and follow it with the enciphered message. However, it was made more complicated than that. The 'WHJ' was *itself* enciphered on the machine. For this purpose a so-called *ground-setting* was also laid down in the instructions for the day. This, like the rotor order, plugboard and ring-setting would be common to every operator in the network. Suppose the ground-setting were 'RTY'. Then the cipher clerk would set up his Enigma with the specified rotor order, plugboard and ring-setting. He would turn the rotors to read 'RTY'. Then his job was to encipher, twice over, his own choice of rotor setting. That is, he would encipher 'WHJWHJ', producing say 'ERIONM'. He would transmit 'ERIONM', then turn the rotors to 'WHJ', encipher the message, and transmit it. The strength was that every message, after the first six letters, was enciphered on a different setting. The weakness was that, for one day, all the operators in the network would be using exactly the same state of the machine for the first six letters of their messages. Worse, those six letters always represented the encipherment of a repeated triplet. It was this element of repetition that the Polish cryptanalysts were able to exploit.

    Their method was to collect each day, from their radio intercepts, a list of these initial six-letter sequences. They knew that in this list, there would be a pattern. For if in *one* message the first letter were A, and the fourth letter was R, then in any *other* message where the first letter was A, the fourth letter would again be R. With enough messages, they could build up a complete table, say:

First letter: A B C D E F G H I J K L M N O P Q R S T U V W X Y Z
Fourth letter: R G Z L Y Q M J D X A O W V H N F B P C K I T S E U

There would be two further tables, connecting the *second* and *fifth* letters, and the *third* and *sixth*. There were a number of ways of using this information to derive the state of the Enigma machine from which all those six-letter sequences had emanated. But particularly significant was a method which responded to the mechanical work of the cipher clerk, with a mechanised form of analysis.

They wrote these tables of letter connections in the form of *cycles*. The cycle notation was common currency in elementary group theory. To bring the specific letter connection above into 'cycle' form, one would start with the letter A, and note that A was connected with R. Then R was connected with B, B with G, G with M, M with W, W with T, T with C, C with Z, Z with U, U with K, and K with A – making a complete 'cycle': (ARBGMWTCZUK). The complete connection could be written out as the product of four cycles:

(A R B G M W T C Z U K) (D L O H J X S P N V I) (E Y) (F Q)

The reason for doing this was that the analysts had noticed that the *lengths* of these cycles (in this example, 11, 11, 2, 2) were independent of the plugboard. They would depend only upon the position of the rotors, the plugboard affecting *which* letters appeared in the cycles, but not *how many*. This observation showed that in a rather beautiful way the rotor positions left their fingerprints upon the cipher-text, when the traffic was considered as a whole. In fact they left just three fingerprints, the cycle-lengths of each of the three letter-connection tables.

It followed that if they possessed a complete file of the cycle-length fingerprints, three for each rotor position, then to determine which rotor position was being used for the first six letters all they would have to do was to search through the file. The snag was that there were 6×17576 possible rotor positions to catalogue. But they did it. To help in the work, the Polish mathematicians devised a small electrical machine which incorporated Enigma rotors, and which automatically produced the required sets of numbers. It took them a year to do the work, the results of which were entered on file cards. But then the detective work was effectively mechanised. It only took twenty minutes to look through the file to identify the combination of cycle-lengths which matched the cipher traffic of the day. This would reveal the positions of the rotors as they stood during the encipherment of the six indicator letters, and from that information, the rest could be worked out and the day's traffic read.

It was an elegant method, but the disadvantage was that it depended entirely on the specific indicator system. It did not last. The naval Enigma was the first to be lost, and[6]

. . . after the end of April 1937, when the Germans changed the naval indicators, they had been able to read the naval traffic only for the period from 30 April to 8 May 1937, and that only retrospectively. Moreover, this small success left them in no doubt that the new indicator system had given the Enigma machine a much higher degree of security . . .

And then on 15 September 1938, as Chamberlain flew to Munich, a greater disaster struck. All the other German systems were changed. It was only a minor modification, but it meant that overnight, all the catalogued cycle-lengths became completely worthless.

In the new system, the ground-setting was no longer fixed in advance. Instead, it would be chosen by the cipher clerk, who therefore had to communicate it to the receiver. This was done in the simplest possible way, by transmitting it as it stood. Thus the clerk might choose AGH, then set the rotors to read AGH. He would then choose another setting, say TUI. He would encipher TUITUI, to give say RYNFYP. He would then transmit AGHRYNFYP as indicator letters, followed by the actual message as enciphered with the rotors starting on the setting TUI.

This method depended for its security upon the fact that the ring-setting would be varying from day to day, for otherwise the first three letters (AG H, in the example) would give the whole thing away. The task of the analyst, correspondingly, was to determine this ring-setting which was common to all the traffic of the network. And amazingly, the Polish analysts were able to bounce back with a new kind of fingerprint, which had the effect of finding this ring-setting, or equivalently, of finding the core-position which corresponded to the openly announced rotor setting such as AGH in the example.

As with the older method, the fingerprint depended upon looking at the entire traffic, and in exploiting the element of repetition in the last six of the nine indicator letters. Without a common ground-setting, there was no fixed correspondence between first and fourth, second and fifth, third and sixth letters, to analyse. But one remnant of this idea, like the grin on the Cheshire cat, survived. Sometimes it would happen that the first and the fourth letters would actually be *the same* – or the second and the fifth, or the third and the sixth. This phenomenon was, for no apparent reason, called 'a female'. Thus, supposing that TUITUI were indeed enciphered as RYNFYP, that repeated Y would be 'a female'. This fact would then give a small piece of information about the state of the rotors as they were when the letters TUITUI were being enciphered. The method depended upon putting enough of these clues together to deduce that state.

More precisely a core-position would be said to have a 'female' letter, if that letter's encipherment happened to be the same three steps later. This was not a rare phenomenon, but would occur on average one time in twenty-five. Some core-positions (about forty per cent) would have the property of possessing at least one 'female' letter, and the rest would not. The property of having a female, or not, would be plugboard-independent, although the identity of the female letter would depend upon the plugboard.

The analysts could easily locate all the observed females in the traffic of the day. They would not know the core-positions which had given rise to them, but from the openly announced rotor settings like AGH in the example, they would know the *relative* core-positions. This information yielded a *pattern* of females. Because only about forty per cent of the core positions had females, there might only be one way in which this pattern could be matched with their known distribution. Here therefore was the new fingerprint – a pattern of 'females'.

But it was not possible to catalogue in advance all possible patterns, as they had been able to do with the cycle lengths. There had to be some other, more sophisticated means of making the match. The method they employed made use of perforated sheets. These were simply tables of all the core-positions, in which instead of printing 'has a female' or 'has no female', there would either be a hole punched, or not. In principle they could first have constructed one such huge table, and then each day could have made a template with the pattern of females observed in the traffic of that day. Passing the template over the table, they would eventually have found a position where the holes matched. But that would have been far too inefficient a method. Instead, they had a method of piling pieces of the table of core positions on top of each other, staggered in a manner corresponding to the observed relative positions of the females. A 'matching' of the pattern would then show up as a place where light passed through all the sheets. The advantage of this staggering system was that 676 possibilities could be examined simultaneously. It was still a long job, requiring 6×26 operations for a complete search. It also required the construction of perforated sheets listing the 6×17576 core-positions. Yet they achieved this within a few months.

This was not the only method they devised. The perforated sheet system required the location of about ten females in the traffic. A second system required only three, but it used not only the mere existence of a female, but the particular letter that appeared as female in the cipher-text. It was essential to the principle of the method that these particular letters had to be among those left unaffected by the plugboard. Since in 1938 the plugboard was being used with only six or seven pairs connected up, this was not too stringent a requirement.

The principle of this method was to match the observed pattern of three particular female letters, against the properties of the core positions. But it was impossible to catalogue in advance all the female letters of 6×17576 positions, and then perform a search, even by staggering sheets. There were far too many possible cases. Instead, they took a radical new step. They would search through the properties of the rotor positions afresh each time, doing no advance cataloguing. But this would not be a human search. It would be done by a machine. By November 1938 they had actually built such machines – six in fact, one for each possible rotor order. They produced a loud ticking sound, and were accordingly called the *Bombes*.

The Bombes exploited the electrical circuitry of the Enigma machine, by using an electrical method of recognising when a 'matching' had been found. The very fact that the Enigma *was* a machine, made mechanical cryptanalysis a possibility. The essential idea was that of wiring up six copies of the basic Enigma, in such a way that a circuit closed when the three particular 'females' occurred. The relative core-positions of these six Enigmas would be fixed by the known relative settings of the 'females' – just as in the 'staggering' of the sheets. Keeping these relative positions constant, the Enigmas would then be driven through every possible position. The complete search could be made in two hours, meaning that several positions could be tested in each second. It was a brute force method, for all it did was naively to try out all the possibilities one after the other. It had no algebraic subtlety. Yet it brought cryptanalysis into the twentieth century.

Unfortunately for the Polish analysts, the Germans were slightly further ahead in the twentieth century, and no sooner had this electromechanical device been brought to bear on the Enigma, than a new complication rendered it powerless again. In December 1938, the German systems all augmented their stock of three rotors, to a repertoire of five. Instead of *six* choices of ordering the rotors, there were now *sixty*. The Polish analysts did not lack for enterprise, and succeeded in working out the new wirings, thanks to cryptographic mistakes made by the self-styled German Security Service, the SD. But the arithmetic was simple. Instead of six Bombes, the method would now require sixty. Instead of six sets of perforated sheets, it would require sixty. They were lost. And this was the position when the British and French delegations went to Warsaw in July 1939. The Poles did not have the technical resources for further development.

This was the story that Alan heard. It was a story that had ground to a halt, but even so, the Poles were years ahead of the British, who were still where they had been in 1932. The British had not been able to work out the wirings, nor had they established the fact that the keyboard was connected to the first rotor in a simple order. Like the Polish cryptanalysts, they assumed that its design would include another jumbling operation at this point, and were amazed to learn that it did not. Nor had GC and CS ever thought of 'the possibility of high-speed machine testing against the Enigma before the July 1939 meeting'. At some level, there had been a failure of will. They had not really wanted to think, had not really wanted to know. Now that particular hurdle was passed, only to confront them with the problem the Poles had found insoluble:[7]

> When the various papers from the Poles – and in particular the wheel wirings – reached GC and CS it was soon possible to decrypt the old messages for which the Poles had broken the keys, but more recent messages remained unreadable.

They would have been unreadable for the same reason as the Poles found them unreadable. They did not have enough Bombes or perforated sheets for the five-rotor Enigma. There was another difficulty, which was that since 1 January 1939 the German systems had used ten pairs on the plugboard, which made the Polish Bombe method unlikely to work. Behind all this lay a deeper problem, which was that the chief Polish methods depended entirely upon the specific indicator system used. Something quite new was required. And this was where Alan played his first crucial part.

The British analysts did immediately embark upon making the sixty sets of perforated sheets that were required for the first 'female' method – now swollen to a colossal task of examining a million rotor settings. But they would have known that if ever the nine-letter indicator system were changed, even slightly, then the sheets would become useless. They needed something more general, something which did not depend upon specific indicator systems.

There did exist such methods, in the case of Enigma machines used without a plugboard. Such, for instance, was the case with the Italian Enigma, and also with that used by Franco's forces in the Spanish Civil War, whose system had been broken by GC and CS in April 1937. One particular attack was based on what Sinkov described as the 'Intuitive' or 'Probable Word' method. For this, the analyst had to guess a word appearing in the message, and its exact position. This was not impossible, given the stereotyped nature of many military communications, and would be helped by the feature of the Enigma that no letter could be enciphered into itself. Assuming the Enigma rotor wirings known, a correctly guessed word could quite easily lead the cryptanalyst to the identity of the first rotor and its starting position.

Such analysis would be done by hand methods. But in principle, a much more mechanical approach could be made, exploiting the fact that even a million possible rotor positions did not constitute a 'tremendously large number'. Like the Polish Bombe, a machine could simply work through the possible positions one by one until it found one that transformed the cipher-text into the known plain-text.

In the following diagrams, we forget the internal details of the basic Enigma, and think of it just as a box which transforms an input letter into an output letter. The state of the machine is represented by three numbers, giving the positions of the rotors. (We also leave aside the problem that the middle and inner rotors may move, and assume them static; in practice this would prove an important consideration in applying the method, but not affecting the principle.)

input

6.16.11

output

Now suppose it is known for certain that UILKNTN is the encipherment of the word GENERAL by an Enigma without plugboard. This means that there exists a rotor position, such that U is transformed to G, and such that the *next* position transforms I to E, the *next* one taking L to N, and so forth. There is no obstacle in principle to making a search through all the rotor positions until this particular position is found. The most efficient way would be to consider all seven letters *simultaneously*. This could be achieved by setting up a line of seven Enigmas, with their rotors in consecutive positions. One would feed in the letters UILKNTN, respectively, and look to see whether the letters GENERAL emerged. If not, all the Enigmas would move on by one step, and the process would be repeated. Eventually, the right rotor position would be found, the state of the machines then appearing as, say,

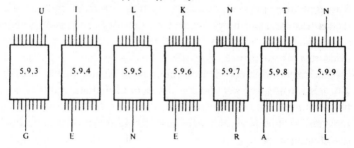

None of this would require technical advances beyond that of the Polish Bombe; it would be easy to attach wires such that a current would flow if and only if all seven letters agreed with GENERAL, and stop the machine.

Even in the early days, such an idea was not particularly far-fetched. Alan's contemporary, the Oxford physicist R. V. Jones who had become scientific advisor to the secret service, was billeted at Bletchley in late 1939. He conversed with Edward Travis, Denniston's deputy, about the current cryptanalytic problems. Travis posed the far more ambitious problem of the automatic recognition not of a fixed text, but of German language in general. Jones inventively proposed various solutions, one of which was[8]

> to mark or punch a paper or film in any one of 26 positions, corresponding to the letter coming out of the machine . . . and to run the resulting record past a battery of photocells, so that each could count the number of times of occurrence of the letter that it was looking for. After a given total count had been achieved, the frequency distribution between the letters could be compared with the one appropriate to the language, which could have been set up on some kind of template.

Travis introduced Jones to Alan, who 'liked the idea'. But with the Enigma, at least, the central method remained on entirely different lines. It kept to the idea of exploiting a known piece of plain-text. The difficulty, of course, was that the military Enigma *did* use a plugboard, which rendered such a naive searching process impossible, there being 150,738,274,937,250 possible ways* of connecting ten pairs of letters. In no way could a machine run through them all.

Yet no serious analyst would be daunted by this frightening number. Large numbers would not in themselves guarantee immunity from attack. Anyone who had solved a puzzle-page cryptogram had succeeded in eliminating all but one of the 403, 291,461,126,605,635,584,000,000 different alphabetic substitutions.[†] It could be done because such facts as that E was common, AO rare, and so forth, would each serve to eliminate vast numbers of possibilities at once.

It can be seen that the sheer number of plugboards is not in itself the problem, by considering an entirely hypothetical machine, in which a plugboard

---

* i.e., $\left(\dfrac{26!}{10!\, 6!\, 2^{10}} \right)$ Actually 11 pairs gives slightly more ways – not that it makes any difference; 12 or 13 pairs rather less

† *i.e.*, 26! This is also the number of possible wirings for each rotor of an Enigma.

swapping is applied only *before* encipherment by a basic Enigma. Suppose that for such a machine, it is known for certain that the cipher-text FHOPQBZ is the encipherment of GENERAL.

Once again, it would be possible to feed the letters FHOPQBZ into seven consecutive Enigmas, and examine the output. But this time, one could not expect the letters GENERAL to emerge, for an unknown plugboard swapping would have been applied to those letters. Nevertheless, something could still be done. Suppose that at some point in the process of running through all the rotor positions, the set-up happens to be:

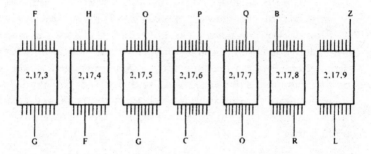

Then it may be asked whether the letters GFGCORL could, or could not, be obtained from GENERAL by the effect of a plugboard swapping. In this example, the answer is 'no', since no swapping could leave the first G unchanged, but swap the second G into an N; no swapping could turn the first E of GENERAL into an F and the second into a C. Furthermore, no swapping could change the R of GENERAL into an O, but then change the A into an R. Any one of these observations suffices to rule out that particular rotor position.

One way of thinking of this question is in terms of consistency. Having fed the cipher-text into the Enigmas, is the output consistent with the known plain-text, in that it differs only by virtue of swapping? From this point of view, the correspondences (OR) and (RA), or (EF) and (EC), are *contradictions*. A single contradiction is enough to eliminate all the billions of possible plugboards, on this hypothetical machine. Sheer numerical size, therefore, may be insignificant, compared with the logical properties of the cipher system.

The crucial discovery was that something like this could be done for the actual military Enigma, with its plugboard swapping taking place both before and after entry into the rotors of the basic Enigma. But the discovery was not immediate, nor was it the product of a single brain. It required a few months, and there were two figures primarily involved. For while Jeffries looked after the production of the new perforated sheets, the other two mathematical recruits, Alan and Gordon Welchman, were responsible for devising what became the British Bombe.[9]

It was Alan who had begun the attack, Welchman having been assigned to traffic analysis, and so it was he who first formulated the principle of mechanising a search for logical consistency based on a 'probable word'. The Poles had mechanised a simple form of recognition, limited to the special indicator system currently employed; a machine such as Alan envisaged would be considerably more ambitious, requiring circuitry for the simulation of 'implications' flowing from a plugboard hypothesis, and means for recognising not a simple matching, but the appearance of a contradiction.

The Turing Bombe

Suppose now that the letters LAKNQKR are known to be the encipherment of GENERAL on the full Enigma with plugboard. This time there is no point in trying out LAKNQKR on the basic Enigmas, and looking at what emerges, for some unknown plugboard swapping must be applied to LAKNQKR before it enters the Enigma rotors. Yet the quest is not hopeless. Consider just one letter, the A. There are only 26 possibilities for the effect of the plugboard on A, and so we can think about trying them out. We may start by taking the hypothesis (AA), *i.e.,* the supposition that the plugboard leaves the letter A unaffected.

What follows is an exploitation of the fact that there is only one plugboard, performing the same swapping operation on the letters going into the rotors as on the letters coming out. (If the Enigma had been fitted with two different plugboards, one swapping the ingoing letters and one the outcoming, then it would have been a very different story.) It also exploits the fact that this particular illustrative 'crib' contains a special feature – a closed loop. This is most easily seen by working out the deductions that can be made from (AA).

Looking at the second letter of the sequence, we feed A into the Enigma rotors, and obtain an output, say O. This means that the plugboard must contain the swapping (EO).

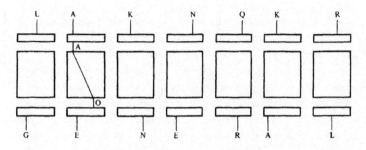

Now looking at the fourth letter, the assertion (EO) will have an implication for N, say (NQ); now the third letter will give an implication for K, say (KG).

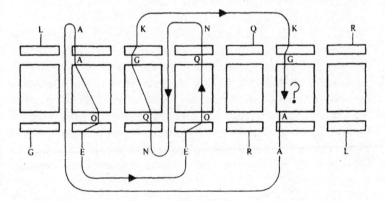

Finally we consider the sixth letter: here the loop closes and there will either be a consistency or a contradiction between (KG) and the original hypothesis (AA). If it is a contradiction, then the hypothesis must be wrong, and can be eliminated.

This method was far from ideal. For it depended absolutely upon finding closed loops in the 'crib', and not all cribs would exhibit this phenomenon.* But it was a method that would actually work, for the idea of completing a closed circuit was one that could be translated naturally into an electrical form. It showed that the sheer number of plugboards was not, in itself, an insuperable barrier.

It was a start, and it was Alan's first success. Like most wartime scientific work, it was not so much that it needed the most advanced knowledge of the day. It was rather that it required the same kind of skill that was used in advanced research, but applied to more elementary problems. The idea of automating processes was familiar enough to the twentieth century; it did not need the author of *Computable Numbers*. But his serious interest in mathematical machines, his fascination with the idea of working like a machine, was extraordinarily relevant.† Again, the 'contradictions' and 'consistency' conditions of the plugboard were concerned only with a decidedly finite problem, and not with anything like Gödel's theorem, which concerned the infinite variety of the theory of numbers. But the analogy with the formalist conception of mathematics, in which implications were to be followed through mechanically, was still a striking one.

Alan was able to embody this idea in the design of a new form of Bombe at the beginning of 1940. Practical construction began, and was pursued with a speed inconceivable in peacetime, under the direction of Harold 'Doc' Keen at the British Tabulating Machinery factory at Letchworth. Here they were used to building office calculators and sorters in which relays performed simple logical functions such as adding and recognising. It was now their task to make relays perform the switching job required for the Bombe to 'recognise' the positions in which consistency appeared, and stop.

---

* Questions about the chance of obtaining loops could be posed and answered in the language of probability theory and combinatorial mathematics – very much what Alan, or indeed any Cambridge mathematician, would be well-placed to tackle. One would be lucky to get a loop within one *word*, as shown in the artificial example; in practice the analyst would have to pick letters out of a longer 'crib' sequence. Furthermore one loop would not be enough – far too many rotor positions would satisfy the consistency condition by chance. Three loops would be required – a taller order.

† The Bombe, nevertheless, had nothing whatever to do with the Universal Turing Machine. It was more general than the Polish Bombe, which worked on a specific indicator system; but otherwise it could hardly have been less universal, being specific to the Enigma wirings and requiring an absolutely accurate 'crib'.

Here· again, Alan was the right person to see what was needed, for his unusual experience with the relay multiplier had given him insight into the problems of embodying logical manipulations in this kind of machinery. Perhaps no one, in 1940, was better placed to oversee such work than him.

Yet Alan had not seen that a dramatic improvement could be made to his design. Here it was Gordon Welchman who played the vital part. He had moved into the Enigma cryptanalytic group with a remarkable achievement to his credit: he had re-invented the perforated sheets method by himself, entirely ignorant of the fact that the Poles had worked it out and that Jeffries already had the production in hand. Then, on studying the Turing Bombe design, he saw that it had failed to exploit Enigma weakness to the full.

Returning to the illustration of the Turing Bombe, we notice that there are other implications which were not followed up, as indicated by the heavier lines:

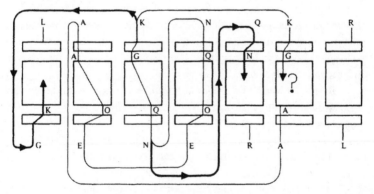

These differ in being implications that could not be foreseen in advance. They arise because (KG) also means (GK), and so at position 1 has an implication for L. Similarly (NQ) also means (QN) and hence at position 5 has an implication for R. This will give rise to a further implication for L at position 7. Clearly there is the possibility of a contradiction arising between these further implications, over and above the question of the loop closing at position 6. Indeed it is not now necessary for the texts to exhibit a loop for a contradiction to arise in this more general way. But this greater power of deduction does depend upon having some automatic means of going from (K G) to (G K), and similarly for every other implication reached, without knowing in advance when or where this may be required.

Welchman not only saw the possibility of improvement, but quickly solved the problem of how to incorporate the further implications into a mechanical process. It required only a simple piece of electrical circuitry – soon to be called the 'diagonal board'. The name referred to its array of 676 electrical terminals in a 26 × 26 square, each corresponding to an assertion such as (KG), and with wires attached diagonally in such a way that (KG) was permanently connected to (GK). The diagonal board could be attached to the Bombe in such a way that it had precisely the required effect. No switching operations were required for this; the following of implications could still be achieved by the virtually instantaneous flow of electricity into a connected circuit.

Welchman could hardly believe that he had solved the problem, but drew a rough wiring diagram and convinced himself that it would work. Hurrying to the Cottage, he showed it to Alan, who was also incredulous at first, but rapidly became equally excited about the possibilities it opened up. It was a spectacular improvement; they no longer needed to look for loops, and so could make do with fewer and shorter 'cribs'.

With the addition of a diagonal board, the Bombe would enjoy an almost uncanny elegance and power. Any assertion reached, say (BL), would feed back into every B and every L appearing in either plain-text or cipher-text. With this four-fold proliferation of implications at every stage it became possible to use the Bombe on any 'crib' of three or four words. The analyst could select a 'menu' of some ten or so letters from the 'crib' sequence – not necessarily including a 'loop', but still as rich as possible in letters bound to lead to implications for other letters. And this would provide a very severe consistency condition, sweeping away billions of false hypotheses with the speed of light.

The principle was startlingly like that of mathematical logic, in which one might seek to draw as many conclusions as possible from a set of interesting axioms. There was also a particularly logical subtlety about the process of deduction. As so far described, the operation would require the trying out of one plugboard hypothesis at a time. If (AA) were brought down by its own contradictions, then (AB) would be tried, and so on, until all 26 possibilities had been exhausted. Only then would the rotors move on by one step and the next position be examined in the same way. But Alan had seen that this was unnecessary.

If (AA) were inconsistent then it would generally lead to (AB), (AC), and so forth, in the process of following all the implications. This would mean that all these were also self-contradictory, and there was no need to try them out. An exception would occur when the rotor position was actually correct. In this case either the plugboard hypothesis would also be correct, and would lead to no contradictions; or would be incorrect and lead to every plugboard statement *except* the correct one. This meant that the Bombe was to stop when the electric current had reached either only *one*, or just *twenty-five*, of twenty-six terminals. It was this rather complicated condition that the relay switches had to test. This was not at all an obvious point, but seeing it made the process twenty-six times faster. Alan would comment on the likeness to mathematical logic, in which a single contradiction would imply any proposition. Wittgenstein, discussing this point, had said that contradictions never got anyone into trouble. But *these* contradictions would make something go very wrong for Germany, and lead to bridges falling down.

Thus the logical principle of the Bombe was the wonderfully simple one of following the proliferation of implications to the bitter end. But there was nothing simple about the construction of such a machine. To be of practical use, a Bombe would have to work through an average of half a million rotor positions in hours rather than days, which meant that the logical process would have to be applied to at least twenty positions every second. This was within the range of automatic telephone exchange equipment, which could perform switching operations in a thousandth of a second. But unlike the relays of telephone exchanges, the Bombe components would have to work continuously and in concert, for hours at a stretch, with the rotors moving in perfect synchrony. Without the solution of these engineering problems, in a time that would normally see no more than a rough blueprint prepared, all the logical ideas would have been idle dreams.

Even with Bombes designed and in the pipeline, the problem of the Enigma was far from solved. A Bombe would not take all the work out of the probable word method. One very important point was that when the consistency conditions were met, and a Bombe came to a halt, this did not necessarily mean that the correct rotor position had been arrived at. Such a 'stop', as it would be called, could arise by chance. (The calculation of how often such chance 'stops' were to be expected was a nice application of probability theory.) Each 'stop' would have to be tested out on an Enigma to see whether it turned the rest of the cipher-text into German, until the correct rotor position was discovered.

Nor was it a trivial matter to guess the probable word, nor to match it against the cipher-text. A good cipher clerk, indeed, could make these operations impossible. The right way to use the Enigma, like any ciphering machine, was to guard against the probable word attack by such obvious devices as prefacing the message with a variable amount of random nonsense, inserting X's in long words, using a 'burying procedure' for stereotyped or repetitious parts of the transmission, and generally making the system as unpredictable, as un-mechanical, as was possible without loss of comprehensibility to the legitimate receiver. If this were done thoroughly the accurate 'cribs' required for the Bombe could never be found. But perhaps it was too easy for the Enigma user to imagine that the clever machine would take care of itself, and there were often regularities for the British cryptanalysts to exploit.

Even when they had overcome subtleties of this kind, and learnt how to guess words with perfect accuracy, the story was far from over. Deciphering *one* message would not help to fight the war. The problem was to solve *every* message, of which there would be thousands every day in each network. The solution of this problem would depend upon the cipher system as a whole. In a system as simple as the pre-war use of repeated triplets of indicator letters, a single solved message could be used to undo the whole process, find the 'ground-setting', and thus reveal the entire traffic. But the enemy would not always be so obliging. Moreover, there was a sort of double bind, since the process of guessing a word with virtual certainty would only become possible when there was a good acquaintance with the traffic as a whole. The Bombe would be of little use unless a break into that traffic were first made in some other way.

With the Luftwaffe signals they did have another way – the method with perforated sheets which worked for the nine-letter indicator system. During the autumn of 1939, the construction of the sixty sets of sheets was completed, and a copy taken to the French cryptanalysts at Vignolles. This was an act of hope. No Enigma messages had been solved since December 1938, so they had no assurance that by the time the sheets were completed they would be of any use. But the hope was justified, for[10]

> 'At the end of the year', GC and CS records, 'our emissary returned with the great news that a key had been broken (October 28, Green)* on the . . . sheets he had taken with him. Immediately we got to work on a key (October 25, Green) . . . ; this, the first wartime Enigma key to come out in this country, was broken at the beginning of January 1940'. The GC and CS account continues: 'Had the Germans made a change in the machine at the New Year? While we waited . . . several other 1939 keys were broken. At last a favourable day arrived . . . The sheets were laid . . . and [the] Red of 6 January was out. Other keys soon followed . . . .'

Their luck had held, and the perforated sheets gave the first entry into the system. It was like the Princeton treasure hunt, in that each success would give the clue to the next goal, that of speedier and more comprehensive decryption. Special methods like those of the sheets – and there were many other algebraic, linguistic, and psychological tricks – could open up the way for something better. But it was never simple, for the rules were changing all the time, and they had to run as fast as they could to keep up. They were only just in time and, had they fallen behind by a few months, might never have caught up. In the spring of 1940 it was particularly precarious, as they held on with a mixture of ingenuity and intuition – or as the military were likely to call it, sheer bloody guesswork.

---

* Welchman, whose first work had been on the identification of the different key-systems, had devised a way of naming them by colours. 'Red' was the general purpose Luftwaffe system; 'Green' the Home Administration of the Wehrmacht. Despite these early breaks, 'Green' turned out to be an example of where an Enigma system was almost totally unbreakable because it was used properly.

Guessing and hoping were entirely characteristic of current British operations. The government had little more idea of how a war could be won, or even of what was happening, than did the public. It seemed that the British and German armed forces had, after all, agreed to have a battle again, but the British Tweedledee was decidedly reluctant to be the one to start, and the German Tweedledum expected it all to be over by six o'clock. Tweedledee's weaponry was still concealed behind Chamberlain's umbrella. The Red King was snoring on the square to the east, and no one (not even at Bletchley) knew what he was dreaming of. The blockade was supposed to bring an already 'stretched' Germany to crack from within, if only Britain could 'hold out'. Half desired, half dreaded by the British rulers, was the reappearance of the Monstrous Crow, currently flapping ambiguously on the other side of the Atlantic.

Appropriately, the Luftwaffe messages so laboriously and expensively deciphered at Bletchley in March 1940 turned out to consist mostly of nursery rhymes sent as practice transmissions. Even there, where at least they were busy with very exciting work, a sense of unreality and anticlimax was often felt. It was the same at Cambridge. Alan would return there occasionally for leave days, to work on mathematics and to see friends. At King's they had all dutifully trooped down into air-raid shelters (all except Pigou, who refused to compromise with the Luftwaffe), but the promised bombardment had not come. Three quarters of the children evacuated to Cambridge had returned home by mid-1940.

Yet the war had not been over by Christmas; Alan had exercised his option to suspend his fellowship for the duration of the war on 2 October 1939, and although his course on the Foundations of Mathematics had been advertised in the lecture list, it was not to be given. And there was Finland. Once during this period there was a party in Patrick Wilkinson's room, where Alan met a third-year undergraduate, Robin Gandy, who was reading mathematics and also rather conscientiously trying to defend the Communist party line. 'Hands off Finland' was double-talk such as Alan despised, but he liked Robin Gandy, and instead of marching off in disgust, led him on gently with Socratic questioning to arrive at a contradiction.

And one thing that was real, even in the phoney war, was the conflict at sea. Just as in the First World War, it was the strength and the weakness of the offshore European island that war with Britain was an attack upon the world trading economy. One third of all merchant shipping was British, and apart from coal and bricks there was scarcely a commodity in which the island was self-sufficient. Despite the blockade Germany could survive by pressing the resources and labour of Europe into its service. But British survival depended upon the ocean lanes. There lay the critical, paramount asymmetry.

It was the sea war that would become Alan's particular province. In early 1940 the different Enigma systems were divided among the chief cryptanalysts, who were allocated huts outside the Bletchley mansion. Welchman took over the army and air force Enigma systems, in Hut 6, joined by a number of new recruits. Dillwyn Knox took the Italian Enigma* and that used by the German SD, again with new recruits. These systems, which did not use plugboards, suited his psychological methods. And Alan was allocated Hut 8 in which to head the work on the naval Enigma signals. Other huts housed sections translating and interpreting the output; thus Hut 3 dealt with the army and air force material issuing from Hut 6, while the naval signals, if and when any were produced, would be interpreted by Hut 4, which was headed by Frank Birch.

Alan probably knew little of the context in which he was working, apart from the general air of urgency that issued from Hut 4. This was probably just as well, for the context was not exactly an encouraging one. He was working for the Admiralty, which only grudgingly had relinquished naval cryptanalysis to GC and CS. Traditionally, the Royal Navy expected autonomy. As possessor of the world's largest fleet, the Admiralty might be supposed capable of organising warfare for itself. Yet it had signally failed to learn the lesson that navies depended not only upon force but upon *information*, for guns and torpedoes were impotent unless in the right place at the right time. Like the giant Cyclops, 'Our Fighting Navy' was decidedly one-eyed. Naval Intelligence was embodied in an organisation that anyone of the new generation would find absurdly Victorian, if not criminally incompetent.

Only in the First World War had any Naval Intelligence Division been set up, and this had declined in peace-time into

---

* Knox's work had a very direct pay-off in the battle of Matapan, in March 1941.

Kafkaesque fantasy. In 1937, the NID was[11] '. . . neither interested in nor equipped to collect or disseminate information about the organization, dispositions, and movements of foreign fleets . . . the situation was very little better than it had been . . . in 1892. . . . Large old-fashioned ledgers were used in which to enter in longhand the last known whereabouts of Japanese, Italian and German warships . . . . These reports were often months old, and only once a quarter [were] the supposed dispositions of foreign navies . . . issued to the Fleet.' The Movements Section of the NID .(consisting of a single part-time officer) 'did not even subscribe to Lloyd's list, which would at least have provided a daily and highly accurate record of all the world's merchant ships. Reports of the movement of warships from the Secret Service were virtually non-existent . . . The possibility of locating ships at sea was . . . even more remote than that of obtaining up-to-date information about them when they were in port.' The admirals did not really want to know.

By September 1939 a new man, Norman Denning, had some-what improved the position. There was a card-index instead of ledgers, a direct telephone link to Lloyd's, and a Tracking Room on which a plot of merchant ships' positions could be up-dated. Links with GC and CS were not so successful. Indeed, the cryptanalytic organisation, captured by the Foreign Office after the First World War, tended to be treated as the enemy. Denning continued to plot its reconquest by the Admiralty until February 1941.

But the forward-looking Denning had also been able to establish the principle that a new sub-section of the NID, the Operational Intelligence Centre, which replaced the old Movements Section, should receive and coordinate information from all sources. This had been impossible in the First World War and represented a revolutionary advance. On the eve of war the OIC stood by with a staff of thirty-six. They had many problems to overcome, but the main problem of 1939 was that they had virtually no information to coordinate. Like Tweedledee, the Admiralty could hit out bravely at anything it could see, but it could see very little.

Occasionally, Coastal Command aircraft would catch sight of U-boats, and the RAF had been persuaded to inform the Admiralty when this happened. Aerial reconnaissance was limited to the hiring of a commercial pilot to take shots of the German coastline.

Information from agents in Europe was 'scanty. The best . . . came from a black market dealer in silk stockings with a contact in the German Naval Post Office, who from time to time was able to give the address of mail for certain ships, thus providing some fragmentary clues to their whereabouts.' When the *Rawalpindi* was sunk in November 1939, the Admiralty were unable to discover even the *class* of ship responsible. And as for signals, not only were the Enigma-enciphered messages indecipherable, but the German Navy[12]

> went over to war-time wireless procedure shortly before the attack on Poland, putting an end to the possibility of following its movements by correlating call-signs with the results of direction-finding, and it was to be months before work on the German naval signals system at GC and CS and in the [OIC] . . . made it possible to produce even tentative deductions on the basis of Traffic Analysis. The first step was to distinguish U-boat from other German naval communications, and it is some indication of the extent of the black-out that this elementary advance was not made until the end of 1939.

Until the outbreak of war, 'the naval sub-section of the German Section' of GC and CS 'which was started with one officer and a clerk as late as May 1938, still had no cryptanalysts.' It was just one aspect of the failure even to try to meet the German challenge. The prospects were better now, with the help of the Poles and with Bombes on the way, but the overall picture was dire:[13]

> Since the outbreak of war GC and CS had continued to give work on the G[erman] A[ir] F[orce] variant of the Enigma priority over its attack on the naval traffic. It had done so for two good reasons. The GAF traffic was more voluminous. Over and above that, those who worked on the naval Enigma had been held up first by the fact that the German Navy used the machine more carefully than the GAF, so that by the beginning of 1940 GC and CS had been able to break the settings for only 5 days of 1938, and then by the discovery that, sometime about the outbreak of war, the naval machine had undergone more radical modification than had the GAF's. During 1940 small amounts of captured naval cypher material had confirmed that, while both still used only three wheels at a time, the naval Enigma's wheels were selected from . . . 8 instead of from 5.

To make any headway, Alan would need something more to go on. 'From December 1939 GC and CS had left the Admiralty in no doubt about the urgency of this . . . requirement, but the Admiralty had had little opportunity to meet it.' But there was (at least at sea) a war on, which meant that the German authorities had to work on the assumption that the Enigma machine itself would be captured. Indeed this was so; the Polish revelations had only given GC and CS seven months' start in this respect, for 'Three Enigma wheels had been recovered from the crew of U-33 in February 1940.' But this 'had not provided a sufficient basis for a further advance.' Possession of the naval machine, while necessary, would have been far from sufficient. If the German navy used the machine 'more carefully', then its key systems were perhaps much less transparent than the foolish repeated triplets exploited by the Poles. And a few days of sparse peacetime traffic would provide a slender basis upon which to mount an attack.

Then the sea war spread to the land, with the German attack on Norway forestalling British designs. The Anglo-French response was not helped by the fact that the German cryptanalytic department, the *Beobachter Dienst*, was able to read a number of their messages, as indeed they had been doing all the time since 1938, and that these were used with great effect. At the end of the campaign, the Commander-in-Chief of the Home Fleet complained that 'it is most galling that the enemy should know just where our ships . . . always are, whereas we generally learn where his major forces are when they sink one or more of our ships.' In the final withdrawal from Narvik, the aircraft carrier *Glorious* was sunk by the *Scharnhorst* and the *Gneisenau* on 8 June. The OIC did not know the position of the *Glorious*, let alone of the German warships, and learnt of the sinking from an open victory broadcast.

Norway brought Bletchley Park into the war, inasmuch as the main Luftwaffe key, and an inter-service key, were read 'by hand methods' throughout the campaign, and revealed a good deal about German movements. Even on the naval side, Hut 4 was able to do work on traffic analysis which could have helped with the *Glorious*. But there were no arrangements for putting the information to use – not that the conditions in Norway itself were such that it could have been used to much advantage. One negative achievement was that the OIC was now obliged to take some notice of Bletchley. The desperate need for better naval intelligence was now clear. 'At the outset of the campaign the Admiralty's own ignorance was complete. When it intervened to give the orders which resulted in the first battle of Narvik on 9 April, it did so in the belief, based on Press reports, that one German ship had arrived there, whereas the German expedition to Narvik had reached the port in ten destroyers.'

It was in this context that an almost miraculous chance of helping Alan's work on naval Enigma was thrown away. For[14]

On 26 April the Navy captured the German patrol boat VP2623, while she was on passage from Germany to Narvik, and took from her a few papers . . . More might have been achieved if VP2623 had not been looted by her captors before she could be carefully searched; and the Admiralty at once issued instructions designed to prevent such disastrous carelessness in the future. As it was, except that they provided some information on the extent of the damage sustained by the German main units during the Norwegian campaign, the decrypts were of no operational use.

The capture of cipher hardware was to be expected and allowed for; the taking of flimsy, water-soluble pages* of current instructions for the use of the machine was a very different matter.

While the parliamentary upheaval meant that Winston Churchill now ceased to be responsible for this and other muddles, and instead took on the far greater muddle which was called the war effort, the 'instructions designed to prevent such disastrous carelessness in the future' were symbolic of an equally significant change. This time it would not do to have the military men behave as in some glorified Footer match, with the old masters giving earnest pep talks from the touchline, and back-room boys running dutiful errands. The lesson of the public schools was obsolete, for patriotism was not enough. They had to apply *intelligence*, at all levels, or they were lost. This was the conflict that would dominate the British war.

Meanwhile the work on the Luftwaffe Enigma, the Bletchley success of early 1940, was taking the first steps towards military usefulness. The steps were faltering, for on 1 May 1940, the 'German authorities introduced new indicators on all Enigma keys except the

---

* Literally, in future parlance, software.

Yellow'.* The perforated sheets had come only just in time to start off the treasure hunt; now they were almost useless. But there were 'German mistakes in the few days after the change of 1 May', very likely the classic one of sending out messages in both old and new systems. So by 22 May, Hut 6 was able to find out the new ('Red') system for the main Luftwaffe signals, and from that date to break it virtually every day thereafter. By that time, however, the German forces were at the Somme and closing on Dunkerque. The Bletchley success did not come soon enough to reveal German intentions during the first phase of the western attack. Indeed, 'For a fortnight ignorance of what the enemy was up to was so great that, in the records of the Cabinet and the Chiefs of Staff, discussions of the fighting continued to be headed "The Netherlands and Belgium".' By the time they found out, it was too late to make any difference.

But it was now that the first Bombes came into operation – probably a Turing prototype in May 1940 and then more with the diagonal board after August. Naturally, the machines 'greatly increased the speed and regularity with which GC and CS broke the daily-changing Enigma keys.' The Bombes were installed not at Bletchley, but at various out-stations such as Gayhurst Manor, in a remote corner of Buckinghamshire. They were tended by ladies of the Women's Royal Naval Service, the Wrens, who without knowing what they were doing and without asking the reason why, loaded the rotors and telephoned the analysts to say when a machine had come to a stop. They were impressive and rather beautiful machines, making a noise like that of a thousand knitting needles as the relay switches clicked their way through the proliferating implications.

Military officers attached to Bletchley were vividly impressed by the Bombes in operation. The secret service officer, F. W. Winterbotham, would refer to the Bombe as[15] 'like some Eastern Goddess who was destined to become the oracle of Bletchley', and at the OIC, too, they spoke of 'the oracle'. It was a usage that would have amused Alan, for he too had conceived of an oracle that would produce answers to unsolvable problems. What they began to discover, however, was that interpretation of the utterances was itself a major enterprise. If cipher machines had brought military communications into the

---

* The Yellow was the temporary inter-service system used in Norway.

Edwardian age, then the effect of the Bombes was to jolt military intelligence into the mass-production era.

In the First World War, Room 40 had worked hidden away in the Admiralty, its productions never coordinated with the results of sightings and interrogation. Only in the autumn of 1917, when the U-boat offensive was at its height, had the officer responsible for tracking them been allowed access to its information. The left hand did not know what the right hand was doing. And although the Navy's cryptanalytic work had been[16] 'incomparably better than that of any other power, or of the British War Office', Room 40 had operated in such a way that 'there were no records, no cross-indexing, and what was not of immediate operational interest went into the waste paper basket.'

It was not until the fall of France, when the war ceased to be a re-run of 1915, that the Room 40 flavour began to give way. The Poles, Welchman, and Alan Turing had put a Bombe under the British establishment, and nothing could ever be the same again.[17] 'Based on a machine and broken on a machine, the Enigma's cyphered mesages were mechanically converted direct into plain language; so that it yielded up its end-product in cornucopian abundance once the daily setting had been solved.' Here was the opportunity to capture not just messages, but the whole enemy communication *system*. Indeed, it was essential to do so, for the 'cornucopian abundance' required a second level of code-breaking in order to interpret it:[18]

> Apart from their sheer bulk, the texts teemed with obscurities – abbreviations for units and equipment, map and grid references, geographical and personal code names, pro-formas, Service jargon and other arcane references. One example is furnished by the fact that the Germans made frequent use of map references based on the CSGS 1:50,000 map of France. This series had been withdrawn from use in the British Army. Unable to obtain a copy of it, GC and CS was obliged to reconstruct it from the German references to it.

The Hut 3 filing systems, therefore, had to mirror the German system as a whole, in order to give meaning to the cipher traffic as a whole. Only when this had been done could the Enigma decrypts yield their real value – not so much in juicy secret messages, but in giving a general knowledge of the enemy mind. Without them,

Europe was an almost complete blank, out of which anything could emerge. With them, they had some insight into what was possible.

No precedent existed for a 'cornucopian abundance', and no means existed for making use of it. In 1940 the immediate problem was that of convincing anyone of the information thereby gained, without explaining its origin. At first it was passed off as the effects of spying. The result was that no military commander could take it seriously, since the offerings of the secret service were regarded as '80% inaccurate'. They had only just begun to consider more satisfactory arrangements for using the Luftwaffe decrypts in France, when events made the oracle irrelevant.

With customary British *sang froid*, those off duty at Bletchley Park played rounders in the afternoon as the news of the armistice came through.* Stern speeches and attitudes were of little help. In any case, it was radar that dominated the British eyes and ears in the coming months, although later in the year, gems of Enigma information provided clues to the Luftwaffe navigation beams. Radar, both in its technical development and in the communications network that it had forced upon the RAF, was three years ahead of Bletchley, an incoherent organisation whose time had not yet come.

Nor was there any pretence at heroism in Bletchley circles. It was not simply that Intelligence traditionally represented the most gentlemanly war work; not simply that the unspoken agreement was that of doing one's bit while making as little fuss as possible. For at the higher levels, the cryptanalytic work was intensely enjoyable. Being paid, or otherwise rewarded, seemed almost a curiosity. It was also something of a holiday even from professional mathematics, for the kind of work required was more on the line of ingenious application of elementary ideas, rather than pushing back the frontiers of scientific knowledge. It was like a solid diet of the hard puzzles in the *New Statesman*, with the difference that no one knew that solutions existed.

Nor was there anything heroic about the scheme that Alan devised in 1940 for protecting his savings against imminent disaster.

---

* They might well have worried that the German occupiers would now learn of the successful start to Enigma decryption from a French source. But no such disclosure or discovery was ever made.

David Champernowne had observed that silver was one thing that had gained in real value during the First World War. Both he and Alan invested accordingly in silver bullion, but while 'Champ' prudently kept his in the bank, Alan typically decided to go the whole hog with a Burying Procedure.

Apparently he imagined that by burying the silver ingots, he could recover them after an invasion had been repelled, or that at least he could evade a post-war capital levy. (In 1920, Churchill and the Labour party had both favoured such a policy.) It was an odd idea. It was logical enough to be pessimistic about the outcome of the war, but if there had been an invasion, then surely some transatlantic evacuation of code-breakers would have taken place (just as the Poles had escaped to France), in which case he would have been better off with his savings in a form more suitable for transport. He bought two bars, worth about £250, and wheeled them out in an old pram to some woods near Shenley. One was buried under the forest floor, the other under a bridge in the bed of a stream. He wrote out instructions for the recovery of the buried treasure and enciphered them. At one point the clues were stuck in an old benzedrine inhaler and left under another bridge. He liked talking about ingenious schemes for coping with the war, and once proposed to Peter Twinn an alternative plan of buying a suitcase full of razor blades. It suggested the curious, but not totally impossible, picture of Alan as a street-corner hawker in a reduced Britain.

In August or September 1940 Alan had a week's holiday, and spent it with Bob, making an effort to give the boy a treat. He had arranged for them to stay at what was for Alan a smart hotel, a renovated castle near Pandy, in Wales. It had been the usual hell for Bob in his first term, but, like Alan, he had survived the year, and at least had not encountered the usual public school anti-semitism. Alan asked a little about the past, and his family, but it was impossible to chat, for Bob had cast out the past as best he could, and Alan had no ability to heal such wounds. In fact, he probably never knew of the scenes which had taken place in Manchester as Bob pleaded unsuccessfully with the H—— family to rescue his mother from Vienna.

They went fishing and for long walks over the hills. After a day or two Alan made a gentle sexual approach, but Bob rebuffed it. Alan did not ask again. It did not affect the holiday. Bob perceived that the possibility had been at the back of Alan's mind from the beginning, but did not feel that Alan had taken advantage of him. He was simply not interested.

None of this was quite what Churchill had in mind when calling the British people to brace themselves to their duties, or speaking of the Empire that might last a thousand years. But duty and empires did not solve ciphers, and Churchill never bargained for an Alan Turing.

If the danger of direct invasion diminished, the attack on shipping was itself an invasion of the British metabolism. In the first year of war the sinkings by U-boats had not been the dominating problem. More significant were the disposition of the merchant fleets of newly occupied and neutral countries, the closure to trade of both the Channel and the Mediterranean, and the reduced capacity of British ports and inland transport to absorb whatever arrived.

From late 1940, however, the position began to clarify. The British-controlled merchant fleet had to supply an island separated by only twenty miles from an enemy continent, and to do so from bases thousands of miles across submarine-infested seas. Britain also had to continue the economic system on which vast populations around the globe depended and, to remain at war at all, had to attack Italy in a Middle East which was now as distant from Britain as was New Zealand. The lessons of 1917 had been applied, and a convoy system introduced since the outbreak of war, but the hard-pressed Navy could not escort convoys far into the Atlantic. And this time, Germany had achieved in a few weeks what four years of machine-guns and mustard-gas had sought to prevent. There were U-boat bases on the French Atlantic coast.

One factor alone weighed against the probability of German victory in the naval war. The U-boat force, so phenomenally success-ful in 1917, had not been built up in time for 1939. The bluffing over Danzig had meant that Hitler blundered into war while Dönitz commanded less than sixty submarines. Short-sighted strategy would keep the numbers at this level until late 1941. Although the sudden increase in U-boat successes after the collapse of France was alarming, it was not in itself a British disaster.

To remain capable of a belligerent policy Britain required imports of thirty million tons a year. A capital stock of thirteen million tons of shipping was available for this purpose. During the year after June 1940, U-boat sinkings were to deplete that stock by an average of 200,000 tons a month. This loss in capacity could just about be replaced. But anyone could see that a U-boat force just three times larger, enjoying a corresponding degree of success, would have a crippling effect both on the level of current supply, and on the total stock of shipping. Each U-boat was sinking more than twenty ships in its life-time, and there was no counter-strategy while the U-boat remained invisible. It was the logical, rather than the physical, power of the U-boat that was its strength. It was the German failure to follow up this tremendous advantage against its only remaining enemy that allowed a period of reprieve in which to counter this logical power with new weapons of information and communication. Radio direction finding and radar had already joined sonar in taking the Admiralty a short way beyond the resources of Nelson. The work of Hut 8 was still far behind.

Alan had begun his investigation of the naval Enigma messages on his own, but then was joined (for a time) by Peter Twinn and Kendrick. Clerical work was done by women who would be called 'big room girls'. Then in June 1940 there was a new mathematical recruit: Joan Clarke, who was one of several 'men of the Professor type' to be a woman. The principle of equal pay and rank being stoutly resisted by the civil service, she had to be promoted to the humble rank of 'linguist' that the pre-war establishment reserved for women, and there was talk by Travis of her being made a WRNS officer so that she could be better paid. But in the Hut itself, a more progressive Cambridge atmosphere prevailed. She had just been reading for Part III, and was recruited for Bletchley by Gordon Welchman, who had supervised her for projective geometry in Part II. Her brother being a Fellow of King's, she had once met Alan in Cambridge.

So in the summer of 1940, Alan Turing found himself in the position of telling other people what to do, for the first time since school. It was like school inasmuch as the WRNS and the 'big room girls' played the role of 'fags', and because it meant meeting, or avoiding, members of the armed services. Alan's methods for dealing with the clerical assistance, and other administrative problems, which gradually grew in scale, were like those of a shy school 'brain' who had been made a prefect by virtue of winning a scholarship. On the other hand, one notable difference from school was that it brought him for the first time into contact with women.

The rest of 1940 saw little progress with naval Enigma. The April U-boat capture, although largely wasted, had given them something to work on[19] – and it was for this reason that Joan Clarke had been directed to Hut 8.

> It had enabled GC and CS to read during May 1940 the naval Enigma traffic for six days of the previous month, and thus to add considerably to its knowledge of the German Navy's [radio] and cypher organisation. GC and CS was able to confirm that, though the Germans resorted to fairly simple hand codes and cyphers for such things as light-ships, dockyards and merchant shipping, their naval units, down to the smallest, relied entirely on the Enigma machine. More important still, it established that they used only two Enigma keys – the Home and the Foreign – and that U-boats and surface units shared the same keys, transferring to the Foreign key only for operations in distant waters.

But only a further five days' traffic, for days in April and May, were broken in the rest of 1940, and 'the advance of knowledge had also confirmed GC and CS's worst fears about the difficulty of breaking even the Home key, in which 95 per cent of German naval traffic was encyphered.' Alan's work showed that they could not hope for progress without further captures. But while they waited, he was not idle. He developed the mathematical theory that would be required to exploit them. There was far, far more to it than the building of Bombes.

Looking at the cipher traffic, an experienced hand might say that such and such a thing 'seemed likely', but now that mass production was the objective, it was necessary to make vague, intuitive judgments into something explicit and mechanical. Much of the mental apparatus required for this had already been constructed in the eighteenth century, although it was new to GC and CS. The English mathematician Thomas Bayes had seen how to formalise the concept of 'inverse probability' – this being the technical term for the likely cause of an effect, rather than the probable effect of a cause.

The basic idea was nothing but the common sense calculation of 'likeliness' of a cause, such as people would use all the time without thinking. The classical presentation of it was like this: suppose there to be two identical boxes, one containing two white balls and one black ball, the other containing one white ball and two black balls. Someone then has to guess which box is which, and is allowed to make an experiment, that of taking just one ball out of either box (without, of course, looking inside). If it turns out to be white, the common-sense judgment would be that it is *twice as likely* that it has come from the box containing two white balls, as from the other. Bayes's theory gave an exact account of this idea.

One feature of such a theory was that it referred not to the happening of events, but to the changes in a state of mind. In fact, it was very important to bear in mind that experiments could only produce relative changes in 'likeliness', and never an absolute value. The conclusion drawn would always depend upon the *a priori* likeliness which the experimenter had had in mind at the beginning.

To give a concrete feel to the theory, Alan liked to think in terms of a perfectly rational person obliged to make bets upon hypotheses. He liked the idea of betting, and put the theory into the form of odds. So in the example, the effect of the experiment would be to double the odds, one way or the other. If further experiments were allowed, the odds would eventually increase to very large numbers although, in principle, certainty would never be attained. Alternatively, the process could be thought of as one of accumulating more and more evidence. From this point of view, it would be more natural to think of *adding* something each time an experiment was made, rather than of *multiplying* the current odds. This could be achieved by using logarithms. The American philosopher C. S. Peirce had described a related idea in 1878, giving it the name 'weight of evidence'. The principle was that a scientific experiment would give a numerical 'weight of evidence' to be added to, or subtracted from, the likeliness of a hypothesis. In the example, the discovery of a white ball would add a weight of log 2 to the hypothesis that the box it came from was the one with two white balls. It was not a new idea, but[20]

Turing was the first to recognise one value of naming the units in terms of which weight of evidence is measured. When the base of logarithms was e he called the unit a natural ban, and simply a ban when the base was 10. . . . Turing introduced the name deciban in the self-explanatory sense of one-tenth of a ban, by analogy with the decibel. The reason for the name ban was that tens of thousands of sheets were printed in the town of Banbury on which weights of evidence were entered in decibans for carrying out an important process called Banburismus.

So a 'ban' of evidence was something that would make a hypothesis ten times as likely as it had been before. Rather like a decibel, a deciban would be 'about the smallest change in weight of evidence that is directly perceptible to human intuition'. He had mechanised guessing, and was ready to put it on machines which would add up decibans to arrive at a rational decision.

Alan developed the theory in several ways. The crucially important application lay in a new procedure for making experiments, later to be called 'sequential analysis'. His idea was to set a target for the weight of evidence required one way or the other, and to continue making observations until that target was attained. This would be a far more efficient method than deciding in advance how many experiments to make.

But he also introduced the principle of judging the value of an experiment by the amount of weight of evidence that it would, on average, produce; and he even went on to consider the 'variance' of the weight of evidence produced by an experiment, a measure of how erratic it was likely to be. In bringing these ideas together, he brought the art of guessing, as employed in cryptanalysis, into the 1940s. Typically, he had worked it all out for himself, either not knowing of earlier developments (as in the case of 'weight of evidence' defined by Peirce) or preferring his own theory to the statistical methods pioneered by R. A. Fisher in the 1930s.

Now, therefore, when they thought that a crib was 'probably' right, or that one message had 'probably' been transmitted twice, or that the same setting had 'probably' been used twice, or that one particular rotor was 'probably' the outermost one, there lay the possibility of adding up the weight of evidence from faint clues in a systematic, rational way, and of designing their procedures so as to make the most of what they had. To save an hour thereby was to gain an hour in which a U-boat gained six miles upon a convoy.

Just after the end of 1940, the theory began to turn into practice. In about December Alan wrote to Shaun Wylie, then teaching at Wellington College, and invited him to join. He arrived in about February 1941. Later Hugh Alexander, the British chess champion, was transferred to Hut 8 from elsewhere in Bletchley. Alexander was also a Kingsman, one who had graduated in 1931 and attributed his failure to become a mathematical Fellow to his having played too much chess. Instead, he had taught at Winchester and then became Director of Research for the John Lewis Partnership, the leading group of department stores. At the outbreak of war he and the other British chess masters had been caught in Argentina at the 1939 Chess Olympiad. It was a matter of some satisfaction that the British team had managed to return while the Germans could not. The next increase in strength in Hut 8 would come when the young mathematician I. J. Good was detached from his Cambridge research work with Hardy to join in May 1941. But by that time everything would have changed.[21]

> When I arrived at Bletchley I was met at the station by Hugh Alexander the British chess champion. On the walk to the office Hugh revealed to me a number of secrets about Enigma. Of course we were not really supposed to talk about such things outside the precincts of the office. I shall never forget that sensational conversation.

For Alan Turing's ideas had been embodied in a working system. The Bombe was at the centre, but there was punched-card machinery, and the 'big room girls' working in a production line, to make the guessing-game as effective and rapid as the improvised conditions allowed. They were beginning to *do* something for the war.

The first planned capture took place on 23 February 1941, during a raid on the Lofoten Islands on the Norwegian coast. It meant that someone died for the Enigma instructions that Alan needed:[22] 'the German armed trawler *Krebs* was disabled, her commanding officer killed before he could complete the destruction of his secret papers, and the ship abandoned by the survivors.' Enough was taken for it to be possible for Hut 8 to read the whole of the naval traffic for February 1941 at various dates from 10 March onwards.

The time lag, for those interpreting the messages, was profoundly frustrating. The naval messages, unlike the bulk of those emanating from the other services, had to carry pieces of first-rate information. One of the first to be decrypted read:[23]

> Naval Attache Washington reports convoy rendezvous 25th February 200 sea miles east of Sable Island. 13 cargo boats, 4 tankers 100,000 tons. Cargo: aeroplane parts, machine parts, motor lorries, munitions, chemicals. Probably the number of the convoy is HX 114.

But on 12 March, when its decipherment was achieved, it was three weeks too late to do anything but wonder how the Naval Attaché had found out so much. Two days later they read a message from Dönitz:

> From: Admiral commanding U boats
> Escort for U69 and U107 will be at point 2 on March 1st at 0800 hours.

which two weeks earlier would have been exactly what the Tracking Room wanted – if only it could be known where Point 2 was. It was necessary to amass the traffic in order to attack such problems of interpretation. Thus:

> English ship *Anchises* lies in AM 4538, damaged from the air.

would, provided it were not slung into the waste-paper basket as in Room 40 days, reveal the location of grid reference AM 4538.

No break was made into the March 1941 traffic. But then came a triumph for Hut 8: the decryption of the April traffic without the benefit of any further captures. Both April and May messages were read 'by cryptanalytic methods'. At last they were beginning to beat the system. Hut 4 was now able to look right into the eye of the enemy, with messages such as:

> From: NOIC Stavanger   [24 April; deciphered 18 May]
> To: Admiral West Coast
> Enemy Report   Offizier G and W
> Supreme Naval Command (First operations division) wires no. 8231/41 re captured Swedish fishing vessels:

1) Operations division believes that it was the task of the Swedish fishing vessels to obtain information about mines in the interests of Britain.

2) Make certain that neither Sweden nor the enemy hears about their capture. The impression should for the time being be allowed to arise that the vessels were sunk by mines.

3) Crews are to be kept under arrest until further notice. You are to forward a detailed report of their interrogation.

Some were even more ironic:

[22 April; deciphered 19 May]

From: C in C Navy

    The U boat campaign makes it necessary to restrict severely the reading of signals by unauthorised persons. Once again I forbid all authorities who have not express orders from the operations division or the Admiral commanding U boats to tune in on the operational U boat wave. I shall in future consider all transgressions of this order as a criminal act endangering national security.

Weeks-old material still had value in building up a knowledge of the system, but of course it was of desperate importance that the time-lag be reduced. By the end of May 1941 they were able to bring down the time to as little as a day. One message that was deciphered within a week read:

[19 May; deciphered 25 May]

From: Admiral commanding U boats
To: U 94 and U 556

    The Fuehrer has decorated both captains with the Ritterkreuz to the Iron Cross. I wish to convey to you, on the occasion of this recognition of the services and successes of the boats and their crews, my sincere congratulations. Good luck and success in future too. Defeat England.

That defeat would now be more difficult than they imagined. For even old messages imperilled German plans. When the *Bismarck* sailed from Kiel on 19 May, the delay of three days or more in decipherment rendered Hut 8 powerless to reveal the secrets of her course. But on the morning of 21 May, some April messages emerged to put it beyond doubt she was making for the trade routes. Thereafter it was left to the Admiralty to derive intelligence in its more traditional way, which included plotting a radio direction on the wrong kind of map projection, though its eventual good guessing was confirmed by a Luftwaffe Enigma message on 25 May. The sequence of events was extremely complicated, and naval Enigma played only a minor role in it. But had the *Bismarck* sailed just a week later, the story would have been very different. New developments in Hut 8 were transforming the picture.

This was because the older material was discovered to have powerful implications:[24]

> After studying the decyphered traffic of February and April, GC and CS was able to show conclusively that the Germans were keeping weather-ships on station in two areas, one north of Iceland and the other in mid-Atlantic, and that, though their routine reports were transmitted in weather cypher and were different in outward appearance from Enigma signals, the ships carried the naval Enigma.

This clever analysis of essentially dull material represented a victory for the new men and the new methods, in which Alan had a personal share. The Admiralty would never have had the time or wit to make the amazing discovery that these vulnerable little weather-ships were supplied with the keys to the Reich. But they were now prepared to act on the prompting of a civilian department, and plotted a series of captures.

The *München* was found and taken on 7 May 1941 and it was with the settings thus obtained that they became able to read the June traffic 'practically currently'. At last, they had a command of the day-to-day tactics. The July settings were captured from another of the weather-reporting trawlers, the *Lauenburg*, on 28 June. Meanwhile on 9 May, an accidental, but brilliantly conducted operation had taken place. A convoy escort detected and disabled the U-110 which had attacked the convoy. In a split-second manoeuvre on the high seas, they boarded the U-boat and took intact its cipher material. The lessons of 1940 had been learnt. The material filled some outstanding gaps, for it included 'the code-book used by the U-boats when making short-signal sighting reports', and 'the special settings used in the Navy for "officer-only" signals'. These *Offizierte* signals were *doubly enciphered* for extra security within the U-boat itself. From the Hut 8 point of view, these were signals which, even after the day's settings had been found and the decryption process applied, remained gibberish while the other messages became German. It required a second stage of attack to recover these, the innermost secrets of the U-boat operations. Now they had what they needed to do it.

The growing body of knowledge was rapidly put to use by the Admiralty. As June 1941 opened, and the naval traffic was read currently, it was able to make an almost clean sweep of the supply ships sent into the Atlantic in advance of the *Bismarck*, disposing of seven out of the eight. This bulldog action, however, provoked a disturbing question. In Hut 8, as they read messages about U-boat rendezvous points and so forth, they assumed quite naively that with the aid of this wonderful information, the U-boats could readily be despatched. In June 1941 this simple view was presumably also taken by the Admiralty, for only afterwards did anyone voice concern that the succession of sinkings, following the loss of the *Bismarck*, might alert the German authorities to the possibility of cipher compromise.

In fact, the operation *had* betrayed Alan's success, for the German authorities decided that the positions of the supply vessels had somehow been disclosed, and set up an investigation. Their experts, however, ruled out the possibility that the Enigma cipher had been broken. Instead, they pinned the blame upon the British secret service, which enjoyed a high reputation in German ruling circles. It was a diagnosis remote from the truth. They had assigned an *a priori* probability of zero to Enigma decryption, and no weight of evidence sufficed to increase it.

It was a blunder, but one easy to make when the implications were so shattering. At Bletchley, where it was explained to Hut 8 that decrypts could not in future be exploited so easily, there was nothing to do but to cross their fingers. The Bombe method, which was central to the system, hung upon a single thread. If, to be on the safe side, the Germans had gone over to a double encipherment of *every* message, then there would have been no more cribs, and all would have been lost. At any time, the mere suspicion that something had gone wrong might stimulate such a change. They walked on a knife-edge.

From mid-June 1941, the Admiralty caught on to the idea

that messages which contained information derived exclusively from Enigma decryption (normally until then, from Luftwaffe Enigma) should go out as ULTRA SECRET on special one-time pads. The other services also began to adapt, setting up Special Liaison Units, attached to headquarters in the field and around the Empire, charged with the reception and control of Bletchley information.

But there was still far to go in the integration of brain and brawn. The Admiralty was the *most* flexible in this regard, but they laboured under the difficulty that while a year before there had been too little information, in mid-1941 they were swamped by its abundance. The OIC could not cope with the new era, in which a vast German system had to be mirrored by a British one.

It had been a revolutionary innovation to place a civilian, a barrister called Rodger Winn, in charge of the OIC Tracking Room at the end of 1940, replacing an ancient naval Paymaster. It was through the mind of Winn that the output of Hut 8 had to be translated into action. Fortunately it was an imaginative mind, one which suggested forecasting where the U-boats were going to be, in time for the convoys to dodge them. Despite great initial resistance, towards the spring of 1941 this entirely novel idea was 'beginning to gain acceptance'. Winn considered[25] that

> it was worth while to 'have a go'. If, as he subsequently said, one beat the law of average [sic] and was right only fifty-one per cent of the time, that one per cent, in terms of lives and ships saved, or U-boats sunk, was surely worth the effort.

However new to the Navy, this was hardly an idea which matched the finesse of 'sequential analysis'. And as the translated decrypts passed down the teleprinter line to the OIC, they travelled back fifty years in time. Even after great improvements,

> ... Winn still had fewer than half a dozen assistants. They had to maintain an Atlantic plot on which were shown not only the latest estimated positions of all U-boats but also the positions and routes of British warships, convoys and independently routed vessels. This of course was on top of their task of dealing with the minute to minute and hour to hour flow of incoming signals concerning attacks, sightings, D/F fixes, and the queries from the Operations, Plans and Trade Divisions in the Admiralty,

from Coastal Command and from headquarters in Ottawa, Newfoundland, Iceland, Freetown, Gibraltar and Cape Town. The situation was beginning to resemble that in Room 40 in 1916 when only the most urgent matters could receive attention. When the flow of decrypts began, Winn, partly for security reasons and partly because of shortage of staff, had to handle and file them all himself. He had no shorthand typist, not even a confidential filing clerk.

Whatever the capacities and dedication of the individuals, the system had not adjusted to the scale and significance of the information it processed. If Bletchley had its successes through traditional British virtues of teamwork and of getting on with the job without a fuss, it suffered from limitations derived from an equally traditional British shabbiness and paltriness. In Hut 4 they had their own tracking charts in order to deduce the meaning of grid references and so forth, and it must have seemed that they could easily take on all the work of plotting and guiding the convoys more effectively than the OIC.

But this was a problem common to the higher reaches of the war effort, as young scientists and academics found themselves confronting the peacetime establishment. In many ways, the war was, for Alan Turing's generation, the continuation of the conflicts expressed in another language in 1933.

They were not taking orders from brainless brass-hats, and, more positively, government was forced to adopt the central planning, scientific methods, and remedies for depression that had been argued for in the 1930s. Bletchley was at the heart of this struggle. It was in 1941 that:[26]

> The staff at GC and CS, recognising no frontiers in research, no division of labour in intelligence work, invaded the field of appreciation.

There were 'unavoidable clashes of priority and personality' as the compartment walls were breached. And such clashes were symptomatic of the difficulty which faced the Services in accepting advice from a peculiar civilian department without name or tradition:

> GC and CS had increased in size four-fold in the first sixteen months of the war. At the beginning of 1941 it was by Whitehall standards poorly organised. This was partly because the growth in its size and in the

complexity of its activities had outstripped the experience of those who administered it. . . .

It was not a single, tidy organisation but 'a loose collection of groups', each pushing ahead in an *ad hoc* manner, doing its best to knock some sense into the relevant military heads before it was too late. The intellectuals, finding themselves in an unprecedented position, virtually ignored the formal structure left over from the peacetime days, and organised one for themselves. This time the war was too important to be left to the generals or to the politicians. They

> inaugurated and manned the various cells which had sprung up within or alongside the original sections. They contributed by their variety and individuality to the lack of uniformity. There is also no doubt they thrived on it, as they did on the absence at GC and CS of any emphasis on rank or insistence on hierarchy.

The Service chiefs were highly indignant at

> . . . the condition of creative anarchy, within and between the sections, that distinguished GC and CS's day-to-day work and brought to the front the best among its unorthodox and 'undisciplined' war-time staff.

Alan was sheltered by Hut 4 from direct contact with the service mentality. But it was his work that was causing the trouble, and he was *par excellence* the 'undisciplined' person who 'thrived' on the 'lack of uniformity' and the 'absence of any emphasis on rank' – a military nightmare.

More precisely, it was the irrelevance of *official* rank that was so striking. The cryptanalysts were highly conscious of differences of talent and speed among themselves. If it was democracy (or 'anarchy', as it would appear to the military mind) it was of the Greek kind, in which the slaves did not count. Hut 8 was an aristocracy of intelligence, a dispensation which suited Alan perfectly. As Hugh Alexander saw it:[27]

> He was always impatient of pompousness or officialdom of any kind – indeed it was incomprehensible to him; authority to him was based solely on reason and the only grounds for being in charge was that you had a better grasp of the subject involved than anyone else. He found unreasonableness in others very hard to cope with because he found it very hard to believe

that other people weren't all prepared to listen to reason; thus a practical
weakness in him in the office was that he wouldn't suffer fools or humbugs
as gladly as one sometimes has to.

The problems came in dealing with the rest of the world.
The civilians tended naively to suppose that the military services
existed to fight the war, and did not appreciate that like almost
all organisations, they expended much of their energy in resisting
change and fending off each other's encroachments. Alan had little
time for Denniston, who never caught up with the change in scale
and vision over which he had presided. Travis, who oversaw the naval
work, and had responsibility for machinery, was a more Churchillian
character, who gave some push to the new ideas; and another man,
Brigadier J. H. Tiltman, won a high respect from the analysts. But
there was a tardy, grudging quality to the administration which to
the new recruits was simply incomprehensible. It was blindingly
obvious how important the miraculous information was, and they
could not understand (Alan least of all) why the system could not
immediately adapt to it. The provision of six Bombes by mid-
1941, for instance, fell far short of the scale he had envisaged; and
parsimony of any kind seemed absurd when frantic efforts were
being made to produce bombers as though all depended on them,
and streams of exhortations to the public issued forth concerning
matters of infinitely less importance to the war effort.

In coping with such problems, Hugh Alexander soon proved
the all-round organiser and diplomat that Alan could never be.
Meanwhile, Jack Good took over the statistical theory, in which he
became more interested. Shaun Wylie and others could be relied
upon to do any pure mathematics which arose. They were all
better than he was at the day-to-day operational work. Yet there
was no question that naval Enigma was Alan Turing's, and that he
was in charge of it inasmuch as anyone was. He had lived with it
from beginning to end, and threw himself into the whole process,
relishing the shift work on the incoming messages as much as any
of the others. This was Snow White's little hut in the forest, where
they all worked together with a will, and whistled as they worked.
Partly his position of leadership was because, like R. V. Jones, he
was one of 'the men who went first'. He just happened to be in

at the beginning. But it was also analogous to his attack on the Hilbert problem. The Turing machine idea had owed nothing to the Mathematical Tripos, and likewise his cryptanalytic ideas stormed ahead without the benefit of books or papers to build upon, for there were none. In the British amateur tradition, he took out his pencil-box, sat down in his Hut, and set to work.

In this respect the war had resolved some of his conflicts. The business of getting to the heart of something, abstracting its meaning, and connecting it with something that worked in the physical world, was exactly what he had been searching for before the war. It was the fault of human history that he found his niche in the intellectual equivalent of filling in holes that others had dug.

If the fighting services were slow to come to terms with the significance of the Enigma decrypts, Winston Churchill was not. He loved them, as one who had been fascinated by cryptanalytic intelligence from 1914 onwards, and who regarded it as of the utmost importance. At first he had asked to read *every* Enigma message, but compromised by receiving each day a special box of the most exciting revelations – in which a resumé of naval Enigma took its place. Since GC and CS officially remained the responsibility of the chief of the secret service, one side-effect of Alan's work was the restored prestige which thereby accrued to the British spying organisation.

It also strengthened prime ministerial government. Churchill alone enjoyed this overall view of Intelligence. At this stage there was no integration of the material except in his head. It was a state of affairs that did not appeal to the military departments or the Foreign Office, especially when the Prime Minister was[28] 'liable to spring upon them undigested snippets of information of which they had not heard', and made 'calls for action or comment from the Chiefs of Staff or the Foreign Office and sent signals direct to the operational theatres and individual commanders.'

War, Churchill had written in 1930, had been 'completely spoilt. It is all the fault of Democracy and Science.' But he still made use of democracy and science when necessary, and did not overlook those who produced the decrypts. In the summer of 1941 he paid a visit to Bletchley, and gave a pep talk to the cryptanalysts as they gathered round him on the grass. He went into Hut 8, and was introduced to a very nervous Alan Turing. The Prime Minister used to refer to the Bletchley workers as[29] 'the geese who laid the golden eggs and never cackled.' Alan was the prize goose.

The last of the German supply ships had been sunk on 23 June 1941. But that day there was something else to think about. Tweedledum had turned upon the slumbering Red King. It was not only Stalin who was caught napping; the Luftwaffe Enigma evidence pointing to an imminent German invasion had been the subject of another fight between GC and CS on the one hand and the service chiefs on the other. They had not been able to believe their ears. But now the world war had begun. From now on the Atlantic lay at Germany's back, and the Mediterranean was a sideshow. The game had changed, and the period of anarchy was over.

In the spring of 1941 Alan developed a new friendship. It was with Joan Clarke, a fact which presented him with a very difficult decision. First they had gone out together to the cinema a few times, and spent some leave days together. Soon everything was pointing in one direction. He proposed marriage, and Joan gladly accepted.

Many people, in 1941, would not have thought it important that marriage did not correspond with his sexual desires; the idea that marriage should include a mutual sexual satisfaction was still a modern one, which had not yet replaced the older idea of marriage as a social duty. One thing that Alan never questioned was the form of the marriage relationship, with the wife as housekeeper. But in other ways he took a modern view, and above all was honest to a fault. So he told her a few days later that they should not count on it working out, because he had 'homosexual tendencies'.

He had expected this to end the question, and was surprised that it did not. He underestimated her, for Joan was not the person to be frightened by a bogey word. The engagement continued. He gave her a ring, and they made a visit to Guildford for a formal introduction to the Turing family, which went well enough. On the way they also had lunch with the Clarkes – Joan's father was a London clergyman.

He must have had thoughts of his own when, for instance, Joan went to Communion with his mother at Guildford. He might well have soft-pedalled the fierceness of his views in a way that in the long run would not have been possible. Again, the nebulous word 'tendency' fell short of the honesty with which he spoke to close male friends. If in fact he had suggested that there was more to it than that, she would have been hurt and shocked. He told Joan about Bob, explaining how he remained for the time being a financial commitment, and said that it was not a sexual matter – again true, but not quite the whole truth. But they were comrades in the aristocracy of talent, even if he was her superior in the work, and he specifically told her that he was glad he could talk to her 'as to a man'. Alan was often lost when dealing with the Hut 8 'girls', not least because he was unable to cope with the 'talking down' which was expected. But Joan's position as cryptanalyst gave her the status of an honorary male.

Alan arranged the shifts so that they could work together. Joan did not wear her ring in the Hut, and only Shaun Wylie was told that there was an actual engagement, but the others could see that something was in the offing, and Alan had managed to find a few bottles of scarce sherry, putting them by for an office party when the time came to announce it. When off duty, they talked a little about the future. Alan said that he would like them to have children, but that of course there was no question of expecting her to leave such important work at such a time. Besides, the outcome of the war, in summer 1941, was far from clear, and he still tended to pessimism. There seemed no stopping the Axis forces in Russia and the south-east.

But when Alan said that he could talk to her as to a man, it certainly did not mean that he had to be solemn. It was the other way round: he was free to be himself, and not conventionally polite. If he came up with some scheme or entertainment then they would both join in with gusto. He had learnt how to knit, and had progressed as far as making a pair of gloves, except for sewing up the ends. Joan was able to explain how to finish them off.

The joy, or the difficulty, was that they enjoyed so easy a friend-ship. They were both keen on chess, and were quite well matched, even though Joan was a novice, whose interest had been drawn by attending Hugh Alexander's course for beginners. Alan used to call their efforts 'sleepy chess', taking place as they did after the nine-hour night shift. Joan had only a cardboard pocket set, and proper chess pieces were unobtainable in wartime conditions, so they improvised their own solution. Alan got some clay from one of the local pits, and they modelled the figures together. Alan then fired them on the hob of the coal fire in his room at the Crown Inn. The resulting set was quite usable, if somewhat liable to breaking. He also tried to make a one-valve wireless set, telling her about the one he had made at school, but this was not such a success.

They had been to see a matinée of a Bernard Shaw play while making their London visit, and besides Shaw, Alan was currently keen on Thomas Hardy, lending *Tess of the d'Urbervilles* to Joan. These were, after all, with Samuel Butler, the writers who had attacked the Victorian codes. But they spent more time taking long country bicycle rides. And because she had studied botany at school, Joan was able to join in one of Alan's enthusiasms which went back to *Natural Wonders*. He was particularly interested in the growth and form of plants.

Before the war he had read the classic work *Growth and Form* by the biologist D'Arcy Thompson, published in 1917 but still the only mathematical discussion of biological structure. He was particularly fascinated by the appearance in nature of the Fibonacci numbers – the series beginning

1, 1, 2, 3, 5, 8, 13, 21, 34, 55, 89 . . .

in which each term was the sum of the previous two. They occurred in the leaf arrangement and flower patterns of many common plants, a connection between mathematics and nature which to others was a mere oddity, but to him deeply exciting.

One day he and Joan were lying on the Bletchley Park lawn – after a game of tennis, perhaps – and looking at the daisies. They started talking about them and Joan explained how she had been taught to record and classify the arrangement of leaves on plants by following them upwards round the stem, counting the number of leaves and the number of turns made before returning to a leaf directly above the starting-point. These numbers would usually appear in the Fibonacci series. Once Alan produced a fir cone from his pocket, on which the Fibonacci numbers could be traced rather clearly, but the same idea could also be taken to apply to the florets of the daisy flower. In this case it was rather harder to see how to count off the petals, and Joan wondered whether the numbers did not then arise merely as a consequence of the method of following them. This was pretty much the view of D'Arcy Thompson, who played down the idea that the numbers had any real significance in nature. They made a series of diagrams to test this hypothesis which did not satisfy Alan, who continued to think about 'watching the daisies grow'.

In 1941 everyone had to knit and glue and make their own entertainments. At the Clock House, where Mrs Morcom died this year, they were eating the young goats, and at Bletchley the shipping crisis was reflected not only in the work of Hut 8, but in the miserable régime of school dinners. Apart from the diet, the siege mentality suited Alan rather well, with matters of social protocol that in the 1930s seemed so important now falling into abeyance. He always liked making things for himself, be they gloves, radio sets or probability theorems. At Cambridge he had a way of telling the time from the stars. Now the war was on his side. In a more self-sufficient England, everyone had to live in a more Turingesque way, with less waste of energy.

This was well understood in the higher realms of Bletchley, in many ways a *New Statesman* readers' establishment, distilling the more creative elements of the ancient universities and leaving behind the upper-class finishing school mentality along with the misogynistic port-passing. By this time, the establishment had sprouted clubs for amateur dramatics and so forth. Alan was as shy as ever of this sort of thing, and never became a figure in the Bletchley social world. To some extent he was a 'character', but without the dominating egoism of the much older Dillwyn Knox. He retained a shy boy-next-door manner which muted his detachment from convention. Among the Hut 8 people, his persona was that of 'the Prof'; for while all the new men were 'men of the Professor type', the word suited him particularly well. It relieved people of the difficulty with forms of address, for women especially, and was a mark of respect while still expressing the amateur quality of his manner – more the *ITMA* professor than an eminent authority. Joan also called him 'Prof' while they were at work, although off-duty Alan commented on this, not actually objecting, but making her promise that she would not ever do so when he really was a professor, or indeed had returned to academic life. There was, in fact, a streak of vulgarity in the usage, which Mrs Turing had been quick to point out to him, comparing it with the lower-middle-class habit of wives referring to husbands by title rather than by name. But it was also that he did not want to sound presumptuous of professorial status.

Pigou was also known as 'Prof' to everyone in King's, and for similar reasons. In fact, they were rather alike. David Champernowne had introduced them before the war, and Pigou became perhaps the only one of the elder King's dons (or 'old fogies', as Alan was liable to call them) to know him well and indeed to find a mutual admiration. Pigou enjoyed a[30] 'sure grasp of logical relations and . . . fanatical intellectual integrity', and he had 'an astonishing capacity for simplifying life and all its important issues', he would 'dispense altogether with pretence as a weapon', and his 'eye for beauty was concerned with mountains and men' – words that would have fitted Alan almost as well.

In Alan's case, there was a suggestion in the nickname of his role at school, as the tolerated 'Maths Brain' with his star globe and pendulum, who had performed the feat of cycling from Southampton. As at school, trivial examples of 'eccentricity' circulated in Bletchley circles. Near the beginning of June he would suffer from hay fever, which blinded him as he cycled to work, so he would use a gas mask to keep the pollen out, regardless of how he looked. The bicycle itself was unique, since it required the counting of revolutions until a certain bent spoke touched a certain link (rather like a cipher machine), when action would have to be taken to prevent the chain coming off. Alan had been delighted at having, as it were, deciphered the fault in the mechanism, which meant that he saved himself weeks of waiting for repairs, at a time when the bicycle had again become what it was when invented – the means of freedom. It also meant that no one else could ride it. He made a more explicit defence of his tea-mug (again irreplaceable, in wartime conditions) by attaching it with a combination lock to a Hut 8 radiator pipe. But it was picked, to tease him.

Trousers held up by string, pyjama jacket under his sports coat – the stories, whether true or not, went the rounds. And now that he was in a position of authority, the nervousness of his manner was more open to comment. There was his voice, liable to stall in mid-sentence with a tense, high-pitched 'Ah-ah-ah-ah-ah' while he fished, his brain almost visibly labouring away, for the right expression, meanwhile preventing interruption. The word, when it came, might be an unexpected one, a homely analogy, slang expression, pun or wild scheme or rude suggestion accompanied with his machine-like laugh; bold but not with the coarseness of one who had seen it all and been disillusioned, but with the sharpness of one seeing it through strangely fresh eyes. 'Schoolboyish' was the only word they had for it. Once a personnel form came round the Huts, and some joker filled in for him, 'Turing A. M. Age 21', but others, including Joan, said it should be 'Age 16'.

He cared little for appearances, least of all for his own, generally looking as though he had just got up. He disliked shaving with a razor and used an old electric shaver instead – probably because cuts could make him pass out with the sight of blood. He had a permanent five o'clock shadow, which emphasised a dark and rough complexion which needed more than the cursory attention it received. His teeth were noticeably yellow, although he did not smoke. But what people noticed most were his *hands*, which were strange anyway, with odd ridges on his fingernails. These were never clean or cut, and, well before the war, he had made them much worse by a nervous habit of picking at the side, raising an unpleasant peeling scar.

To some extent, his lack of concern for appearances, like his low-budget mode of life, was an intensification of what people meant by 'donnish', and as such was far more striking to those outside university circles, than to those long familiar with bicycling dons eking out their stipends. It departed from the 'don' typology in his peculiar youthfulness of manner, but Alan Turing still presented the world outside Oxford and Cambridge with a crash course in King's College values, and the reaction to his oddness was mostly a concentrated form of the mixture of baffled respect and head-shaking suspicion with which English intellectuals were traditionally regarded. This was particularly true at Guildford, where the engagement was perceived in terms of types, he as the don shy of women, and she as the 'country vicar's daughter'* and bluestocking 'female mathematician'. It was demeaning, but the repetition of

---

* She was *not* a country vicar's daughter, but that was what they thought.

superficial anecdotes about his usually quite sensible solutions to life's small challenges served the useful purpose of deflecting attention away from the more dangerous and difficult questions about what an Alan Turing might think about the world in which he lived. English 'eccentricity' served as a safety valve for those who doubted the general rules of society. More sensitive people at Bletchley were aware of layers of introspection and subtlety of manner that lay beneath the occasional funny stories. But perhaps he himself welcomed the chortling over his habits, which created a line of defence for himself, without a loss of integrity. He, this unsophisticated outsider at the centre, could be left alone at the point where it mattered.

In the summer of 1941 that much more worldly observer Malcolm Muggeridge had cause to visit Bletchley and notice that[31]

> Every day after luncheon when the weather was propitious the cipher-crackers played rounders on the manor-house lawn, assuming the quasi-serious manner dons affect when engaged in activities likely to be regarded as frivolous or insignificant in comparison with their weightier studies. Thus they would dispute some point about the game with the same fervour as they might the question of free-will or determinism. . . . Shaking their heads ponderously, sucking air noisily into their noses between words – 'I thought mine was the surer stroke', or: 'I can assert without contradiction that my right foot was already . . .'

Alan did indeed have that way of sucking in his breath before speaking, while in Hut 8 they were, when off-duty, talking about games, free will and determinism.

He was currently reading a new book by Dorothy Sayers, *The Mind of the Maker*.[32] It was not his usual taste in reading, this being Sayers' attempt to interpret the Christian doctrine of divine creation through her own experience as a novelist, but he would have enjoyed the challenge of her sophisticated attitude to free will, which she saw from God's point of view, in the light of her knowledge that fictional characters had to find their own integrity and unpredictability, and were not determined by a master plan at the outset. One image which caught Alan's fancy was that of Laplacian determinism suggesting that 'God, having created his Universe, has now screwed

the cap on His pen, put His feet on the mantelpiece and left the work
to get on with itself.'

This was not so new, but it must have made striking reading while
the Bombes ticked away, getting on with the work by themselves
– and while the Wrens did their appointed tasks, without knowing
what any of it was for. He was fascinated by the fact that people
could be taking part in something clever, in a quite mindless way.

Machines, and people acting like machines, had replaced a good
deal of human thought, judgment, and recognition. Few knew how
the system worked, and for anyone else, it was a mystic oracle,
producing an unpredictable judgment. Mechanical, determinate
processes were producing clever, astonishing decisions. There was
a connection here with the framework of ideas that had gone into
Computable Numbers. This, of course, was far from forgotten. Alan
explained the Turing machine idea to Joan, and gave her an off-print
of one of Church's papers, though she perhaps disappointed him
in her response. He also gave a talk on the subject of his discovery.
Meanwhile Turing machines, reading and writing, had sprung into
an exceedingly practical form of life, and were producing a kind of
intelligence.

A subject closely analogous to cryptanalysis, and which could be
spoken of when off-duty, was chess. Alan's interest was not limited to
chess as recreation; he was concerned to abstract a point of principle
from his effort to play the game. He became very interested in the
question of whether there was a 'definite method' for playing chess –
a machine method, in fact, although this would not necessarily mean
the construction of a physical machine, but only a book of rules that
could be followed by a mindless player – like the 'instruction note'
formulation of the concept of computability. In such discussions
Alan would often jokingly refer to a 'slave' player.

The analogy between chess and mathematics had already been
employed and in each case the same problem arose, that of how to
choose the right move to reach a given goal – in the case of chess,
to achieve checkmate. Gödel had shown that in mathematics there
was *no* way at all to reach some goals, and Alan had shown that there
was no mechanical way to decide whether, for a given goal, there
was a route or not. But the question could still be asked as to how
mathematicians, chess-players or code-breakers did in practice make

those 'intelligent' steps, and to what extent they could be simulated by machines.

Although his solution of the *Entscheidungsproblem* and his work on ordinal logics had focussed attention upon the *limitations* of mechanical processes, it was now that the underlying materialist stream of thought began to make itself more clear, less interested in what could *not* be done by machines, than in discovering what *could*. He had demolished the Hilbert programme, but he still exuded the Hilbert spirit of attack upon unsolved problems, and enjoyed a confidence that nothing was beyond rational investigation – including rational thought itself.

Jack Good, like Alan, had the Bletchley mind, not being simply 'a mathematician', but a person who enjoyed exploring the connections between logical skills and the physical world. Chess interested him too, and unlike Alan he was a Cambridgeshire county player. He had already in 1938 published a light-hearted article on mechanised chess-playing in the first issue of *Eureka*, the house magazine of the Cambridge mathematics students. Besides playing chess, Alan taught Jack Good the game of Go, and before long found himself being beaten at that as well.

Over meal-times on night shift they would talk about the problem of mechanising chess. They latched on to a basic idea, which they agreed to be obvious. It was that a chess player might often see wonderful moves that could be made if only the opponent would do such-and-such, but in serious play, White would assume that Black would always exploit the situation to maximum advantage. White's strategy therefore would be to make the move *least advantageous* to Black – the move making Black's best move the least successful of all the possible best moves – the minimum maximum, in fact.

This was not a new idea. The theory of games had been studied mathematically since the 1920s, and this principle, second nature to chess-players,* had been abstracted and formalised in the manner of modern mathematics. The word 'minimax' had been coined for the idea of the least bad course of action. It applied not only to games like chess, but also to those which involved guessing and bluffing.

---

* Though any but a very naive player would be able to do better than this and play so as to exploit the particular weaknesses of the opponent.

Much of the mathematical work had been done by von Neumann, taking up ideas first published by the French mathematician E. Borel in 1921. Borel had defined 'pure' and 'mixed' strategies in game-playing. Pure strategies were definite rules, setting out the proper action in any contingency; a mixed strategy would consist of two or more different pure strategies, to be chosen at random, but with specified probabilities for each strategy according to the contingency.

Von Neumann had been able to show that for any game with two players, with fixed rules, there would exist optimal strategies, usually mixed, for each player. Alan would very likely have attended his talk on the game of Poker, given at Princeton in 1937, which illustrated the result.[33] It was von Neumann's beautiful, if depressing theorem, that in *any* two-person game* both players would be locked into their 'minimax' strategies, both finding that all they could do was to make the best of a bad job, and to give the opponent the worst of a good job, and that these two objectives would always coincide.

Poker, with its bluffing and guessing, was a better illustration of the von Neumann theory than chess.† A game without concealment, such as chess, von Neumann called a game of 'perfect information', and he proved that any such game would always possess an optimal 'pure strategy'. In the case of chess, this would be a complete set of rules for what to do in every contingency. There being far more possible chess positions than plugboard positions for the Enigma, however, the general von Neumann theory had nothing of practical value to say about the game. It was an example of where a high-powered, abstract approach failed to be of use. Alan and Jack Good's approach was quite different in nature, being for one thing concerned not so much with a theory of the game, but with a discussion of human thought processes. It was an *ad hoc* discussion, 'dull and elementary' by pure-mathematical standards, and pursued quite independently of the existing theory of games. They could have done it at school.

---

* More strictly, any 'zero-sum' game, one in which the loss of one player would always be the gain of the other.

† Less complicated than poker (which in fact is much too complex for a full mathematical analysis), the game of 'stone, paper, scissors' illustrates the idea. In this game the optimal strategy, for both players, is a 'mixed' strategy, that of choosing the three options randomly and with equal probability. For clearly, if one player departs from randomness, the other can exploit the departure to gain an advantage.

In their analysis it first had to be assumed that there was some sensible scoring system, awarding a numerical value to the various possible future positions on the basis of pieces held, pieces threatened, squares controlled and so forth. With this agreed, the most crude 'definite method' would be simply to make the move that maximised the score. The next level of refinement would take the opponent's reply into account, using the 'minimax' idea to choose the 'least bad' move. In chess there would normally be about thirty possible moves for each player, so that even this crude system would require about a thousand separate assessments. A further step of looking ahead would require thirty thousand.

Reducing the figure of thirty to a mere two for the sake of a diagram, the player (White) making a three-move look-ahead is confronted by a 'tree' such as:

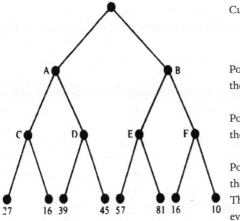

Current position, W to play

Position resulting from the 2 possible W moves

Position resulting from the possible B replies

Position resulting from the possible W replies. These positions are evaluated by scoring.

A human playing White might reason that it would be good to reach position E, but Black will not be so obliging, and will respond to move B by reply F. A second best, for White, would be position D, but again it must be assumed that Black will play C to prevent this. Of the two evils, positions C and F, C is the lesser one, since at least it guarantees for White a position of value 27. So White plays move A.

A 'machine' could simulate this train of thought by a method of 'backing up' the tree. Having worked out all the scores for three moves ahead, it would then label the intermediate positions on a minimax basis. It would assign 27 to C, 45 to D, 81 to E, 16 to F (the *best* in each case), then 27 to A, 16 to B (the *worst* in each case), and finally select move A for White.

This basic idea created a 'machine' that could effect a decision procedure bearing some relation to human intelligence. It was small beer compared with the Hilbert problem, which had required thinking about decision procedures for the whole of mathematics. But on the other hand, it was something that could actually work. As a practical model for mechanical 'thinking' it fascinated Alan to the point of obsession.

Such a three-move analysis would be hopelessly ineffective in real chess, in which players would think not in terms of moves but of *chains* of moves, as for example when a sequence of obligatory captures was set in motion. Alan and Jack Good saw this and decided that the 'look-ahead depth' would have to be variable, continuing as far as any capture was possible at all, so that evaluations would only be made on 'quiescent' positions. Even so, such a scheme would fail to cope with more subtle play, involving traps leading to pins or forks, a fact which they discussed. It was a crude, brute force attack on chess-playing, but it was a first step in mechanising a fairly sophisticated thought process – at least, a first non-secret step.

They thought these ideas too obvious to be worth publishing. Alan did, however, continue to pursue his own mathematical work and to submit it for publication in America. A true intellectual, he would have been ashamed to let human crime and folly defeat him. 'Before the war my work was in logic and my hobby was cryptanalysis,' he once said, 'and now it is the other way round.' He had to thank Newman for stimulating his thoughts on this 'hobby' of mathematical logic, for they corresponded[34] in 1940 and 1941, in which latter year Newman again gave Cambridge lectures on Foundations of Mathematics.

Most of Alan's efforts were directed towards a new formulation of the theory of types. Russell had regarded types as rather a nuisance, adopted *faute de mieux* in order to save Frege's set theory. Other logicians had felt that a hierarchy of logical categories was really quite a natural idea, and that it was the attempt to lump together every conceivable entity into 'sets' that was strange. Alan inclined to the latter view. He would prefer a theory which agreed with the way in which mathematicians actually thought, and which worked in a practical way. He also wanted to see mathematical logic used to make the work of mathematicians more rigorous. In a less technical essay written in this period,[35] 'The Reform of Mathematical Notation', he explained that despite all the efforts of Frege and Russell and Hilbert

> . . . mathematics has profited very little from researches in symbolic logic. The chief reason for this seems to be a lack of liaison between the logician and the mathematician-in-the-street. Symbolic logic is a very alarming mouthful for most mathematicians, and the logicians are not very much interested in making it more palatable.

His own effort to bridge the gap began with an attempt

> . . . to put the theory of types into a form in which it can be used by the mathematician-in-the-street without having to study symbolic logic, much less use it. The statement of the type principle given below was suggested by lectures of Wittgenstein, but its shortcomings should not be laid at his door.
>
> The type principle is effectively taken care of in ordinary language by the fact that there are nouns as well as adjectives. We can make the statement 'All horses are four-legged', which can be verified by examination of every horse, at any rate if there are only a finite number of them. If however we try to use words like 'thing' or 'thing whatever' trouble begins. Suppose we understand 'thing' to include everything whatever, books, cats, men, women, thoughts, functions of men with cats as values, numbers, matrices, classes of classes, procedures, propositions . . . Under these circumstances what can we make of the statement 'All things are not prime multiples of 6'. . . . What do we mean by it? Under no circumstances is the number of things to be examined finite. It may be that some meaning can be given to statements of this kind, but for the present we do not know of any. In effect then the theory of types requires us to refrain from the use of such nouns as 'thing', 'object' etc., which are intended to convey the idea 'anything whatever'.

The technical work of separating mathematical 'nouns' from 'adjectives' was based upon that of Church, whose lectures he had followed at Princeton, and who published a description of his type theory in 1940. Part of Alan's work was done in collaboration with Newman through correspondence; their joint paper[36] being received at Princeton on 9 May 1941. It must have crossed the Atlantic just as the *München* was captured. Alan produced a further paper[37] of a highly technical nature, 'The Use of Dots as Brackets in Church's system', and submitted it just a year later. This promised as forthcoming two more papers, but these never emerged.

He did not allow the war to expunge the idea that, for mathematics, the world should be a single country. In a letter to Newman of autumn 1941, concerned with arrangements for sending out reprints of their paper, he commented: 'Also expect they might send a copy to Scholz, but I expect that will be impossible by then.'

This was not the only thing that became impossible in the course of 1941. The engagement had continued during the summer, but there had been signs of inner conflict on Alan's part. Once they had a weekend together in Oxford, visiting Joan's brother there; Alan went off by himself for a while, apparently thinking it over again, though deciding to carry on. Then in the last week of August they were able to have a whole week together. (They were allowed a week of leave each quarter.) They took their holiday in North Wales, going up by train from Bletchley with bicycles and rucksacks, and arriving at Portmadoc when it was already dark. Alan had arranged for them to stay at a hotel, but the management had made a mess of it and overbooked; only by making a fuss could they be accommodated for the night, and they lost half a precious day finding another place in the morning. Food was a problem too, Alan not having a temporary ration card. But they had brought some margarine, and managed on bread and a surprise discovery of unrationed meat paste. They walked the lesser mountains, the same as he had tramped as a boy: Moelwyn Bach, Cnicht, and others, suffering only the usual troubles with punctures and rain.

It was soon after they returned that Alan made up his mind and burnt his bridges. It was neither a happy nor an easy decision. He quoted to her Oscar Wilde's words, the closing lines from *The Ballad of Reading Gaol*, which bore both an immediate and a prophetic interpretation:

> Yet each man kills the thing he loves,
> By each let this be heard,
> Some do it with a bitter look,
> Some with a flattering word,
> The coward does it with a kiss,
> The brave man with a sword!

There had been several times when he had come out with 'I do love you'. Lack of love was not Alan's problem. The break created a difficult situation in the Hut. Alan told Shaun Wylie that the engagement had been broken off, but not the true reason. In fact he used a dream to make up an explanation, saying that he had dreamt that they had gone to Guildford together and that Joan had not been accepted by his family. Alan dropped out of shift work so that he and Joan did not have to meet more often than necessary at first. It was very upsetting for both of them, but he had behaved in such a way that she knew she had not been rejected as a person. The break was a barrier, but the understanding of it continued as a link.

As they talked about games at Bletchley, something of that locked-in minimax logic of combat was developing out on the Atlantic, strategy forcing counter-strategy, weapon forcing counter-weapon, detection forcing counter-detection. Less tidy than poker or chess, these real conflicts involved rules which were always changing, strategies whose consequences could not be foreseen, and losses which were more than marks on paper. But like poker, the U-boat war was a game of imperfect information, with bluffing and guessing. It was also a game in which, by August 1941, the British had placed a mirror behind the opponent's hand, and was able to cheat by looking at almost all the German cards.* There was no need for further captures in the rest of 1941, and

---

* The 'Foreign' key-system, used by German vessels in waters such as the Indian Ocean, was *never* broken. Furthermore the 'Home' key-system no longer covered the communications of the Mediterranean surface vessels. These, from April 1941, had gone on to a new system which remained immune from decryption for another year.

decryption was performed by Hut 8 within a period of thirty-six hours – this despite the fact that the eight rotors of the naval Enigma had 336 possible orders, as compared with the 60 of the other services' machines.

But this perfection of methods was not the responsibility of Hut 8 alone. It had brought into play the operation of the Bletchley establishment as a whole, attacking the German communication system as a whole:[38]

> ... From the spring of 1941, assisted first by a captured document and then by the discovery that some of the signals were repetitions of decrypted Enigma messages, it broke a dockyards and fairways hand cypher ('Werft'). From August 1941, as a result of the fact that some of its signals were re-encyphered in the Enigma and re-transmitted, and of GC and CS's ability to isolate these signals, the 'Werft' decrypts made, in return, an invaluable contribution to the daily cryptanalytical assault on the naval Enigma settings. At the same time it was as a result of breaking into the Enigma that GC and CS was able to complete its mastery of the dockyard cypher ...

And besides this Rosetta stone, the 'naval meteorological cypher' also 'turned out to be of especial importance'.

> It was first broken in February 1941 and in May of that year the Meteorological Section at GC and CS discovered that it carried weather reports from U-boats in the Atlantic which had originally been transmitted in the naval Enigma. Thereafter its decrypts were no less useful than those of the dockyard cypher in helping to break the Enigma keys.

These developments, while a triumph for Bletchley, were something of a personal blow for Alan. He had worked out subtle mathematical methods for the cryptanalytic attack earlier in the year, only to have an almost insultingly direct method thrust upon him by the dockyard and weather 'cribs'. But he had to give way to the events which his own pioneering work had made possible.

The key to the development of GC and CS now lay in the *integration* of its work, rather than in individual brilliance. These new discoveries were the final vindication for all that the new men had been fighting for. The dockyard messages held nothing of operational value, and according to Room 40 standards, would never have been touched. But at GC and CS they had established a principle of attacking *everything*, however apparently insignificant, and thinking big had now paid off. It was also crucial that a single organisation handled all the decrypts, and was allowed to use them as it saw fit. Had the Admiralty been allowed to recapture naval cryptanalysis, this might never have been possible. But these were considerations of a kind which did not call for Alan Turing's expertise as much as for administrative and political skill. He could well appreciate what was being done, but his own strength lay in the more self-contained problems.

In a wider sense, too, the cryptanalytic work only gained meaning through a coordination of many different kinds of activity, of which puzzle-solving, though critical, was but one. The audacious U-boat captures, the painstaking labour on dreary dockyard lists, the comparisons with aerial reconnaissance and current incidents, the filing systems whereby duplication of material could be exploited, the engineering of new machinery – all had to be fitted together, and all rested too upon the gruelling transcription of dim, fading, meaningless Morse signals, scrupulously effected during month after bewildering month by those invisible devoted servants at the radio receivers.

Again, the reading of German signals was but one of the many factors in the Atlantic game which changed in mid-1941. The assault on Russia drew off the Luftwaffe, and British aircraft were better able to control the Western Approaches. The U-boats moved out into a new battlefield in mid-Atlantic. Both escort ships and aircraft were being fitted with radar for short-range submarine detection. The Huff Duff system for automated, accurate direction finding was beginning to work. More significantly, trade links were, as in the First World War, bringing about an undeclared American war. The US Navy was escorting convoys half-way across the Atlantic, and its official neutrality was of advantage to Britain in that U-boats were instructed not to attack American vessels.

But it was the Enigma which lay at the heart of the British recovery of summer 1941; not only in the tactical routeing of convoys, but in making possible action against U-boats, especially their supply system. Above all the British now had a clear and virtually complete picture of what was happening. It was through Alan Turing's work that 'in July and August, by which time Winn had got into his stride', losses fell to under 100,000 tons a month. Overall, the second half of 1941 saw German successes halved despite an increase in the number of U-boats to 80 in October. By the end of the year it was being claimed that the shipping problem was solved.

Yet the fight was very far from over. The British improvements were only just keeping pace with the continually growing U-boat strength, and they were at the mercy of the Enigma enciphering system. September 1941, in particular, saw a dramatic increase in sinkings for the few weeks after a small sophistication was introduced into the U-boat signals. They had all along been indicating positions by means of the grid references on their maps, and[39]

> . . . not by latitude and longitude. Thus position AB1234 would indicate point, say 55 degrees 30 minutes North, 25 degrees 40 minutes West. This of course presented no problem to us once a portion of a gridded chart had been captured and the whole reconstructed. But in [September] 1941 the Germans started transposing these letters, square AB becoming, for example, XY while a figure would have to be added to or subtracted from the numerals, so that 1234 would appear in the text of the signal as, say, 2345. These transpositions were changed at regular intervals.

Yet either the Enigma was being read, in which case these precautions were a feeble response, or it was not, in which case they were a waste of time. The rationale lay not in foiling the British cryptanalysts, but in creating a defence against imagined spying and treachery. The cumbersome disguise succeeded in confusing their own officers:

> On one occasion we successfully solved a disguised grid reference and diverted a convoy clear of a waiting patrol line, only to find that the C.O. of one of the U-boats involved had not been as clever as we had and had misinterpreted the disguised grid reference given in his orders and blundered into the convoy in consequence.

In November 1941 the system was made even more complicated, and gave rise to long periods of uncertainty at Bletchley. They were still on a knife-edge, and were never allowed to forget it.

It was in the autumn of 1941 that the cryptanalysts finally rebelled against the administrative system. As one of the very few who had the vision, it fell to Alan Turing to force the British government into the modern world. He and the others broke all the rules by writing directly to another man who knew how to break all the rules, and who now had the power to change them:[40]

Secret and Confidential
Prime Minister only
                                                    Hut 6 and Hut 8,
                                                    (Bletchley Park)
                                                    21st October 1941

Dear Prime Minister,

Some weeks ago you paid us the honour of a visit, and we believe that you regard our work as important. You will have seen that, thanks largely to the energy and foresight of Commander Travis, we have been well supplied with the 'bombes' for the breaking of the German Enigma codes. We think, however, that you ought to know that this work is being held up, and in some cases is not being done at all, principally because we cannot get sufficient staff to deal with it. Our reason for writing to you direct is that for months we have done everything that we possibly can through the normal channels, and that we despair of any early improvement without your intervention. No doubt in the long run these particular requirements will be met, but meanwhile still more precious months will have been wasted, and as our needs are continually expanding we see little hope of ever being adequately staffed.

We realise that there is a tremendous demand for labour of all kinds and that its allocation is a matter of priorities. The trouble to our mind is that as we are a very small section with numerically trivial requirements it is very difficult to bring home to the authorities finally responsible either the importance of what is done here or the urgent necessity of dealing promptly with our requests. At the same time we find it hard to believe that it is really impossible to produce quickly the additional staff that we need, even if this meant interfering with the normal machinery of allocations.

We do not wish to burden you with a detailed list of our difficulties, but the following are the bottlenecks which are causing us the most acute anxiety.

1. *Breaking of Naval Enigma (Hut 8)*
Owing to shortage of staff and the overworking of his present team the Hollerith section* here under Mr Freeborn has had to stop working night

---

* A reference to punched-card machine work employed on other stages of the process.

shifts. The effect of this is that the finding of the naval keys is being delayed at least twelve hours every day. In order to enable him to start night shifts again Freeborn needs immediately about twenty more untrained Grade III women clerks. To put himself in a really adequate position to deal with any likely demands he will want a good many more.

A further serious danger now threatening us is that some of the skilled male staff, both with the British Tabulating Company at Letchworth and in Freeborn's section here, who have so far been exempt from military service, are now liable to be called up.

2. Military and Air Force Enigma (Hut 6)
We are intercepting quite a substantial proportion of wireless traffic in the Middle East which cannot be picked up by our intercepting stations here. This contains among other things a good deal of new 'Light blue'* intelligence. Owing to shortage of trained typists, however, and the fatigue of our present decoding staff, we cannot get all this traffic decoded. This has been the state of affairs since May. Yet all that we need to put matters right is about twenty trained typists.

3. Bombe testing, Hut 6 and Hut 8
In July we were promised that the testing of the 'stories' produced by the bombes† would be taken over by the WRNS in the bombe hut and that sufficient WRNS would be provided for this purpose. It is now late in October and nothing has been done. We do not wish to stress this so strongly as the two preceding points, because it has not actually delayed us in delivering the goods. It has, however, meant that staff in Huts 6 and 8 who are needed for other jobs have had to do the testing themselves. We cannot help feeling that with a Service matter of this kind it should have been possible to detail a body of WRNS for this purpose, if sufficiently urgent instructions had been sent to the right quarters.

4. Apart altogether from staff matters, there are a number of other directions in which it seems to us that we have met with unnecessary impediments. It would take too long to set these out in full, and we realise that some of the matters involved are controversial. The cumulative effect, however, has

---

* Luftwaffe key-system used in Africa.
† A reference to the problem of testing the positions at which the Bombe stopped, to eliminate those which had arisen by chance.

been to drive us to the conviction that the importance of the work is not being impressed with sufficient force upon those outside authorities with whom we have to deal.

We have written this letter entirely on our own initiative. We do not know who or what is responsible for our difficulties, and most emphatically we do not want to be taken as criticising Commander Travis who has all along done his utmost to help us in every possible way. But if we are to do our job as well as it could and should be done it is absolutely vital that our wants, small as they are, should be promptly attended to. We have felt that we should be failing in our duty if we did not draw your attention to the facts and to the effects which they are having and must continue to have on our work, unless immediate action is taken.

We are, Sir, Your obedient servants,

A. M. Turing
W. G. Welchman
C. H. O'D. Alexander
P. S. Milner-Barry

This letter had an electric effect. Immediately upon its receipt, Winston Churchill minuted[41] to General Ismay, his principal staff officer:

ACTION THIS DAY

Make sure they have all they want on extreme priority and report to me that this has been done.

On 18 November the chief of the secret service reported that every possible measure was being taken; though the arrangements were not then entirely completed, Bletchley's needs were being met.

Meanwhile, another profound change was beginning to affect their work. America's phoney war, preceding rather than following an official declaration, was reflected not only in the judicious aspirations of the Atlantic Charter, but in the more substantial negotiations with Britain over the sharing of intelligence. Already in 1940 a limited disclosure of cryptanalytic success had been made. This had involved work for Alan, who had gone to extraordinary lengths to devise methods that could be used to explain their decryption of Engima messages at a time when the Bombe was being retained as a British secret. The British doubted the ability of the Americans to keep secrets – and for all that Churchill spoke of the American republic as a rather bigger and better Dominion, the fact was that it was a very different country, one apparently lacking the habits of deference, secretiveness and deviousness, and with powerful elements inimical to British interests. But in the course of 1941, arrangements were made for liaison officers to be attached to Bletchley, and the charade was dropped. The Turing eggs were now for export.

Germany declared war on the United States on 11 December 1941, four days after the attack on Pearl Harbor. 'So we had won after all! . . . England would live, Britain would live; the Commonwealth of Nations and the Empire would live . . .', reflected Churchill. But the first effects were disastrous for Britain. The Pacific drew off the American naval vessels which had protected the convoys. And it proved to be an even tougher job selling intelligence to the US Navy than it had been in the British Admiralty. Naval Enigma information had indicated the operation of fifteen U-boats off the American coast at the declaration of war, but the warning had been spurned, and no precautions taken. Enormous shipping losses marked an unhappy start to the Grand Alliance. Then on 1 February 1942, a far greater blow was dealt. The U-boats switched over on to a new Enigma system. The Bombes failed to deliver their prophecies. There was no more ULTRA.

The black-out of February 1942 meant that the U-boat Enigma analysis had to begin all over again, with the past two years serving as warming-up practice. It was symbolic of the war effort as a whole, the British situation being one that in 1939 would have seemed disastrous beyond belief. The loss of all European allies, the reversal of early gains from Italy, the surrender of Singapore – these and other blows were still offset only by the promise of help from an ill-prepared, inexperienced America. For what it was worth, which was very little, the RAF was gaining superiority over the Luftwaffe in crude bombing capacity. Yet this did not prevent the *Scharnhorst* and *Gneisenau* passing Dover in broad daylight. Meanwhile the economy of German Europe, hitherto complacently assessed as 'taut', was in reality only just beginning to adapt to full-scale war production. And its principal enemy had only just staved off defeat at the gates of Moscow.

They had to think impossible things, and to think them before breakfast. An American army had to be created out of almost nothing, and conveyed across the Atlantic to invade a heavily fortified continent dominated by a no less advanced industrial power. But even the preparations for that invasion, let alone its success, were impossible while the Atlantic U-boat fleet was allowed to operate. With Hitler now taking the war seriously, the U-boat force had swollen to a fleet of a hundred larger vessels by January 1942, and was increasing every week. After February, with invisibility restored, they were able to inflict damage that approached disaster levels – half a million tons a month, exceeding the construction rate of the new Allies combined. It would be hard even to stay in the same place, let alone build up to the possibility of a victory.

Everything had changed. There was no unemployment in Britain now, as there had been in 1940, and now everything was being planned. Indeed, Britain and the United States found themselves planning the entire trading economy of the world outside Axis and Soviet jurisdiction. At Bletchley, the country-house party spirit was gone, replaced by a conscription of the intelligentsia, to be ferried round Buckinghamshire in fleets of buses. The chaos of 1940 and the floundering of 1941 had been sorted out just in time to make use of the 'cornucopian abundance'. Now the military departments had been forced to swallow their pride and adapt to its output: not sporadic 'golden eggs', but the production of an intelligent, integrated organisation which mirrored every level of the enemy system. In 1941 the supply of resources to Bletchley was still regarded as a concession, one that subtracted from the real men's war of aircraft and guns. Even at the end of the year, the cryptanalysts had been obliged to manage with no more than sixteen Bombes – and this when the breaking into a number of German Army key-systems had rapidly increased their requirements. But their desperate letter to Churchill had brought about a change of attitude. Travis took over the direction from Denniston, and presided over an administrative revolution which at last brought the management of Intelligence into line with its mode of production. Meanwhile the services, recognising the hard fact that Intelligence was dominating Churchill's control of the war, began to slacken their resistance to Bletchley's claims.

But however wonderfully their minds were concentrated, the fact remained that the U-boat Enigma problem now surpassed their means. In 1941, Hut 8 had given sight to the blind, and if this experience had been traumatic, the taking of sight away was an even more cruel blow. More precisely, the Admiralty again became, like Nelson, one-eyed. For only the ocean-going U-boats had adopted the new system. Surface vessels and U-boats in coastal waters continued to use a 'Home' key which could still be broken. They therefore had information on the departures from port of U-boats, and knew how many U-boats were at large – data which could be correlated with sightings and Huff-Duff findings. But this was very poor stuff compared with the operational commands and position reports to which they had now become accustomed.

Within Hut 8 the black-out had a different meaning. The game had been going very enjoyably, and now the Germans had spoilt it by changing the rules. The temptation was to regard the Atlantic problem as a tiresome interruption, and to carry on with the fascinating work of decrypting the signals from European waters. But as they read about the sinkings, and saw the dim, dismal charts, reality penetrated the mathematical game. And much of the fun went out of it.

What had happened was not only a change in the system for using the Engima machine. It was a change in the machine itself. It had been modified in such a way that it now possessed a space for a *fourth rotor*. Hitherto, the naval Enigma settings had continued to involve a choice of three rotors out of eight, for which there were 336 possibilities. Had the machine been modified to allow a free choice of four rotors out of nine, the figure would have gone up to 3024 (a ninefold increase) and the setting of the new rotor would have introduced a further twenty-six-fold increase on top of that. But this was not done. There was indeed a new ninth rotor, but it had to stay in its place. It was only the old machine, but with a new rotor attached to the end, capable of twenty-six different settings. It was equivalent to having twenty-six different reflector wirings. Accordingly, the problem had become not 234 times worse, but only twenty-six times worse.

It was a half-hearted measure, like the encoding of the map references, and was undertaken for the same misguided reason: the internal protection of the U-boat messages. It was not that the Germans feared British cryptanalysis. But even if half-hearted, it was a change that pushed Hut 8 off the knife-edge and into almost total blindness. It was already a fluke that the numbers worked out at all, allowing Bombes that worked in hours rather than weeks. Already the naval Enigma had strained every nerve in the process of achieving decryption in the day or two that was necessary for it to be of convoy-diverting use. Now the twenty-six-fold increase turned every hour into a day, or would require twenty-six Bombes for every one that they had used in 1941, unless ingenuity found another way.

There was one point of success: they knew the wiring of the new fourth wheel. This was because the new four-rotor Enigmas were not new machines, but modified versions of the old ones. The fourth wheel had been sitting in a 'neutral' position on the U-boat machines during late 1941. Once in December a U-boat cipher operator had carelessly let it move out of this position while enciphering a message; Hut 8 had noted the ensuing gibberish, and also spotted the re-transmission of the message on the correct setting. This elementary blunder of repetition, so easy to make while the Germans held complete trust in their machines, had allowed the British analysts to deduce the wiring of the wheel. Armed with this information, they were in fact able to break the traffic for 23 and 24 February and for 14 March – days for which they had particularly clear 'cribs' from messages that had also been enciphered in other, breakable, systems.* But it took twenty-six times too long: six Bombes working for seventeen days were employed. This development well illustrated the chanciness of the whole endeavour. Had this enlarged Enigma been adopted from the start, the treasure hunt might never have got off the Polish ground.

'Faster, faster!', the White Queen now cried. But nothing could make Bombes go twenty-six times faster overnight. There had, in fact, been an opportunity to prepare against the dreadful day, since as early as the spring of 1941 there had been references in the decrypts to the addition of a fourth rotor. The Hut 8 analysts afterwards

---

* The 'crib' for 14 March came from a special message sent out both on the (broken) Home key-system, and on the U-boat system, announcing the no doubt vital news that Dönitz had been promoted to the rank of Admiral.

rebuked themselves for not having impressed this fact with greater force upon the administrators. But in the conditions of 1941 it was quite unrealistic to think in terms of finding resources for bigger and better Bombes to cover a possible future development, when they had to fight to get enough Bombes merely to keep up with the existing traffic. The authorities had thrown away this advantage of foreknowledge. But with the shake-up of late 1941 a more dynamic approach had been taken, and one very important effect of the impending naval Enigma crisis was that at the turn of the year it brought in fresh expertise on the engineering side.

One obvious approach was that of enlarging the Bombe to include the new fourth rotor spinning through its twenty-six possible positions at extremely high speed. The task of devising such a high-speed rotor system was entrusted to the inventive Cambridge physicist, C. E. Wynn-Williams, who in 1941 was working for the radar research laboratory, the Telecommunications Research Establishment as it became on its move to Malvern in May 1942.

One aspect of this assignment was that, with the high speeds proposed, the logical system for following through the proliferating implications of each rotor hypothesis could no longer be embodied in a network of electromagnetic relays. These would be too slow. Instead, an *electronic* system would be necessary. In this way the first suggestions arose for applying the new and arcane technology of electronics to Bletchley work.

It must have pleased Alan Turing that 'electronics' took its name from the word 'electron' coined by his remote relative George Johnstone Stoney. (He used to comment disparagingly on the fact that Stoney was famous merely for inventing a *name*.) The point was that electronic valves could respond in a millionth of a second, there being no moving parts but the electrons themselves, while the electromagnetic relay had to make a physical click. Here lay the possibility of a thousand-fold increase in speed at a time when they were frantic. But electronic valves were notoriously prone to failure, besides being hot, cumbersome, and expensive. There were few people with the knowledge and skills required for using them.

More precisely, applications to Bletchley problems required the use of electronic components in *logical* systems, to act as switches in place of relays. But the pre-eminent use of the electronic valve had remained that of acting as an amplifier in radio reception. It was a quite different matter to think of electronic components as providing on-or-off switches, although the principle had been demonstrated in 1919. In this respect Wynn-Williams had the advantage of having pioneered electronic Geiger counters, and so was one of the even fewer people aware that electronics could be applied to a discrete problem.

While radar research had created a fund of high-powered electronic expertise, TRE was not the only establishment with electronic engineers. There was also the Post Office Research Station, located in the London suburb of Dollis Hill. It had been established to protect the Post Office, in its installation of a modern telephone system, from the monopolistic practices of the equipment manufacturers. It was the vanguard of what was the only state-owned industry of the 1930s, and despite being run on a shoestring, had maintained a high level of research. Its young engineers, picked out by fierce competition, had ambitions and skills which went far beyond the opportunities offered in the 1930s economic climate. The senior of these, T. H. Flowers, had[42]

> . . . joined the Research Station as a probationary engineer in 1930, after serving his apprenticeship at Woolwich Arsenal. His major research interest over the years had been long distance signalling, and in particular the problem of transmitting control signals, so enabling human operators to be replaced by automatic switching equipment. Even at this early date he had considerable experience of electronics, having started research on the use of electronic valves for telephone switching in 1931. This work had resulted in an experimental toll dialling circuit which was certainly operational in 1935 . . .

Here then was world leadership in the field of electronic switching.

That a TRE expert could collaborate on a GC and CS project was already a reflection of the breakdown of frontiers induced by the conditions of 1942. It was even more remarkable that a third organisation, the Post Office Research Station, could be brought in as well. In fact, its engineers took on two different projects arising from the naval Enigma crisis. Wynn-Williams was assisted with the development of the high-speed fourth rotor by W. W. Chandler, a young man recruited by the Post Office in 1936, who had gained expertise in the new use of electronic valves for trunk-line switching. Meanwhile Flowers himself was assisted by S. W. Broadhurst, an electromechanical engineer who in the slump of the 1920s could only be employed in the grade of 'labourer', but had worked his way up through a command of automatic telephone switching to this advanced position at Dollis Hill. They worked on a quite different machine which was to automate the testing of 'stops'. It was intended that the large number of false 'stops' (to be expected with the increased number of rotor positions) would be eliminated very much more rapidly than would be possible if they continued to try each one out by hand on an Enigma.

These developments were begun in the spring of 1942, but were disappointing in their outcome. Wynn-Williams often seemed about to succeed with the high-speed rotor, but never managed it in that year. The work done on designing an associated electronic network was therefore not of any use. The stop-testing machine, in contrast, was designed, built and working by the summer of 1942. But it turned out not to be of practical application after all. Meanwhile Flowers and his colleagues had made suggestions to Keen for the improvement of the Bombes by the inclusion of some electronic components, but these were rejected.

Thus the summer of 1942 saw an unhappy state of affairs and highly frustrated young engineers. No use had been made of their electronics and Alan, who told them what was required, had not achieved anything either. It was a step in the right direction, but the Atlantic remained as opaque as it had become in February.

Hut 8, meanwhile, had acquired more high-level cryptanalytic staff, although the total never rose much above seven. At the end of 1941 Hugh Alexander had brought in Harry Golombek, the chess master, who also had returned from Argentina but had been obliged to serve two years in the infantry. Then in January 1942 arrived Peter Hilton, who had done just one term at Oxford reading mathematics and was only eighteen. He would describe his initiation thus:[43]

> . . . this man came over to speak to me and he said, 'My name is Alan Turing. Are you interested in chess?' And so I thought, 'Now I am going to find out what it is all about!' So I said, 'Well, I am, as a matter of fact'.

He said, 'Oh, that is very good, because I have a chess problem here I can't solve.'

A whole day passed before Peter Hilton found out what he was there to do. But as 1942 went grimly on, this idiosyncratic style of organisation gave way to something more smoothly business-like. Alan remained 'the Prof', but gently, subtly, Hugh Alexander became more and more the *de facto* head. In the nicest possible way, Alan found the rug being pulled from under his feet. He had brought the naval Enigma into being, but it needed a more adroit person to foster its development. He lacked the attention to detail, as well as other skills in managing people. Hugh Alexander was, for instance, the kind of person who could compose and write out a perfectly expressed memorandum without a single crossing-out – not at all a Turing strength. Inevitably Alan felt the loss, as one whose baby had been taken away. But he could not have disputed that Alexander was the better organiser, even though this upset the more cosy arrangements of 1941. Jack Good noticed [44]

> . . . one example of Hugh Alexander's technique as an administrator. Since the section worked 24 hours a day, we had a three-shift system, so the 'girls' had three shift-heads. One of them made herself unpopular because she was always getting in a flap, although she behaved well in ordinary social relationships. Hugh said he'd like to experiment with a complicated five-shift system, so two new shift-heads were required. After a few weeks he decided that the experiment had failed, and he returned to the three-shift system. Two of the shift-heads had to be dropped and you can guess who one of them was.

Alan would never have dreamt of so devious a ploy, although he would have laughed loudly enough when it was explained to him. He had in fact been quite helpful to the 'girls', in matters of leave and working hours, in the old 1940 days. But now a more professional approach to management was required.

Gradually he was being eased out of the immediate problems, and into longer-term research. It was personally disappointing, for he had enjoyed the shift work as much as any of them, and loved having the feel of the whole thing from beginning to end of the process. But it was a rational way of using his abstract mind.

Though technically still attached to Hut 8, he now worked in a room of his own, becoming in effect the chief consultant to GC and CS. While the others worked on a 'need to know' system, not allowed to know anything beyond the specific sphere in which they were engaged, his role became unlimited. The Prof was let into everything, drawn deeper and deeper into the enormously expanded communications that now reflected a world in total war. It was not the same any more, but he could not complain. There was a war on, and he had a unique ability to put his country in the picture.

There was one picture emerging in fragmented form which was nearly as exciting as that of the Atlantic U-boats. The analysts had begun to intercept a small amount of traffic which was entirely different in character from the Enigma signals. It was not in Morse code, having instead the features of a teleprinter signal. Teleprinter transmissions, which had rapidly developed during the 1930s, employed the Baudot-Murray code, not Morse code, the point being that this was a system whose operation could be made automatic. The Baudot-Murray code represented letters of the alphabet by using the thirty-two different possibilities offered by five-hole paper tape. A teleprinter could translate the resulting pattern of holes directly into pulses; at the receiving end the pulses could be translated back into a written message without human intervention. This idea had been developed in Germany to create cipher machine systems, in which encipherment, transmission and decipherment were made automatic – systems more convenient, and making much more effective use of contemporary technology, than the Enigma.

From a logical point of view, the 'hole' in the tape might as well be a 1, and the 'no hole' a 0. So the transmissions were in the form of five sequences of binary digits, 0s and 1s. It had long since occurred to cryptographers that the Baudot-Murray code could be used as the basis for an 'adding on' type of cipher. The principle was dignified with the name of the American inventor Vernam. In fact a Vernam cipher was based on the simplest possible kind of adding, since 'modular' addition with binary digits would use nothing but the rules shown in the figure.

key plain	0	1
0	0	1
1	1	0

In other words, a plain-text teleprinter tape could be 'added' to a key teleprinter tape, according to the rule that a 'hole' in the key-tape would *change* the plain-text-tape (from 'hole' to 'no hole' or vice versa), while a 'no hole' would leave the plain-text unchanged,*as below.

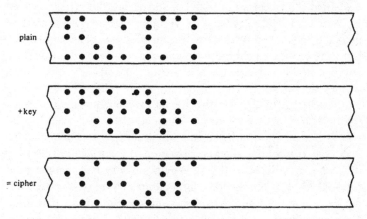

If the key were produced truly randomly, and used on the one-time principle, then such a system would be secure, the same for binary digits as for decimal digits. If all keys were equally likely, then no weight of evidence could accrue to any particular possible plain-text. But that was not the case with these German transmissions. The key was generated by the action of a machine. There were several different kinds of teleprinter-enciphering machines in use, but they shared common features, the key being a pattern generated

---

* They thought of the tape as reading from left to right, as shown, and so thought of it as having five 'rows'. This is not the usual terminology, but for consistency it will be used throughout.

by the irregular motion of ten or so wheels.[45] Such machines did not actually produce a paper key-tape, but from the cryptanalyst's point of view it came to the same thing.

As with Enigma, there was always the possibility of an eventual hardware capture, and the German cryptographers should have allowed for that possibility. But the important break into this type of traffic did not, as it happened, come about in this way. In 1941 someone at Bletchley guessed that a certain message had been sent out twice, and in a very particular way. Some fault in the system for using the machine had allowed an elementary blunder. The message had been sent out each time enciphered by the *same* key, but with the key advanced by one character in one transmission. Once this guess had been made, the recovery of both key and plain-text was a simple step.

With a machine designed for complete security, it should have proved impossible to make any further progress. The sequence of key thus elucidated should have appeared to be random, without any discernible pattern. But it was not. The decisive observation was made by W. T. Tutte, a young Cambridge chemist turned mathematician. This was the breakthrough equivalent to what the Poles had achieved with the Enigma in 1932. Like that work, it constituted a logical rather than a physical capture of the machine, and again it was only the very first step, the *sine qua non*. But one difference was that this time German industry had made a more serious effort. This was not a jumped-up version of a commercial device as was the Enigma. Another difference lay in its part in the German military system. This traffic was sparse but juicy, carrying high-level reports and appreciations. It brought Bletchley much closer to Berlin, at a time when Hitler was taking over personal direction of the war.

Even with an effective capture of the machine, further cryptanalysis should have been impossible: it was the rule of good cryptography. And the 'period' of the enciphering mechanism was not 17576 but a truly 'tremendously large number'. Yet these problems proved not to be entirely insuperable, and by 1942 the analysts were beginning slowly to find ways to exploit the knowledge they had. The work on this particular type* of machine-enciphered

---

* Other types of teleprinter-enciphering machine systems remained unbroken.

traffic became known as *Fish*. One of the most important and general methods was developed by Alan on the basis of Tutte's work in the course of his months of work during 1942 on Fish. It became known as 'Turingismus'.

A new Bletchley industry was sprouting up. It was another way in which 1942 meant starting all over again from the beginning. But this was not to become Alan Turing's game, as naval Enigma had been. For one thing, he had not been the one to start it off. For another, it was someone else who took the step of mechanising its analysis. This person was Newman, who arrived in summer 1942.

Newman had been recruited by his friend P. M. S. Blackett, the Cambridge physicist (and King's College Fellow), who had been Professor of Physics at Manchester since 1937, and was currently applying statistical analysis to the problem of convoy organisation. (For at last the Admiralty was allowing scientific interference with its operations as well as with its intelligence.)

Newman was assigned to the Research section to work on the Fish signals but found himself not particularly adept with the hand methods. He was thinking of going back to Cambridge, when he conceived of an approach that could be automated. The theoretical rationale was founded upon the statistical methods that Alan had developed during 1940 and 1941. These ideas, indeed, were crucial to Newman's plans. But the implementation of them would require the construction of entirely new machines to do very fast counting operations. Newman persuaded Travis to approve this development, and the existing links with the Post Office Research Station were brought into play as early as autumn 1942. This recognition of their frustrated skills meant that the electronic engineers were able to make an impact after all. For the rest of 1942 the project was floundering on the engineering side, but this was not so much because of the electronics, as because of the mechanical difficulties associated with passing paper tape very rapidly through a reader.

Alan knew all about this project, but his active part in the Fish analysis was confined to 'Turingismus'. This, in the autumn of 1942, was taken up by the section called the Testery,* where they

---

* Nothing to do with testing, but named after its head, a Major Tester.

tried out hand-methods on the Fish traffic – just as, years before, the Enigma decryption had painfully begun. Peter Hilton had been moved there from Hut 8, and then another arrival, in autumn 1942, was that of an even younger man, straight from Rugby School. This was Donald Michie, who had won an Oxford classics scholarship, and while waiting to do a Japanese course had entered elementary cryptological training. His talent recognised, he was thrust into the deep end at Bletchley. Both he and Peter Hilton developed Turingismus and reported their ideas back to its originator.

Although the news was so unremittingly bleak, and the prospects so uncertain, 1942 could be a wonderful, liberating year for the young, given opportunities and ideas that would never have been possible in peacetime. Alan's own youthfulness much endeared him to the younger recruits. It was in fact his thirtieth birthday when the fall of Tobruk came as the latest in a series of disasters, but to those fresh from school it would be hard to decide whether one so 'schoolboyish' himself could be as much as thirty, or whether one carrying so much intellectual standing could be so young. A conversation with him was like being invited into some older boy's study where House Colours and Chapel Parade gave way to illicit jazz and D. H. Lawrence novels, but where the housemaster had to turn a blind eye because a precious scholarship was being won.

Peter Hilton was a racy raconteur, and his favourite Turing story concerned the Home Guard. The authorities quaintly insisted upon the Bletchley analysts doing soldierly work in their spare time. The heads of sections were exempt, but Alan conceived a passion for becoming proficient with a rifle, which amazed Harry Golombek, who after two years in the army had no such enthusiasm. Alan enrolled in the infantry section of the Home Guard, and to do so[46]

had to complete a form, and one of the questions on this form was: 'Do you understand that by enrolling in the Home Guard you place yourself liable to military law?' Well, Turing, absolutely characteristically, said: 'There can be no conceivable advantage in answering this question "Yes"' and therefore he answered it 'No'. And of course he was duly enrolled, because people only look to see that these things are signed at the bottom. And so . . . he went through the training, and became a first-class shot. Having become a first-class shot he had no further use for the Home Guard. So he ceased to attend parades. And then in particular we were approaching a time when the danger of a German invasion was receding and so Turing wanted to get on to other and better things. But of course the reports that he was missing on parade were constantly being relayed back to Headquarters and the officer commanding the Home Guard eventually summoned Turing to explain his repeated absence. It was a Colonel Fillingham, I remember him very well, because he became absolutely apopletic in situations of this kind.

This was perhaps the worst that he had had to deal with, because Turing went along and when asked why he had not been attending parades he explained it was because he was now an excellent shot and that was why he had joined. And Fillingham said, 'But it is not up to you whether you attend parades or not. When you are called on parade, it is your duty as a soldier to attend.' And Turing said, 'But I am not a soldier.' Fillingham: 'What do you mean, you are not a soldier! You are under military law!' And Turing: 'You know, I rather thought this sort of situation could arise,' and to Fillingham he said: 'I don't know I am under military law.' And anyway, to cut a long story short, Turing said, 'If you look at my form you will see that I protected myself against this situation.' And so, of course, they got the form; they could not touch him; he had been improperly enrolled. So all they could do was to declare that he was not a member of the Home Guard. Of course that suited him perfectly. It was quite characteristic of him. And it was not being clever. It was just taking this form, taking it at its face value and deciding what was the optimal strategy if you had to complete a form of this kind. So much like the man all the way through.

This Looking Glass ploy of taking instructions literally was one that created a similar fuss when his identity card was found unsigned, on the grounds that he had been told not to write anything on it. It came to light when he was stopped and interrogated by two policemen as he took a country walk. His awkward appearance and habit of examining wild flowers in the hedgerows had excited the imagination of a spy-conscious citizen.[47]

But besides sharing in such victories over blimps and bureaucrats, there was the experience of free association with people who were among the best in British mathematics, in a sort of secret university, one in which tradition and form, together with rank, age, degrees and all such superficialities were ignored. All that mattered was the ability to think. And they had a mathematical Flash Gordon, a logical Superboy, to encourage them – someone who refused to admit defeat, or any limitations on their capacities to succeed. To Peter Hilton, Alan was

> . . . a very easily approachable man – though you always felt there was lots more you did not know anything about. There was always a sense of this immense power and of his ability to tackle every problem, and always from first principles. I mean, he not only . . . did a lot of theoretical work, but he actually designed machines to help in the solution of problems – and with all the electrical circuitry that would be involved, as well.

He did, for instance, design a special machine to help Harry Golombek with the analysis of the particular Enigma system employed by the German motor torpedo boats. There was another designed for use on the main naval Enigma problem; there was far more to it than the Bombe. The technology was not always new; thus the Banburismus process involved the use of paper sheets on which cipher-text messages were represented as punched holes. These had to be moved against each other and coincident holes laboriously counted before the sophisticated statistical methods could be used. There was a hint of irony in the way that Alan chose to call the process ROMsing – a reference to that progressive slogan, the Resources of Modern Science. But it also represented the essential truth about the Bletchley work, and Alan Turing was at the heart of it, never too proud to get his hands dirty with the 'dull and elementary':

> In all these ways he always tackled the whole problem and never ran away from a calculation. If it was a question of wanting to know how something would in fact behave in practice, he would do all the numerical calculations as well.
>
> We were all very much inspired by him, his interest in the work but the simultaneous interest in almost everything else . . . And he was a delightful person to work with. He had great patience with those who were not as gifted as himself. I remember he always gave me enormous encouragement when I did anything that was at all noteworthy. And we were very very fond of him.

Alan's 'great patience' was not usually his most conspicuous characteristic, nor his approachability. But Peter Hilton was in fact the fastest thinker of the new Fish group, and drew out the most rewarding aspects of the 'creative anarchy' that was Alan Turing. It was pure joy to achieve something new, and show him, and have him grunt, gasp, brush back his hair, and exclaim, stabbing with his strange fingers, 'I see! I see!' But then it was down to earth again with the rules and regulations:

> But there again, he began to be beset by the bureaucrats who wanted him to be in at a certain time and work till five o'clock and leave. His procedure – and that of many others of us, let me say, not only he, who were really fascinated by the work – would be maybe to come in at midday and work until midnight the next day. And then, the problem being essentially solved, go off and rest up and not come back for 24 hours perhaps . . . they were getting much more work out of Alan Turing that way. But, as I say, the bureaucrats came along and wanted forms to be filled in and wanted us to clock in, and so on.

Once he ordered a barrel of beer for the office, but it was 'disallowed'. Such questions were trivial, but behind them lay more serious confrontations with the old mentality, which little by little and nearly too late had been obliged to give way to intelligence. Alan's role in this process, however annoying to authority, was not entirely unrewarded. One day in 1942 he, Gordon Welchman and Hugh Alexander were suddenly summoned to the Foreign Office and awarded £200 each. Alan told Joan that they could not be given decorations, so had been given money instead. He probably found it more useful.

In September 1942 the British position was a little less hopeless, but only inasmuch as there had been no serious loss since that of Tobruk. Rommel's eastward advance on Egypt had been checked by Auchinleck in July and by Montgomery in August, the latter being particularly helped by deciphered signals. The desert war was more like a naval war than a conventional front, and was particularly dependent upon information. It desperately required the effective integration of the three services, who had been obliged to swallow a bitter pill indeed by allowing Bletchley information and interpretation to be transmitted over the heads of the London chiefs directly to an intelligence centre in Cairo. But a more centralised system was forced upon them by the cornucopia of north Buckinghamshire.

By May 1942 they were breaking every Enigma key system of the African theatre. In August this was joined by a new Hut 8 success, the breaking of the system used by Mediterranean surface ships. Rommel was now losing one quarter of his supplies through British attacks which were almost totally dependent on detailed Enigma information – sometimes enabling them to pick out the more important cargoes for destruction. News of this triumph was passed back to the analysts in Hut 8 to encourage them in their work.

But the Mediterranean was, ultimately, an Anglo-German diversion. In the world struggle there had been a major setback for Japan at the battle of Midway, where the US Navy proved it could put its own Intelligence to devastating effect. But in Europe there was no such hint of a reverse. The Axis attack on Russia had reached Stalingrad, and the Dieppe raid had ended lingering fantasies of an easy victory in the west. More frightening than either of these developments, however, for Churchill and for everyone else, was the state of the fragile Atlantic bridge. Without it Britain was nothing.

Although the first American troops had arrived in Britain early in 1942, it was the stream of war materials, tanks and aircraft in particular, that alone could make the reconquest of western Europe conceivable. That stream had to face the Atlantic U-boat fleet, which by October amounted to 196 in number. Since 1940 the numbers had trebled, and the sinkings had trebled too. Until mid-1942, American reluctance to provide coastal convoys had diverted U-boats to easy pickings off the eastern seaboard, but in August counter-measures had remedied this gap in defence. Accordingly, the U-boats had turned back to the Atlantic convoys, exploiting the area in mid-ocean where air cover was not supplied, and were now accounting for over half the merchant fleet required to supply Britain within a year. The revived American shipyards were turning out new vessels at top speed, only to have each sunk after three or so ocean voyages. But now the United States had its own pressing demands in the Pacific. The total Allied stock of shipping was actually declining, while the number of U-boats was increasing: there would be 212 at the end of 1942, with another 181 on trial.

Fast approaching was the crisis of the western war. 1943 might see either Britain stocked up as the forward base of an impregnable American industry, or might see it slowly sink. Though more diffuse a crisis than that of the air war of September 1940, it likewise stood to see a make-or-break resolution. Ten years earlier, Alan had conceived a model of action: 'We have a will which is able to determine the actions of the atoms probably in a small portion of the brain . . . The rest of the body acts so as to amplify this.' Now he was one of the clustered nerve-cells, and around him a colossal system which had translated his ideas into concrete form: a British brain, an electric brain of relays clicking through the contradictions, perhaps the most complex logical system ever devised. Meanwhile the two years of the reprieve had rendered the rest of the body more prepared and coordinated to use its intelligence. In the Middle East it was amplifying the dim Morse signals into the sinking of Rommel's army. But the Atlantic was different; here Eisenhower and Marshall might be cut off on a far greater scale than Rommel unless the brain could awaken into life again.

But those two years had seen another momentous change. The ten-fold increase in rotor positions had forced Poland to turn to the technically superior West. And now the twenty-six-fold increase had brought the United States into the electromagnetic relay race. Its Admiral King, who had been more obstinate than the British Admiralty, resisted the setting up of a tracking room until mid-1942. But the US Navy's cryptanalysts had been quick off the mark to see what was needed. Their department had been using modern machinery since 1935, and when the black-out came in February 1942 they were not content to stand and wait until the British caught up: they could do it themselves. This did not accord at all with the British view, which held that the Americans should concentrate on the Japanese ciphers, and not duplicate the work done at Bletchley. But the US Navy was particularly insistent. Already in June, its relations with GC and CS were 'strained' by complaints at the delay they experienced in obtaining a promised Bombe, and then[49]

> in September the Navy Department announced that it had developed a more advanced machine of its own, would have built 360 copies of it by the end of the year, and intended to attack the U-boat Enigma settings forthwith.

These were figures to send Bletchley minds reeling. The whole of GC and CS Enigma work depended upon managing with just thirty

Bombes in summer 1942, although another twenty were on the way. The Americans were proposing a take-over bid for the Atlantic work, by the brute-force expedient of building twenty-six times as many Bombes as the British had available, and using them in parallel.

> But in October a second deputation from GC and CS to Washington negotiated another compromise. GC and CS 'acceded to US desires to attack the German naval and submarine problems' and agreed to supply the Navy Department with the intercepts and with technical assistance. In return the Navy Department . . . undertook to construct only 100 Bombes, accepted that GC and CS should be responsible for co-ordinating the work done by the American machines with that done by the British, and agreed to the complete and immediate exchange of the cryptanalytic results.

There was only one person who knew everything about the methods and the machines, and who was free from day to day responsibility. The responsibility for the detailed coordination thus promised now fell to the Prof. Sorting out the tensions between American bluster and British arrogance was not at all the kind of work he enjoyed, but the Anglo-American liaison had to be made concrete. There was a war on. Accredited as an official with the British Joint Staff Mission in Washington, he was issued with a visa[50] on 19 October. He told Joan, 'The first thing I shall do is to buy a Hershey bar.'

This was not to be the only purpose of his visit. Now that joint operations were being planned, the Allied authorities were in need of new technology to communicate the more subtle aspects of the Grand Alliance. Telegraph communication was not enough. They required proper means for speech signals. There being no submarine Atlantic telephone cables, all speech transmissions had to go by short-wave radio. But as a June 1942 Foreign Office memorandum had stated:[51]

> The security device has not yet been invented which is of any protection whatever against the skilled engineers who are employed by the enemy to record every word of every conversation made.

No one could speak except on the understanding that what they were saying was being overheard in Berlin. There was a row in September 1942, when Prince Olaf of Norway had to be refused permission to talk to his five-year-old daughter, lest it set a precedent for the transmission of uncensored messages by governments-in-exile.

The essential difficulty of providing speech secrecy lay in the overwhelming *redundancy* of speech, as compared with writing. While the modular sum of two written messages would require painstaking work to unravel, the ear and brain could, almost without thinking, analyse and separate a sound signal into conversation, music and background noise. This was only possible because speech signals carried so much more information than was necessary for comprehension. Cryptanalysis thrived on redundancy, whether on routine 'Probable Words', repeated triplets in indicators, or re-enciphered messages. Any secure form of speech encipherment would have to remove it. The systems in use in 1942 did not attempt to satisfy that requirement. Systems existed which split up speech into pitch levels and then permuted them, preventing casual overhearing. But such 'scramblers' were easily penetrated by doing jigsaw puzzle work on a sound spectrograph of the resulting signal. They did not tackle the essential problem. Some efforts were being made at Dollis Hill to create a more sophisticated system, but American developments were much further ahead, and part of Alan's assignment was to investigate them. It marked a shift from the cryptanalytic to the cryptographic side, reflecting the demands of the more offensive Allied war now being planned.

Speech secrecy was in another sense also a problem to the British authorities. In 1940 it had not been too difficult: a few jolly clever chaps in a country house were having a go at breaking German codes. In 1941 it changed: Churchill was getting the most important information from a source which only a select few knew about. The problem was to throw a ring of secrecy around this mushrooming department which lay outside the normal structure of the state. But by 1942 the problem had changed again. Bletchley Park was no longer outside the ordinary channels: it dominated them. Its productions were not the spice added to some other body of knowledge. It was nearly all that they had – photo-reconnaissance and POW interrogation adding points of important detail but never matching in scale what they had fresh from the horse's mouth. There were sixty key-systems broken, producing fifty thousand decrypted messages a month – one every minute. The old days of 'Red' and

'Yellow' were long over and the soaring imagination of the analysts, exhausting the colours of the rainbow, had plundered the vegetable and animal kingdoms: *Quince* for the SS key, *Chaffinch* for Rommel's reports to Berlin, *Vulture* for the Wehrmacht on the Russian front. Certain key-systems were used with the proper safeguards, and with these Bletchley was powerless. The *Shark* system, as they called the U-boat key, still remained intact but for those three days in February and March 1942. But except for these gaps, the German radio communication system had become an open book – to an élite.

It meant that a cloud of mystery and obscurity was settling over the whole British war. All of its documentation had to be falsified, acting out a charade in which 'the old procedures', as Muggeridge saw it,[52]

> like the setting of agents, the suborning of informants, the sending of messages written in invisible ink, the masquerading, the dressing-up, the secret transmitters, and the examining of the contents of waste paper-baskets, all turned out to be largely cover for this other source; as one might keep some old-fashioned business in rare books going in order to be able, under cover of it, to do a thriving trade in pornography and erotica.

It was the ability of the British system to absorb the necessary innovations in an *ad hoc* fashion, when forced to do so, that constituted its real secret weapon. Without that flexibility, all the mathematical and linguistic skill would have been of no avail. Here, perhaps, the habit of paternal Empire enjoyed its triumph. While A. V. Alexander, the trade union man who followed Churchill as First Lord of the Admiralty, was never allowed to know about naval intelligence, let alone cryptanalysis, the prefect layers of the stratified British system were characterised by a trust that allowed them to keep in control and in communication with each other. There were bitter conflicts at every level over what was the most important and exciting development ever to be thrust into the unready hands of British government. But the conflict took place within a club where the rules were agreed tacitly, not enforced by legalities or by handcuffs. Alan Turing could hardly have survived in any other system – certainly not in the German, obsessed by spying and treachery, and perhaps not in the American. He was no team man himself, preferring always to have something tidily his own, but they were able to use him as the sixth-form Maths Brain, distinguished, as the headmaster had said, within his sphere.

For those in charge, the fruits of his labours presented logical problems no less difficult than those of Bertrand Russell. Who was to know what, and who to know they knew? Liaison with the very differently organised American system was just one problem; there was the deception of Dominions, free forces and Russians. The capture of cipher material had to be avoided when it was *not* needed; they had to prevent the 'indoctrinated' from ever falling into enemy hands, and above all, they had to arrange convincing explanations for the foreknowledge that successful operations might betray. But how could this be done without vast numbers of people knowing that something strange was happening, and how could the information be used without giving itself away?

It could not. Bletchley's continued successes depended upon the willingness of German authorities to believe that ciphers were *proved* secure, instead of asking whether they actually were. It was a military Gödel theorem, in which systematic inertia rendered German leadership incapable of looking at their system from the outside. Nor was the principle of 'need to know' one that worked like a complete and consistent logical system. At Cambridge – and at Sherborne School – people guessed the nature of work being done at Bletchley. In 1941 the *Daily Mirror* had carried an article[53] headed SPIES TAP NAZI CODE, which dwelt proudly upon the work of amateur radio operators who were 'taking down the Morse messages which fill the air'. In the hands of 'code experts', the article explained, 'they might produce a message of vital importance to our Intelligence Service'. 'A letter of thanks from headquarters telling us that we have been able to supply some useful information is all the reward we ask,' said the radio spies themselves. And more significantly, on another part of the board, the initiative of the Red Queen was allowing Soviet authorities access to Enigma decrypts. Yet the system still held together.

The traffic in signals was indeed like 'a thriving trade in pornography and erotica', inasmuch as what mattered was not so much the guarding of any particular fact, or thing, as the maintenance of the whole subject as something improper for discussion, like 'dirty talk'. A deeply ingrained fear and embarrassment about the unmentionable was the keynote of all that depended upon the Bletchley work, rather than this or that rule. This was how it worked so well, but it left Alan Turing in an extreme position. It was difficult enough being a mathematician, this being the frightening subject of which even educated people knew nothing, not even what it was, and of which they might proudly boast ignorance. His sexuality might at best elicit a similar condescension, but more likely the associations of evil, tragedy and disease. Above all, it was a matter on which society still demanded silence. Such silence was for him tantamount to an uneasy game of deceit, and he loathed pretence. But as chief consultant to GC and CS, he was living at the heart of yet another imitation game, doing work that did not officially exist. Now there was almost nothing in his life that he could talk about but chess-playing and fir cones.

Fragments of ordinary life continued. Occasionally he saw his friend David Champernowne, himself now working in the Ministry of Aircraft Production, but of course they never spoke of their work. He had continued to concern himself with Bob's future, being concerned that he should try for a Cambridge scholarship. This he did, cramming for Latin, but reaching only the standard for an ordinary place. In the circumstances this was quite an achievement, but Bob felt that he had disappointed Alan, in not possessing a feel for abstract ideas. Alan could not possibly afford to send him to Cambridge, so Bob went in autumn 1942 to take the telescoped chemistry course at Manchester University, earning his own keep by stoking the boilers in the Friends Meeting House.

Bob had a sharp eye, and guessed that Alan, 'Champ' and Fred Clayton formed a team working together in Intelligence. He was mistaken in that respect, but right about Alan, though all he knew was that Alan worked at a place called Bletchley. Other people likewise were able to put two and two together. John Turing, while serving in Egypt, found that his superior officer had a brother working at the same town, and they guessed it had to do with ciphers. Mrs Turing guessed correctly too, remembering Alan's letter about 'the most general code or cipher' in 1936, and knowing he worked for 'the Foreign Office'. She liked the idea of him being assigned duties once more, although perhaps it was a disappointment that these did not oblige a military haircut. Her long letters sometimes found their way unread into the Hut 8 waste-paper basket, Alan telling Peter Hilton 'Oh, she's all right.' She also visited him in autumn 1941. He tried to hint that he was doing an important job, with 'about a hundred girls' working for him, but neither she nor indeed anyone else had the faintest idea of how important it was. How could they? The concept of an information processing system, one matching the organisation of an advanced industrial power, had only just been invented.

Which was now the ordinary and which the extraordinary? Which the reality and which the illusion? The pure mathematician of 1938 had wandered into an amazing position on the board, so that his brain concentrated ideas upon which the struggle for Europe depended. Miranda was playing chess when she saw the brave new world – and here it was in this organisation, a progressive scientist's dream in which the experts had confounded the Blimps, and forced them to play the modern tune. Far above the heads of its slaves, far above the heads of the British people, the secret technocracy was working like an intelligent machine. There at the centre was the brain of the Alpha Plus who had breathed life into it and nurtured its growth – the unhappy Alpha, cursed with the ability to think for himself, and beginning to be edged out by his own creation.

The broken Enigma and Fish systems were only just decipherable, and were stretching the fastest minds and the resources of modern science to the limit. They also depended upon luck and sudden, brilliant observations. On 30 October another stroke of fortune, the capture of U-559 off Port Said, at last gave Bletchley the key to the blank Atlantic, just as Alan was preparing to cross it. Thus elements of pure chance, amplified by a youthful will that had thrown off the 'old fogies' of the 1930s, had implanted into the British state a fantastic new element. In his central control of the war, Churchill now relied totally upon an unmentionable department, in which no one knew what anyone else was doing, and which made deceit into second nature. Starting with those early discoveries in the outhouses of Bletchley Park, a great explosion of implications had, in silence, proliferated through level after level of military and political organisation. It was a logical chain reaction, whose after-effects no one had time or inclination to consider.

General Montgomery always put his troops 'in the picture'. Indeed he was prone to give too much of the ultra-modern picture away, and had to be reprimanded by Churchill. But with that picture integrated effectively into Montgomery's plans, his troops inflicted defeat at last upon the Afrika Korps. It was the first decisive British victory over German arms in three years of war. On 6 November 1942, General Alexander signalled 'Ring out the bells!'. The British occupation of Egypt was preserved, its puppet regime saved, the southern German pincer on the Middle East destroyed. Then on 8 November, the Allied forces landed in Morocco and Algeria, achieving complete surprise. It was a first victory for planning and the coordination of Intelligence. The Americans were now back in the Old World – and to British consternation, would treat with the Vichyiste Darlan. But the British could not complain, for they had handed over the torch.

Alan Turing had boarded the *Queen Elizabeth* on 7 November.[54] As the refitted Monster zigzagged alone towards America, the fighter escort left behind, the King's First Minister was explaining that he did not intend to preside over the liquidation of the British Empire. Churchill also said that it was only the end of the beginning. But for the goose that laid the greatest golden egg, it was already the beginning of the end.

Bridge Passage

Aboard at a ship's helm,
A young steersman steering with care.

Through fog on a sea-coast dolefully ringing,
An ocean-bell – O a warning bell, rock'd by the waves.

O you give good notice indeed, you bell by the sea-reefs ringing,
Ringing, ringing, to warn the ship from its wreck-place.

For as on the alert O steersman, you mind the loud admonition,
The bows turn, the freighted ship tacking speeds away under her
   gray sails,
The beautiful and noble ship with all her precious wealth speeds
   away gayly and safe.

But O the ship, the immortal ship! O ship aboard the ship!
Ship of the body, ship of the soul, voyaging, voyaging, voyaging.

While the Atlantic remained in the dark, November 1942 proved to be the worst month yet for Allied shipping. But the North African landings drew off part of the U-boat force, and the *Queen Elizabeth*, faster than any U-boat, made her way in safety. Alan disembarked at New York on 13 November, but according to a story he told his mother,[1] was very nearly refused entry to the United States:

> He had on arrival some difficulty over admission as he had been told on no account to take any papers other than those in the Diplomatic Bag which he carried. The triumvirate who confronted him on landing talked of despatching him to Ellis Island. Alan's laconic comment was, 'That will teach my employers to furnish me with better credentials.' After further deliberation and passing of slips of paper, two of the triumvirate outvoted the third member and he was admitted.

Such problems were supposed to be kept under control by W. Stephenson, the Canadian millionaire who directed 'British Security Coordination' from Rockefeller Center. Stephenson, originally installed to liaise between the British secret service and the FBI, had made a considerable effort to advance British interests in America by undercover manipulation. Since 1941 his office had expanded to take in the more serious work of channelling Bletchley's productions to Washington. But perhaps Alan's tiresome habit of taking instructions literally had defeated even him.[2] It was certainly a curious greeting for a person who in so many ways was bridging old and new worlds. His primary assignment took him to the enormously expanded capital city, much changed since the sleepy days of 1938, where his opposite numbers in the Navy's cryptanalytic service, 'Communications Supplementary Activities (Washington)', were based.

From Bletchley's point of view, America was the miraculous land across the rainbow bridge, possessing resources and skilled labour in quantities that desperate Britain could not supply. The CSAW was already closely connected with the most advanced sections of American industry, using Eastman Kodak, National Cash Register and IBM to plan and build its machinery. As in other ways, Hitler had the effect of adding British ideas to the massive capacity of American business. Again It was Alan Turing's role to connect the logical and the physical.

But CSAW was certainly not without its own brains, and one of its staff was the brilliant young Yale graduate mathematician, Andrew Gleason. He and another member of the organisation, Joe Eachus, looked after Alan during his period in Washington. Once Alan was taken by Andrew Gleason to a crowded restaurant on 18th Street. They were sitting on a table for two, just a few inches from the next one, and talking of statistical problems, such as that of how best to estimate the total number of taxicabs in a town, having seen a random selection of their licence numbers. The man on the next table was very upset by hearing this technical discussion, which he took to be a breach of 'security', and said, 'People shouldn't be talking about things like that.' Alan said, 'Shall we continue our conversation in *German?*' The man was insulted and told them in no uncertain terms how he had fought in the First World War.

They were all spy-conscious in Washington now, but such anecdotes apart, the central event of Alan's visit was the breakthrough back into the U-boat Enigma. This was achieved without the possession of faster Bombes; it depended upon a precarious thread of luck, ingenuity, and a German blunder. It went back to the weather signals that in mid-1941 had given them an almost unfairly simple crib each day, thanks to the fact that they were transmitted both in the Enigma, and in the special meteorological cipher. But early in 1942 a change in the system had denied this method to Hut 8. Not until the U-boat capture of 30 October could it be regained. This gave them the cribs, but the difficulty remained that it would take three weeks to work through all the rotor settings, just for one day's traffic. Here, however, they were saved by a German blunder which, in effect, threw away all the advantage that the fourth wheel offered. For the weather reports, and other routine short signals, the U-boats used their Enigma with the fourth wheel in 'neutral' position, thus reducing the cryptanalysts' problem to the one they had mastered in 1941. This in itself was not fatal for Germany; the greater mistake lay in the fact that the three rotor settings used for the weather reports were also used for all the other traffic of the day. For this the analysts now needed only to work through 26 possibilities for the fourth wheel, rather than the $26 \times 336 \times 17576$ possibilities that would otherwise have been the case. As a result of this slip, Hut 8 was able to supply decrypted messages from 13 December. It was not a sudden restoration of sight, but more like a return to the period in the spring of 1941. They had weeks in which nothing worked out. But it was sufficiently copious a flow of information for the Tracking Room in the OIC to have by 21 December a clear idea of the location of all eighty-four U-boats at large in the North Atlantic. And this time Hut 8 was not alone. In Washington, Alan Turing was indoctrinating the American analysts into all their methods. Now, when the rotor settings were discovered, they were passed back and forth across the Atlantic, the analysts beginning to communicate directly as indeed the two tracking rooms were also doing.

The decrypts flooded in, at an average rate of 3000 a day, like a newspaper filled with nothing but concise, up-to-date news about the Atlantic operations. Just as the flow began, in early December, the[3] 'irreplaceable' Winn collapsed from 'total mental and physical exhaustion' and 'what was not of immediate operational importance had to be put on one side for later study and usually, before this could be undertaken . . . the next crisis was upon us and the study had to be abandoned', and so forth. But somehow the joint system managed to keep going, and in the new year became able once more to divert convoys away from known U-boats. The result was that on the other side, they could not understand why their success in sinking Allied shipping had suddenly been reduced to the level of September 1941. Indeed they were sure that U-boat positions were somehow known to the enemy. But the head of the German naval intelligence service at Naval High Command adhered to the opinion that it would be impossible for the enemy to have deciphered the signals. They continued to assume that there was a spy network operating in their bases in occupied France, although nothing could have been further from the truth. And so their faith in machines and experts continued to be matched by distrust of men. There were in fact many other factors involved besides cryptanalysis – the provision of escorts and aircraft patrols, the development of radar and of counter-radar measures, and the fearsome weather of this fourth winter of the war. But the crucial change was that Allied authorities once more knew where the U-boats were.

His liaison job accomplished, Alan left Washington at the end of December. He had been working at the nerve-centre of the alliance, at its point of equilibrium. The British contribution had not yet been overtaken by the American. The Casablanca conference, from 14 to 24 January, saw Churchill as Roosevelt's equal. For the last time, the Americans supported a British strategy to regain the Mediterranean. For the first time, Britain was to act as an American base. It was also the equilibrium point of the war. It was taking far longer than expected to clear North Africa, Montgomery missing some ultra-fine chances, with disastrous consequences all over the globe. The Russian front was still undecided. Nothing was clear, despite the demand for 'unconditional surrender'. The crude 'strategic bombing' was endorsed, for want of anything better. But the Atlantic battle, agreed at Casablanca to remain the top priority, had taken a turn. For the first time, new Allied ship construction was exceeding losses.

Alan went to Saunderstown, Rhode Island, to visit Jack and Mary Crawford again, as he had done from Princeton. But Jack had died on 6 January, a few days before Alan arrived. His widow asked him to stay on a few days nonetheless. Then his direction turned too.

He went down to New York City, arriving[4] at the Bell Laboratories building on West Street, by the piers, on the afternoon of 19 January 1943. And for two months he soaked himself in the electronic technology of speech encipherment.

Like most organisations devoted to secret work, Bell Labs operated in a cellular fashion so that people never knew what was happening outside their own department. Alan, however, was free to move into any 'cell' he wished, although he had to be careful not to transfer information himself. It came through to the Bell engineers with whom he worked that his 'clearance' had come not from the Army or the Navy, but from the White House itself. Most of his time, however, was spent in one particular 'cell' which had the responsibility of trying to crack speech encipherment systems which had been proposed. He made an impression from the start, for within an hour of arrival he had solved a problem. It involved a scrambling system in which time segments were permuted, by means of nine magnetic heads simultaneously reading a magnetic tape. 'That ought to give you 945 codes,' said Alan when this was explained. 'It's only $9 \times 7 \times 5 \times 3$.' It had taken one of their technicians a week to work it out.

For his first week, Alan acquainted himself with all the projects on which they were working, and became particularly keen on taking up one of them himself. This was the challenge presented by an RCA engineer, who had devised a system in which a speech signal would be *multiplied* by a key signal. It presented quite an unusual problem. On 23 January, Alan announced he had begun to think of a way of attacking it, and then he came in after the weekend convinced of its possibilities. His idea involved the use of the Vocoder.

Alan had probably already learnt about the Vocoder when in Britain, since Dollis Hill had received information on it in 1941. It was a very advanced piece of communications technology, which had been patented by the Bell engineer H. W. Dudley in 1935 and developed since then at the Bell laboratories. The idea of the Vocoder was to abstract the essential elements of speech, throwing away much of its redundancy, and conversely to reconstruct the speech signal from its essential components. One way of thinking of this process was to regard it as reducing the *bandwidth*, or frequency range, of a speech signal.

Any Bell Labs engineer would be familiar with the idea of reducing the frequency range of speech, since the telephone cut off sound above 4000 Hz. The resulting lack-lustre tones were still perfectly comprehensible, the point being that higher frequencies were redundant in ordinary applications. But to reduce that frequency limit much further would produce a sort of miserable grunting, which would not do at all. The Vocoder did something far more sophisticated. It collected information about the amplitude of the speech signal at each of ten frequencies up to 3000 Hz, and also took an eleventh component which coded either the fundamental pitch of the sound or (during unvoiced sounds like *ssss*) an absence of pitch. Each of these eleven signals required a frequency range of only 25 Hz. In this way, sufficient information was abstracted for intelligible speech to be reconstructed, and yet the total bandwidth was confined to less than 300 Hz.

Alan had already suggested that the principle of the Vocoder, taking samples at ten different frequency levels, could be applied to an attack on the time-segment permuting type of speech scrambler – perhaps with the idea of recognising the neighbouring segments automatically. His idea for applying the Vocoder to the RCA multiplying speech cipher was something much more sophisticated, and he said it would require at least a week of computational work to see if it was feasible. In his second week at Bell Labs he settled down to this work, which involved calculations with Hermite polynomials, and in his third week he had some assistance with the calculations.

But Alan was also involved in a quite separate 'cell', which was devoted to creating the world's first totally unbreakable speech encipherment system. This was the most advanced work in progress at Bell Labs, and its best guarded secret. The original goal had been that of finding a way to encipher speech on the Vernam principle, so that if a one-time key were employed, the result would be as unbreakable for speech as it was for telegraph signals. With this end in view, they had attacked the quite novel problem of representing speech by the discrete 0's and 1's used in a Vernam cipher system.

They had begun in 1941 with the Vocoder, and tried to adapt it to their purpose by approximating its eleven outputs as being either 'on' or 'off'. This, however, resulted in a 'badly mutilated' speech signal.

Accordingly they had abandoned the simple binary 'on or off' of the Vernam cipher, and instead approximated the Vocoder outputs not by two possible levels, but by *six*. The eleventh signal required a finer tuning than the others, and was allowed thirty-six levels. The effect was to encode the speech-signal as a total of twelve streams of 'base six' digits, like 041435243021353. . . . Each such stream would then be added in modular fashion* to a similar but random key sequence, and the result transmitted. At the other end the identical key would be subtracted, and the speech reconstituted. The speech signal was to be sampled for its 'levels' fifty times a second, which meant that the transmission was roughly equivalent to sending 300 teleprinter characters a second. They had succeeded in devising the equivalent of a one-time pad system for speech.†

The development was given the intriguing name of 'Project X' or the 'X-system'. By November 1942 an experimental model had been installed at New York and tested with[5] 'a synthetic set of signals from a signal generator that had previously been sent to England'. In January 1943 they were beginning to assemble the first model intended for operational use. There were tremendous technical obstacles. Not only was the basic Vocoder already very complicated, but it required a large number of further components for taking the discrete (or 'quantised') levels. It also required the allocation of seventy-two different frequencies, for the twelve streams of digits were to be played like music, with a different frequency, not a different amplitude, for each possible digit. The system also required perfect synchrony between sender and receiver, and had to allow for the fading and time delays in the Atlantic ionosphere.

The result was a roomful of electronic equipment at each end of the system:

> A terminal occupied over 30 of the standard 7-foot relay rack mounting bays, required about 30kW of power to operate, and needed complete air conditioning in the large room housing it. Members working on the job occasionally remarked about the terrible conversion ratio – 30kW of power for 1 milliwatt of poor-quality speech.

---

* Curiously, they had not found this by any means an obvious idea, although it was just like the base-10 modular addition used in one-time pad ciphers. They had invented it afresh.
† They had also independently invented a form of pulse-coded modulation.

But it worked, which was the main thing. For the first time, secret speech could cross the Atlantic. Alan's inspection of the apparatus on behalf of the British government preceded a formal Anglo-American agreement on the subject. The somewhat disgruntled minutes[6] of the meeting of the Chiefs of Staff Committee of the War Cabinet, on 15 February 1943, explained the position:

> THE COMMITTEE had before them a Memorandum by the British Joint Communications Board on a proposal for the installation by the Americans of a highly secret apparatus for telephone communication between the United States and London.
>
> THE COMMITTEE were informed that Major Millar, a US officer specially sent over for the purpose of installing the apparatus, had now arrived. His instructions were to place it in some building where it would be exclusively under American control, though it could be used by high officials in the British Government. There were only two other sets, one of which was being installed in the White House, and one in the War Department in Washington. No more sets could be produced for eight or nine months.
>
> The following were the main points in the discussion:
>
> (a) *Security.* It was noted that the only Englishman who had yet been able to examine the apparatus was Dr Turing of the Government Code and Cypher School. In view of the fact that conversations relating to British operations would undoubtedly take place over the secret telephone, we had a legitimate interest in finding out whether the new apparatus could really be considered one hundred per cent secure. It was thought that this could best be cleared up by the Joint Staff Mission in Washington, where ample technical talent was available.
>
> (b) *Site for installation.** In view of the fact that the apparatus would undoubtedly be used by the Prime Minister, and that no extension telephone involving an outside line could be permitted, it seemed that the only practicable site for the installation would be the new Government Office building, Great George Street. It was noted that the Americans hoped to have the installation complete by the 1st April.
>
> (c) *Control of the Apparatus.* Although the secrecy being observed by the

---

* The British did not get their way regarding the location of the London terminal. In April the X-system was installed in the American headquarters and only later was a line run to Churchill's war room.

Americans about the apparatus, and their desire for exclusive control, might be open to criticism, it was thought better not to raise objections at this stage.

The Committee went on to tell the Joint Staff Mission in Washington to 'approach the Americans with the object of making a thorough examination of the new secrecy apparatus, so that we could be satisfied that its security could be relied upon.' Alan left Bell Labs for a week in Washington from 17 to 25 February, and this might well have been in connection with these negotiations. Apparently he found room for improvement, since according to a later minute of the Chiefs of Staff:[7]

LT. GENERAL NYE recalled that Dr Turing had not been completely satisfied as to the security of the equipment and had suggested certain alterations.

Meanwhile Alan's work on the RCA cipher seemed to show that his method would not work. He joined in the work of the 'cell' on another approach to the problem. Despite the great technical secrecy, there were enough straws in the wind for his colleagues to realise that he was doing other, top-level, work. Thus it was noted that while speaking with H. Nyquist, one of the top Bell consultants working on the X-system, Alan had met William Friedman, who was the chief American cryptanalyst. It got back to the 'cell' that Alan was 'the top cryptanalyst in England'. One of his colleagues there, Alex Fowler, heard this and saying, 'Oh, you can help me,' produced a newspaper puzzle. 'That's one of those *Herald Tribune* cryptograms,' replied Alan. 'I've never been able to do *those*.' He sometimes mentioned his previous period in America, and his connection with Church, and some of the mathematicians at Bell were aware of the Turing machine. But he still found it difficult to adjust to American civilities. New acquaintances at Bell Labs complained of Alan giving no sign of recognition or greeting when he passed them in the halls; instead, he seemed to 'look straight through them'. Alex Fowler, who was an older man of just over forty, was able to take Alan to task. He was abject, but made an explanation hinting at why he found so many aspects of life difficult. 'You know at Cambridge,' he said, 'you come out in the morning and it's *redundant* to keep saying hallo, hallo, hallo.' He was too conscious of what he was doing, to slip into conventions without thinking. But he promised to do better.

There was no time for social relaxation. It was the peak of the war effort, and they were all working for up to twelve hours a day. Alex Fowler would have liked to find the time and energy to entertain Alan, but it was out of the question. Like many people, he was also afraid of boring him. Alan meanwhile was accommodated in a hotel. He told a joke about how he tried going to read in the toilet when the black-out was on, but found to his chagrin that the lights went out there too.

Greenwich Village in 1943 was perhaps more exciting than Princeton in 1938. Alan later told a story about a man in the hotel having made a sexual approach, amazing him by its casualness. No hint of anything like this was heard in Bell Labs, though Alan once said, 'I've spent a considerable portion of time in your *subway*. I met someone who lived in your *Brooklyn* who wanted me to play *Go.*' Another time he said: 'I had a *dream* last night. I dreamt I was walking up your Broadway carrying a flag, a *Confederate* flag. One of your *bobbies* came up to me and said, "See here! You can't do that," and I said, "Why *not*? I fought in the *War between the States*".' Alan's curious English voice, like the X-system encoding his information by frequency rather than by amplitude, made a vivid impression on his temporary colleagues.

By the end of February Alan had gained more familiarity with the electronic equipment that was used in the laboratory. Although his work was primarily theoretical, he asked many questions about oscilloscopes and frequency analysers, such as they were using for breaking speech encipherment systems, and left them impressed with the amount of knowledge he had picked up. He also took advantage of the theoreticians at Bell Labs, for instance learning from Nyquist his theory of feedback, which was a new departure making use of the complex numbers.

But another significant interaction of his visit was the one he made every day at teatime in the cafeteria. Here he met a person who had been able to take the part of an academic, philosophical engineer, the role that Alan might have liked had the English system allowed for it. This was Claude Shannon, since 1941 working for

Bell Laboratories and producing ideas of a breadth which would find scant encouragement in any British company. While Friedman was Alan's opposite number in terms of direct responsibility for cryptanalytical work, Friedman was an older and more old-fashioned figure: a code and cipher fanatic, rather than someone who had looked at cryptology through the eyes of modern science as Alan had. In intellectual depth it was Shannon who was Alan's opposite number, and they found a good deal in common.

People had thought about machines since the dawn of civilisation, but *Computable Numbers* had come up with a precise, mathematical definition of the concept of 'machine'. People had thought equally long about communication, but here again it needed a modern mind, in this case Claude Shannon's, to provide a precise definition of the concepts involved. These were somewhat parallel developments. Shannon had completed his first paper[8] in this direction in 1940, and by 1943 his fundamental ideas were beginning to be used in Bell Laboratories, in whose mathematical department he was now employed. He was consulted about the design of the X-system, which posed some of the questions answered by his work.

The transmitter, the ionosphere, and the receiver, were a *communication channel*, in his terms; a channel with a limited capacity, and a channel plagued by noise. Into this channel had to be squeezed a signal. Shannon found ways to define channel capacity, noise, and signal, in terms of a precise measure of information. The problem of the communication engineer was that of encoding the signal in such a way as to make best use of the channel, and to prevent it being distorted by the noise; Shannon found new precise theorems which placed limits upon what could be achieved.

There was not only a parallel between his work and Alan Turing's; there was a sort of reciprocity. On his side Alan, although his main strength was in the logic of machines, had dipped into the study of information. Not only was this true in general terms of all his cryptological work, but there was an even more specific point of contact. Shannon's measure of information was essentially the same as the Turing 'decibans'. A *ban* of weight of evidence made something ten times as likely; a binary digit or *bit* of information made something twice as definite. There were fundamental connections between the theories, although they were not free to discuss them. Shannon only knew by implication why Alan was at Bell Labs at all.

Then on his side, Shannon had also independently thought about logical machines. From 1936 to 1938 he had been working on the differential analyser at MIT, and had designed a logical apparatus with relays in connection with a particular problem. This in turn had led him to write a paper[9] in 1937 which drew the connection between the 'switching' operations of electromagnetic relays, and Boolean algebra – hence doing this just as Alan was designing his electric multiplier at Princeton.

Alan showed *Computable Numbers* to Shannon, which he read, immediately impressed. They also discussed the idea implicit in *Computable Numbers*, an idea of which they were independently convinced. Shannon had always been fascinated with the idea that a machine should be able to imitate the brain; he had studied neurology as well as mathematics and logic, and had seen his work on the differential analyser as a first step towards a thinking machine. They found their outlook to be the same: there was nothing sacred about the brain, and that if a machine could do as well as a brain, then it *would* be thinking – although neither proposed any particular way in which this might be achieved. This was a back-room Casablanca, planning an assault not on Europe, but on inner space.

Here at least was something they could speak of freely. Once Alan said at lunch, 'Shannon wants to feed not just *data* to a Brain, but *cultural* things! He wants to play *music* to it!' And there was another occasion in the executive mess, when Alan was holding forth on the possibilities of a 'thinking machine'. His high-pitched voice already stood out above the general murmur of well-behaved junior executives grooming themselves for promotion within the Bell corporation. Then he was suddenly heard to say: 'No, I'm not interested in developing a *powerful* brain. All I'm after is just a *mediocre* brain, something like the President of the American Telephone and Telegraph Company.' The room was paralysed, while Alan nonchalantly continued to explain how he imagined feeding in facts on prices of commodities and stock, and asking the machine the question 'Do I buy or sell?' All afternoon the phone was ringing in his laboratory, with people asking who on earth it was.

On 2 February 1943, the German surrender at Stalingrad had marked the turning of the tide. But while the eastern front was turned by sheer brute force, the western powers had the space and time for developments in which force was not the only element. The intricacy and sophistication of their cryptanalysis was the most extreme example, but this was not the only sphere in which machinery was taking the war out of the old world of duty and self-sacrifice. In November 1942 the ground had been cleared at Los Alamos, and by March 1943 the first scientists were already moving in. The atomic bomb they planned to make would not release greater energy than the raids of 1943 already deployed against Germany. But it would make thousands of bombers redundant, thus mechanising the discipline and coordination of air offensives. The Manhattan Project would still depend upon an aircraft pilot – but then he, too, was being automated at Peenemünde, where the long-predicted 'monster cannons' were on the way. The V-weapons would lack sufficient accuracy – but such problems of guidance were also being attacked at Germany's back by the new techniques of proximity fuses, automatic celestial navigation, and automatic fire control. People easily understood powerful guns, fast ships, impenetrable tanks, which extended human limbs. By now the secret of radar was out, and it could be understood how its manifold applications extended human eyesight to the longer wavelengths of the electromagnetic spectrum. But rapidly developing, and not only at Bletchley and Washington, was a new kind of machinery, a new kind of science, in which it was not the physics and chemistry that mattered, but the logical structure of information, communication, and control.

This development was not confined to warfare. In Dublin, Schrödinger was lecturing under the title 'What is Life?', and advancing the conjecture that the information defining a living organism must somehow be encoded in molecular patterns. In Chicago, two neurologists had read *Computable Numbers*, and were publishing[10] an idea that connected the definition of the logical machine with the actual physiology of brains. They had applied Boolean algebra to the properties of nerve-cells. As Hilbert died at Göttingen on 14 February 1943, a new kind of applied logic was taking shape. Against the distant thunder of the east, there were the first glimpses of a post-war science. This first half-serious, half-joking talk of 'thinking machines' reflected both the immensely wider horizon that the war had opened to science, and the fact that an end at last seemed possible.

By 4 March Alan had completed a report on his suggestions regarding the RCA speech cipher, and had studied in great detail all the speech systems with which they were working. The head of the section had expressed concern that Alan might invent something which would create a tangle over patent rights, but Alan pooh-poohed this, saying that he wanted Bell Telephone to have anything that he thought of. 'Hands across the sea', he said. But what idea could possibly compare with those he had already handed across the sea, ideas far too important for any patent office to know existed? From 5 to 12 March he had to spend another week in Washington at the request of the Navy, to look after this side of his mission again. It was another critical point for the U-boat Enigma, for on 10 March the codebook for the short weather report signals, on which the December breakthrough had been based, was withdrawn. But the three months of successful decryption had allowed analysts to develop alternative methods in time, finding in particular that other 'short signals', allowed to remain in force, were also enciphered with the fourth wheel in its 'neutral' position. Again the German force threw away its advantage, and with more than sixty Bombes at Bletchley now, the Allied dependence upon this special trick was lessening. The change of 10 March was overcome in just nine days.

Returning to Bell Labs, Alan worked for a few more days on the RCA cipher. He wanted them to keep him informed on their progress after he had returned to England, and indicated two possible means of communication: either through Friedman, or through Professor Bayly, a Canadian engineer attached to British Security Coordination. At a quarter past four on 16 March, he had a telephone call from BSC, telling him it was time to embark. He stopped work and left the West Street building within half an hour. His ship[11] was not a *Queen* this time, but the 26,000-ton troop transporter *Empress of Scotland*. This British vessel could sustain 19½ knots, while packed with 3867 enlisted men, 471 officers – and just one civilian.

*

After a week's delay, the *Empress of Scotland* left New York harbour on the night of 23 March. She steamed due east into mid-Atlantic, and then swung up to the north. Only one of the thousands carried into the midst of the battle knew the precarious system upon which so much depended – but that knowledge made no difference now. For a week Alan was ordinary again, having to take the risks and trust the authorities like everyone else. The danger was real enough: on 14 March the similar *Empress of Canada* had been sighted and sunk. Briefly he was on the receiving end of the system, and briefly he was relieved of responsibility for it.

In a sense he had stood upon the burning deck since 1939; but the sentiments of *Casabianca*, those of doing a patriotic duty contrary to inclination, were very remote from the spirit of Alan Turing's war. He was doing what he had chosen to do, and was expressing himself, not subjecting himself. His mind continued to work away, fascinated by the problems, even during this voyage home. While briefly sharing in the helplessness, confinement and danger of the war, he spent his time studying a twenty-five-cent handbook on electronics, the *RCA Radio Tube Manual*, and invented a new way of enciphering speech.

If he dreamt about fighting in the War between the States, it was the reverse of the truth. He was committed to the Yankee side, and had seen no fighting. His struggle had all been behind the lines. But that was not quite the whole story. Once in conversation with his friend Fred Clayton, the question arose as to how scientists could have continued to work for Germany. With both personal honesty and political realism, Alan pointed out that in scientific research it was inevitable that one became absorbed in the work, and did not think of the implications. It was, in this respect, a Looking Glass war, with the *B. Dienst* analysts no less fascinated by their work.* It could be an entrancing dream world, without connection with the issues of the war. But Fred did get him to admit that Germany raised other questions.

For Alan Turing's generation, the First World War had been a War between the States, meaningless as Tweedledum and Tweedledee. The mirror symmetry of nationalism had disgusted

---

* At least one of Scholz's students was working directly against him.

Russell and Einstein, Hardy and Eddington, who saw only human beings with labels pinned upon them, destroying each other. They longed to jump out of the system of *La Grande Illusion*, and in 1933 the new generation had voiced this longing openly. But Russell and Einstein came round to support of *this* war, the war for the Anti-War, the war that could be imagined not as the 'national war' but the world Civil War, a crusade against slavery. That it was primarily a war between two tyrannies; that it had massively reinforced national governments; that it had made mass slaughter respectable again; that it had militarised the advanced economies – these did not countervail. Against this enemy anything could be justified. In 1933 they had reviled the arms manufacturers above all others. But they were all arms manufacturers now.

There had been British atrocities in an Ireland now so obdurately neutral, but not with filing systems, medical experiments, and industrial cyanide. At Bletchley they had already deciphered some of the figures that Germans did not know, or want to know. That sheer explicit single-mindedness, in following ideas to a logical conclusion, was what lay beyond the grasp of English minds. But that Nazi definiteness had helped to stimulate the scientific consciousness without which the western Allies, at least, would have been helpless.

This dimension of the war went without saying, and did not need talking about. Yet in Alan Turing's case there was a sharp irony,* in that Himmler had sneered at British Intelligence for making use of

---

* German security policy was more advanced than the British. In a letter[12] of 9 October 1942, Himmler replied to a memorandum from the Consultant Physician to the *Reichsicherheitshauptamt* (Supreme State Security Office) on the subject of *die Homosexualität in der Spionage und Sabotage*. 'I grant you . . . that the British have found some rather promising (*passender*) material for their purposes here,' he wrote, but decreed that there was no question of a remission in the vigorous prosecution of homosexuality for the sake of gaining recruits, in view of the risk of homosexual vice rampaging unpunished amongst the *Volk*, and whole sections of the youth being seduced. Anyway, he said, if one of these degenerates and crooks (*Pathologen und Gauner*) were set on betraying his country, he would do so whether punished according to Paragraph 175 or not. Prosecution, in 1942, meant consignment as a 'pink triangle' prisoner to a concentration camp. The doctors were sharply rebuked by Himmler on 23 June 1943 for their suggestion of retraining (*Erziehungsversuche an anormalen Menschen*) as a waste of effort at a time when Germany struggled for its existence, and because the outcome of such efforts was so dubious (*höchst zweifelhaft*).

homosexuals, and had specifically directed that in Germany useful talents could not exempt those so identified from the general rule. Few indeed could have appreciated that irony, and fewer still have believed that this strange civilian on the *Empress of Scotland* was playing as much of a part as any in bringing Himmler to his own poison.

In 1939 Forster[13] had expressed the numbing conviction that to defeat fascism it would be necessary to become fascist. It had not happened like that, and in many ways the channels of communication had been opened up. Yet far more subtly, the logic of the game was reflecting something inhuman into what was called democracy: not just in the bombing raids alone, but in a deep internal way. As the Allied war turned from defence to offence, from innocence to experience, from thinking to doing, an undefinable naiveté was going with the wind. The very success and efficiency of its scientific solutions was bringing this about. In 1940 there had been a feeling, quite illusory perhaps, of individual contact with the course of events. But now even a Churchill was dwarfed by the scale and complexity of operations. In the 1930s it had seemed that there were simple choices to be made between good and evil. But after 1943, as the Allies prepared to join the Russians in biting on the Nazi apple, nothing would be simple again. Nothing could even be properly known.

In the cold dawn of 31 March, a British escort was waiting for the *Empress of Scotland* in the Western Approaches. The danger was passed, no U-boat having sighted the ship, and the odd civilian returned safely to his country. For three years now he had helped to stem the tide by thinking, and they had built a colossal machine around his brain. But they could not fight the war by knowing about it. Intelligence was not enough; it had to be embodied in a savage world. Nor would its engineer escape that general rule.

PART TWO

THE PHYSICAL

5 Running Up

One's-self I sing, a simple separate person,
Yet utter the word Democratic, the word En-Masse.

Of physiology from top to toe I sing,
Not physiognomy alone nor brain alone is worthy for the Muse,
I say the Form complete is worthier far,
The Female equally with the Male I sing.

Of Life immense in passion, pulse and power,
Cheerful, for freest action form'd under the laws divine,
The Modern Man I sing.

The surrender at Stalingrad had marked the beginning of the end for Germany. The war had turned. Yet in the south and west there was little evidence of progress for the Allies. The African war dragged on, the *Luftwaffe* still mounted raids on Britain. And the ports were sheltering the survivors of what had been the most damaging convoy battle of the war, fought in mid-Atlantic while Alan had waited in New York.

When Churchill and Roosevelt had conferred at Casablanca, they had good reason to suppose that, with Atlantic U-boat Enigma restored, the sinkings could be kept down to the level of late 1941. In January they were. But in February they had doubled, nearly back to 1942 levels. And then in March they were the worst of the war: ninety-five ships, amounting to three-quarters of a million tons. Massed U-boats had been able to sink twenty-two out of the 125 ships that had set out in convoy on the eastbound Atlantic passage that month. There was a reason for the deteriorating Allied control of events, one scarcely credible. It was not just that the convoys had sailed during the nine days' blackout caused by the change to the U-boat weather report system. It was that all the time, and to an ever-increasing degree, the convoy routeing cipher, among others, was being broken by the *B. Dienst*.

Convoy SC.122 had started out on 5 March, HX.229 on 8 March, and the smaller and more fortunate HX.229A the next day. On 12 March, SC.122 was re-routed to the north to avoid what was thought to be the position of a U-boat line, the *Raubgraf.* This signal was intercepted and deciphered. On 13 March the *Raubgraf* attacked a westward-bound convoy, thus openly betraying its position; SC.122 and HX.229 were both diverted again. Both diversion signals were intercepted and deciphered within four hours. The *Raubgraf* group could not catch up with SC.122 but 300 miles to the east, the forty-strong *Stürmer* and *Dränger* lines were sent to intercept them. There was ill luck on the German side – they were confused as to which convoy was which – but good luck too, for one of the *Raubgraf* happened to sight HX.229 by chance, and beckoned the others on. In London they could see the two convoys moving into the midst of the U-boat lines – but it was too late to do anything but to have them fight it out. On 17 March, they were surrounded by U-boats, and over the three days that followed twenty-two vessels were sunk, for the loss of one U-boat. Chance had played its part in this particular action, but underlying these and other current engagements lay the systematic failure of Allied communications.

In London and Washington, the first suspicion that this was so had been aroused in February 1943, when it was noticed that three U-boat line diversions were ordered within thirty minutes to operate successfully against a convoy on the 18th. Clear proof came only in mid-May when three doubly-enciphered Enigma messages showed evidence of the decipherment of particular Allied transmissions. Identifiable Enigma information had gone since 1941 into one-time-pad messages, and so had not been directly compromised. But it was implicit in the daily U-boat Situation Report, which by February 1943 was being decrypted. Yet again, the German authorities imputed Allied knowledge to a combination of airborne radar and the treachery of their officers. In a futile gesture, they reduced the number of people allowed to know about U-boat traffic. Again and again, only an *a priori* faith in the machine prevented them from seeing the truth. The Allies had very nearly given their own game away.

It was a dismal story, not perhaps one of individuals, but of the system. Neither in London nor in Washington was there a section in a position to do the very difficult detailed work of sorting out what the German command must have known, from what it could have known. The cryptanalysts were not given access to Allied dispatches – of which, in any case, there was no complete record. At the OIC they were still understaffed, underequipped, and under great strain with the convoy battles.

The cryptographic and operational authorities were working to standards which to Hut 8 eyes would seem criminally negligent. For one thing, the convoy routeing cipher, introduced as a joint Anglo-American system, was in fact an old British book cipher which the *B. Dienst* were able to recognise. Although in December 1942 a 'recipherment of indicators' had caused a setback for the *B. Dienst*, every kind of mistake was still being made. According to the American post-mortem:[1]

> USN-British Naval Communications were so complex, and often repetitious, that no-one seemed to know how many times a thing might not be sent and by whom – and in what systems. It is possible that the question of cipher compromise might have been settled earlier than May had the Combined Communications system been less obscure and had there been closer cooperation between the British and the US in such matters

while according to Travis's German counterpart,[2]

> The Admiral at Halifax, Nova Scotia, was a big help to us. He sent out a Daily Situation Report which reached us every evening and it always began 'Addressees, Situation, Date', and this repetition of opening style helped us to select very quickly the correct code in use at that time. . . .

All the time, while minds and technology were being pushed to the limit at Bletchley in the attack on German signals, the most elementary blunders were being made in the defence of their own. The result was that since late 1941, the German successes had been owed not only to the growing numerical strength of the U-boat fleet, but to their knowledge of Allied convoy routes; and during 1942 the effect of the Enigma blackout was only half the story.

Unlike the German authorities, the British were capable of recognising a mistake. The error was not that of the Admiralty alone, for GC and CS had exercised that part of its remit which called it to advise upon cipher security. But it was a part of GC and CS which had been left untouched by the revolutions elsewhere, and whose timescale still ran in terms of years. In 1941 it had devised a new system, which in 1942 the Admiralty had agreed to introduce in June 1943. Even allowing for the fact that it took six months simply to equip the Navy with new tables, this was a story of delays normal in peacetime, but bearing no connection with the new standards applied to anything considered essential for the war. If it were the decipherment of exciting messages, or airborne radar to make German cities visible for night raids, or the atomic bomb, then new industries could be conjured up in months. The less glamorous work of convoy protection called forth no such effort. Although the principle of integration had been applied so powerfully at Bletchley, it had not been extended to match up the two sides of its work.

They had learnt, but it was a painful way to learn, and those who had suffered most were unable to benefit from the lesson. They were at the bottom of the sea. Fifty thousand Allied seamen died in the course of the war, trying to mind their own business in the most gruelling conditions of the western war; 360 in the March 1943 convoy battle alone. Nor were their trials then over; the Merchant Navy cipher system continued to be breakable for the rest of 1943, long after the Navy was protected by the introduction of its new system on 10 June. Peculiarly vulnerable, and given the lowest priority, the merchant shipping ran a danger of which few knew, and whose enormity even fewer could appreciate.

In retrospect the failure of Allied naval communications vindicated the policy urged before the war by Mountbatten, and rejected by the Admiralty, that cipher machines should be employed. After 1943 the Navy joined the other services in an increasing use both of the Typex and of the equivalent American machine. Against these the B. Dienst made no headway. And yet the modernists such as Mountbatten might have been right for the wrong reason. Machine ciphers were not inherently secure, as the Enigma proved. The Foreign Office continued to use a hand system based on books; it remained unbroken. Bletchley deciphered the Italians' naval machine system; but was increasingly powerless against their book ciphers. What was enciphered on a machine might all the easier be deciphered on a machine. It was not the machine, but the whole human system in which it was embodied, that mattered. Behind the mis-match of Allied cryptanalytic and cryptographic standards there lay another question: were the Typex transmissions really more secure than those of the Enigma? Perhaps the most salient fact was the negative one: that the B. *Dienst* made no serious effort against them, just as in 1938 no serious effort had been made against the Enigma. If an attack upon the Typex had been made with the resources mobilised at Bletchley, the story might have been very different.[3] But perhaps they had no Alan Turing – nor a system in which an Alan Turing could be used.

Such was the background against which Alan returned to base himself in Hut 8. The game had gone sour. The cryptanalysts tended to assume that their productions were being fed into a system that knew what it was doing, and it was a shock when they were told of the convoy cipher fiasco. Hut 8 itself had been taken over by Hugh Alexander during Alan's absence. There was a story that a form had come round, asking for the name of the head of the section. Alexander had said, 'Well, I suppose I am,' and thereafter he remained in smooth control of the naval Enigma. There were no further crises, despite the later proliferation of German naval key-systems. The introduction of an alternative fourth wheel in July 1943 for the U-boat system caused them no problem; they were able to deduce its wiring without a capture. None of this needed Alan any more; indeed several of the high-level analysts were moved to more innovative work on Fish. Nor, indeed, did the U-boat Enigma need the British effort now. Although the British[*] produced the first working high-speed four-wheel Bombe in June 1943, the Americans produced more and better Bombes after August. By the end of 1943 they had taken over the U-boat work entirely, and had spare capacity for other Enigma problems.

If they did not need Alan Turing on what had become a routine task, his help might well have been of use in the cryptographic context, where 1943 saw a slightly greater degree of cooperation and coordination prevail. He had already been introduced to the job of inspecting speech cipher systems, and to the delicate work of Anglo-American liaison. The Allies now had the problem of recovering from the delays and narrow vision of 1942, at a time when

---

[*] Wynn-Williams did make some progress, but this machine was probably the work of Keen and BTM.

communications were expanding enormously and growing towards their great climax. The times had been out of joint, something they could not afford to allow in the intricate plans for 1944. For Alan Turing this would be dull and dispiriting work compared with the excitement of the relay race; but it was the job crying out for expert attention.[4]

After June 1943 the Atlantic war turned dramatically in favour of the Allies, with ship sinkings reduced to tolerable levels. In retrospect, March 1943 had seen 'the crisis' of the battle of the Atlantic, and thereafter it could be claimed that 'the U-boat was defeated'. But more truly, 1943 saw a continuing state of crisis, one in which it was not the boat but the system that was beaten from day to day by a superior system. At last they introduced long-range air patrols to cover the mid-Atlantic gap. And the logical advantage held by the U-boats in 1940 had been reversed. They were now visible from afar through the Enigma (by the end of 1943 the British had a clearer idea of where they were than did their own command), and at close range through the airborne radar work of TRE. Meanwhile the convoy communications became secure. The combination was a winning one, and the Atlantic poker game appeared as a quiet front, only noticed when occasionally the cheating failed to work. But from the German point of view it was not a quiet front at all. For them, 1943 saw a tremendous stepping up of the attack. At the end of the year they would have over 400 U-boats to deploy, equipped with elaborate measures to counter the radar detection they believed responsible for all their failure to find convoys. The fleet was still alive and aggressive, even if individual U-boats were increasingly short-lived. It was a game of perfect information – or *Sigint*, as it became in the new language of 1943 – for one player. But the other did not admit defeat. The Second World War was not a game.

The introduction of the fourth rotor in February 1942 thus had effects unknown in Germany. That it was employed half-heartedly and foolishly, allowing it to be mastered after December 1942, meant the loss of the battle of the Atlantic. But that it had been employed at all meant that it had introduced electronic engineers to Bletchley and hence to the Fish problem. And while 1943 saw a general resolution of Anglo-American friction over Intelligence, by means of an agreement to divide the world between them – Britain taking

Europe and America taking Asia – the US Navy retained its more aggressive stance. Their rapid development of Bombes reflected the fact that the Atlantic was now an American sea. Alan Turing's work had denied the ocean lanes to Germany, and secured them for the United States.

Alan had written to Joan while away in America, asking her what she would like as a present, but in her reply she had not been able to answer this question because of the censorship. In the event he brought back a good-quality fountain pen for her – and others too had presents. There were Hershey bars among the sweets he left for general consumption in Hut 8, and he also brought an electric shaver for Bob, making a transformer to convert it from the American to the British voltage. He told Joan how seeing Mary Crawford in January, just after Jack had died, had affected him with a sense of how much they had meant to each other. He hinted that they should 'try again', but Joan did not take up the hint; she knew that it was over.

He showed her a book on Go, and lay on the floor in his room at the Crown Inn demonstrating some of the situations in the game. And he also lent her a remarkable new novel. It was by his friend Fred Clayton, though under a pseudonym,[5] and had been published in January 1943. *The Cloven Pine*, as it was called, in a cryptic reference to Ariel's imprisonment by Sycorax in *The Tempest*, vented groans about politics and sex which were closer to Fred's experiences and problems than to Alan's, Fred having set his plot in the Germany of 1937 and 1938, and drawing upon his complex and conflicting reactions to the Vienna and Dresden of a little earlier.

He had tried to understand the collapse of the ideals of 1933. On one level, he showed German individuals, no less and no more lovable than English individuals. On another, he showed the system, the Nazi system. And while he portrayed himself as the Englishman, asking how Germans could believe such things, he tried to see himself and English attitudes through German eyes. In an internationalist gesture, *The Cloven Pine* was dedicated jointly to George, his younger brother, and to Wolf, one of the boys in Dresden that he had known. 'Freedom and consistency of mind', he had the German boy of his story think to himself, analysing English liberalism. 'They were illusions! What freedom or consistency was there in this Self, a thing of moods that did not understand one another. . . .' It was the conclusion of a King's liberal, trying hard to comprehend the absolute denial of Self.

There was a second thread to his story, that of the English schoolmaster's friendship with the German boy which remained 'suspended in an atmosphere of semi-Platonic sentimentality'. This for Joan represented a quality of self-restraint that deserved admiration, but Alan, who had often teased Fred in terms rather like these, would probably have taken a different view. The book was saved from the obvious danger, one that Evelyn Waugh had mocked in *Put Out More Flags*, by the stringency and sophistication with which it examined the contradictions. The personal realities were ever questioning, and questioned by, a political background which included the late-1930s Nazi propaganda about boy-corrupting Jews and Catholic clergy. On this level it served Alan as a way of saying that his 'tendencies' could not be separated from his place in society, nor regarded as peripheral to his own freedom and consistency of mind.

Although he had dropped away from the direct cryptanalytic work, Alan remained within the Bletchley fold, and was to be seen in the cafeteria off duty. Conversation at these times often revolved round mathematical and logical puzzles, and Alan was particularly good at taking some quite elementary problem and showing how some point of principle lay behind it – or conversely, illustrating some mathematical argument with an everyday application. It was part of his special concern for connecting the abstract and the concrete, as well as a pleasure in demystifying the higher mathematician's preserve. It might be wallpaper patterns for an argument about symmetries. His 'paper tape' in *Computable Numbers* had the same flavour, bringing an 'abstruse branch of logic' down to earth with a bump.

One person who appreciated this approach was Donald Michie, to whom as a classicist it all came as fresh and new. He became very friendly with Alan, and in 1943 they began to meet every Friday evening in a pub at Stony Stratford, just north of Bletchley itself, to play chess and talk – or more often, for Donald to listen. The Prof's chess had always been something of a joke at Bletchley, being all the more exposed to invidious comparison when the chess masters arrived. Harry Golombek had been able to give him queen odds, and still win; or when Alan resigned he was able to turn the board round and win from the position given up as hopeless. He complained that Alan had no idea how to make the pieces work together, and it might well be that as in his social behaviour, he was too conscious of what he was trying to do, to play with fluency. As Jack Good saw it, he was too intelligent to accept as obvious the moves that others might make without thinking. He always had to work it out from the beginning. There had been an amusing moment when Alan had come off a night shift (this would have been in late 1941) and then played a game with Harry Golombek in the early morning. Travis had looked in and was embarrassed to find, as he thought, his senior cryptanalyst playing while on duty. 'Er . . . er . . . want to see you about something, Turing,' he said awkwardly, like the housemaster catching a sixth-former with a cigarette in the toilet. 'Hope you can beat him,' he added to Golombek as they left the room, assuming quite wrongly that the master cryptanalyst was the champion player. But young Donald Michie was a player of Alan's standard.

These meetings were an opportunity for Alan to develop the ideas for chess-playing machines that had begun in his 1941 discussion with Jack Good. They often talked about the mechanisation of thought processes, bringing in the theory of probability and weight of evidence, with which Donald Michie was by now familiar. The development of machines for cryptanalytic work had in any case stimulated discussion as to mathematical problems that could be solved with the mechanical aid – that of finding large prime numbers, for instance, was a topic that came up in lunchtime conversations, rather to the amazement of Flowers, the electronic engineer, who could see no point in it. But Alan's talk went in rather a different direction. He was not so much concerned with the building of machines designed to carry out this or that complicated task. He was now fascinated with the idea of a machine that could *learn*. It was a development of his suggestion in *Computable Numbers* that the states of a machine could be regarded as analogous to 'states of mind'. If this were so, if a machine could simulate a brain in the way he had discussed with Claude Shannon, then it would have to enjoy the faculty of brains, that of learning new tricks. He was concerned to counter the objection that a machine, however intricate its task, would only be doing what a person had explicitly designed it to do. In these off-duty discussions they spent a good deal of time on what would be said to count as 'learning'.

Implicit in these discussions was the materialist view that there was no autonomous 'mind' or 'soul' which used the mechanism of the brain. (He had perhaps hardened his stance as an atheist, and his conversation was more free with anti-God and anti-church jokes than it would have been before the war.) To avoid philosophical discussions about what 'mind' or 'thought' or 'free will' were supposed to be, he favoured the idea of judging a machine's mental capacity simply by comparing its performance with that of a human being. It was an *operational* definition of 'thinking', rather as Einstein had insisted on operational definitions of time and space in order to liberate his theory from *a priori* assumptions. This was nothing new – it was an entirely standard line of rationalist thought. Indeed in 1933 he had seen it on the stage, for in *Back to Methuselah* Shaw had a future scientist produce an artificial 'automaton' which could show, or at least imitate, the thought and emotions of twentieth-century people. Shaw had the 'man of science' assert that he had no way of drawing a line between 'an automaton and a living organism'. Far from it being a novelty, Shaw was trying to make this argument appear a dated piece of Victoriana. Again, his *Natural Wonders* book had accepted the rationalist view, with a chapter called 'Where some of the Animals do their Thinking' which treated thought, intelligence and learning as differing only in degree as between monocellular animals and human beings. It was no new idea, therefore, when Alan talked in terms of an imitation principle: that if a machine appeared to be doing as well as a human being, then it *was* doing as well as a human being. But it gave a sharp, constructive edge to their discussions.

Meanwhile Donald Michie had been plucked from the Testery, and Jack Good from Hut 8, to work as Newman's first staff on a very exciting development of the Fish analysis. Donald Michie had continued to work on refinements of the Turingismus method, reporting informally to Alan on their progress – advances reflected in the fact that at the beginning of 1943 a proportion of the Fish signals were being read regularly and with little delay. The Turing theory of statistics, with its formalisation of 'likeliness' and 'weight of evidence', and with its 'sequential analysis' idea, were also playing a general part in the Fish work, in which it found greater application than in the Enigma methods. But by the spring of 1943, Newman's ideas for mechanisation had begun to bear fruit. Here the new developments with electronic technology, in which the crucial steps had been taken while Alan was in America, were in themselves very significant.

The Post Office engineers had been able to install a first electronic counting machine in Hut F, where Newman and his two assistants worked, in about April 1943. This and its successors were called the 'Robinsons'.* Although they had overcome some of the engineering problems associated with passing paper tape very rapidly through an electronic counter, these 'Robinsons' still suffered from many defects. They were prone to catch fire; the paper tapes were always breaking; and the counts were unreliable. This was because the slower parts of the counting process were performed by the old relays, and these produced an electrical interference effect upon the electronic components. But the fundamental technological problem was that of synchronising the ingestion of the two separate paper tapes demanded by the method. For all these reasons, the Robinsons proved too unreliable and too slow for effective cryptanalytic use. They were employed only for research purposes. There was also another fundamental difficulty, not so much physical as logical, which made the machine method too slow. In using it for the cryptanalytic process, the operator would constantly have to produce fresh tapes, resorting for this purpose to an[6] 'auxiliary machine that was used to produce the tapes which formed one of the two inputs to the Heath Robinson.'

But even before the first Robinson was finished, Flowers had made a revolutionary proposal which both solved the problem of tape synchronisation, and did away with the laborious production of fresh tapes. The idea was to store the Fish key-patterns *internally*, and in electronic form. If this were done, only one tape would be

---

* There was more than one kind of Robinson: a 'Peter Robinson' and a 'Robinson and Cleaver' after London department stores, and a 'Heath Robinson' after the famous cartoonist specialising in elaborate machines to perform absurdly simple tasks.

required. The difficulty lay in the fact that such internal storage would require the more extensive use of electronic valves. It was a suggestion regarded with deep suspicion by the established experts, Keen and Wynn-Williams. But Newman understood and supported Flowers' initiative.

By any normal standards, this project was a stab in the technological dark. But they were not in normal times, but in the conditions of 1943. What happened next was a development unthinkable even two years before. Flowers simply told Gordon Radley, director of the Post Office laboratories, that it was necessary for Bletchley work. Under instructions from Churchill to give Bletchley's work absolute priority, without questions or delay, Radley had no decision to make, although the development consumed half the resources of his laboratories. Construction began in February 1943 and the machine that Flowers had envisaged was completed after eleven months of night-and-day working. No one but Flowers, Broadhurst and Chandler who together had designed the machine had been permitted to see all the parts, let alone to know what the machine was for. There were no drawings for many of the parts, only the designers' originals; there were no manuals, no accounts, nor questions asked about materials and labour consumed. In the laboratory the machine was assembled, wired and made to work in separate sections which did not come together until the whole machine was installed and working at Bletchley in December 1943.

In three years, they had caught up with half a century of technological progress. Dillwyn Knox died in February 1943, passing just before the Italian empire he had done a good deal to undermine, and with him the pre-industrial mind. They had been forced into one scientific revolution by the Enigma, and already they were in the throes of a second. The all-electronic machine proved to be much more reliable than the Robinsons, as well as faster. They called it the *Colossus*, and it demonstrated that the colossal number of 1500 electronic valves, if correctly used, could work together for long periods without error. This was an astonishing fact for those trained in the conventional wisdom. But in 1943 it was possible both to think and do the impossible before breakfast.

Alan knew about all these developments, but declined the invitation to play a direct part.[7] Newman built up an increasingly large and powerful group, drawing in the best talent from the other huts and from the mathematical world outside. Alan moved in the opposite direction; he was not a Newman, skilled in overall direction, and still less was he a Blackett, moving in political circles. He had not fought to retain control of naval Enigma, but had retreated before Hugh Alexander's organising power. If he had been a quite different person, he could now have made his position one of great influence, it being the time for sitting on coordinating committees, Anglo-American committees, future policy committees. But he had no concern for finding a place in anything but scientific research itself. Other scientists were finding the war to be awarding them a power and influence denied in the 1930s, and thrived upon it. For Alan Turing, the war had certainly brought new experiences and ideas, and the chance to do something. But it had given him no taste for organising other people, and it had left his axioms unchanged. A confirmed solitary, he wanted something of his own again.

It would also take more than the Second World War to change his mother's ideas, which in December 1943 focussed as always upon the duty of choosing Christmas presents. Alan wrote to her[8] on 23 December:

My dear Mother,

Thank you for your enquiry as to what I should like for Christmas, but really I think we had better have a moratorium this year. I can think of a lot of things I should like which I know to be unobtainable, e.g. a nice chess set to replace the one that was stolen from here while I was away, and which you gave me in 1922 or so: but I know it is useless to try at present. There is an old set here that I can use till the war is over.

I had a week's leave fairly recently.* Went up to the lakes with Champernowne and stayed in Prof. Pigou's cottage on Buttermere. I had no idea it was worth while to go amongst the mountains at this time of year, but we had the most marvellous weather, with no rain at all and snow only for a few minutes when we were up on Great Gable. Unfortunately Champernowne caught a chill and retired to bed half-way through. This was in the middle of November, so I don't think I'll be taking Christmas leave until February . . . .

Yours,   Alan

---

* It was from 16 to 22 November 1943.

But by Christmas 1943, as the *Scharnhorst* was sunk with Enigma help, Alan had set out on a new project, this time something of his own. He handed over his files on American machinery to Gordon Welchman, who left Hut 6 at this point to take on an overall coordinating role. Welchman had lost interest in mathematics, but found a new life in the study of efficient organisation, and was particularly attracted to American liaison. But Alan, since returning from America, had spent a good deal of time on the devising of a new speech encipherment process. And while other mathematicians might be content to *use* electronic equipment, or to know about it in general terms, he was determined to build upon his Bell Labs experience and actually create something that worked with his own hands. In late 1943 he became free to devote time to some experiments.

Speech encipherment was not now regarded as an urgent problem. On 23 July 1943, the X-system had been inaugurated for conversations at top level between London and Washington (the extension to the War Room was not completed for another month). The Chiefs of Staff memorandum[9] of that date stated that 'The British experts, who were appointed to examine the secrecy [*sic*] of the equipment, have expressed themselves as completely satisfied'; it also listed the twenty-four British top brass, from Churchill downwards, who were allowed to use it, and the forty Americans, from Roosevelt downwards, whom they might call. This solved the problem of high-level Atlantic communications, although it meant that the British had to go cap in hand to use it, and found themselves outdone by the links to the Philippines and Australia that the Americans were busily installing. Nor did they necessarily want to have all their transmissions recorded by the Americans, the alliance never being so close that the British government confided all to the United States. There was every incentive, from the point of view of future policy, to develop an independent British high-grade speech system. It was Britain, not the United States, that was supposed to be the centre of a world political and commercial system.

But this was not undertaken, and nor did Alan's new idea have the potential for such a development. The principle he had in mind would be impossible to apply with the variable time-delays and fading experienced with short-wave transatlantic radio transmissions. It would never be the rival of the X-system, which overcame these problems, and this was made clear at the start. It bore the mark of something that he wanted to achieve for himself, rather than of something he had been asked to provide. The war was no longer calling for his original attack on problems, and he found himself almost redundant after 1943. Neither was his idea backed with more than a token share of resources. It was like going back to early, grudging days. To pursue it he had to move to a rather different establishment. While the Bletchley industry continued, with ten thousand people now at work on rolling secrets off the production line, not only deciphering and translating but doing a great deal of the interpretation and appreciation over the heads of the services, Alan Turing gradually transferred himself to nearby Hanslope Park.

While the Government Code and Cypher School had expanded to dimensions quite unimaginable in 1939, the secret service had also mushroomed in a variety of directions. Just before the war, it had recruited Brigadier Richard Gambier-Parry to improve its radio communications. Gambier-Parry, a veteran of the Royal Flying Corps and a genial paternalist to whom junior officers would refer as 'Pop', had thereafter spread his wings much further. His first chance had come in May 1941 when the secret service managed to detach from MI5 the Radio Security Service, then responsible for tracking down enemy agents in Britain. It was he who had taken it over. With all such enemy agents soon under control, the role of the RSS had been diverted into that of intercepting enemy agents' radio transmissions from all over the world. Now known as 'Special Communications Unit No. 3', this organisation used a number of large receiving stations centred on the one at Hanslope Park, a large eighteenth-century mansion in a remote corner of north Buckinghamshire.

Gambier-Parry had also acquired the responsibility for other aspects of secret-service work. These included providing the transmitters for the black broadcasting organisation, which began its 'Soldatensender Calais' broadcasts on 24 October 1943. (The studios where journalists and German exiles concocted their ingenious falsities were at Simpson, another Buckinghamshire village.) SCU3 had further taken under its wing the manufacture of the cryptographic system Rockex, which was to be used for top-grade British telegraph signals. Such traffic now amounted to a million words a day to America alone – pre-eminently, of course, carrying Bletchley's productions. The Rockex represented a technical improvement to the Vernam one-time teleprinter cipher system.

One problem with the Vernam principle was that the ciphertext, regarded as a teleprinter input in the Baudot code, was bound to include many occurrences of the operational or 'stunt' symbols, those which produced not letters but 'line feed', 'carriage return' and so forth. For this reason, the cipher-text could not be handed over to a commercial telegraph company for transmission in Morse, as was often desirable. It was Professor Bayly, the Canadian engineer at Stephenson's organisation in New York,* who had devised a method of suppressing the unwanted characters and replacing them in such a way that the resulting cipher-text could be printed neatly on a page. This required the development of electronics which could automatically 'recognise' the unwanted telegraph symbols. The problem involved logical circuits such as appeared in the Colossus, albeit on a much more modest scale, using electronic switching for Boolean operations applied to the holes of telegraph tape.

By late 1943, the research had been completed. An inventive telegraph engineer, R. J. Griffith, had been borrowed from Cable and Wireless Ltd to do the detailed design. The manufacture was now going ahead at Hanslope Park, where Griffith was also at work on the problem of generating the key-tapes automatically, by using electronic random noise.

Hanslope Park, with its web of connections with secret enterprise and electronic cryptographic work, was therefore a natural place for the Turing speech encipherment project to be based. The Post Office Research Station might have housed it, but it was a great deal further from Bletchley than was Hanslope, which was only ten miles to the north. It was a rather strange place, strange for its very appearance of being an ordinary military station, with all the accoutrements

---

* Hence the name Rockex, coined by Travis and inspired by that better-known feature of the Rockefeller Center, the Rockettes.

of military ranks and language. Quite different from Bletchley
Park, where the military had been obliged to adjust to the young
Cambridge intelligentsia, here a service mentality was unaffected
by the advent of modern technology. Here there was not a civilian
cafeteria but an officers' mess, where framed in passepartout lay the
clue, a quotation from *Henry V*:

> The King hath note of all that they intend,
> By interception which they dream not of

But in fact Gambier-Parry's staff were working in a dream war themselves, one in which they knew neither the significance of what they
were doing, nor what anyone else did. The newcomer would spend
many months before being able to work out that the organisation
came under the direction of the secret service.

Alan's first contact with Hanslope Park came in about September
1943, when he cycled the ten miles from Bletchley to inspect the
possibilities. A senior ex-Post Office man, W. H. 'Jumbo' Lee, was
deputed to look after his requirements. Hanslope was not exactly a
model of spit-and-polish smartness; some of its uniformed personnel
were 'real soldiers' but many were of an unmilitary disposition,
transferred straight from the Post Office, Cable and Wireless, and
similar organisations. There was, however, a sufficiently military
air at Hanslope for a misunderstanding to arise when 'Jumbo' Lee
introduced Alan to his superior, Major Keen. 'Dick' Keen was the
top British expert on radio direction-finding, who had written the
only textbook on the subject during the First World War, and spent
much of the Second on writing a new edition. Alan and 'Jumbo' Lee
stood together at his door and Keen waved them away, assuming
from his appearance that Alan was a cleaner or delivery boy.

Hanslope Park had a precedent for the arrival of a cryptographic
project, but whereas Griffith had demanded, and received, a new
workshop and adequate staff, Alan simply took what he was given,
which was not very much. In fact his project was granted bench-
space in a large hut where a number of other research projects
were being conducted, and he was offered some mathematical
assistance in the form of Mary Wilson, who did direction-finding
analysis with Keen. She was a graduate from a Scottish university,
and working with Keen had considerably raised the standard from

the early days when people said 'Two fixes are better than three – there is no triangle of error.' Instead, they were offering to the analysts ellipses on the map which represented the area in which the point of transmission could be asserted to be with such-and-such a probability. But she did not have enough mathematics to understand what Alan wanted when he explained his idea. (He helped her later with the direction-finding work, though expressing a somewhat dim view of her training.) So over the next six months he had to work alone on the project, coming in a couple of days a week, not every week. Two army signalmen were assigned to assemble pieces of electronic equipment under his direction, but that was all.

In mid-March 1944 there was a distinct change in the Hanslope staffing, with an influx of mathematical and engineering expertise. Such a change was needed. There was, for instance, an occasion when 'Jumbo' Lee showed Alan a problem on which they were stuck. It was no more than a trigonometrical series (in connection with aerial design) easily within the grasp of a Cambridge scholarship candidate, but he was most impressed when Alan immediately produced the answer, the more so as Post Office engineers had been laboriously summing it term by term. The authorities had chosen five new young officers, selected from those taking courses at the Army Radio School near Richmond in Surrey. Two of them would take special places in Alan Turing's life. Indeed, this was a fresh start for him. In 1943, he had met Victor Beuttell over lunch in London, with some of their personal troubles coming out. (Victor had finally rebelled against his father, and joined the RAF.) They would never see each other again; but the personal *rapport* that thereby lapsed was to be found within new friendships.

The first was Robin Gandy, the undergraduate who in 1940 had stoutly maintained 'Hands off Finland' at Patrick Wilkinson's party in the face of Alan's quizzical scepticism. His arrival brought to Hanslope a breath of the King's spirit. He had been conscripted into the ranks in December 1940, with six months on a coastal defence battery, until his mathematical mind had enjoyed more recognition, as he became a radar operator, and then an instructor. After being commissioned into REME, a series of courses, sandwiched with practical experience, had taught him about all the radio and radar equipment used by the British forces.

The second was yet another Donald. This was Donald Bayley, who came from a quite different background, that of Walsall Grammar School (where Alan's friend James Atkins had taught him mathematics) and Birmingham University, where he had graduated in electrical engineering in 1942. He also had been commissioned into REME and had likewise shot ahead in all the courses.

Both were introduced to the large 'laboratory' hut where the research projects were in progress, and found Alan at work there. If civilians from Cambridge were apt to find him unusually careless in appearance, his divergences from respectability were very much more noticeable at military Hanslope. With holes in his sports jacket, shiny grey flannel trousers held up with an ancient tie, and hair sticking out at the back, he became the cartoonist's 'boffin' – an impression accentuated by his manner of practical work, in which he would grunt and swear as solder failed to stick, scratch his head and make a strange squelching noise as he thought to himself, and yelp when shocked by the current that he forgot to turn off before soldering the joints in his 'bird's nest' – so they called it – of electronic valves.

But Robin Gandy was struck in another way on about the first day that he set to work investigating the effectiveness of high-permeability cores in the transformers of the radio receivers. There were two engineers in his section, who started the tedious task of testing the things, when Alan pottered in, and decided that it should all be solved from theoretical principles – in this case, it being an electromagnetic problem, from Maxwell's equations. These he wrote down at the top of his paper, just as though it were some contrived Tripos question instead of one from real life, and eventually performed a *tour de force* of partial differential equations to get an answer.

Donald Bayley was impressed in a similar way by the speech encipherment project, which at Hanslope became known as the *Delilah*. Alan had offered a prize for the best name, and awarded it to Robin for his suggestion of Delilah, the biblical 'deceiver of men'. It made full use of his experience in cryptanalysis, and as Alan would explain, was designed to meet the basic condition that even if the equipment were compromised, it would still provide complete security. Yet the system he had conceived on board the *Empress of Scotland* a year earlier was essentially very simple.[10] It was a mathematician's design, and one which had depended upon Alan asking 'But why not?'

What he had done was to consider the roomful of equipment which made up the X-system, and to ask what were the crucial features which made it into a secure speech cipher. The Vocoder was *not* essential, although it had been the starting point of the project. Nor was the business of quantising the output amplitudes into a number of discrete levels. By jettisoning these he reduced the number of ideas involved to two: the fact that it *sampled* the speech at a succession of moments in time, and the fact that it used *modular* addition, like a one-time pad.

The Delilah was based on these two ideas from the beginning, while in the X-system they had arrived by a back-door route. The point about sampling was that it removed the redundancy of the continuous sound wave. Any sound signal could be represented by a curve such as:

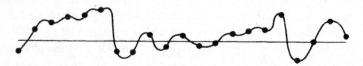

The point was that it would be unnecessary to transmit the whole curve. It would be sufficient to communicate the knowledge of certain points on the curve, provided that the recipient could thereupon perform the exercise of 'joining the dots' to reconsitute the curve. This could be done, at least in principle, provided that it was known how sharply the curve was allowed to wiggle in between the points. Since sharp wiggles would correspond to high frequencies, it followed that provided there was a limit on the frequencies contained in the signal, then a sequence of discrete points, or samples, of the curve, taken at regular intervals, would contain *all* the information of the signal. Since telephone channels did in any case cut off high frequencies, the restriction on allowed 'wiggling' of the curve was no real restriction at all, and in fact a rather small number of samples could be shown to suffice to convey the signal.

The idea was well-known to communication engineers. In the X-system, it was the practice to sample each of the twelve 25-Hz channels fifty times a second. These figures were illustration of a general result, that it was necessary to sample at a rate of twice the maximum variation in frequency of the sound, or bandwidth. There was an exact mathematical result to this effect, proved as early as 1915 but which Shannon had re-stated[11] and discussed with Alan at Bell Labs. If, for instance, the sound signal were restricted to frequencies less than 2000 Hz, then a sample taken 4000 times a second would be exactly enough to reconstitute the signal. There would be precisely one curve of the stated frequency restriction that passed through all the sampled points. Alan described and proved this result to Don Bayley as the 'Bandwidth theorem'. His 'Why not?' had come in asking why this elegant fact could not be made the pivot on which to turn the whole encipherment process.

The figure of 2000 Hz was in fact the one he intended to use, and his encipherment process would start with the speech signal being sampled 4000 times a second. The Delilah would then have to effect the addition of these sampled speech amplitudes to another stream of key amplitudes. The addition would be done in modular fashion, meaning that while speech sample amplitude of 0.256 units and key amplitude 0.567 units would be added to give 0.823 units, the addition of 0.768 and 0.845 would give 0.613, not 1.613. The result of all this would be a train of sharp 'spikes', of heights varying between zero and one unit:*

The next problem was that of how to transmit the information of these 'spike' heights to the receiver. In contrast to the X-system, Alan planned no quantisation of amplitudes here. He wanted to transmit them as directly as possible. In principle the 'spikes'

---

* Technically, of course, there was more to it than this. The speech would first be filtered to remove frequencies above 2000 Hz, and to restrict it to a specific range in amplitude so that it could be described at any point by a number between 0 and I. Then in fact the encipherment was done by adding a continuous key signal first, and then taking the sample by making the resulting speech-plus-key signal modulate a pulse train. The 'remaindering' process would then be performed, chopping down a 'spike' by one unit if it exceeded one unit in amplitude.

themselves could be transmitted, but being of such short duration, a few microseconds in fact, they would require a channel which could carry very high frequencies. No telephone circuit could do this. To use a telephone channel the information of the 'spikes' would have to be encoded into an audio-frequency signal. Alan's proposal was to feed each 'spike' into a specially devised electronic circuit with an 'orthogonal' property. This meant that its response to a spike of unit amplitude would be a wave with unit height after one time interval, and with zero height at every other unit time interval:

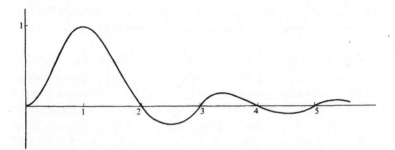

Assuming the circuit to be 'linear', meaning that the input of a 'spike' of say half a unit would produce exactly half of this response, the effect of feeding into it the train of 'spikes' would be to 'join the dots' in a very precise way. The information of each 'spike' would go exactly into the amplitude of the response of this circuit one unit of time later, and nowhere else.

The transmission would then be straightforward, and could be achieved by perfectly standard means; the decipherment process did not require any further new ideas.* Apart from the question of supplying a key-system, this was all that was required for the Delilah to effect an 'adding-on' encipherment of speech, the analogy of what agents like Muggeridge, or the machines which produced the Fish signals, or the Rockex, were all alike doing for telegraph

---

* The output of the 'orthogonal' circuit would have the characteristics of random noise in the frequency range up to 2000 Hz. It would be deciphered by performing the sampling process (in strict synchrony, of course, with the sender), and a modular subtraction of the identical key. This would yield the samples of the original speech signal, and it was then a standard procedure, requiring only a low-frequency filter, to recover the speech itself.

or teleprinter. If the key were truly 'random', or without any discernible pattern, such a speech-cipher system would be as secure as the Vernam one-time cipher for telegraph tape, and on exactly the same argument. From the enemy's point of view, if all keys were equally probable, then all messages would be equally likely. There would be nothing to go on.*

The disadvantage of the simple Delilah system, as compared with the X-system, was that its output signal would be one of bandwidth 2000 Hz, rather than a stream of digits, and that it would have to be communicated perfectly, or all would be lost. In particular, any variation in time delay, or distortion in amplitudes, would ruin the decipherment process. Sender and receiver would have to keep in time to the microsecond for the same reason. This was why it could never be used for long-range short-wave transmissions. But it could be used for local short-wave, for VHF and for telephone communication. For tactical or domestic purposes, therefore, it held considerable potential.

Don Bayley was very eager to work on the Delilah, but it was not at first allowed. He was assigned to other tasks, and it only gradually became possible for him to spend time on Alan's project. It was several months before formal permission was granted for his participation, and even then it was only on the understanding that he would have to do other jobs from time to time.

Alan's waiting for help coincided with the time when everyone was waiting for the rather more important question of the Second Front. And it was, after all, the enterprise for which all his efforts, fascinating and depressing alike, had helped to secure the conditions. But back at Bletchley Park there was quite a different reason for excitement in Newman's section. They had shown that even in the time of intricate planning and coordination, there was room for initiative. In fact there had been a scramble at the last minute in the latest development of the treasure hunt. Again it was the fresh generation that had done it, disproving an assumption that

---

* As Alan would stress in explaining the system, this depended crucially upon the use of *modular* addition. If ordinary unremaindered addition were used, then there would be a correlation between the speech amplitudes and the speech-plus-key amplitudes, and this the cryptanalyst could exploit. Indeed, this is precisely what the ear does in sorting out speech from background noise.

something could not be done. It was something they could be proud to tell Alan Turing about.

Using the new electronic Colossus, installed since December, Jack Good and Donald Michie had made the marvellous discovery that by making manual changes while it was in operation, they could do work that hitherto it had been assumed would have to be done by hand methods in the Testery. The discovery meant that in March 1944 an order had been placed with Dollis Hill for six more Colossi by 1 June. This demand could not possibly be met, but with desperate efforts one Mark II Colossus was finished on the night of 31 May, and others followed. The Mark II included technical improvements, was five times faster, and also incorporated 2400 valves. But the essential point was that it incorporated the means for performing automatically the manual changes that Jack Good and Donald Michie had made. The original Colossus, by recognising and counting, was able to produce the best match of a given piece of pattern with the text. The new Colossus, by automating the process of varying the piece of pattern, was able to work out which was the best one to try out. This meant that it performed simple acts of decision which went much further than the 'yes or no' of a Bombe. The result of one counting process would determine what the Colossus was to do next. The Bombe was merely supplied with a 'menu'; the Colossus was provided with a set of instructions.

This greatly extended the role of the machine in bringing Fish to a state of 'cornucopian abundance'. As with the Bombe, it was not that the Colossus did everything. It was at the centre of an extremely sophisticated and complex theory, in which far from being 'dull and elementary', the mathematics involved was by now at the frontiers of research. There were in fact many ways in which the Colossus could be used, exploiting the flexibility offered by its variable instruction table. It took the analysts' work into a quite new realm of enchantment. In one of the main uses, the human and the machine would work together:[12]

> . . . The analyst would sit at the typewriter output and call out instructions for a Wren to make changes in the programs. Some of the other uses were eventually reduced to decision trees and were handed over to the machine operators.

These 'decision trees' were like the 'trees' of the mechanical chess-playing schemes. It meant that some of the work of the intelligent analyst had been replaced by the electronic hardware of the Colossi; some went into the devising of instructions for them; some into the 'decision trees' that could be left to uncomprehending 'slaves'; and some retained for the human mind. When off duty, they had talked about the machines playing chess, taking intelligent decisions automatically. In their work, in this new extraordinary phase, the arbitrary dispensations of the German cryptographic system had brought something like this into being – and even more uncanny for those who did it, a sense of dialogue with the machine. The line between the 'mechanical' and the 'intelligent' was very, very slightly blurred. Whatever its application to the great surprise that awaited the Germans, they were having a wonderful time in seeing the history of the future.

No one at Hanslope, seeing the strange civilian boffin cycling across with a handkerchief around his nose (it was his hay fever period) could have connected him with the success of the assault on Normandy. And by now his part in the necessary conditions for making it was something that lay in the past; the success he wanted was of something truly and more wholly his own. As ten years before, it was his privilege to continue in his own way, with the least waste of energy, the civilisation which demanded harsher sacrifice from others. And it was another kind of invasion that he had in mind, one not yet ready for announcement.

The successful passage of 6 June 1944 roughly coincided with the point at which it became possible for Alan and Don Bayley to get down to work on constructing the Delilah equipment, clearing up the rather messy efforts that the Prof had made on his own. The main task was that of building the circuit to produce the highly accurate 'orthogonal' response. It was the design of this circuit that had absorbed most of Alan's earlier thought and experiment. He had realised that it could be synthesised out of standard components. This was an entirely new idea to Don Bayley, as was the mathematics of Fourier theory* that had been used to attack it. It was a tough

---

* Fourier theory very naturally involved the use of the 'complex numbers', and so did other aspects of the analysis of electronic circuits. The mathematics that he needed was at

problem, which Alan said had involved spending a whole month in working out the roots of a seventh-degree equation. Although he was an amateur and self-taught electronic engineer, he was able to tell his new assistant a good deal about the mathematics of circuit design, and for that matter was by now able to show most of them in the laboratory hut a thing or two about electronics. But it needed Don to bring his practical experience to bear on the problem, and to tame the straggly bird's nest. He also kept beautifully neat notes of their experiments, and generally kept Alan in order.

Alan came in most mornings on his bicycle – sometimes in pouring rain, of which he appeared oblivious. He had been offered the use of an official car to take him to and fro, but declined it, preferring the use of his own motive power. Once, very unusually, he was late in, and even more dishevelled than usual. As explanation he produced a grubby wad of £200 in notes, explaining that he had dug them up out of their hiding-place in a wood and that there were two silver bars still to be recovered.

But late in the summer, as the bridgehead was finally established and the Allied armies swept their way across France, he gave up his lodgings with Mrs Ramshaw at the Crown Inn, and moved to the Officers' Mess at Hanslope Park. At first he occupied a room on the top floor of the mansion (he had one to himself, enjoying a rather more privileged status than the junior officers), and then later moved into a cottage in the walled kitchen garden, which he shared with Robin Gandy and a large tabby cat. The cat was called Timothy, and had been brought to Hanslope by Robin on his return from staying with friends in London. Alan was well-disposed towards Timothy, even though (or perhaps because) it had a way of launching playful swats at the typewriter keys when he was at work.

As a place to hibernate while they waited for the war to end, Hanslope enjoyed one special advantage. The mess officer was Bernard Walsh, the proprietor of Wheeler's, the smart oyster restaurant in Soho. As if by magic, fresh eggs and partridges found their way on to the Hanslope dining table, while the rest of Britain

---

the undergraduate level – nothing as advanced as the work on the Riemann zeta-function before the war. As with the statistical theory he developed at Bletchley, this was a very good example of how quite elementary nineteenth-century mathematics had applications to the technology of the 1940s that no one had seen, or had tried to see.

chomped its way through Woolton pie and reconstituted egg. These might be supplemented by a rabbit from the copse, or a duck egg from the pond which lay at the bottom of the meadow surrounding the house, and Alan was also able to have the apple that as a rule he would always eat before going to bed. He would go out for walks or runs round the fields, and might be seen thoughtfully chewing on leaves of grass as he loped along, or perhaps burrowing around for mushrooms. During the year, a timely Penguin guide[13] to edible and inedible fungi had appeared, of which he made use to bring back amazing specimens for Mrs Lee (who organised the day-to-day catering) to cook for him. He particularly relished the name of the most poisonous of all, the Death Cap or *Amanita phalloides*. He would roll the name off his tongue with glee, and they would all search for it, but they never found a specimen.

Once of an evening he went out for a run and managed to break his ankle by slipping on a slime-covered brick which formed part of the garden path, just as he was getting into his stride. He had to be sent off in an ambulance to have it set in hospital. But at other times the Prof delighted everyone by winning the sports-day race, and by defeating the young Alan Wesley (another of the March intake) who rashly challenged him to a circuit of the large field. He was accepted (with due respect to his differences) as one of a crowd of junior officers. At lunchtimes, they would gather in the mess, and look at the newspapers: the *Daily Mirror* first, for the comic strip *Jane*. Don Bayley, who was very keen on military matters, might tell them about the developing strategy of the eastward-moving armies, while Alan might hold forth on some topic of a scientific or technical flavour, such as why water was opaque to electromagnetic radiation of the radar wavelength, or how a rocket could accelerate its own fuel quickly enough. They would sometimes go for lunchtime walks together, Timothy the cat accompanying them. Robin Gandy was learning Russian – not because of his earlier membership of the Communist party (which had lapsed in 1940), but because of his admiration for the Russian classics. Robin still considered himself a fellow-traveller, and in this respect 1941 had not changed Alan's view that he was quite misguided. But there was little discussion of politics at Hanslope, where the prevailing attitude was that of doing the job without asking questions.

Every month or so there was a Mess Night at which dress uniform, or in Alan's case a dinner jacket, would be required, and at which pheasant might be on the menu. This he would enjoy, for although generally so austere, he liked to live it up occasionally, dancing vigorously with the ladies of the ATS afterwards. There was plenty of social gossip and intrigue, which Alan rather liked hearing about, and discussing with Mrs Lee and Mary Wilson. Sometimes his own rather glamorous position as the mysterious Prof, coupled with his unthreatening friendly way with the women members of staff, aroused a mild jealousy. In this respect, he kept his own secret.

It was the first time in his life that he had mixed with ordinary people for any length of time, people picked out neither by social class nor by a special kind of intellect. It was a typical Turing irony that this should happen at an establishment working for the secret service. He liked its unpretentiousness, and perhaps the escape from the intellectual pressure at Bletchley. He certainly had the pleasure of being a large fish in a small pond. This liking was reciprocated. There was an occasion when he was invited to a drinking-party organised by the Other Ranks. For some reason it did not come off, but he was still very pleased: partly at breaching social class barriers, but surely also because of the allure, felt almost inescapably by a homosexual of his background, of that vast unknown England of working-class men.

In the evenings most of the officers would play billiards or drink in the bar, and sometimes Alan did too. But Donald Bayley, Robin Gandy and Alan Wesley had the idea of doing something more mind-improving, and asked Alan to give a course of lectures on mathematical methods. They found a place upstairs in the mansion, which in the winter of 1944 was a singularly cold classroom, and retired thither, somewhat to the amazement of the less zealous. Alan wrote out notes, which the others would copy, mainly on Fourier analysis and related material using the calculus of complex numbers. He illustrated his discussion of the idea of 'convolution' – the blurring or spreading out of one function in a way defined by some other function – with the example of a mushroom fairy ring.

It was not only the mushroom which currently reflected his interest in biological form; on his return from runs he would often show examples of the Fibonacci numbers to Don Bayley, producing

's father, Julius Turing, in about
7.

On the beach: Alan and his brother
John at St Leonard's in 1917.

he cliffs: Alan and his mother at St Lunaire, Brittany, in 1921 (see page 14).

Colonel Morcom, Mrs Morcom and Christopher on holiday, summer 1929.

George Maclure, Peter Hogg and Alan setting out on their hike, Easter 1931 (see page 75).

Alan Turing in a Guildford street, a chance snapshot taken in 1934.

A brushed-up Alan with his father and mother and a family friend (right) outside 8 Ennismore Avenue in 1938.

Boy and buoys at Bosham, August 1939 (see page 200). From the front: Alan, Bob, Karl and Fred Clayton.

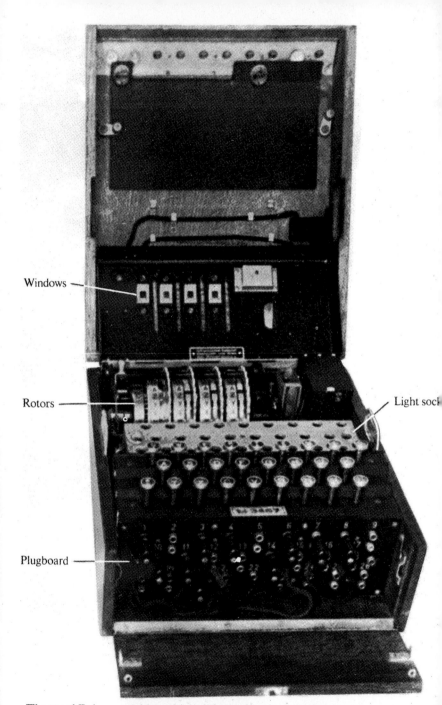

Windows

Rotors

Light sock

Plugboard

The naval Enigma machine with its lid raised to show the four rotors.

Above: The Colossus in operation at Bletchley Park, 1944–5, showing the rapid punched tape input mechanism.

Right: In contrast, the complete Delilah terminal fitted easily onto a table top (note the scale).

Left: The key unit, with lid removed to show both rotors and multivibrators.

Above: Robin Gandy in Summer 1953, on holiday in France.

Left: A sporting second in a three-mile race: possibly the event of 26 December 1946 (see page 444).

Below: The pilot ACE computer, on show at the NPL in November 1950. Jim Wilkinson is the right hand figure.

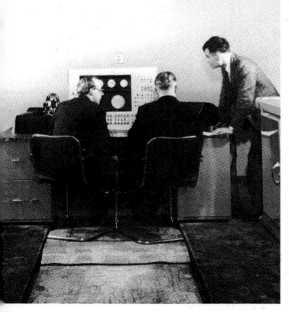

Above: The prototype Manchester computer. The six racks shown are essentially those of the 'baby machine' which ran its first program in June 1948; by June 1949 when this photograph was taken, the machine had roughly doubled in size.

Left: Alan Turing with two Ferranti engineers at the console of the 'Mark I' computer at Manchester, 1951.

Alan Mathison Turing on election to Fellowship of the Royal Society, 1951.

fir cones as he had in 1941. He was still sure there had to be a reason for it. And he found time for mathematical study of his own, taking up von Neumann's *Mathematical Foundations of Quantum Mechanics* again. In the evening there might also be chess, or card games, which he enjoyed, although these also brought out his most childish side in which – just as when a little boy – if he thought that someone else had cheated or changed the rules, he would storm out and slam the door. Such behaviour also typified his dealings with authority, which he still naively expected to maintain literal truth-telling and constancy of policy.

It was like the last two terms at school, staying on after having won a scholarship, without any clear function but accorded a gratifying respect. In August 1944, at about the same time as he came to live at the Hanslope mess, a small extension was built on to the large laboratory hut, and one of the four ten-foot by eight-foot rooms in it was allocated for work on Delilah. This gave him a more self-contained world in which to experiment, read, and think for the future. It was an odd position for 'the top cryptanalyst in England', waiting for his opponent to concede the game now dragging on and on. The Delilah project made more sense now that he had a qualified engineer to sort it out, but even this had been something of an accident. Don Bayley had not been assigned to it, but had been obliged to enveigle his way into participation, and there were always pressures on him to forsake it for other duties. Alan was irritated when this happened and would sometimes help in getting them out of the way.

Once, for instance, his advice was sought on the question of whether the 'wide-band' amplifiers, used in the process of distributing signals from a single large aerial to several different receivers, were introducing an element of noise into the system. He devised some experiments for testing them and did a little theoretical analysis. For this there was an outing to Cambridge, to search for appropriate literature on thermal noise. They had the privilege of an official car, and Don Bayley was rather pleased at being able to make a first visit to Cambridge. Before they went, Alan told the others not to call him 'Prof' while they were there.

Alan certainly enjoyed working together with his assistant in this way, but it meant being involved in what was very small stuff compared with his role in naval Enigma, or in Anglo-American liaison. Don knew that he had worked in cryptanalysis, and had been to America, but virtually nothing else. Alan did not supply any more to go on, and it was particularly striking at Hanslope, where with most people a few leading questions and a suggestion of already knowing more than was really the case would usually prise out further details. This method did not work with the Prof.* It was not simply the government's secrets that he protected with a uniquely rigid silence, but all personal confidences too. He treated all promises with a perhaps rather annoying literalness, as sacrosanct pieces of his own mind. (He often complained of politicians that they never kept their promises.) It left his colleague very puzzled as to his status. Alan showed himself slightly put out when after a short while he was taken to be one of the SCU3 staff, and made it clear he considered himself something rather better. But he had no discernible superior to whom to report, and no one ever came to see the progress of the Delilah.

There were a few social visits by his Bletchley colleagues, and evidence of one piece of Bletchley work on which he was consulted. This was to do with the design of a new Enigma-type machine which Gordon Welchman was currently organising. It was to encipher Baudot-code messages, and so had rotors with thirty-two rather than twenty-six contacts. This he also described to Shaun Wylie, explaining how he had been shown the proposed machine and complained that it had a period of only $32 \times 32 \times 32$. Meeting resistance, he had embarked on cranking through the settings by hand, only to discover that it was even worse – the period was only 32×32. His algebraic work on this problem stimulated some pure-mathematical offshoots which he kept for himself.[14]

There was some cryptographic consultation at Hanslope too, perhaps more typical of the work he had been doing since returning from America. He was asked to check that the Rockex key-tapes that were being generated by electronic noise were, in fact, sufficiently random. Unprotected by the buffer of a Hut 4 or a Hugh Alexander in such dealings with the military, there were often failures of

---

* He once expressed himself as shocked by the indiscreet talk at a college dinner of a certain eminent wartime scientist.

communication. Speaking too technically about 'the imaginary part of the error', he found the top brass had stopped listening. What he perceived as incompetence and stupidity would often send him into a black mood. In that case he would often take off for a run round the large field at the south of the Hanslope Park mansion to work off his feelings.

There was another issue which created argument and frustration, this time in the Delilah hut itself. Alan suddenly dropped into the conversation, with apparent casualness, the fact that he was a homosexual. His young Midlands assistant was both amazed and profoundly upset. He had heard mention of homosexuality only through jokes at school (which he was not the sort of person to find amusing) and through the vague allusions to 'grave charges' employed by the popular Sunday newspapers reporting court cases. It was not only what Alan told him that he found repellent, but the unapologetic attitude.

But it was the attitude of a Cambridge background that was as different from Don Bayley's as mathematics from engineering. Alan's assistant had equally firm, clear views, and said rather sharply that he had never before met anyone who not only admitted to what he considered at best distasteful and at worst disgusting propensities, but who seemed to think it perfectly natural and almost to be proud of it. Alan in turn was upset and disappointed by this reaction, which he described as only too typical of society at large. But this was, perhaps, one of the very few times that he had ever directly sampled the opinions of society at large. The reality, whether he liked it or not, was that most ordinary people would think of his feelings as alien and nauseating. His own attitudes having hardened since before the war – perhaps since breaking the engagement, but surely also because of increased confidence in himself after the work he had done – he did not drop the subject in embarrassment, but continued to argue in such a way that the exchange became quite heated. The progress of the Delilah was jeopardised.

Alan had ridden roughshod over fundamental differences. But he managed to overcome the difficulties in a way that did not mean either of them backing down. Don Bayley was able to cope by regarding it as another Turing eccentricity, and by weighing it against the advantage of working on such high-level ideas with a

Systematic search for exceptional groups. Theory

   In examining all possible uprights for a given T the main
difficulty lies in the large number of uprights involved. Once it
has been proved that a particular upright is unexceptional
the same will follow for a great number of others. ~~Xxxkxxx~~
~~ihxxxfxrxxixxxkryxxxmkxxixxxxifxximgxthxrxthxvxxrixhtxyxhxxxwitt~~
More generally given any upright we can find a great number of
others which generate either the same group H or an
isomorphic group. If we can classify these uprights together
in some way we shall enormously reduce the labour, since we
shall only need to investigate one member of each class. The
chief principles which enable us to find equivalent uprights are

   (i) If $U' = R^m U R^n$ then $H(U') = H(U)$

   (ii) If V commutes with(R)then $H(V U V^{-1}) \cong H(U)$. (N.B. if
V commutes with(R) then $V \wedge V^{-1} = R^k_{\quad})$

   (iii) If $U' = U^m$ ....on... U, U'^n then $H(U) = H(U')$.

The principle (i) is the one of which we make the most
systematic use. Our method depends on the fact that there are
very few U for which none of the permutations $R^m U R^n$ leave
two letters invariant (In other words there are very few U
without a beetle) and none of T is even. We therefore investigate
separately the U with no beetles and the U without beetles.

   U with no beetle. We can find an expression which
determines the classes of permutations obtainable from one
another by multiplic tion right and left by powers of R as
follows. Let $U R^{m+l} Z = R^{f(n)} U R^k Z$ (here Z represents
the last letter of the alphabet however many characters there
may be in it). Then we take the number $f(n)$ as describing
$$f(1) f(n) \ldots f(T)$$

A few lines from Alan Turing's attack on the problem of designing rotor
wirings, showing his use of group theory. The extract might be thought to
show also the influence of Timothy the cat on his work, but in fact this was
typical of his typewriting standard.

person whom in other ways he had come to like, and thought he knew quite well. So the Delilah survived the revelation. By the end of 1944, the equipment which did the sampling of the speech signal, and which processed the enciphered samples, was finished. They had proved it to work satisfactorily by setting up both a transmitting and a receiving end within the laboratory, and feeding both with an identical 'key', in the form of random noise from a radio receiver working with its aerial removed. It remained to design and build a system for feeding an identical key to terminals that would in practice be far apart.

In principle the Delilah could have used a one-time key recorded on gramophone records as the X-system did, analogous to the 'one-time pad' for telegraph transmissions. But Alan had chosen to devise a system which, though as good as 'one-time', would not require the shipping of thousands of tapes or records, but would instead allow sender and receiver to generate the identical key simultaneously at the time of transmission.

It was in this aspect of the Delilah that his cryptanalytic experience came into play. The work they had done so far constituted the mechanism for 'adding on' to speech. The crucial question of *what* to add was the one on which he had spent much of his time since 1938. In this he could act as the mathematical Cambridge and Bletchley figure, rather than someone who had joined in, somewhat awkwardly and embarrassingly, with the expanding world of electronic engineering.

Although he could not say, nor even hint, the task amounted to creating something like the Fish key generator. It had to be deterministic, for otherwise it could not be produced identically at two independent ends. But it had to be sufficiently devoid of pattern or repetition, to be as secure as something truly 'random', such as electronic noise. Any kind of mechanism would, inevitably, have *some* pattern to it; the job was to make sure that it was one that the enemy cryptanalyst could not possibly detect. So in doing this for the Delilah, he was finally scoring off the half-hearted efforts of German cryptography. In fact, he was doing very much better, for the Delilah key would have to be supplied in sequences of hundreds of thousands of numbers. It was like enciphering not telegraph messages, but *War and Peace*.

The idea of generating a key for speech encipherment in this way was not entirely new. The X-system was not always used with one-time gramophone records for the key. There was an alternative, called 'the threshing machine'. But this only had to produce a stream of digits at a rate of 300 a second, and was only used for testing or for low-status signals. The Delilah was much more demanding.

The generator had to be electronic, and the basic unit he used was the 'multivibrator' – a pair of valves possessing the property of locking into an oscillation between 'on' and 'off', with a length which would be some integral multiple of a basic period. His key generator made use of the outputs of eight such multivibrators, each locked into a different mode of oscillation. But that was just the beginning. These outputs were fed into several circuits with non-linear elements, which combined them in a complicated way. He had worked out a circuit design which ensured that the energy of the output would be spread as evenly as possible over the whole frequency range, and he explained to Donald Bayley with the aid of Fourier theory that this would endow the amplitude of the resulting output with the necessary degree of 'randomness' for cryptographic security.

There had to be some variability in the circuits, or else the generator would produce the same noise over and over again. This was allowed for by making the interconnections required for the combination of the eight multivibrator outputs to pass through the wirings rather like those of an Enigma, with rotors and a plugboard. So a setting of this 'Enigma' would serve to define a particular sequence of key, in a way that both sender and receiver could agree upon in advance. With the rotors fixed in position, the key would not repeat itself for about seven minutes. In practical operation, speech in one direction of transmission could be limited to this time and a new key sequence started off on reversing the direction of transmission. This could be done simply by stepping on the rotors. There were enough rotor and plugboard positions for the resulting system to be as safe – according to his theory – as a genuinely random one-time key.

Getting the Delilah system to work as a whole was a job which pushed their resources to the limit. The system was useless unless sender and receiver could keep their multivibrators in time to the microsecond. They spent most of the first half of 1945 in achieving the necessary precision. They also had to test the output of the Delilah key generator, when they had built it, for the evenness over the frequency range which the calculations predicted for it. It was typical of the conditions in which they worked that they had no frequency analyser. Alan would have seen one at Bell Labs, and there was one known to exist at Dollis Hill, but at Hanslope they had to make one for themselves. It provided a challenge of the desert-island sort, which Alan as usual enjoyed. After a lot of work they had a device, but when they first tried it out, Alan had to confess 'It's a bit of an abortion, isn't it!' – so they called it the ABORT Mark I.

To requisition anything at all required an act of skilful and aggressive diplomacy in handling the administrators. All they could obtain was a double-beam oscilloscope and an audio-frequency Hewlett-Packard oscillator. They even had to fight for this, first being fobbed off with an inferior one, and having to demand something better from SCU3's Controller, Colonel Maltby. For Alan, the process was as baffling as it ever was for Alice in the Looking Glass shop, trying to locate what she needed on the White Queen's shelves. Dealing with Maltby on the telephone drove Alan to extremes of nervousness, and people remarked upon how his speech, halting at the best of times, was on these occasions an almost indecipherable scrambling operation itself. He hated the showmanship that was required in negotiating for equipment. It was forever his bitter complaint that more adept players – 'Charlatans', 'Politicians', 'Salesmen' – would get their way not through expertise but through clever talk. He still tended to expect reason to prevail as if by magic.

It was a small-scale, homely example of the conflict that permeated the British war effort. But the Delilah project, manifestly too late for the German war, could not possibly expect a high priority, as he must have known. It was not like the work at Bletchley. And so even if angry over what he saw as incomprehensible waste and stupidity, he could also afford to stand aside and see the establishment in a more detached way. In this respect he and Robin Gandy would see things in very much the same light, and they both enjoyed reading Nigel Balchin's novel *The Small Back Room*, which had appeared in 1943. This presented with unconcealed bitterness, yet also with a mordant wit, the frustrations felt by young scientists trying to get the war won and over with, and hamstrung by games of one-upmanship and empire-defending. At Hanslope a number of amusing stories were told, fairly or unfairly, about the plots and coups of the upper echelons, but Alan certainly did not suffer from all the tribulations Balchin described, and in particular he was spared the problem of dealing with dead-wood 'eminent scientists' who stifled initiative in the name of efficiency. In fact no one took any interest, scientific or otherwise, in the Delilah project. This remained true even when the addition of the key generator showed that he had a means of giving complete speech security with two small boxes of equipment.

The army officers emerged in Balchin's book as 'red-tabbed stooges' who had entered a 'profession for fools', but to Alan the army system was less pernicious than ludicrous. He was very fond of Trollope's novels and kept a stock of them in the Hanslope cottage. He would hold forth on the similarities between the organisation of the Church of England and that of the army; with the help of Robin Gandy and Don Bayley he would seek out parallels between the Barchester machinations and those of the Hanslope hierarchy. They worked out a correspondence between the respective ranks, so that a lieutenant-colonel became a dean, and a major-general a bishop, while a brigadier was pegged at the status of suffragan bishop (the *cheapest* kind of bishop, Alan explained).

Occasionally there would be episcopal visitations, when Gambier-Parry and Maltby would look in to pay their respects and listen to the Delilah output, but this would be for reasons of form rather than out of real interest, as they had no direct responsibility for the work and only the haziest of notions of what Alan and Don Bayley were up to. Nor was it much use asking, since Alan was quite incomprehensible to them, a fact which was somewhat uncongenial since they claimed some scientific knowledge. The visitors were liable to be treated to a rendering of Winston Churchill's voice, since they used a gramophone recording of one of his speeches to test the Delilah. It was the broadcast made on 26 March 1944 in which after dwelling rather uncharacteristically on the subject of post-war housing policy, the Prime Minister had turned to more immediate prospects:[15]

> . . . The hour of our greatest effort is approaching. We march with valiant
> Allies who count on us as we count on them. The flashing eyes of all our

soldiers, sailors and airmen must be fixed upon the enemy on their front. The only homeward road for all of us lies through the arch of victory. The magnificent Armies of the United States are here or are pouring in. Our own troops, the best trained and the best equipped we have ever had, stand at their side in equal numbers and in true comradeship. Leaders are appointed in whom we all have faith. We shall require from our own people here, from Parliament, from the Press, from all classes, the same cool, strong nerves, the same toughness of fibre, which stood us in good stead in those days when we were all alone under the blitz.

With the help of the ABORT Mark I they could check that the Delilah had enciphered Churchill's phrases into a White Noise – a perfectly even and uninformative hiss. And then, by passing the output into the decipherment process, they could recover:

And here I must warn you that in order to deceive and baffle the enemy as well as to exercise the forces, there will be many false alarms, many feints, and many dress rehearsals. We may also ourselves be the object of new forms of attack from the enemy. Britain can take it. She has never flinched or failed. And when the signal is given, the whole circle of avenging nations will hurl themselves upon the foe and batter out the life of the cruellest tyranny which has ever sought to bar the progress of mankind.

It was not very good policy to test the Delilah with the same recording over and over again, because even when in the spring of 1945 they got it to work, it still helped to know the words when listening to the output. The deciphered speech had to compete with a noisy background,* and a 4000 Hz whistle. The latter arose from a 4000 Hz signal used to synchronise sender and receiver, and which was only imperfectly filtered out of the final speech output. But the Delilah *actually worked* – that was the joy of it, for all its deficiencies. Alan had created a sophisticated piece of electronic technology out of nothing, and it worked. They did go as far as making a proper sixteen-inch disc recording of the effect, which entailed a trip to the black broadcasting studios at Simpson, since Hanslope lacked the requisite equipment. While they were there, Alan's braces burst. Harold Robin, the chief engineer of the organisation, produced

---

* The signal-to-noise ratio was only 10dB, meaning that the speech had only ten times the power of the noise.

some bright red cord from an American packing case. Alan used it every day thereafter as his normal way of keeping his trousers up.

As the chief goose would surely have guessed, Churchill's prophecies had owed something to the continued supply of golden eggs, and so did the fact that by the time the Delilah was enciphering them, the words had come true. The pre-invasion 'feint' had succeeded in outwitting the German command, as they had been able to tell by listening in. At the critical points of the Normandy campaign, they had enjoyed the advantage of hearing the story from the other side. But he might well have wondered why it was taking so appallingly long to finish the war.

As month after over-confident month passed, the technological developments at Bletchley became less and less relevant to fighting the war. If Sigint continued to help with the general knowledge, it failed resoundingly at critical points. For all the wonders of the electronic revolution, the Allies had been taken completely by surprise in December 1944 when the front, already held far longer than anyone expected, threatened to settle into another and more gruesome 1917. There had been a radio silence. It was perhaps more the fault of the military that no serious assessment had been made of the German forces at Arnhem. But there was a limit to what Sigint could do. Knowledge of the 'new forms of attack' from the pilotless V1 and rocket-propelled V2 had not sufficed to stop them. And most remarkably, even the U-boat war, pre-eminently the war of information, was no walkover for the Allies. One factor was a political one: the RAF insisted on its role as an independent war-winning organisation, and devoted itself to the devastation of German cities rather than to the careful elimination of U-boats. But an increasing use of radio silence made cryptanalysis almost irrelevant at the end. The extraordinary fact was that when Dönitz took over from Hitler in April 1945, he still commanded a powerful, if suicidal, force. More U-boats were patrolling the American coasts than at any time since the mid-war winter, and new types of true submarines – rather than submersible boats – were in service. They were too late, like the new Enigmas that were ready but never came into service, but not very much too late.

The tapes whizzed round, the rotors spun, the Wrens followed their decision trees, but in the last months the mathematicians, finally given everything they wanted, had whizzed on into a world of their own. (Though what was real, and what absurd, would now be hard to tell.) Brute force, rather than wit and ingenuity, characterised the final Allied effort. This was not Alan Turing's war. The achievement of his work was more the defensive one, rather appropriate for him, who only wanted to be left in peace. There had been no repetition of the Atlantic of 1917, and the almost impossible had been made possible in time, just before German science and industry was being used seriously. As for the Europe of 1945, the Dresden of his friend, the Warsaw where it had begun, was this the victory of anyone's intelligence? Had the poker game of 1941 done anything for this? It did not bear thinking about.

Indeed, hardly anyone was permitted to think about it. The 'cracking from within' of 1918 had often filled British strategists of the Second World War with comforting delusions of easy victory, but it had also created a betrayal myth on which the Nazi party had capitalised. The great crack in logical control achieved by Bletchley Park no doubt had its influence upon the strategists of its aftermath, but this time there was no popular impact. It was completely hushed up. The victorious western governments had a common interest, for obvious reasons, in concealing the fact that the world's most sophisticated communication system had been mastered.

No one questioned that this had to be so. Those who knew a part of their story transferred it to a sealed compartment of the mind, so that the whole war became a blank out of which only stories about bicycles could emerge. The Bletchley vision had taken a few people on a time-travel trip, into a world of Our Ford where science had an answer for everything. Now they had to come back to the mid-1940s. Some, of course, had been grappling with the grimy reality of the 1940s all the time, and knew how almost impossible it had been to bridge the gap. Alan Turing, however, had been able more than most to protect himself from the abrasion. It would not be easy for him to adjust. And as one acquainted with a wider span of knowledge than any other of the 'men of the Professor type', it meant a particularly acute mind-splitting operation. On VE day, 8 May 1945, he joined Robin Gandy, Don Bayley and Alan Wesley for a walk in the nearby woods at Paulerspury. 'Well, the war is over, now you can tell all,' said Don, not very seriously. 'Don't be bloody silly,' said Alan – and that was the last word.

*

The Delilah was finished at about the same time as the German surrender. There was no particular drive to improve standards for the Japanese war, nor for future purposes, and its fundamental advance met with little enthusiasm. Radley and another engineer, R. J. Halsey, came to Hanslope to inspect it in somewhat perplexed fashion. The Post Office did have some system of their own under development – possibly based on the Vocoder, on which they had requested and received information in 1941. Their main concern was that the crackly output of the Delilah was too poor to be commercially acceptable, as indeed it was. They showed no sign of interest in the potential of the principle. Alan then spent some time himself at Dollis Hill in the summer of 1945, where he explained his system to a somewhat sceptical Flowers.

It was all over except the details – and Alan was never good at bothering with the last details. He was happy to leave it for Don Bayley to work on. For he had other ideas in mind. He had several times discussed with Don the question of his plans for peacetime, and had said that he was expecting to return to his King's fellowship and a cut back to £300 per annum. There were eighteen months of his 1938 fellowship still to run, but beyond this he now had a longer period assured, since on 27 May 1944, making a rather special gesture of confidence, King's had prolonged the tenure of his fellowship by a further three years. He could go back as if the war had never happened, and continue from where he had left off in 1939. A university lectureship might soon come his way.[16] And yet the war *had* happened, and everything had changed. It had not simply been an interruption to the course of his intellectual career, as for some of the 'professor type' it might have been. It had mobilised his inner life. His ideas had been enmeshed with its critical developments, and they had been able to grow with the scale of the war itself. The world had learned to think big, and so had he. For though expecting to return to Cambridge, Alan had also told Don Bayley from the start of their collaboration that he wanted 'to build a brain'.

His use of the word 'brain' was entirely consistent with his bold appeal to 'states of mind' ten years before. If the states of a Turing machine could be compared with 'states of mind', then its physical embodiment could be compared with a brain. One important aspect of this comparison, important to anyone who was concerned with the mystery of mind, the apparent paradox of free will and determinism, was that the Turing machine model was one independent of physics. The argument from Laplacian physical determinism could be shrugged aside with the observation that no such prediction could ever be performed in practice. This rebuttal could not be applied to a Turing machine, in which everything that happened could be described in terms of a finite set of symbols, and worked out with complete precision in terms of discrete states. Later he would articulate this himself:[17]

> The prediction which we are considering is however rather nearer to practicability than that considered by Laplace. The system of the 'universe as a whole' is such that quite small errors in the initial conditions can have an overwhelming effect at a later time. The displacement of a single electron by a billionth of a centimetre at one moment might make the difference between a man being killed by an avalanche a year later, or escaping. It is an essential property of the mechanical systems which we have called 'discrete state machines' that this phenomenon does not occur.

To understand the Turing model of 'the brain', it was crucial to see that it regarded physics and chemistry, including all the arguments about quantum mechanics to which Eddington had appealed, as essentially irrelevant. In his view, the physics and chemistry were relevant only in as much as they sustained the medium for the embodiment of discrete 'states', 'reading' and 'writing'. Only the *logical* pattern of these 'states' could really matter. The claim was that whatever a brain did, it did by virtue of its structure as a logical system, and not because it was inside a person's head, or because it was a spongy tissue made up of a particular kind of biological cell formation. And if this were so, then its logical structure could just as well be represented in some other medium, embodied by some other physical machinery. It was a materialist view of mind, but one that did not confuse logical patterns and relations with physical substances and things, as so often people did.

In particular it was a different claim from that of behaviourist psychology, which spoke of reducing psychology to physics. The Turing model did not seek to explain one kind of phenomenon, that of mind, in terms of another. It did not expect to 'reduce' psychology to anything. The thesis was that 'mind' or psychology could properly be described in terms of Turing machines because they both lay on the *same* level of description of the world, that of discrete logical systems. It was not a reduction, but an attempt at transference, when he imagined embodying such systems in an artificial 'brain'.

Alan probably did not know much in 1945 about the actual physiology of human brains: quite possibly no more than from jolly pictures of the brain as a humming telephone exchange in the *Children's Encyclopaedia*, or from the passage in *Natural Wonders* describing the 'small thinking place in the brain':

> Directly over the ear, a place that you can almost cover with your thumb, lies the most important part of all, the place where we remember and handle words. At the bottom of this word spot, we remember how words sound. An inch farther up and toward the back, we remember how words look in print. A little farther up and forward lies the 'speech center' from which, when we want to talk, we direct the tongue and lips what to say. Thus we get our word-hearing, our word-seeing, and our word-speaking centers close together, so that when we speak we have close by and handy our memory of what we have heard in words, and of what we have read.

But that would have been quite sufficient. He would have seen pictures of nerve cells (there were a few in *Natural Wonders*), but at the level at which he was approaching the description of mind, the details were not important. In speaking of 'building a brain' he did not mean that the components of his machine should resemble the components of a brain, or that their connections should imitate the manner in which the regions of the brain were connected. That the brain stored words, pictures, skills in *some* definite way, connected with input signals from the senses and output signals to the muscles, was almost all he needed. But ten years before, he had also had to fight his own way through to the crucial idea that Brewster glossed over; he had rejected the idea of a 'we' behind the brain that somehow 'did' this signalling and organising of the memory. The signalling and the organisation had to be all that there was.

But in describing the Turing machines ten years before, he had also justified his formalisation of the idea of 'mechanical' with a complementary argument, that of the 'instruction note'. This put the emphasis not on the internal workings of the brain, but upon the explicit instructions that a human worker could follow blindly. In 1936 such 'instruction notes' had entered his experience through the rules of Sherborne School, other social conventions, and of course in the mathematical formulae that one could apply 'without thinking'. But in 1945 a great deal of water had flowed under the bridge, and the 'instruction notes' that had been somewhat fanciful in 1936, just as were the theoretical logical machines, had become exceedingly concrete and practical. The cornucopian abundance was one of messages 'based on a machine and broken on a machine', and these machines were *Turing* machines, in which the logical transformation of symbols was what mattered, not physical power. And in designing such machines, and in working out processes that could be given to people acting like machines – the 'slaves' – they had effectively been writing elaborate 'instruction notes'.

This was a different, but not incompatible, approach to the idea of 'brain'. It was the interplay between the two approaches that perhaps fascinated Alan most – just as at Bletchley there had been a constant play between human intelligence and the use of machines or 'slave' methods. His 'weight of evidence' theory had shown how to transfer certain kinds of human recognition, judgment and decision into an 'instruction note' form. His chess-playing methods did the same thing – as did the games on the Colossi – and posed the question as to where a line could be drawn between the 'intelligent' and the 'mechanical'. His view, expressed in terms of the imitation principle, was that there was *no* such line, and neither did he ever draw a sharp distinction between the 'states of mind' approach and the 'instruction note' approach to the problem of reconciling the appearances of freedom and of determinism.

All these questions remained to be explored, for the exigencies of the German cipher machines had barely scratched the surface of what could be done. It was yet to be seen how much could be achieved by writing 'instruction notes', and yet to be seen whether a machine could behave like a brain in developing 'thinking spots' for itself. As he had stressed in his discussions with Donald Michie, it had to be shown that a machine could *learn*. To explore these questions it would be necessary to have machines on which to experiment. But the almost incredible fact was that it would require only *one* machine, for the performance of any and all such experiments. For a *universal* Turing machine could imitate the behaviour of any Turing machine whatever.

In 1936 the Universal Turing Machine had played a purely theoretical part in his attack upon the Hilbert *Entscheidungsproblem*. But in 1945 it had a very much more practical potential. For the Bombes and Colossi and all the other machines and mechanical processes were parasitic beasts, dependent upon the whims and blindness of the German cryptographers. A change of mind on the other side of the Channel would mean that all the engineering that had been required to construct them would suddenly become useless. It had happened right from the start, with the Polish 'fingerprint' file, their perforated sheets and their simple Bombe, and it had nearly led to catastrophe in the blackout of 1942. The construction of special machines had led the cryptanalysts into one problem after another with the acquisition and application of new technology. But a *universal* machine, if only it could be realised in practice, would require no fresh engineering, only fresh tables, encoded as 'description numbers' and placed upon its 'tape'. Such a machine could replace not only Bombes, Colossi, decision trees and all the other mechanical Bletchley tasks, but the whole laborious work of computation into which mathematicians had been conscripted by the war. The zeta-function machine, the calculation of roots of seventh-order equations, the large sets of equations arising in electrical-circuit theory – they could all alike be performed by a single machine. It was a vision beyond the comprehension of most people in 1945, but not beyond Alan Turing. As he would write later in 1945:[18]

> There will positively be no internal alterations to be made even if we wish suddenly to switch from calculating the energy levels of the neon atom to the enumeration of groups of order 720.

Or as he would put it in 1948,[19]

> We do not need to have an infinity of different machines doing different jobs. A single one will suffice. The engineering problem of producing various machines for various jobs is replaced by the office work of 'programming' the universal machine to do these jobs.

From this point of view, a 'brain' would not be just some bigger or better machine, some superior kind of Colossus. It did not develop out of an experience of things, but out of a consciousness of underlying ideas. A universal machine would not just be a machine; it would be *all* machines. It would replace not only the physical Bletchley machinery, but all that was routine – almost all that those ten thousand people had been doing. And not even the 'intelligent' work of the high-level analysts would be sacrosanct. For a universal machine could also play out the workings of human brains. Whatever a brain did, *any* brain, could in principle be placed as a 'description number' on the tape of a Universal Machine. This was his vision.

But there was nothing in the paper design of the Universal Turing Machine that suggested it could be made a practical proposition. In particular, there was nothing about its speed of operation. The tables of *Computable Numbers* could be realised by people sending postcards to each other, without affecting the theoretical argument. But if a universal machine were to be of any practical use, it would have to be able to run through millions of steps in a reasonable time. This demand for speed could only be met by electronic components. And this was where the revolution of 1943 had made all the difference in the world.

More precisely, the point was that electronic components could be regarded as operating upon *discrete*, on-or-off, quantities, and so could realise a Turing machine. This he had learnt in 1942, and thereafter he had known all about the Robinsons, the X-system, and the Rockex; he had also picked up a fund of radar knowledge from his new friends at Hanslope. But above all there were the two developments that had begun in 1943. Whatever its usefulness to the war effort, the technical success of the Colossi told him that thousands of electronic valves could be used in conjunction – something that few could have believed in until it had been done. And then he had worked with his own bare hands on the Delilah. There had been a method in his madness all along. By working in these second-rate conditions, on a device that officialdom had not called for, he had proved that he could carry off an electronic project of his own. Coordinated with his theoretical ideas and his experience of mechanical methods, this direct knowledge of electronic technology formed the last link in his plans. He had learned how to build a brain – not an *electric* brain, as he might possibly have imagined before the war – but an *electronic* brain. It was thus that 'round about 1944', Alan's mother heard him talking[20] about 'his plans for the construction of a universal [machine] and of the service such a machine might render to psychology in the study of the human brain.'

There was a further fundamental consideration besides that of discreteness, reliability and speed: that of sheer size. There would have to be room on the 'tape' of a universal machine both for the 'description numbers' of the machines it had to imitate, and its workings. The abstract universal machine of 1936 was equipped with a 'tape' of *infinite* length, meaning that although at any stage the amount of tape used would be finite, it was assumed that as more space was required it could always be made available. In a practical machine, space would always be limited in extent – and for that reason no physical machine could actually realise a truly universal machine. Still, Alan had suggested in *Computable Numbers* that human memory was finite in extent. If this were so then the human brain itself could hold only a limited number of 'tables of behaviour', and a sufficiently large tape could contain them all. The finiteness of any practical machine would not, on this argument, debar it from having a brain-like quality. The question was, however, how much 'tape' would be required for a machine that could actually be built: enough to make it interesting, but not more than would be technically feasible. And how could such storage be arranged without inconceivable expense in terms of electronic valves?

This practical question was one more up Don Bayley's street. As the European war ground to its end, and the problems of the Delilah were essentially solved, it became clear that Alan's interest had turned to 'the brain'. He described to his assistant the universal machine of *Computable Numbers*, and its 'tape' on which instructions would be stored. They began thinking together about ways in which to realise a 'tape' that could store such information. And thus it was that in this remote station of the new Sigint empire, working with one assistant in a small hut, and thinking in his spare time, an English homosexual atheist mathematician had conceived of the *computer.**

That was not how the world was to see it, and the world was not being entirely unfair. Alan Turing's invention had to take its place in a historical context, in which he was neither the first to think about constructing universal machines, nor the only one to arrive in 1945 at an electronic version of the universal machine of *Computable Numbers.*

There were, of course, all manner of thought-saving machines in existence, going back to the invention of the abacus. Broadly these could be classed into two categories, 'analogue' and 'digital'. The two machines on which Alan worked just before the war were examples of each kind. The zeta-function machine depended on measuring the moment of a collection of rotating wheels. This physical quantity was to be the 'analogue' of the mathematical quantity being calculated. On the other hand, the binary multiplier had depended upon nothing but observations of 'on' and 'off'. It was a machine not for measuring quantities, but for organising symbols. In practice, there might be both analogue and digital aspects to a machine. There was not a hard-and-fast distinction. The Bombe, for instance, certainly operated on symbols, and so was essentially 'digital', but its mode of operation depended upon the accurate physical motion of the rotors, and their analogy with the enciphering Enigma. Even counting on one's fingers, by definition

---

* That is, he had arrived at the *automatic electronic digital computer with internal program storage.* In what follows, the word 'computer' will be reserved for machines satisfying all these conditions. But in 1945 the word 'computer' meant what it had meant in 1935: either a person who did computations, or any type of machine (in anti-aircraft artillery, for instance) which mechanised that computation. It was not for about ten years that 'computer', or even 'digital computer', took on the new meaning. Meanwhile a variety of more cumbersome terms were used, and correspondingly the concept was less clear at the time than it later became, particularly regarding the internally stored program. Alan Turing had not invented a *thing*, but had brought together a powerful collection of *ideas*; since the ideas condensed into exactly what became 'the computer' it does not do too grave an injury to history to employ the word anachronistically. In fact the anachronism reflects quite well the difficulty that he faced in communicating to the 1940s a picture that belonged to the 1960s.

'digital', would have an aspect of physical analogy with the objects being counted. However, there was a practical consideration which provided the effective distinction between an analogue and a digital approach. It was the question of what happened when increased accuracy was sought.

His projected zeta-function machine would have well illustrated the point. It was designed to calculate the zeta-function to within a certain accuracy of measurement. If he had then found that this accuracy was insufficient for his purpose of investigating the Riemann Hypothesis, and needed another decimal place, then it would have meant a complete re-engineering of the physical equipment – with much larger gear-wheels, or a much more delicate balance. Every successive increase in accuracy would demand new equipment. In contrast, if the values of the zeta-function were found by 'digital' methods – by pencil and paper and desk calculators – then an increase in accuracy might well entail a hundred times more work, but would not need any more physical apparatus. This limitation in physical accuracy was the problem with the pre-war 'differential analysers', which existed to set up analogies (in terms of electrical amplitudes) for certain systems of differential equations. It was this question which set up the great divide between 'analogue' and 'digital'.

Alan was naturally drawn towards the 'digital' machine, because the Turing machines of *Computable Numbers* were precisely the abstract versions of such machines. His predisposition would have been reinforced by long experience with 'digital' problems in cryptanalysis – problems of which those working on numerical questions would be entirely ignorant, by virtue of the secrecy surrounding them. He was certainly not ignorant of analogue approaches to problem-solving. Apart from the zeta-function machine, the Delilah had an important 'analogue' aspect. It depended crucially on accurate measurement and transmission of the amplitudes, in contrast to the X-system, which made them 'digital'. He might well have admitted that for certain problems, the analogue solution could not remotely be rivalled by a digital method. Putting a model aircraft in a wind tunnel would immediately produce a picture of stresses and vortices that centuries of calculation would never obtain. In 1945 there was plenty of scope for debating the relative practical usefulness of analogue and digital devices, and the priorities for construction. But so far as

Alan Turing was concerned, this was a debate for other people. He was committed to the digital approach, flowing out of the Turing machine concept, and with its potential universality at the centre. No analogue machine could lay claim to universality, such devices being constructed to be physical analogies of particular problems. It followed that his ideas had to find their place among, and compete with, the prevailing developments of digital calculators.

There had been machines to add and multiply numbers, the digital equivalents of the slide rule, since the seventeenth century. Alan had a desk calculator at Hanslope, and used it for the calculation of circuit properties. It was a very long step indeed from such devices to the idea of a practical universal machine. But as Alan knew by this time, that step had been made a hundred years before, by the British mathematician Charles Babbage (1791–1871). He used to speak to Don Bayley of Babbage, and knew something of what Babbage had planned.

After working on a 'Difference Engine', to mechanise the particular numerical method used in the construction of mathematical tables, Babbage had conceived (by 1837) of an Analytical Engine, whose essential property was that of mechanising *any* mathematical operation. It embodied the crucial idea of replacing the engineering of different machines for different tasks, by the office work of producing new instructions for the same machine. Babbage did not have a theory like that of *Computable Numbers*, to argue for universality, and his attention was focussed upon operations using numbers in decimal notation. Yet he did perceive that its mechanism could serve to effect operations upon symbols of any kind whatever,* and in this

---

* Certainly Ada, Countess of Lovelace, writing her interpretation[21] of Babbage's ideas in 1842, expressed this idea in a passage of prophetic insight:

The bounds of *arithmetic* were, however, outstepped the moment the idea of applying the cards had occurred; and the Analytical Engine does not occupy common ground with mere 'calculating machines'. It holds a position wholly its own; and the considerations it suggests are most interesting in their nature. In enabling mechanism to combine together *general* symbols, in successions of unlimited variety and extent, a uniting link is established between the operations of matter and the abstract mental processes of the *most abstract* branch of mathematical science. A new, a vast and a powerful language is developed for the future use of analysis, in which to wield its truths so that these may become of more speedy and accurate practical application for the purposes of mankind than the means hitherto in our possession have rendered possible. Thus not only the mental and the material, but the theoretical and the

and other ways the Analytical Engine came close in its conception to the Universal Turing Machine. Babbage wanted a 'scanner', in effect, working on a stream of instructions, and putting them into operation. He hit on the idea of coding the instructions on punched cards, such as then were used for the weaving of complicated patterns in brocade. His plans also called for storing numbers in the form of positions of gear wheels. Each instruction card would cause an arithmetical operation such as 'subtract the number in location 5 from that in location 8, and put the result in location 16.' This required machinery he called the 'mill' to do the arithmetical operations, but the crucial innovation of Babbage's plans did not lie in the efficient mechanisation of adding and multiplying. It lay in his perception that it was the mechanisation of the organising or *logical control* of the arithmetic that mattered.

In particular, Babbage had the vital idea that it must be possible to move forwards or backwards among the stream of instruction cards, skipping or repeating, according to criteria which were to be tested by the machine itself in the course of its calculation. This idea, that of 'conditional branching', was his most advanced. It was equivalent to the freedom allowed to Turing machines, that of changing 'configuration' according to what was read on the tape, and it was this that made Babbage's planned machine a universal one, as he himself was well aware.

Without 'conditional branching', the ability to mechanise the word IF, the grandest calculator would be no more than a glorified adding machine. It might be thought of as an assembly line, everything being laid down from start to finish, and there being no possibility of interference in the process once started. The facility of 'conditional branching', in this model, would be analogous to specifying not only the routine tasks of the workers, but the testing, deciding and controlling operations of the *management*. Babbage was well-placed to perceive this idea, his book *On the Economy of Machinery and Manufactures* being the foundation of modern management.

---

practical in the mathematical world, are brought into more intimate and effective connexion with each other. We are not aware of its being on record that anything partaking of the nature of what is so well designated the *Analytical* Engine has been hitherto proposed, or even thought of, as a practical possibility, any more than the idea of a thinking or of a reasoning machine.

These ideas were a hundred years ahead of their time, and were never to be embodied in working machinery during Babbage's lifetime. Government funding did not solve the problems created by his grossly over-ambitious specifications; the project was not advanced by Babbage's contempt for committees, administrators, and other scientists; nor did his own efforts to bring mechanical engineering to an entirely new standard, and his absorption in every theoretical and practical aspect of the work, overcome these difficulties.

Indeed, it was *exactly* a hundred years from the conception of the Analytical Engine, until there were substantially new developments either in the theory or in the construction of such a universal machine. On the side of theory, 1937 saw the publication of *Computable Numbers*, which made all these ideas precise, explicit, and conscious. On the practical side, there had been the inevitable Looking Glass war as the revived and expanding electrical industry of the 1930s provided rival powers with new opportunities.

The first development had, in fact, occurred in 1937 Germany, at the Berlin home of K. Zuse, an engineer who had rediscovered many of Babbage's ideas, though not that of conditional branching. Like the Babbage machine his first design, which was actually built in 1938, was mechanical and not electrical. But he had avoided the thousands of meshing ten-spoke gear wheels that Babbage had demanded, by the simple expedient of having his machine work in binary arithmetic. This was not a deep theoretical advance, but from any practical point of view it was an immense simplification. It was also a liberation from the usual engineer's assumption that numbers had to be represented in the decimal way. Alan had used the same idea at the same time in his 1937 electric multiplier. Zuse had quickly moved on to construct further versions of his machine which made use of electromagnetic relays rather than mechanical elements, and with collaborators experimented with electronics before the end of the war. Zuse calculators were used in aircraft engineering but not in code-breaking; it was argued that the war would be over too soon. Short-term Nazism left Zuse in 1945 desperately trying to save his work from destruction.

These events were unknown on the Allied side, where roughly parallel but larger developments had been taking place. In Britain there was no such digital calculator, controlled by a sequence of instructions, except the Colossi. This was in marked contrast with the situation in the United States. The British success, frantic but triumphant, had been achieved at the last moment by individuals giving their all to wartime public service. The Americans, so much richer in capitalist enterprise, were years ahead in pursuing two different, perhaps slightly unimaginative, approaches to the Babbage idea – and in doing so even in peacetime, just as they were ahead in the analogue differential analyser of the earlier 1930s. For again it was 1937 when at Harvard the physicist H. Aiken began to realise it in terms of electromagnetic relays. The resulting machine was built by IBM and turned over to the US Navy for secret work in 1944. It was lavish and impressive in scale, but like Zuse's machines, did not embody conditional branching, even though Aiken knew of Babbage's plans. The instructions had to be followed rigidly from beginning to end. Aiken's machine was also more conservative than Zuse's in that it based the arithmetical machinery on the decimal notation.

The second American project was under way at Bell Laboratories. Here the engineer G. Stibitz had first only thought of designing relay machines to perform decimal arithmetic with complex numbers, but after the outbreak of war had incorporated the facility to carry out a fixed sequence of arithmetical operations. His 'Model III' was under way in the New York building at the time of Alan's stay there, but it had not drawn his attention.

There was another person, however, who did make a thorough examination of these two advanced projects, and who like Alan had the mind with which to form a more abstract view of what was happening. This was that other mathematician of the wizard war, John von Neumann. He had been connected with US Army ballistic research as a consultant since 1937. From 1941 most of his time had been spent upon the applied mathematics of explosions and aerodynamics. In the first six months of 1943 he was in Britain, conferring on those subjects with G. I. Taylor, the British applied mathematician. It was then that he had first become involved in programming a large calculation, in the sense of organising how it could best be done by people working on desk calculators. Back in the United States, his entry in September 1943 into the atomic bomb project had taken him into similar problems with shock waves, whose prediction by numerical computation required months of slogging work. In 1944 he toured the available machines in search of help. W. Weaver, at the Office of Scientific Research and Development, had put him in touch with Stibitz, and on 27 March 1944 von Neumann wrote[22] to Weaver:

> Will write to Stibitz: my curiosity to learn more about the relay computation method, as well as my expectations concerning possibilities in this direction, are much aroused.

On 10 April he wrote again to say that Stibitz had shown him 'the principle and working of his relay counting mechanisms'. On 14 April he wrote to R. Peierls at Los Alamos about the 'shock decay problem', saying that it could probably be mechanised, and adding that he was now also in touch with Aiken. In July 1944 there were negotiations to use the Harvard-IBM machine. But then everything changed. For the pressure of wartime demands had brought about the same technological revolution as had happened at Bletchley, and at exactly the same time. In quite another place, namely the engineering department of the University of Pennsylvania (the Moore School), work had begun on yet another large calculator in April 1943. This was the ENIAC – the *Electronic* Numerical Integrator and Calculator.

The new machine was designed by the electronic engineers J. P. Eckert and J. Mauchly, although von Neumann's first knowledge of it, apparently something of an accident, came through talking on a railroad station with H. H. Goldstine, a mathematician associated with the project. Von Neumann seized upon the possibilities opened up by a machine that when built would perform arithmetical operations a thousand times faster than Aiken's. From August 1944 he was regularly attending ENIAC team meetings, writing on 1 November 1944 to Weaver:

> There are some other things, mostly connected with mechanised computation, which I should like a chance to talk to you about. I am exceedingly obliged to you for having put me in contact with several workers in this field, especially with Aiken and Stibitz. In the meantime I have had a very extensive exchange of views with Aiken, and still more with the group at the Moore School . . . who are now in the process of planning a second electronic machine. I have been asked to act as their adviser, mainly on the matters connected with logical control, memory, etc.

The ENIAC project was immensely impressive, quite enough to give people allowed to see it the sensation of seeing the future. It employed no fewer than 19,000 electronic valves. As such it surpassed the Colossi, with which in many ways it was comparable, though one difference was that in the summer of 1945 the ENIAC was still incomplete, and would come too late to have any use whatever in the war.

It required a greater total number of valves than the Colossi because it stored long decimal numbers – the more so because of the primitive system the designers had employed whereby ten valves were allocated to each decimal digit required, a '9' being represented by the ninth of those valves being 'on'. In contrast the Colossi operated on single pulses which represented the logical 'yes' or 'no' of holes in telegraph tape.

But this was a fairly superficial point of difference. Both alike demonstrated that thousands of valves, hitherto regarded as too unreliable for operation *en masse*, could be kept in simultaneous use.* And the ENIAC project was embodying the idea that Zuse, Aiken and Stibitz had missed. Like the Mark II Colossi, with their ability to automate acts of decision, the results of one counting operation automatically deciding what step was taken next, the ENIAC would have a form of conditional branching. It was designed so that it could be made to hop to and fro within the stock of instructions supplied to it, repeating sections as many times as the progress of the calculation showed was necessary, without the interference of human management. None of this went beyond what Babbage had envisaged – except that electronic components were so much faster, and that the ENIAC was (or was nearly) a reality.

Like the Colossus, the ENIAC had been designed for a specific task, that of calculating artillery-range tables. Essentially it was to simulate the trajectories of shells through varying conditions of air resistance and wind velocity, which involved the summation of thousands of little pieces of trajectory. It had external switches which would be set to store the constant parameters for a trajectory calculation, and further external devices for setting up the

---

* But neither was the first in this respect. At Iowa State University, J. V. Atanasoff had been using electronics for mechanising arithmetical operations since 1939.

instructions on how to calculate the segments of motion. Then there would be valves to hold the intermediate working figures. In these arrangements it resembled the Colossi. But in both cases, people had quickly discovered the possibility for using the machines for a wider range of tasks than those for which they had been designed. The role of the original Colossus had been much extended by Donald Michie and Jack Good, and then the Mark II had once been set up to punch out a deciphered message, although this was done out of interest, not for the sake of efficiency. Even though a parasite on the German cipher machine, the flexibility offered by its instruction table facility was such that it could even be 'almost' set up to perform numerical multiplication. The ENIAC was flexible in a more serious way, and von Neumann had already discovered that it could be used, when ready, for Los Alamos problems.*

But the ENIAC had not been conceived as a universal machine, and in one important respect the designers had departed from Babbage's line of development. Babbage had been proud of the fact that the planned Analytical Engine would be able to ingest an *unlimited* number of instruction cards. The Aiken relay machine enjoyed the same feature, although cards had been replaced by a sort of pianola roll. But on the ENIAC the state of affairs was different. Its operations, being electronic, would be so fast as to make it impossible to supply cards or tape quickly enough. The engineers had had to find a way to make the instructions available to the machine in electronic times of a few millionths of a second.

On the ENIAC they were arranging for this by a system of external devices which would set up the instructions for each job. It took the form of making connections with plugs as on a manual telephone exchange. (The Colossi had something very similar.) The advantage of this solution was that the instructions would in effect be available instantaneously, once the plugging work was done. The disadvantage was that the sequence of instructions was limited in length, and that it would take a day or so to do the plugging. It would be like building a new machine for each different task. Both ENIAC and Colossus were like kits out of which many slightly different machines could be

---

* Indeed its first serious use, in late 1945, would be on a trial calculation for the hydrogen bomb.

made. Neither sought to embody the true universality of Babbage's conception, in which the machinery would be entirely unchanged, and only the instruction cards rewritten.

But even when von Neumann joined the ENIAC team as 'adviser' in late 1944, Eckert and Mauchly had perceived a quite different solution to their problem. This was to leave the hardware alone, and to make the instructions available at electronic speeds by storing them *internally*, in electronic form. The ENIAC was designed to store its arithmetical working internally, and the point of the original Colossus had been that it would hold Fish key-patterns internally. It was a quite different matter to consider holding the *instructions* to the machine internally. Instructions were naturally thought of as coming from the outside, to act upon an inside. But the 'second electronic machine' mentioned in von Neumann's letter to Weaver was intended to incorporate this new idea.

Every tradition of common sense and clear thinking would tend to suggest that 'numbers' were entirely different in kind from 'instructions'. The obvious thing was to keep them apart: the data in one place, and the stock of instructions to operate *on* the data, in another place. It was obvious – but wrong. In March and April 1945, the ENIAC team had prepared a proposal, the *Draft Report on the EDVAC*. The EDVAC – the Electronic Discrete Variable Calculator – was the planned 'second electronic machine'. The report was dated 30 June 1945, and signed by von Neumann. It was not his design, but the description of it bore the mark of his more mathematical mind rising above the technicalities.

In particular, it articulated the very cautious, circumspect, but quite new idea, at which the ENIAC team had arrived in planning a better machine. It discussed the different kinds of storage that existing machines required: intermediate results, instructions, fixed constant parameters, statistical data, and then stated that[23]

> The device requires a considerable memory. While it appeared, that various parts of this memory have to perform functions which differ somewhat in their nature and considerably in their purpose, it is nevertheless tempting to treat the entire memory as one organ.

But such a proposal, that of the 'one organ', was equivalent to adopting the 'one tape' of the Universal Turing Machine, on which everything – instructions, data, and working – was to be stored. This was the new idea, different from anything in Babbage's design, and one which marked a turning point in proposals for digital machines. For it threw all the emphasis on to a new place – the construction of a large, fast, effective, all-purpose electronic 'memory'. And in its way it made everything much more simple, less cluttered in conception. Von Neumann might well have seen it as 'tempting', because it was almost too good an idea to be true. But it had been in *Computable Numbers* all the time.

So the spring of 1945 saw the ENIAC team on the one hand, and Alan Turing on the other, arrive naturally at the idea of constructing a universal machine with a single 'tape'. But they did so in rather different ways. The ENIAC, now already shown to be out-of-date in principle even before it was finished, had been something of a sledgehammer in cracking open the problem. And von Neumann had been obliged to hack his way through the jungle of every known approach to computation, assimilating all the current needs of military research and the capabilities of American industry. The result was something close to the Lancelot Hogben view of science: the political and economic needs of the day determining new ideas.

But when Alan Turing spoke of 'building a brain', he was working and thinking alone in his spare time, pottering around in a British back-garden shed with a few pieces of equipment grudgingly conceded by the secret service. He was not being asked to provide the solution to numerical problems such as those von Neumann was engaged upon; he had been thinking for himself. He had simply put together things that no one had put together before: his one-tape universal machine, the knowledge that large-scale electronic pulse technology could work, and the experience of turning cryptanalytic thought into 'definite methods' and 'mechanical processes'. Since 1939 he had been concerned with little but symbols, states, and instruction tables – and with the problem of embodying these as effectively as possible in concrete forms. Now he could consummate it all.

And now the war was over, his motives were much nearer to those of G. H. Hardy than to the practicalities of the world's work. They had more to do with the paradox of determinism and free will, than with the effecting of long calculations. Of course, no one was likely to pay for a 'brain' that had no useful application. And in this respect Hardy would have found justification of his views regarding the applications of mathematics. On 30 January 1945 von Neumann had written[24] that the EDVAC was being designed for three-dimensional 'aerodynamic and shock-wave problems . . . shell, bomb and rocket work . . . progress in the field of propellants and high explosives'. Such, in Churchill's phrase, would be the 'progress of mankind'. Alan Turing too would have to come a long way from the logic of Hilbert and Gödel if he was really to build a brain.

The *Draft Report on the EDVAC* did carry a more theoretical burden (one reflecting von Neumann's interests) in that it drew attention to the analogy between a computer and the human nervous system. The use of the word 'memory' was an aspect of this. In its way, it was 'building a brain'. However, its emphasis was placed not on an abstract thesis concerning 'states of mind', but on the similarities between input and output mechanisms, and afferent and efferent nerves respectively. It also drew upon the 1943 paper written by the Chicago neurologists W. S. McCulloch and W. Pitts, which analysed the activity of neurons in logical terms, and used their symbolism for describing the logical connections of electronic components.

McCulloch and Pitts had been inspired by *Computable Numbers*, and so in a very indirect way, the EDVAC proposal owed something to the concept of the Turing machine. One of its curiosities, however, was that it made no mention of *Computable Numbers*, nor made precise the universal machine concept. Yet von Neumann had been familiar with it before the war, and must surely have recognised the connection when he liberated himself from the assumption that data and instructions had to be stored in different ways. According to S. Frankel,[25] who worked at Los Alamos on the atomic bomb and was one of the first to use the ENIAC,

> in about 1943 or '44 von Neumann was well aware of the fundamental importance of Turing's paper of 1936 'On computable numbers' . . . Von Neumann introduced me to that paper and at his urging I studied it with care. . . . he firmly emphasised to me, and to others I am sure, that the fundamental conception is owing to Turing – insofar as not anticipated by Babbage, Lovelace and others.

So the Wizard might well have learnt something from Dorothy. However, the essential point about these two initiatives, American and British, was not the rather tenuous connection between them. It was their very marked independence.[26]

Whatever ideas had flowed westwards, the *Draft Report on the EDVAC* was the first to put them together in writing. So once again, British originality had been pipped at the post by an American publication – and at a time when everyone was looking to the west. The Americans had won, and Alan was a sporting second. This time, however, American priority was nothing but an advantage to the Turing plans, for it provided the political and economic impetus that his own ideas alone could never have enjoyed.

Indeed, it was probably only the existence of the ENIAC and the EDVAC idea that made possible the next stage of Alan Turing's life. For in June he had a telephone call at Hanslope. It was from J. R. Womersley, Superintendent of the Mathematics Division at the National Physical Laboratory.

Womersley was a new man in a new post in a new organisation. The National Physical Laboratory was not new; it had been set up in shabby, suburban Teddington in 1900, the British response to state-sponsored German scientific research. Its site was carved out of Bushy Park, itself largely given over to Supreme Headquarters, Allied Expeditionary Forces. It was the most extensive government laboratory in the United Kingdom, enjoying a high reputation within its traditional sphere, that of setting and maintaining physical standards for the benefit of British industry. Its current Director, installed in 1938, was Sir Charles Galton Darwin, grandson of the theorist of evolution and himself an eminent Cambridge applied mathematician. His major contribution had been in the field of X-ray crystallography, and like Humpty Dumpty, who was able to explain the *Jabberwocky*, he was regarded as the[27] 'interpreter of the new quantum theory to experimental physicists'. Large, awesome, and remote, during the war he had spent a year as director of what became the British Central Scientific Mission in Washington, and had been the first scientific adviser to the British Army.

The Mathematics Division, however, was new. Indeed, it was a computational equivalent of the planned welfare state, the product of a calculators' Beveridge Report. In about March 1944 a proposal[28] had been mooted for an independent Mathematical Station, and this suggestion, a fine example of wartime planning for peace, went to a large interdepartmental committee, itself a manifestation of cooperation and coordination unthinkable in peacetime days. The government accepted the principle of continuing the funding found necessary in war, and a centralised, rationalised institution was planned to take over the various *ad hoc* offices which had done the mostly dreary work of numerical computation for military purposes. Sir Charles Darwin had persuaded the committee to establish it as a division of the NPL.

But the telephone call to Hanslope was not made at Darwin's behest. It was made on the initiative of his subordinate, Womersley, who had been selected as head of the new Division on 27 September 1944. Womersley, a bulky Yorkshireman then attached to the Ministry of Supply and a member of the interdepartmental committee, was probably the nominee of D. R. Hartree, who in mathematical matters was a power behind the Darwin throne. Womersley had appeared as joint author with Hartree of a 1937 paper on the application of the differential analyser to partial differential equations.

The official research programme for the new Division in October 1944 included 'Investigation of the possible adaptation of automatic telephone equipment to scientific computing' and 'Development of electronic counting device suitable for rapid computing'. Behind these words there were more definite intentions to imitate the American developments. Hartree, with his differential analyser at Manchester University, already had an interest in computing machinery, and had a finger in many wartime scientific projects. In the high levels where he moved, some details of the secret Aiken and ENIAC machines filtered through. Such knowledge was reflected in Womersley's report in December 1944, which, though it placed emphasis upon building a large differential analyser, remarked on the speed of electronics, and suggested that 'a machine can be made to perform certain cycles of operations mechanically . . . the instructions to the machine [could] depend on the result of previous operations . . . the problem is already being tackled in the USA.' The press release[29] in April 1945, when the new Division was officially inaugurated, only made mention of 'analytical engines, including the differential analyser and other machines both existing and awaiting invention . . . it is certain that this field is capable of great developments, but it is more difficult to predict in what directions they will lie.' But it seemed that the direction in which to look was westwards, and in February 1945, Womersley had been packed off on a two-month tour of the computing installations of the United States, where on 12 March he was the first non-American to be allowed access to the ENIAC, and to be informed of the EDVAC report.

By 15 May, Womersley was back at the NPL 'revising his plans'. The American revelations were enough to give anyone pause for thought. But they held a particular meaning for Womersley, who held one very unusual card up his sleeve. Before the war, while employed at Woolwich Arsenal on practical computation, he had learnt of Turing machines. Even more remarkably, for a mathematician-in-the-street, he had not been daunted by the abstruse language of mathematical logic. According to his claim:[30]

> *1937–38* Paper Computable Numbers seen by JRW and read. JRW met C. I. Norfolk, a telephone engineer who had specialised in totalisator design and discussed with him the planning of a 'Turing Machine' using automatic telephone equipment, Rough schematics prepared, and possibility of submitting a proposal to NPL discussed. It was decided that machine would be too slow to be effective.

> *June 1938* JRW purchased a uniselector and some relays on Petty Cash at RD Woolwich for spare-time experiments. Experiments abandoned owing to pressure of work on ballistics.

On seeing Aiken's machine at Harvard, Womersley had written home to his wife that he saw it as 'Turing in hardware'. And thus it was that in June 1945, according to his account:

> JRW meets Professor M. H. A. Newman.* Tells Newman he wishes to meet Turing. Meets Turing same day and invites him home. JRW shows Turing the first report on the EDVAC and persuades him to join NPL staff, arranges interview and convinces Director and Secretary.

---

* Here one must assume that the secrecy surrounding Bletchley operations had been breached sufficiently for Womersley (through Darwin and Hartree, perhaps also through Blackett) to learn of the existence of the electronic Colossus and also of Alan's general whereabouts.

Alan would be appointed a Temporary Senior Scientific Officer at £800 per annum. Don Bayley, when told of this, did not think much of the rank, but Alan told him that it was the highest to which they could recruit, and that he had been assured of a promotion within a few weeks. It was not quite 'Fivepence farthing for one – twopence for two', as of the eggs that the Sheep offered Alice in the Looking Glass shop, but at £600 for naval Enigma, and £800 for the digital computer, the British government had certainly acquired a bargain in Alan Turing. Alan claimed that Womersley had asked him whether he knew 'the integral of cos x', which as Don Bayley immediately said, was a ludicrously trivial question to ask any prospective SSO, let alone a B-star Wrangler. 'Ah', said Alan, in a joke against his own capacity for carelessness, 'but what if I had got it *wrong*?'

On his side, Womersley expressed himself to colleagues as delighted with the capture of Alan Turing for his new department. And for Alan, who was careless about rank and terms of appointment, it was still the exciting prospect of having the British government itself support the realisation of a Universal Turing Machine. He had done his bit for them; now they could return the compliment. The NPL had been founded 'to break down the barrier between theory and practice', and this was exactly what Alan proposed to do. Whatever his doubts about the Civil Service, it offered him a chance. When he bid farewell to Joan Clarke and the others tidying up Hut 8, he talked excitedly of the future of automatic computers, and reassured them that mathematicians would not be put out of work.

In the general election of July 1945 he voted for the Labour party. 'Time for a change', he said vaguely later. It was hardly remarkable for one who belonged to a generation that had been champing at the bit while the 'broad-bottomed', as G. H. Hardy used to call them, ruled the roost. The conflicts of 'the small back room' were reflected in the hustings. The war had obliged the planning and state control that had been urged upon deaf ears in the 1930s, and the Labour party offered to preserve what Churchill proposed to dismantle – as Lloyd George had done in 1919. But Alan Turing was no Labour party stalwart, no more than he had been a 'political' person in the 1930s.

The Abdication had roused him more than the Beveridge Report. As an admirer of Bernard Shaw, a reader of the *New Statesman*, and a wartime scientist up against the blinkered inertia of the old regime, he would approve of reform. But organisation and reorganisation did not really interest him.

His attitudes still had more in common with the democratic individualism of J. S. Mill than with the planners of 1945. But he did not share Mill's interest in commercial competition. Indeed, he did not know anything about it. His life had been in interlocking schools, universities and government service. In his undergraduate days business had been a holiday, and the small Beuttell and Morcom firms were themselves exceptions to the twentieth-century trend, representing a spirit which had largely died with Gladstone. And during the war, the contractors for equipment had been working on *carte blanche* government contracts in which ordinary considerations of profit had no application.

Money, commerce and competition had played no obvious part in the central developments in which Alan Turing was enmeshed, developments which had allowed him in many ways to remain the idealistic undergraduate. His retention of a primitive liberalism, his 'championing of the underdog' as it was seen at Hanslope, like his obsession with the absolutely basic, had the flavour of more Utopian thinkers than Mill. Tolstoy was a figure that he brought to one person's mind,[31] and Claude Shannon had perceived him as like Nietzsche, 'beyond Good and Evil'. But perhaps closer in spirit than either of these, and certainly closer to home, was another late nineteenth-century figure who had lurked more in the back room of political consciousness. That awkward figure Edward Carpenter, while sharing much in common with each of these European thinkers, had criticised Tolstoy for a restrictive attitude to sex and Nietzsche for overbearing arrogance. And in Carpenter, at a time when socialism was supposed to be about better organisation, lay the example of an English socialist not interested in organisation but in science, sex and simplicity – and with bringing these into mutual harmony. Born in 1844, he had written words[32] during the First World War that already fitted a small boy at St Leonards-on-Sea, and which Alan Turing had continued to act out quite regardless of the opinions of more respectable persons:

I used to go and sit on the beach at Brighton and dream, and now I sit on the shore of human life and dream practically the same dreams. I remember about the time that I mention – or it may have been a trifle later – coming to the distinct conclusion that there were only two things really worth living for – the glory and beauty of Nature, and the glory and beauty of human love and friendship. And to-day I still feel the same. What else indeed *is* there? All the nonsense about riches, fame, distinction, ease, luxury and so forth – how little does it amount to! It really is not worth wasting time over. These things are so obviously second-hand affairs, useful only and in so far as they may lead to the first two, and short of their doing that liable to become odious and harmful. To become united and in line with the beauty and vitality of Nature (but, Lord help us! we are far enough off from that at present), and to become united with those we love – what other ultimate object in life *is* there? Surely all these other things, these games and examinations, these churches and chapels, these district councils and money markets, these top-hats and telephones and even the general necessity of earning one's living – if they are not ultimately for that, *what are they for?*

Behind all the foibles and funny stories, and behind all the fuss made about his appearance and manners, lay the fact that as a boy he had never been able to understand how anyone could see life in any other way than this, and that at thirty-three only the war against Nazi Germany had dented that clinging to first principles.

There was a closer parallel than this, for Carpenter had been a Cambridge mathematician, and one fascinated by the same theme of mind in a deterministic world. He had the same upper-middle-class background, a parallel interest in biological growth. And he had abandoned Christian belief, while identifying himself as a homosexual. His book *Homogenic Love*, appearing in 1895, had been the first English work to place homosexual desire within a contemporary psychological and social context (rather than in that of ancient Greece), and to do so as part of a wholesale attack on 'fixed moral codes' – very close to the 'general rules' that Keynes, much more privately, rejected. And although not entirely relinquishing the idea that homosexuals had some special part to play, an idea not found in Alan Turing's book, the burden of his argument was that 'homogenic love' should be part of the general give and take, the creative anarchy of life – neither good nor evil in itself, but as sociable, as selfish, as messy as anything else.

By 1945 this could be taken as Alan Turing's view, for if sometimes he had seen his sexuality as a cross to bear, it was more and more a fact of life, one as much at the heart of what he was as that equally unasked-for, equally amoral, love of natural science. But in defending this view in 1945 he was taking a line that in fifty years hardly anyone had dared say in public more clearly than Carpenter. Not all the modernity of the war had changed this fact, and since his consciousness of sex had become clearer in 1933, there had been little but 'semi-platonic sentimentality' allowed.* Others, of course, were more content than he to remain the deceivers of men. And they were wise in worldly terms, for it remained a taboo more dangerous than the Soviet heresy. Untouched by electronic revolutions, unmentioned in the debates of 1945, it was not a subject for political people. But Alan Turing was not one of those.

The early Labour Party had been open to Carpenter's ideals of a more simple life, and even of a New Morality. His naively lucid questioning of what life was for, and of what socialism was going to make it, had played a part in its more innocent days. Even when in power, in 1924, the first Labour cabinet had sent him a letter of thanks on his eightieth birthday. But the thirties had put paid to that. In 1937 George Orwell had ridiculed what remained of such impractical, distracting naiveté in *The Road to Wigan Pier:*

> One sometimes gets the impression that the mere words 'Socialism' and 'Communism' draw towards them with magnetic force every fruit-juice drinker, nudist, sandal-wearer, sex-maniac, Quaker, 'Nature Cure' quack, pacifist and feminist in England.

In 1944 it was E. M. Forster, a liberal rather than a socialist, who remembered Carpenter's centenary, and his cautious eulogy[33] was that of a man forgotten. Both he and Lowes Dickinson had been influenced by Carpenter's ideas – the happy pastoral ending in the 'greenwood' of the still unpublished *Maurice* was drawn from Carpenter's rather scandalous life near Sheffield with a younger

---

* Indeed the second Waugh, in 1945, had largely revisited the semi-platonic sentimentality of the first.

working-class man – and the more democratic side of King's College, its lack of 'stuffiness' as well as its comparative openness to sexual dissent, owed something to his legacy.

The Labour party still had Carpenter's song *England Arise!* as well as *The Red Flag*, but the shift of support to it in 1945 reflected the success of new men and modern methods rather than the sentiments of either anthem. There was now a political consciousness of the importance of science (though hardly in the way Carpenter would have wanted), but not of sex or simplicity. In 1937 – just as the first large calculators began the arithmetical relay race – George Orwell had been as repelled by the creed of 'mechanization, rationalization, modernization', as by the vegetarians and simple-lifers. But the war had brought it to power, and for good reason: Britain would have succumbed without it.

Orwell escaped from the dichotomy by appealing to an England of 'ordinary, decent' people. Alan Turing might have liked to do the same, but he was now hopelessly saddled with a mind full of extraordinary, indecent contradictions. It contained the greatest development of 'mechanization, rationalization, modernization' of the war, and another which was the greatest ever conceived, while still longing for 'the commonest in nature', and while still being precisely what Orwell meant by a 'sex maniac'. He could not avoid these things, and having surrendered half of his mind to the government, was not free to try. He had done something, in a way that Orwell had not, and had passed the point of no return.

The paradox was not his alone, although his life took it to a peculiar intensity. The war had dealt a sharp blow to 'fixed moral codes', with social changes accelerated, old authorities questioned and new talents employed. Everyone had been made conscious of the defects of the old system, and more insidiously, of the fact that systems could be changed when survival made it necessary. To the dismay of conservative forces, British society had undergone a second and more thorough shaking up, this time with knowledge and ideas communicated to those excluded from participation in peace – ordinary men, the young, and even women. Bletchley Park had seen this happen as much as anywhere else. It had not been all a story of 'men of the Professor type'; there had been boys of eighteen, 'female mathematicians', and Post Office engineers who had risen from the bottom of the ladder, all playing crucial parts.

In other ways, too, the consciousness of being a community, sharing a very limited common wealth, had brought people closer to the 'least waste of energy' of Alan Turing, spartan but not joyless. Even at a place such as Hanslope, enmeshed in the technical machinations of the secret service, the pleasures of Mess Nights, mountain-walking holidays, mushroom-cooking, games and self-education had taken on that enhanced value that Carpenter had rather laboriously tried to explain as 'the Simplification of Life'.

There was a new spirit, and yet it was a spirit within a machine. A much enlarged state apparatus, and the more centralised economy, were the legacy of the great battle for intelligence and coordination. This time it would not be undone. And it was the machine, rather than glimpses of workers' control, that inspired Ernest Bevin:[34] 'Calculation of profits and all the other things that have cluttered up progress in the past has to go and the great genius of our managers and technicians is being given full play. . . .' It was true. Unhindered by wasteful competition and by the false economies of public parsimony, the Government Code and Cypher School and the Post Office had proved capable of managing fantastic feats. Now the development of the electronic computer was being taken over by the National Physical Laboratory for the public good. It deserved two cheers, as Forster would say, two cheers for managerial socialism. But management and techniques had not been the whole story, important as they might be. There had been something else, something now fading away while they waited for the other war to finish.

With Hitler out of the way, the games of Red and White could resume. Attlee replaced Churchill at the Potsdam conference when the results of the British election were known. Alan Turing went to Germany at the same time, in a party made up of five British and six American experts to report on German progress in communications. Flowers was one of the other British members. They left on 15 July, and arrived in Paris on a fine hot day. Here they were to meet the Americans, but the military headquarters had no idea who they were, so they took the day off. Late in the afternoon the telegrams from London came through, and they were assigned to the military transit camp, a hotel near the Madeleine. The same thing happened next day at Frankfurt when they reported to the American army headquarters in the I. G. Farben building. It was Patton's area, and they were warned not to continue into Bavaria without permission from his staff, or they would be arrested by the military police. After another day they set off in a jeep along the pot-holed roads, going 'hell for leather' over 200 miles to make their destination before nightfall. They were stopped thirty-seven times by the MPs because, being civilians, they had no tin hats.

So Alan Turing re-entered the ruined land of Gauss and Hilbert under watchful American eyes and in a military jeep. The party stayed at a communications laboratory at Ebermannstadt, near Bayreuth, which they had to reach by trudging up a thousand feet of mountain. It had been a hospital, and still bore a red cross on its roof, so they simply slept in the hospital beds. Women from the village came and did their washing, in return for a fragment of soap. Only he and Flowers had any cryptological interest, and the other members of the party did not (as far as they knew) know that they had. One of the captured German scientists proudly produced a machine of the Fish type, and explained how many billions of steps it would go through without repeating the key. Alan and Flowers just blinked and said, 'Really!' when he went on to tell them that none the less their mathematicians had reckoned it impregnable only for two years, and that then there would be a chance of it being broken.

While they were there, the mushroom cloud fulfilled the wilder prophecies of 1939. The quantum mechanics that Hardy had so recently pronounced gloriously useless, had come of age. It was the outward sign of the new men's work. Maurice Pryce had played an early part in the British research, and the final touch had been added by von Neumann, calculating the height at which it should explode to effect the maximum destruction. The clouds rolled over the second enemy, that would-be latecomer to old-fashioned empire, and warned a potential new one. The Americans had solved the final problem of the war. Yet without the sequence of events that had kept the Enigma on the Allied side in 1943, the war of 1945 might have been very different, with the first atomic weapons reserved for the concrete pens of raiding U-boats.

The great secret was out – or rather, it was known that there *was* a secret, which made it so very different from the other one. American soldiers came up to the Ebermannstadt station with the news, which did not surprise Alan. He had known of the possibility before the war, and was good at picking up straws in the wind. After his return from America, he had posed both to Jack Good and to Shaun Wylie a question about a chain reaction, expressed in terms of barrels of gunpowder. He had also spoken of a possible 'U-bomb' at Hanslope lunchtimes. He gave a talk on the basic physical principle to the others at Ebermannstadt.

He remained in Germany until about the middle of August, and then returned to write his report on the visit. After six years, the war was officially finished. He had given his help in breaking the slave states, and in giving victory to the Yankees. Perhaps, less directly, his work had played a part in deciding the new boundaries of Animal Farm. But in 1945 few were dwelling upon the state of the eastern menagerie, although at Bletchley Park the new men had left the means with which to resume the politics of the 1920s.

No longer responsible for the world, they could get things right at home. In this respect Alan Turing was as fortunate as anyone. Even if his work had often been wasted, he had made the most of the war for himself, and emerged ready to contribute to the peace. The British had avoided defeat, and owed America for that. The ending of Lend-Lease was only the beginning of new problems. The power of British capital had shrunk, and its empire was to melt away. Yet arising in the mind were seeds of other kinds of growth.

6 Mercury Delayed

As I walk these broad majestic days of peace,
(For the war, the struggle of blood finish'd, wherein, O terrific Ideal,
Against vast odds erewhile having gloriously won,
Now thou stridest on, yet perhaps in time toward denser wars,
Perhaps to engage in time in still more dreadful contests, dangers,
Longer campaigns and crises, labors beyond all others,)
Around me I hear that eclat of the world, politics, produce,
The announcements of recognized things, science,
The approved growth of cities and the spread of inventions.

I see the ships, (they will last a few years,)
The vast factories with their foremen and workmen,
And hear the indorsement of all, and do not object to it.

But I too announce solid things,
Science, ships, politics, cities, factories, are not nothing,
Like a grand procession to music of distant bugles pouring,
    triumphantly moving, and grander heaving in sight,
They stand for realities – all is as it should be.

Then my realities;
What else is so real as mine?
Libertad and the divine average, freedom to every slave on the face
    of the earth,
The rapt promises and luminé of seers, the spiritual world, these
    centuries-lasting songs,
And our visions, the visions of poets, the most solid announcements
    of any.

Alan Turing did not wait to take up the NPL post before thinking about the practical construction of his universal machine. In particular he discussed with Don Bayley the problem which dominated its engineering – that of the storage mechanism, or 'tape'. They discussed every form of discrete storage that they could think of. For instance, they considered magnetic recording. They had seen

a captured German Army 'Magnetophon', the first successful tape
recorder, but rejected the idea essentially because magnetic tape was
too much like the tape of the theoretical Universal Turing Machine
– it would require so much physical moving to and fro.* Instead,
they favoured another solution with which Alan was by now well
acquainted – that of the 'acoustic delay line'.

The idea was based on the fact that the time taken for a sound wave
to travel along a few feet of pipe was of the order of a thousandth of
a second. The pipe could be regarded as *storing* the sound wave for
that period. The principle had already been applied in radar, using
information stored in the delay line to cancel out all the radar echoes
which had not changed since the last scanning. In that way the radar
screen could be made to show only new, or changing, objects. It was
Eckert of the ENIAC team who had suggested the use of a delay line
to store the pulses of an electronic computer. There were several
ideas involved. The pipe, or delay line, would have to cope with
pulses separated by only a millionth of a second, and to transmit
them unsmudged. It was also necessary that the pulses should be
stored not just for a thousandth of a second, but indefinitely, which
required recirculating them through the delay line again and again.
If this were done naively then the pulses would soon become too
blurred to be distinguishable. So electronics had to be devised to
detect the existence of a (somewhat degenerated) pulse arriving at
the end of the line, and to start off a clean pulse at the beginning – the
electronic equivalent of the relay used as a telegraph repeater. This
would have to be combined with the facility to accept pulses from
the rest of the computer and to feed them back in as required. It was
well known that it was advantageous to use a medium other than
air for the sound waves, and mercury was already being employed

---

* Curiously, they did not think of what later became magnetic core storage. They knew
all about the properties of toroidal magnetic cores, wound with current-carrying wire,
because they were used in the wide-band radio frequency transformers which Donald
Bayley often had to break away from work on Delilah in order to design. The cores for
this work were chosen for their low hysteresis, meaning that they would respond rapidly
and accurately without loss of signal. It never occurred to them that the 'less satisfactory'
types shown in the manufacturers' catalogue, whose response was not so linear, and
which would tend to remain either 'north' or 'south', could be used in the discrete 'on or
off' manner required for storage.

in radar applications. This appropriate element, associated with the classical deity of speed and communication, was to haunt the developments of the next few years.

This was an attractively cheap solution within the existing technology, and had been provisionally adopted in the *Draft Report on the* EDVAC. In this September 1945 period, they tried out the principle in the Hanslope hut. Don Bayley rigged up a cardboard tube, eight inches across and the whole ten feet of the hut in length, and Alan designed a super-regenerative amplifier (a particularly sensitive form of amplifier fashionable at the time). They connected the amplifier to a microphone at one end of the tube and a loudspeaker at the other. The idea was simply to get a feel for the problem by recycling a sound wave in air on the delay-line principle, clapping at one end and hoping to set up a hundred artificial echoes thereafter. They did not get it to work before Alan left Hanslope to take up his NPL post, which officially began on 1 October 1945. But it meant that he arrived full of ideas both logical and physical, and was far from being the pure mathematician of 1938.

In setting up the new Mathematics Division, Womersley had been able to recruit from the experts in the field of numerical computation, as it had been developed for the war effort. His division took over the highly regarded Admiralty Computing Service as the nucleus of what was the most high-powered group in the western world, the rival being the American equivalent at the National Bureau of Standards. It was not that they were good at doing large sums, although indeed they *were* doing sums on desk calculators. Their problems were roughly analogous to those that faced Alan in calculating the Riemann zeta-function in 1938. When the resources of pure mathematics had been fully exploited, there might still remain a formula, or system of equations, into which actual numbers had to be substituted. Actually doing such substitutions on desk calculators was not very interesting; but the problem of how best to organise the work was a more abstract question, one constituting the branch of mathematics called 'numerical analysis'.*One particular problem

---

* As a branch of mathematics, however, numerical analysis probably ranked the lowest, even below the theory of statistics, in terms of what most university mathematicians found interesting.

was that although equations and formulae would generally relate 'real numbers' of infinite precision, practical computation would work with quantities defined only to so many decimal places, thus introducing an error into every step. Deducing the effect of such errors and minimising them was an important aspect of numerical analysis. The existence of such problems was partly what made Alan say that automatic computers would not make mathematicians redundant.

The section doing such work was headed by E. T. 'Charles' Goodwin, a fellow B-star of 1934 who recognised Alan from undergraduate days. Two other sections, 'Statistics' and 'Punched cards' were also related to the Turing interest, and the existence of punched-card machinery on the premises was to decide the choice of input mechanism for his machine. A fourth section consisted of the staff of the Hartree differential analyser, and remained for the time being at Manchester. The fifth section consisted of Alan Turing alone. By the end of the year there were twenty-seven staff in the Division, which was thus roughly equivalent to a large university department.

Two Victorian houses, Teddington Hall and Cromer House, on the perimeter of the existing NPL territory, had been acquired in March, and in October the whole new Division was housed in Cromer House, where Alan had a little room in the north wing. Charles Goodwin and his colleague Leslie Fox were across the way and working on the problem of how best to find the 'eigenvalues' of matrices, in connection with the problem of finding the resonant frequencies of an aircraft design. In these autumn months they would hear his typewriter banging jerkily away.

Alan lived in a guest house in nearby Hampton village, near the Thames, and generally continued to live out of a suitcase just as in wartime. The transition from war to peace was marked by the fact that now, instead of being under the administration of military officers, he was under the direction of scientists. This was not as much of a change as he might have expected. For Womersley, to whom he would grimly refer as 'my boss', as indeed he was, had turned out to be the epitome of what Alan despised as 'bogus'. Although a man of some dynamism and vision, he lacked the solid grasp of scientific knowledge that Alan considered essential for a person in his position. Thus it transpired that Womersley's lengthy and expensive tour of the United States earlier in 1945 had been a technical failure, since he had lacked the expertise to make detailed notes on what he had been allowed to see. Flowers and Chandler had been obliged to make a visit of their own in September and October to see the ENIAC in connection with work they were doing on special-purpose calculators for the military, instead of using Womersley's notes. Womersley's gifts of management: a mastery of name-dropping, a genial enthusiasm, a pleasant office manner to important visitors, a diplomatic sense of what to report, were not skills that Alan Turing ranked highly; not just because he lacked them himself, but because he still could not understand why anyone should need weapons other than rational argument. Before long, Alan was openly rude to Womersley in the office, saying 'What do *you* want?' and turning his back if Womersley dared to intrude upon some discussion. Later on there was a bet arranged among the staff of the Division, which depended upon someone coming out of Womersley's office with 'an equation, no matter how trivial'; it was abandoned and conceded, Alan reported, 'for lack of entries'. Conversely, Womersley would show visitors round Cromer House, pointing at the Turing office from afar with exaggerated awe, and saying 'Ah, that's Turing, we mustn't disturb *him*,' as of some rare zoological exhibit.

A stronger scientific intellect, with an independent view of how computers should be built, might have hindered rather than helped Alan's plans, which at least found in Womersley no technical resistance. On the contrary, Womersley was all too liable to agree with whatever had last been suggested. Womersley also coined a more happy acronym for the Turing electronic computer project than the soulless ENIAC and EDVAC. It was to be called the Automatic Computing Engine – a reference to Babbage's 'engine'. It would be the ACE. Alan was fond of saying that this was Womersley's *only* contribution to the project. It reminded him of old George Johnstone Stoney, who had not discovered the electron, but gave it its name. In fact, Womersley had displayed considerable political skill in getting the project approved. It was not for nothing that he had a copy of *How to Win Friends and Influence People* on his desk. But Alan was blind to that. He was still the least political person.

*

Alan's first task was to write a report[1], setting out a detailed design of an electronic universal machine, and an account of its operation. Surprisingly, the report that he submitted did not contain mention of *Computable Numbers*. Instead, it referred to the *Draft Report on the* EDVAC, with which his report was to be read 'in conjunction'. However, the ACE proposal was effectively self-contained, and its roots lay not in the EDVAC, but in his own universal machine. Some fragmentary notes[2], dating from this early period, made this clear:

> . . . In 'Computable Numbers' it was assumed that all the stored material was arranged linearly, so that in effect the accessibility time was directly proportional to the amount of material stored, being essentially the digit time multiplied by the number of digits stored. This was the essential reason why the arrangement in 'Computable Numbers' could not be taken over as it stood to give a practical form of machine.

It was also implicit in an opening paragraph of the report he now wrote, which explained how new problems would be 'virtually only a matter of paper work', with examples,* and said:

> It may appear somewhat surprising that this can be done. How can one expect a machine to do all this multitudinous variety of things? The answer is that we should consider the machine as doing something quite simple, namely carrying out orders given to it in a standard form which it is able to understand.

But he considerably amplified this idea in a talk given a year later in February 1947,[3] in words which explained the origin of the ACE as he himself perceived it:

> Some years ago I was researching on what might now be described as an investigation of the theoretical possibilities and limitations of digital computing machines. I considered a type of machine which had a central mechanism and an infinite memory which was contained on an infinite tape. This type of machine appeared to be sufficiently general. One of my conclusions was that the idea of a 'rule of thumb' process and a 'machine process' were synonymous. The expression 'machine process' of course means one which could be carried out by the type of machine I was

---

* In the passage already quoted on page 368.

considering. . . . Machines such as the ACE* may be regarded as practical versions of this same type of machine. There is at least a very close analogy.

Digital computing machines all have a central mechanism or control and some very extensive form of memory. The memory does not have to be infinite, but it certainly needs to be very large. In general the arrangement of the memory on an infinite tape is unsatisfactory in a practical machine, because of the large amount of time which is liable to be spent in shifting up and down the tape to reach the point at which a particular piece of information required at the moment is stored. Thus a problem might easily need a storage of three million entries, and if each entry was equally likely to be the next required the average journey up the tape would be through a million entries, and this would be intolerable. One needs some form of memory with which any required entry can be reached at short notice. This difficulty presumably used to worry the Egyptians when their books were written on papyrus scrolls. It must have been slow work looking up references in them, and the present arrangement of written matter in books which can be opened at any point is greatly to be preferred. We may say that storage on tape and papyrus scrolls is somewhat *inaccessible*. It takes a considerable time to find a given entry. Memory in book form is a good deal better, and is certainly highly suitable when it is to be read by the human eye. We could even imagine a computing machine that was made to work with a memory based on books. It would not be very easy but would be immensely preferable to the single long tape. Let us for the sake of argument suppose that the difficulties involved in using books as memory were overcome, that is to say that mechanical devices for finding the right book and opening it at the right page, etc. etc. had been developed, imitating the use of human hands and eyes. The information contained in the books would still be rather inaccessible because of the time occupied in the mechanical motions. One cannot turn a page over very quickly without tearing it, and if one were to do much book transportation, and do it fast, the energy involved would be very great. Thus if we moved one book every millisecond and each were moved ten metres and weighed 200 grams, and if the kinetic energy were wasted each time, we should consume 10^{10} watts, about half the country's power consumption. If we are to have a really fast machine then we must have our information, or at any rate a part of it, in a more accessible form than can be obtained with books.

---

* A typical circumlocution necessary for 'the computer'.

After this flight of fancy, very typical of his conversation, he discussed various more serious proposals for storage, and commented that 'the provision of proper storage is the key to the problem of the digital computer.'

> In my opinion this problem of making a large memory available at reasonably short notice is much more important than that of doing operations such as multiplication at high speed. Speed is necessary if the machine is to work fast enough for the machine to be commercially valuable, but a large storage is necessary if it is to be capable of anything more than rather trivial operations. The storage capacity is therefore the more fundamental requirement.

He then continued to give a succinct definition of 'building a brain':

> Let us now return to the analogy of the theoretical computing machines with an infinite tape. It can be shown that a single special machine of that type can be made to do the work of all. It could in fact be made to work as a model of any other machine. The special machine may be called the universal machine; it works in the following quite simple manner. When we have decided what machine we wish to imitate, we punch a description of it on the tape of the universal machine. This description explains what the machine would do in every configuration in which it might find itself. The universal machine has only to keep looking at this description in order to find out what it should do at each stage. Thus the complexity of the machine to be imitated is concentrated in the tape and does not appear in the universal machine proper in any way.
>
> If we take the properties of the universal machine in combination with the fact that machine processes and rule of thumb processes are synonymous, we may say that the universal machine is one which, when supplied with the appropriate instructions, can be made to do any rule of thumb process. This feature is parallelled in digital computing machines such as the ACE. They are in fact practical versions of the universal machine. There is a certain central pool of electronic equipment, and a large memory. When any particular problem has to be handled the appropriate instructions for the computing process involved are stored in the memory of the ACE and it is then 'set up' for carrying out that process.

His priorities were a large, fast memory, and then a hardware system that would be as *simple as possible*. The latter requirement was reminiscent of his 'desert island' mentality, doing everything with the least waste. But both features were to exploit the universality of the machine. His idea was always that anything in the way of refinement or convenience for the user could be performed by thought and not by machinery, by instructions and not by hardware.

In his philosophy it was almost an extravagance to supply addition and multiplication facilities as hardware, since in principle they could be replaced by instructions applying only the more primitive logical operations of OR, AND, and NOT. Indeed, the Colossus, when it was 'almost' programmed to do multiplication, had done just that. Since these primitive logical operations (absent from the EDVAC draft design) were incorporated in his plan for the ACE, he could indeed have omitted adders and multipliers, and still have had a universal machine. In reality, he did include special hardware to perform arithmetical tasks, but even there he decomposed the arithmetical operations into small pieces so that he could economise on hardware at the cost of more stored instructions. The whole conception was very puzzling to his contemporaries, to whom a computer was a machine to do sums, and a multiplier the very essence of its function. To Alan Turing, the multiplier was a rather tiresome technicality; the heart lay in the logical control, which took the instructions from the memory, and put them into operation.

For similar reasons, his report placed no great emphasis upon the fact that the ACE would use binary arithmetic. He stated the advantage of the binary representation, namely that electronic switches could naturally represent '1' and '0' by 'on' and 'off'. But that was all, apart from a terse statement that the input and output to the machine would be in ordinary decimal notation, and that the conversion process would have 'virtually no outward and visible form'. In his 1947 talk he would elaborate on this briefest of all possible comments. The point was that the universality of the machine made it possible to encode numbers in any way one wished within the machine – in binary form, if that happened to suit the technology. It would be inappropriate to employ binary numbers in a cash register, because it would be more trouble than it was worth to make the conversions for input and output. On the universal ACE, however, no such conversion was required –

This last statement sounds quite paradoxical, but it is a simple consequence of the fact that these machines can be made to do any rule of thumb process by remembering suitable instructions. In particular it can be made to do binary decimal conversion. For example in the case of the ACE the provision of the converter involves no more than adding two extra delay lines to the memory. This situation is very typical of what happens with the ACE. There are many fussy little details which have to be taken care of, and which according to normal engineering practice would require special circuits. We are able to deal with these points without modification of the machine itself, by pure paperwork, eventually resulting in feeding in appropriate instructions.

Logical as this was, and certainly comprehensible to mathematicians, who had been familiar with binary numbers for three hundred years at least, the fact was that 'fussy little details' such as this were more of a headache for other people. To an engineer, in particular, it would come as a revelation that the concept of number could be separated from its representation in decimal form. Many people would see the 'binary' arithmetic of the ACE as itself a weird and wonderful innovation. Whilst he was perfectly correct in seeing this as a detail, it was a good example of his difficulties in communication with the kind of people who might fund, organise, and build his machine.

But with such details disposed of, he concentrated his report upon the two really important things: the memory and the control.

Considering the storage problem, he listed every form of discrete store that he and Don Bayley had thought of, including film, plugboards, wheels, relays, paper tape, punched cards, magnetic tape, and 'cerebral cortex', each with an estimate, in some cases obviously fanciful, of access time, and of the number of digits that could be stored per pound sterling. At one extreme, the storage could all be on electronic valves, giving access within a microsecond, but this would be prohibitively expensive. As he put it in his 1947 elaboration, 'To store the content of an ordinary novel by such means would cost many millions of pounds.' It was necessary to make a trade-off between cost and speed of access. He agreed with von Neumann, who in the EDVAC report had referred to the future

possibility of developing a special 'Iconoscope' or television screen,*
for storing digits in the form of a pattern of spots. This he described
as 'much the most hopeful scheme, for economy combined with
speed.' But in a prescient paragraph of the ACE report he also
suggested an approach more on the home-made, 'least waste of
energy' line:

> It seems probable that a suitable storage system can be developed without
> involving any new types of tube, using in fact an ordinary cathode ray tube
> with tin-foil over the screen to act as a signal plate. It will be necessary to
> furbish up the charge pattern from time to time, as it will tend to become
> dissipated . . . . It will be necessary to stop the beam from scanning in
> the refurbishing cycle, switch to the point from which the information
> required is to be taken, do some scanning there, replace the information
> removed by the scanning, and return to refurbishing from the point left
> off. Arrangements must also be made to make sure that refurbishing does
> not get neglected for too long because of more pressing duties. None of
> this involves any fundamental difficulty, but no doubt it will take time to
> develop.

Lacking such cathode ray tube storage, he had to plump for the
mercury delay lines, not with any great enthusiasm, but because
they were already working. They held the obvious disadvantage,
from the point of view of accessibility, of involving a delay. His plan
was for a delay line to hold a sequence of 1024 pulses, so it was
like chopping up the 'tape' of the Universal Turing Machine into
segments each of 1024 squares in length. It would take an average
of 512 units of time to reach a given entry. However, this was an
improvement upon the 'papyrus scroll'.

As for the other most important aspect of the machine, this
was the 'Logical Control'. It corresponded to the 'scanner' of the
Universal Turing Machine. The principle was simple: 'The universal
machine has only to keep looking at this description' – that is, at
the instructions on its tape – 'to find out what it should do at each
stage.' So the Logical Control was a piece of electronic hardware

---

* The work done at RCA on the 'Iconoscope' was closely related to the American commercial development of television, and as such was much more technically ambitious than
the use of 'ordinary' cathode ray tubes familiar in radar displays, such as Alan favoured.

which would contain two pieces of information: where it was on the 'tape', and what instruction it had read there. The instruction would take up thirty-two 'squares' or pulses in a delay line store, and might be of two kinds, in the design that he proposed. It might simply cause the 'scanner' to go on to another point of the 'tape' for its next instruction. Alternatively, it might prescribe an operation of adding, multiplying, shifting or copying, of numbers stored elsewhere on the 'tape'. In the latter case, the 'scanner' was to move to the next point on the 'tape' for its next instruction. None of this involved anything but the reading, writing, erasing, changing of state, and moving to left and right, that was to be done by the theoretical Universal Turing Machine working on the description numbers on its tape – except that there were special facilities added so that addition and multiplication could be achieved in only a few steps, rather than with thousands of more elementary operations.

Of course, there was to be no physical motion when the 'scanner' went to fetch an instruction, or operate upon numbers stored on the 'tape'; no motion but that of electrons. Instead, the control of the ACE would work by a process rather like that of dialling a telephone number, to reach the right place. Most of the complexity of the electronic circuits arose from the demands of this 'tree' system. There was also a complexity in the way that thirty-two half-way houses, 'temporary storage' locations consisting of special short delay lines, were provided for the shunting around of pulses. This was quite different from the EDVAC conception, in which all the arithmetic would be done by shunting numbers in and out of a central 'accumulator'. In the ACE design the arithmetical operations were 'distributed' around the thirty-two 'temporary storage' delay lines in an ingenious way.

The point of this complexity was that it increased the speed of operation. Speed took a slightly higher priority than simplicity. This was reflected also in the fact that Alan planned the pulse rate of the ACE to be a million a second, straining electronic technology to the full.* His emphasis on speed was natural enough, granted his Bletchley experience, in which speed was of the very essence, a few hours marking the difference between usefulness and pointlessness. It

---

* Or as he put it, 'it is not thought wise to design for higher speeds than this as yet.'

was also related to the universality of the electronic computer. In 1942 they had tried to get faster Bombes to cope with the fourth rotor; they had been saved by the German slip-up with the weather-reporting signal system, but without this stroke of fortune they would have had to wait over a year for the machinery to match the problem. One virtue of a universal machine would lie in its ability to take on any new problem immediately – but this meant that it must be as fast as possible from the start. It would hardly be desirable to re-engineer a universal machine for the sake of a special problem. The whole point was to get the engineering out of the way once and for all, so that all the work thereafter could go into the devising of instruction tables.

Although the ACE was based upon the idea of the Universal Turing Machine, in one way it departed from it. The departure lay in the feature, at first sight an extraordinary one, that it had no facility for conditional branching. In this respect it might seem to lack the crucial idea that Babbage had introduced a hundred years earlier. For the 'scanner', or Logical Control, of the ACE could only hold one 'address', or position on the tape, at a time. It had no way of holding more than one, and then of selecting a next destination according to some criterion.

The omission, however, was only on the surface. The reason for it was that it was a case where the hardware could be simplified, at the cost of more stored instructions. Alan set out a way in which conditional branching could be done without needing the logical control to hold more than one 'address' at a time. It was not the best technical solution, but it had the merit of a brutal simplicity. Suppose that it were wished to execute instruction 50 if some digit D were 1, and to execute instruction 33 if it were 0. His idea was to 'pretend that the instructions were really numbers and calculate D × (Instruction 50) + (1 – D) × (Instruction 33).' The result of this calculation would be an instruction which would have the effect that was required. The 'IF' was effected not by hardware, but by extra programming. It was a device which had led him to mix up data (the digit D) with instructions. This in itself was of great significance, for he had allowed himself to *modify the stored program*. But this was only the beginning.

Von Neumann had also seen that it was possible to interfere with the stored instructions, but had done so in only one very particular way. If a stored instruction had the effect of 'taking the number at address 786', then he had noticed that it would be convenient to be able to add 1 into the 786, so that it gave the effect of 'taking the number at address 787'. This was just what was needed for working along a long list of numbers, stored in locations 786, 787, 788, 789 and so forth, as would so frequently occur in large calculations. He had programmed the idea of going to the 'next' address, so that it did not have to be spelt out in explicit form. But von Neumann went no further than this. In fact, he actually proposed a way of ensuring that instructions *could not* be modified in any other way than this.

The Turing approach was very different. Commenting on this feature of modifying instructions, he wrote in the report: 'This gives the machine the possibility of constructing its own orders. . . . This can be very powerful.' In 1945 both he and the ENIAC team had hit upon the idea of storing the instructions inside the machine. But this in itself said nothing about the next step, that of exploiting the fact that the instructions could now themselves be changed in the course of running the machine. This was what he now went on to explain.

It was an idea that had arrived somewhat by chance. On the American side, they had thought of storing instructions internally because it was the only way of supplying instructions quickly enough. On the Turing side, he had simply taken over the single tape of the old Universal Turing Machine. But neither of these reasons for adopting stored instructions said anything about the possibility of interfering with the instructions in the course of a computation. On the American side, it was not pointed out as a feature of the new design until 1947.[4] Equally, the Universal Turing Machine, in its paper operation, was not designed to change the 'description number' that it worked upon. It was designed to read, decode, and execute the instruction table stored upon its tape. It would never *change* these instructions. The Universal Turing Machine of 1936 was like the Babbage machine in the way that it would operate with a fixed stock of instructions. (It differed in that that stock was stored in exactly the same medium as the working and the input and output.) And so Alan Turing's own 'universality' argument showed that a Babbage-like machine was enough. In principle there was nothing that could be achieved through modifying the instructions in the course of an operation that could not be achieved by a universal machine without this facility. The faculty of program modification could only *economise* on instructions, and would not enlarge upon the theoretical scope of operations. But that economy could, as Alan said, be 'very powerful'.

This original perception flowed from the very universality of the machine, which might be used for any kind of 'definite method', not necessarily arithmetical. That being so, the pulses '1101' stored in a delay line, might well not refer in any way to the number 'thirteen', but might represent a chess move or a piece of cipher. Or, even if the machine were engaged upon arithmetic, the '1101' might be representing not 'thirteen', but indicating perhaps a possible *error* of thirteen units, or a thirteen in the floating-point representation* of numbers, or something else altogether, at the choice of the user of the machine. As he saw it from the start, there would be far more to adding and multiplying than putting pulses through the hardware adder and multiplier. The pulses would have to be organised, interpreted, chopped up, and put together again, according to the particular way in which they were being used. He dwelt in particular upon the question of doing arithmetic in floating-point form, and showed how the mere addition of two floating-point numbers required a whole instruction table. He wrote some tables of this kind. MULTIP, for instance, would have the effect of multiplying two numbers which had been encoded and stored in floating-point form, encoding and storing the result. His tables made use of the 'very powerful' feature of the machine, for he had it assemble bits of the necessary instructions for itself, and then execute them.

But if a simple operation like multiplying floating-point numbers would require a set of instructions, then a procedure of any useful scale would involve putting many such sets of instructions together. He envisaged this not as a stringing together of tables, but as a *hierarchy*, in which subsidiary tables like MULTIP would serve a 'master' table. He gave a specific example of a master table called CALPOL which was to calculate the value of a fifteenth-order polynominal in floating-point. Every time a multiplication or addition was required,

---

* In decimal floating point, the sequence 2658 13 would be used to represent the number 2.658×10^{13} or 26580000000000; the computer would normally use a binary equivalent of this.

it had to call upon the services of a subsidiary table. The business of doing this calling up and sending back of subsidiary tables was something which *itself* required instructions, as he saw:

> When we wish to start on a subsidiary operation we need only make a note of where we left off the major operation and then apply the first instruction of the subsidiary. When the subsidiary is over we look up the note and continue with the major operation. Each subsidiary operation can end with instructions for the recovery of the note. How is the burying and disinterring of the note to be done? There are of course many ways. One is to keep a list of these notes in one or more standard size delay lines . . . with the most recent last. The position of the most recent of these will be kept in a fixed TS [short delay line] and this reference will be modified every time a subsidiary is started or finished. The burying and disinterring processes are fairly elaborate, but there is fortunately no need to repeat the instructions involved each time, the burying being done through a standard instruction table BURY, and the disinterring by the table UNBURY.

Perhaps he drew the imagery of burying and unburying from the silver bar story.* This was an entirely new idea. Von Neumann had thought only in terms of working through a sequence of instructions.

The concept of a hierarchy of tables brought in further applications of program modification. Thus he imagined 'keeping the instruction tables in an abbreviated form, and expanding each table whenever we want it' – the work being done by the machine itself, using a table called EXPAND. The further he progressed with this idea, the more he saw that the ACE could be used to prepare, collate, and organise its own programs. He wrote:

> Instruction tables will have to be made up by mathematicians with computing experience and perhaps a certain puzzle-solving ability. There will probably be a good deal of work of this kind to be done, for every known process has got to be translated into instruction table form at some stage. This work will go on whilst the machine is being built, in order to avoid some of the delay between the delivery of the machine and the production of results. Delay there must be, due to the virtually inevitable

---

* The standard image, more illuminating, came to be that of the *stack*, holding the 'return addresses' of the 'sub-routines' entered.

snags, for up to a point it is better to let the snags be there than to spend such time in design that there are none (how many decades would this course take?) This process of constructing instruction tables should be very fascinating. There need be no real danger of it ever becoming a drudge, for any processes that are quite mechanical may be turned over to the machine itself.

It is not surprising that he looked forward to the process of writing instruction tables as 'very fascinating'. For he had created something quite original, and something all of his own. He had invented the art of computer programming.* It was a complete break with the old-style calculators – of which, in any case, he knew little. They assembled adding and multiplying mechanisms, and then had the job of feeding in a paper tape to make them work in the right order. They were machines to do arithmetic, in which the logical organisation was a rather tiresome burden. The ACE was quite different. It was to be a machine to play out programs 'for every known process'. The emphasis was placed on the logical organisation of the work, and the hardware arithmetic added only to short-cut the more frequently used constituent operations.

On desk calculators, the figures 0 to 9 would appear visibly in the registers and the keyboard, and the operator would be led to feel that in some way the calculator had the numbers themselves stored within it. In reality, it had nothing but wheels and levers, but the illusion would be strong. The illusion carried over to the big relay calculators, the Aiken and Stibitz machines, and to the ENIAC. Even the EDVAC proposals carried the feeling that the pulses in the delay lines would somehow actually be numbers. But the Turing conception was rather different, and took a more abstract view. In the ACE, one might regard pulses as representing numbers, or as representing instructions. But it was really all in the mind of the beholder. The machine acted, as he put it, 'without understanding', and in fact operated not on numbers nor on instructions, but on electronic pulses. One could 'pretend that the instruction was really a number', because the machine itself knew nothing about either. Accordingly, he was free in his mind to think about mixing

---

* But Zuse, in Germany, had also worked out some highly advanced ideas by this time, under the name *Plankalkül*.

data and instructions, about operating on instructions, about tables of instructions being inserted by other instructions of 'higher authority'.

There was a reason for his facility to take this liberated approach. Ever since he first thought about mathematical logic, he was aware of mathematics as a game played with marks on paper, to be manipulated by chess-like rules, regardless of their 'meaning'. This was the outlook that the Hilbert approach encouraged. Gödel's theorem had cheerfully mixed up 'numbers' and 'theorems', and *Computable Numbers* had represented instruction tables as 'description numbers'. His proof of the existence of unsolvable problems depended upon this mixing up of numbers and instructions, by treating them all alike as abstract symbols.* It was therefore a small step to regard instructions, and instruction tables, as grist to the ACE's own mill. Indeed, since so much of his war work had depended upon indicator systems, in which instructions were deliberately disguised as data, there was no step for him to make at all. He saw as obvious what to others was a jump into confusion and illegality.

This vision of the ACE's function was also tied in with the imitation argument. The ACE would never really be 'doing arithmetic', in the way that a human being would. It would only be *imitating* arithmetic, in the sense that an input representing '67 + 45' could be made to guarantee an output representing '112'. But there were no 'numbers' inside the machine, only pulses. When it came to floating-point numbers, this was an insight of practical significance. The whole point of his development was that the operator of the ACE would be able to use a 'subsidiary table' like MULTIP, *as if* it were a single instruction to 'multiply'. In actual fact, it would have the effect of much shunting and assembling of pulses inside the machine. But that would not matter to the user, who could work *as if* the machine worked directly on 'floating-point numbers'. As he wrote, 'We have only to think how this is to be done once, and then forget how it is done.' The same would apply if the machine

---

* Perhaps, however, he had been made aware that symbols could indifferently represent data or instructions when at his first school he misunderstood the instruction to omit 'the'.

were programmed to play chess: it would be used *as if* it were playing chess. At any stage it would only be outwardly imitating the effect of the brain. But then, who knew how the brain did it? The only fair use of language, in Alan's view, was to apply the same standards, the standards of outward appearance, to the machine as to the brain. In practice, people said quite nonchalantly that it was 'doing arithmetic'; they should also say it was playing chess, learning, or thinking, if it could likewise imitate the function of the brain, quite regardless of what was 'really' happening inside. Even in his technical proposals, therefore, there lay a philosophical vision which was utterly beyond the ambition of building a machine to do large and difficult sums. This did not help him to communicate with other people.

Although he had shifted the emphasis from the building of a machine to the construction of programs, there was nothing nebulous about his engineering plans for the ACE. The delay lines, he wrote,

> have been developed for RDF purposes to a degree considerably beyond our requirements in many respects. Designs are available to us, and one such is well suited to mass production. An estimate of £20 per delay line would seem quite high enough.

He did in fact make a visit to the Admiralty Signals Establishment to see T. Gold, who was working there on delay lines. His plan was for two hundred mercury delay lines, each with a capacity of 1024 digits. But the figures, dimensions, costs, and the choice of mercury as the medium, were not taken off the radar engineers' shelf. He worked out the physics himself. On the basis of his calculation, mercury was only marginally to be preferred to a mixture of water and alcohol, which he observed would have the same strength as gin. He rather hankered after using gin, which would come cheaper than mercury. However, he did not propose doing the development work himself. He wanted this to be done by the Colossus engineers, at the Post Office Research Station. Flowers was already familiar with delay lines, having been shown Eckert's model in October 1945.

As for the engineering of the logical control and arithmetical circuits ('LC and CA'), he wrote:

Work on valve element design might occupy four months or more. In view of the fact that some more work needs to be done on schematic circuits such a delay will be tolerable, but it would be as well to start at the earliest possible moment. . . .

In view of the comparatively small number of valves involved the actual production of LC and CA would not take long; six months would be a generous estimate.

A great many of the 'schematic circuits' were already planned out in the report. He included a detailed design of the arithmetical circuits, making use of (and extending) von Neumann's notation. Perhaps he had the pleasure of drawing upon his experience in designing a binary multiplier before the war. There was also another feature of the design which might well have harked back to an earlier experience. There was to be a facility for plugging in special circuits, if these were required, for operations over and above the arithmetical and Boolean functions that were permanently incorporated in the machine. This idea was rather at variance with the principle of putting as much as possible into the instructions, but it might be appropriate if some exceedingly efficient special-purpose circuit were at hand. This had been the case, for instance, with the Bombes. Here the steps which depended upon relay clicks were slow by electronic standards. But the steps which depended upon electricity flowing through the internal Enigma wirings were effectively instantaneous. These would have taken longer to achieve by means of an instruction table played out on an electronic computer. Alan's design allowed for such short-cuts to be taken, if desired. But no one could possibly have guessed that he was writing on the basis of such experience with mechanical methods.

Nor was it planned out only at the logical level of circuit diagrams. There were many pages also on the specific electronic equipment required. One section owed a direct debt to the Delilah, another unknown aspect of his expertise:

Unit delay. The essential part of the unit delay is a network, designed to work out of a low impedance and into a high one. The response at the output to a pulse at the input should preferably be of the form indicated in Fig. 50,* i.e. there should be maximum response at time 1μs after the initiating pulse, and the response should be zero by a time 2μs after it, and should remain there. It is particularly important that the response should be near to zero at the integral multiples of 1μs after the initiating pulse (other than 1μs after it).

A simple circuit to obtain this effect is shown on Fig. 51a. . . . It differs from the ideal mainly in having its maximum too early. It can be improved at the expense of a less good zero at 2μs by using less damping, i.e. reducing the 500 ohm resistor. It is also possible to obtain altogether better curves with more elaborate circuits.

He also considered the practical requirements of the project as a whole:

It is difficult to make suggestions about buildings owing to the great likelihood of the whole scheme expanding greatly in scope. There have been many possibilities that could helpfully have been incorporated, but which have been omitted owing to the necessity of drawing a line somewhere. In a few years time however, when the machine has proved its worth, we shall certainly want to expand and include these other facilities, or more probably to include better ideas which will have been suggested in the working of the first model. This suggests that whatever size of building is decided on we should leave room for building-on to it.

He had learnt from the mushrooming of Bletchley. He proposed a total of 1400 square feet for the machine and its accessory equipment, and estimated the total capital cost of a machine with two hundred delay lines at £11,200. Changes would have to be made as they went along, but his plans allowed for that: the main thing was to get started.

His February 1947 talk would expand on how the machine would 'prove its worth', by giving

a picture of the operation of the machine. Let us begin with some problem which has been brought in by a customer. It will first go to the problems preparation section where it is examined to see whether it is in a suitable form and self-consistent, and a very rough computing procedure made out.

He gave a specific example of a problem, namely the numerical

---

* A figure similar to that on page 345.

solution of a differential equation involving Bessel functions. (This would be a very typical problem arising in applied mathematics and engineering.) He explained how the instruction table for the Bessel function would be already 'on the shelf', and so would one for the general procedure for solving a differential equation. So

> The instructions for the job would therefore consist of a considerable number taken off the shelf together with a few made up specially for the job in question. The instruction cards for the standard processes would have already been punched, but the new ones would have to be done separately. When these had all been assembled and checked they would be taken to the input mechanism, which is simply a Hollerith card feed. They would be put into the card hopper and a button pressed to start the cards moving through. It must be remembered that initially there are no instructions in the machine, and one's normal facilities are therefore not available. The first few cards that pass in have therefore to be carefully thought out to deal with this situation. They are the initial input cards and are always the same. When they have passed in a few rather fundamental instruction tables will have been set up in the machine, including sufficient to enable the machine to read the special pack of cards that has been prepared for the job we are doing. When this has been done there are various possibilities as to what happens next, depending on the way the job has been programmed. The machine might have been made to go straight on through, and carry out the job, punching or printing all the answers required, and stopping when all of this has been done. But more probably it will have been arranged that the machine stops as soon as the instruction tables have been put in. This allows for the possibility of checking that the content of the memories is correct, and for a number of variations of procedure. It is clearly a suitable moment for a break. We might also make a number of other breaks. For instance, we might be interested in certain particular values of the parameter a, which were experimentally obtained figures, and it would then be convenient to pause after each parameter value, and feed the next parameter value in from another card. Or one might prefer to have the cards all ready in the hopper and let the ACE take them in as it wanted them. One can do as one wishes, but one must make up one's mind.

These proposals were eminently practical, and indeed prophetic, foreseeing the need for flexible interaction between operator and machine. But of course he had already seen the future, in the use of the Colossus. Another passage brought in the possibility of using remote terminals:

> . . . the ACE will do the work of about 10,000 [human] computers. It is to be expected therefore that large scale hand-computing will die out. [Human] computers will still be employed on small calculations, such as the substitution of values in formulae, but whenever a single calculation may be expected to take a [human] computer days of work, it will presumably be done by an electronic computer instead. This will not necessitate everyone interested in such work having an electronic computer. It would be quite possible to arrange to control a distant computer by means of a telephone line. Special input and output machinery would be developed for use at these out-stations, and would cost a few hundred pounds at most.

He also perceived the demand for computer programmers:

> The main bulk of the work done by these computers will however consist of problems which could not have been tackled by hand computing because of the scale of the undertaking. In order to supply the machine with these problems we shall need a great number of mathematicians of ability. These mathematicians will be needed in order to do the preliminary research on the problems, putting them into a form for computation. . . .

And indeed he could foresee a new field of industry and employment:

> It will be seen that the possibilities as to what one may do are immense. One of our difficulties will be the maintenance of an appropriate discipline, so that we do not lose track of what we are doing. We shall need a number of efficient librarian types to keep us in order.

Twenty years ahead of his time, in his conception of the organisation of a computer installation, he had drawn his picture from his experience at Bletchley. There they had likewise employed ten thousand human operators, and had worked as a *system*, involving remote out-stations, telephone communications, an élite who transformed the problems into programs, and plenty of 'librarian types'. But he could never say anything directly about it, and no one could picture what had never officially existed, so his analysis came as news from nowhere.

The ACE report was also the first account of the *uses* to which a universal computer could be put. The ACE was to solve 'those problems which can be solved by human clerical labour, working to fixed rules, and without understanding', restricted by the size of the machine to cases where 'the amount of written material which need be kept at any one stage is limited to . . . 50 sheets of paper' and where 'the instructions to the operator' could be described in 'ordinary language within the space of an ordinary novel'. It would be able to do so in one hundred thousandth of the time taken by 'the human operator, doing his arithmetic without mechanical aid.'

The implication was that the ACE could have taken over all the routine mental work of the British war effort. Here he made what was for him an unusually good 'political' point: in a list of possible applications, 'Construction of range tables' came first. This was the job for which the ENIAC had been specifically designed. There followed four further examples of calculations of practical importance, which currently required months or years of work on desk machines. But another four were non-numerical, reflecting his much wider view of the nature of the computer, and indeed being much closer to his experience.

The first would have the computer interpreting a special language for describing electrical problems:

> Given a complicated electrical circuit and the characteristics of its components, the response to given input signals could be calculated. A standard code for the description of the components could easily be devised for this purpose, and also a code for describing connections.

It would mean the automatic solution of circuit problems such as he had spent weeks over at Hanslope. The second was a more mundane example of the world's work:

> To count the number of butchers due to be demobilised in June 1946 from cards prepared from the army records.

The machine, he wrote, 'would be quite capable of doing this, but it would not be a suitable job for it. The speed at which it could be done would be limited by the rate at which cards can be read, and the high speed and other valuable characteristics of the calculator would never be brought into play. Such a job can and should be done with standard Hollerith equipment.' The third non-numerical problem* was described thus:

---

* W. T. Tutte had worked on this pure-mathematical problem.

A jig-saw puzzle is made up by cutting up a halma-board into pieces each consisting of a number of whole squares. The calculator could be made to find a solution of the jig-saw, and, if they were not too numerous, to list all solutions.

This was the nearest he came to a mention of cryptanalysis, although it was latent in a number of his ideas, such as including the logical operations of AND and OR from the start, and regarding them as on an equal footing with arithmetical operations. Alan wrote that this particular problem was 'of no great importance, but it is typical of a very large class of non-numerical problems that can be treated by the calculator. Some of these have great military importance, and others are of immense interest to mathematicians.' But at the end came the suggestion that he was most interested in himself, although he could hardly expect it to motivate the government:

Given a position in chess the machine could be made to list all the 'winning combinations' to a depth of about three moves on either side. This is not unlike the previous problem, but raises the question, 'Can the machine play chess?' It could fairly easily be made to play a rather bad game. It would be bad because chess requires intelligence. We stated at the beginning of this section that the machine should be treated as entirely without intelligence. There are indications however that it is possible to make the machine display intelligence at the risk of its making occasional serious mistakes. By following up this aspect the machine could probably be made to play very good chess.

This was not so much a report, as a plan of campaign in which tactical and strategic were jostling as close on paper as they did in his own mind. The promise of an electronic 'brain' was as fanciful as space travel, and this report was like explaining the advantages of colonising Mars, while in next breath describing the design of a fuel pump. The naive, colloquial style was not calculated to appeal to the authorities, and its detailed considerations went far beyond their absorptive capacities. No one was going to work through the example programs or the circuit diagrams, nor resolve the baldly stated paradox of a machine 'entirely without intelligence' nevertheless 'displaying intelligence'. Even Hartree found it very hard going.

Although undated, the ACE report was completed by the end

of 1945,[5] representing an amazing burst of energy. It then went to Womersley, who wrote both a memorandum[6] for Darwin and an introductory report[7] for the Executive Committee meeting of 19 February 1946. Womersley was quick to perceive the opportunities offered by a universal machine, and whatever the intellectual limitations noted by Alan and the other mathematicians, he certainly wrote an able defence of what he claimed as 'one of the best bargains the DSIR* has ever made.' 'The possibilities inherent in this equipment are so tremendous that it is difficult to state a practical case . . . without it sounding completely fantastic . . .' Optics, hydraulics, and aerodynamics could be 'revolutionised'; the plastics industry could be 'advanced in a way that is impossible with present computing resources.' Besides the range table problem already noted by Alan, a problem expected to require three years of work by the existing Mathematics Division, Womersley claimed that 'The machine will also grapple successfully with problems of heat-flow in non-uniform substances, or substances in which heat is being continuously generated' – of explosives old and new, in fact. Womersley also claimed that 'the promised support of Commander Sir Edward Travis,† of the Foreign Office, will be invaluable.'

On the more theoretical side, Womersley stressed that 'this device is not a calculating machine in the ordinary sense of the word. One does not need to limit its functions to arithmetic. It is just as much at home in algebra . . .' And on the more political side he drew attention to the large sums already expended in America on machines whose capacities would be far outstripped by the ACE. More subtly, Womersley pointed to the advantage to be gained from having such a machine installed in the NPL:

> . . . we in this country, and particularly in this Division, have a unique contribution to make to world progress. I can say quite definitely that in the *use* of such equipment we shall be far more resourceful and cunning than the Americans. . . . All the USA machines are in Electrical Engineering Departments. In this Division the machine will be in the hands of the user, rather than the producer . . .

---

* The Department of Scientific and Industrial Research, through which the NPL was funded.

† Travis had already received a knighthood.

Discussion of this visionary cooperation of British brain and hand was postponed until the meeting of the Executive Committee on 19 March. This time Alan was invited to attend, and after being dauntingly introduced by Womersley as 'an expert in the field of mathematical logic', he did his best to explain the ACE as simply as possible. It was a particularly clear account, in which he began by saying

that if a high overall computing speed was to be obtained it was necessary to do all operations automatically. It was not sufficient to do the arithmetical operations at electronic speeds: provision must also be made for the transfer of data (numbers, etc.) from place to place. This led to two further requirements – 'storage' or 'memory' for the numbers not immediately in use, and means for instructing the machine to do the right operations in the right order. There were then four problems, two of which were engineering problems and two mathematical or combinatory.

Problem (1) (Engineering) To provide a suitable storage system.

Problem (2) (Engineering) To provide high speed electronic switching units.

Problem (3) (Mathematical) To design circuits for the ACE, building these circuits up from the storage and switching units described under Problems 1 and 2.

Problem (4) (Mathematical) To break down the computing jobs which are to be done on the ACE into the elementary processes which the ACE is designed to carry out. . . . To devise tables of instructions which translate the jobs into a form which is understood by the machine.

Taking these four problems in order, Dr Turing said that a storage system must be both *economical* and *accessible*. Teleprinter tape provided an example of a highly economical but inaccessible system. It was possible to store about ten million binary digits at a cost of £1, but one might spend minutes in unrolling tape to find a single figure. Trigger circuits incorporating radio valves on the other hand provided an example of a highly accessible but highly uneconomical form of storage; the value of any desired figure could be obtained within a microsecond or less, but only one or two digits could be stored for £1. A compromise was required; one suitable system was the 'acoustic delay line' which provided storage for 1000 binary digits at a cost of a few pounds, and any required information could be made available within a millisecond.

But explaining excitedly to the committee how the delay line was to work, he rapidly became too technical and was cut off before even touching upon the question of devising 'tables of instructions'. Darwin was therefore sceptical, reasonably enough.

> The Director asked what would happen, in cases where the machine was instructed to solve an equation with several roots. Dr Turing replied that the controller would have to take all the possibilities into account, so that the construction of instruction tables might be a somewhat 'finicky' business.

Hartree came to the rescue with an argument which appealed less to science than to post-war patriotism:

> . . . It requires only 2000 valves as against 18000 in the ENIAC, and gives a 'memory' capacity of 6000 numbers compared with the 20 numbers of the ENIAC . . . if the ACE is not developed in this country the USA will sweep the field. . . . this country has shown much greater flexibility than the Americans in the use of mathematical hardware. He urged that the machine should have every priority over the existing proposal for the construction of a large differential analyser.

This was a genuinely far-sighted and generous recommendation, coming as it did from a person who had spent much of his own time and energy on the differential analyser. It was a remarkably smooth victory for the digital over the analogue approach. Hartree had, of course, seen the ENIAC when near to completion, and might also have seen the Colossus after the war had ended. He was also a particularly cooperative and helpful person.

Darwin was still not convinced and

> enquired whether the machine could be used for other purposes if it did not fulfil completely Dr Turing's hopes. Dr Turing replied that this would depend largely on what part of the machine failed to operate, but that in general he felt that many purposes could be served by it.

He was probably gritting his teeth at Darwin's failure to grasp the principle of universality. Now Womersley infiltrated a new concept into the discussion, one which had played no role in the Turing report. It was that of a *pilot* machine.

There was next some discussion as to the possible cost of the machine and Mr Womersley said that a pilot set-up could possibly be built for approximately £10000, and it was generally agreed that no close estimate of the overall cost of the full machine could be made at this stage.

Not much notice was taken of Alan's estimate of capital cost. Womersley had said it should be multiplied by a factor of four or five. In fact they were probably annoyed that he had trespassed across the demarcation line into the administrative province; the more so as he wrote as though he could shop around for the equipment himself. Recommendations were passed on, notably that of the Ministry of Supply which handled all military contracts.*

Then the

> Committee resolved unanimously to support with enthusiasm the proposal that Mathematics Division should undertake the development and construction of an automatic computing engine of the type proposed by Dr A. M. Turing, and Director agreed to discuss the financial and other aspects of the matter with Headquarters.

Alan Turing held a great and burning dislike for committee meetings such as this, resenting the fact that the decisions were made not because of a clear understanding of his ideas, but for political and administrative reasons. The report he had submitted was, in fact, largely irrelevant, and played the role of pleasing Humpty Dumpty by having something down on paper, no matter what it was. But Sir Charles Darwin did take swift action. Indeed already, on 22 February, he had written[8] to the Post Office about this 'electronic mathematical machine of a novel type, which should be immensely superior in every way to anything hitherto constructed anywhere'.

> Very broadly it works using principles developed by your staff during the war for a certain Foreign Office project, and we want to be able to take advantage of this, enlisting the help . . . in particular of Mr Flowers who has had much experience in working out the electronic side of it.

---

* They wanted the ACE for 'shell, bombs, rockets and guided missiles'. The MoS was assured by E. S. Hiscocks, Secretary of the NPL, on 20 March that 'we certainly hope that it will be freely available for the types of purpose mentioned in your letter. . . .'

The initial response from the Post Office was encouraging, and on 17 April Darwin was able to present to the Advisory Council of the DSIR a cogent plan of action, which showed that he had by this time marshalled the essential ideas:

... The possibility of the new machine started from a paper by Dr A. M. Turing some years ago, when he showed what a wide range of mathematical problems could be solved, in idea at any rate, by laying down the rules and leaving a machine to do the rest. Dr Turing is now on the staff of the NPL and is responsible for the theoretical side of the present project, and also for the design of many of the more practical details.

It gave three examples of large calculations that it would be able to perform, and explained:

The complete machine will naturally be costly; it is estimated that it may call for over £50,000, but probably not twice as much. A smaller one, containing the essential characteristics, could be constructed first, perhaps for a cost of £10,000, but its chief function would be to reveal some of the details of design that cannot be planned without trial, and its scope would be too limited to be worth constructing for its own sake. This would involve development work on delay lines and trigger circuits, and this part of the work would be undertaken by the Post Office, where facilities and specially trained staff exist, with the collaboration of Dr Turing and his assistants. . . .

The small machine would not be a miniature substitute for the large machine but would later constitute a part of the full scale machine in due course. It is hoped that the complete machine can be constructed in three years. . . . It is proposed to proceed immediately, and with high priority, in the design and construction of this preliminary machine, but in doing so it is important to know that if it fulfils its promise there will be full backing for the greater sums required for the real operating machine. In view of its rapidity of action, and of the ease with which it can be switched over from one type of problem to another it is very possible that the one machine would suffice to solve all the problems that are demanded of it from the whole country . . .

Darwin requested up to £10,000 to be allocated to the 'small machine' in the current financial year. On 8 May 1946 the DSIR agreed to support his application, and also that if the 'small machine' fulfilled expectations then they would recommend the expenditure of up to £100,000 on a full-scale machine. On 15 August the Treasury sanctioned the £10,000, although according to standard procedure refused further commitment. Meanwhile by 18 June the NPL had committed itself by sending a letter to the Post Office asking for development work on delay lines. The ACE was under way. It might now be couched in terms more like those of a Five Year Plan than those of getting on with it in a hut, but it still promised a machine that would solve all the problems demanded of it from the whole country. The legacy of a total war, and of the capture of a total communication system, could now be turned to the construction of a total machine.

After submitting the report, Alan continued to improve the design and to write 'instruction tables' for the paper machine. In this he gained some assistance, Darwin having decided that 'high priority' meant the allocation of two Scientific Officers to the project. First came J. H. Wilkinson, one of the pure-mathematical Part III class of 1939, by now experienced through six years of work in the numerical analysis of explosives problems. It was Charles Goodwin who had sought to bring him to the NPL, to work in numerical analysis; but when Wilkinson made a visit he found himself being told by Alan about the exciting ACE plans.[9] It was the ACE that made him decide to remain in government service, rather than to return to Cambridge mathematics. It was agreed that he should work half-time for Goodwin on desk machines and half-time on the ACE. It was a position fraught with possible inter-departmental frictions, but fortunately Jim Wilkinson was the most equable and diplomatic person. He joined on 1 May 1946. A second assistant joined a little later: this was the young Mike Woodger, son of the theoretical biologist and philosopher of science J. H. Woodger. He was immediately attracted by the Turing vision of a universal machine. Unfortunately, after training on desk machines in June, he caught glandular fever and was away until September.

Mike Woodger was there when Alan's official honour was announced in June. For his war service he was appointed to the Order of the British Empire – standard for civil servants of the rank that he had officially held. The letters OBE were added to his office door, which made him furious, perhaps because he did not want to be asked what it was for, perhaps at the absurdity of advertising a token recognition. The King was ill, so the OBE medal, bearing the inscription 'For God and the Empire', arrived by post. It was lodged in his tool-box.

Already by May, when Jim Wilkinson joined in work on the ACE design, it had reached a Version V. One difference was that this incorporated a hardware facility for conditional branching. It was quickly replaced by a transitional Version VI and then a Version VII. By this time Alan was devoting more attention to speed of operation than he had done in the original report, at the cost of more hardware. In Version VII enough equipment was added to make it possible for an instruction to have the effect of a complete arithmetical operation, taking two numbers from store, adding them, and putting the result back into store. Again in the interest of speed, it was necessary to program the instructions so that as far as possible each instruction would be leaving its delay line just as the previous one had been executed. Since different operations would take different lengths of time to complete, this made the construction of tables into a crossword-puzzle operation. It also led to a decision that every instruction should specify which instruction the 'head' was to take next, abandoning the idea of a natural stream of consecutive orders, flowing with only occasional interruptions. It also increased the length of the instruction words from thirty-two to forty pulses, again requiring more equipment. In Version VII each such operation would take forty microseconds, but it would then take another forty microseconds to assemble the next instruction in the control circuits. Again in the interest of speed, Alan wished to eliminate this period by duplicating part of the equipment so that each instruction could be assembled while the last was being performed.

It was entirely in line with his original report that as experience was gained with writing instruction tables, he should modify aspects of the hardware design. It was consistent, too, that he should sacrifice a certain amount of simplicity for the sake of greater speed. All the same, the assembling of actual components could not begin too soon for him; the paper 'tables' were intended for a real machine, not as theoretical exercises.

In this respect, however, development was very far from speedy. There was an overwhelming problem with the ACE project. This was that although the NPL had a Radio Division, it did not have a single electronic engineer capable of appreciating, let alone implementing, Alan's ideas. Back in December 1944, Womersley had told the Executive Committee that plans for new machines 'could be put into effect only by cooperation between the Division and certain industrial organisations – possibly along the lines of extra-mural development contracts.' But no progress had been made towards making this cooperation possible. There was one solid medium-term possibility, that of having English Electric, a firm with which Darwin had connections, manufacture a machine to commercial standards. Its director, Sir George Nelson, had attended the March meeting. But it was not at all clear who was to do the immediate development work. This was no small impediment to the confident hope expressed in Alan's report of being able to produce the equipment for the logical control within six months.

A further problem lay in the internal structure of the NPL. At Hanslope, Alan had worked happily with Don Bayley, putting their different kinds of skill together. In this respect the Delilah project had been practice for the ACE, for he would have liked to work in much the same way again, if on a larger scale. (He had vaguely suggested that Don himself should come to the NPL, but Don was not due for release until February 1947, and he also felt he was rather more than the 'amplifier king' that Alan said he would need.)

But the National Physical Laboratory did not encourage any such *ad hoc*, informal collaboration. Its division of labour drew a firm line between brain and hand. Alan was cast firmly in the role of the theoretical designer, and was not expected to know anything about practical engineering. The bureaucratic outlook of the NPL also showed in the form-filling and permission-seeking necessary for the requisition of equipment, Meanwhile, to make matters worse, the post-war chaos, presenting something akin to a black market in war-surplus materials, called for the skills of the spiv in locating hardware of any kind. For these reasons, there was no immediate prospect of having engineers qualified to build the circuits, and there was no facility for Alan to conduct practical experiments himself.

There was a more general difficulty facing the ACE project. It was one analogous to the arrival of the Enigma information in 1940. Alan had poured out a decade's worth of ideas into the lap of an organisation which had not been noted for innovation. In 1940 the cryptanalysts had naively expected that the government could take the deciphered messages and put them to good effect. In fact it had needed two years of painful adjustment before this could be done, even with all the pressure of war. Now again the cornucopian abundance of the ACE design would require a change in attitudes. But there was no precedent for it, and although the administrators talked knowingly about what was needed, they had no clear idea of how to proceed. There was no tradition of major innovation at the NPL, and the ACE project brought out the conservative and negative character of the institution. Just as in 1941, Alan was impatient and uncomprehending of the organisational inertia.

Nor, perhaps, did he appreciate that the very high priorities enjoyed by Bletchley, and the willingness of other organisations to sacrifice their independence, could no longer be expected in 1946. Thus at Dollis Hill, while there was no question about the competence of the Colossus engineers, their director, Radley, did not attach great importance to the work being done on delay lines for the NPL. The Post Office had manifold tasks of its own in connection with the wartime backlog, and there was now no higher authority, no national policy, to coordinate the priorities of the various state enterprises. Alan and Womersley made an official visit to Dollis Hill on 3 April 1946, and thereafter work began – but in a desultory way, with the effect of creating an unforeseen delay, and a sense of uncertainty as to direction.

Alan had written in his report about the possibility of using cathode ray tubes as a quite different kind of storage system, and it was probably at his prompting that on 8 May 1946 Womersley wrote to TRE with an enquiry about the state of research at the radar establishment into the use of such tubes for storage, explaining that it might 'form a suitable alternative to, or possibly even an improvement on, the mercury delay line which we are at present intending to use in our automatic computing engine . . .' The response was not unhelpful, and on 13 August Darwin was able to write to Sir Edward Appleton at the DSIR:[10]

... As I told you Womersley was down at TRE to see whether they could do any work about the ACE machine. He tells me that it looks a most promising chance, and I think we should go ahead on it. Their lay-out for the job looks good, and I gather it appealed strongly to F. C. Williams as a job he would like to do, so that it should get a good chance. I am kicking myself for not having thought of it months ago as a possibility.

As to what comes next Womersley has got to use some tact in exploring how we stand with the Post Office who have started to give us help, which would be very good but that they are not in a position to plunge very deep. It will also be necessary to square TRE officially over priorities, and on this I should like to bring high power to bear if necessary, because we have got a splendid chance of jumping in ahead of America.

F. C. Williams was one of the top electronic experts at TRE, and he was enthusiastic not because he was interested in computing, which he was not, but because he was anxious to find peacetime applications for the electronic skills developed in wartime radar. In finding such applications he had a mind of his own, and had an option open to him that did not accord with Darwin's 'high power'. It derived from Manchester.

Newman had left Bletchley to become Professor of Pure Mathematics at Manchester University, taking Jack Good and David Rees with him as junior lecturers. (Rees was another of the 1939 Part III class, who had joined Welchman's hut in December 1939, and later joined the Newmanry.) In moving from a Cambridge lectureship to a Manchester University chair, he was following Blackett, then Professor of Physics. Together they constituted a formidable team, and one which did not see why Darwin should monopolise electronic computers. As the first reader of *Computable Numbers* and the co-author of the Colossus, Newman appreciated the potential of computers as well as anyone, and if he lacked Alan's emotional thrust towards 'building a brain', and the experience of assembling components himself, he compensated with a greater skill in the art of the possible.

Writing to von Neumann[11] on 8 February 1946, Newman had explained that he was

hoping to embark on a computing machine section here; having got very interested in electronic devices of this kind during the last two or three years. By about 18 months ago I had decided to try my hand at starting up a machine unit when I got out. It was indeed one of my reasons for coming to Manchester that the set-up here is favourable in several ways. This was before I knew anything of the American work, or of the scheme for a unit at the National Physical Laboratory. Later I heard of the various American machines, existing and projected, from Hartree and Flowers.

Once the NPL project was started it became questionable whether a further unit was wanted. My view was that in this, as in other branches of technology, basic research is wanted, which can go on without worrying about getting into production . . .

Newman 's intention was that

mathematical problems of an entirely different kind from those so far tackled by machines might be tried, e.g. testing out (say) the 4-colour theorem or various theorems on lattices, groups, etc. . . .

He explained that

Anyhow, I have put in an application to the Royal Society for a grant of enough to make a start. I am of course in close touch with Turing. After discussion with him, and hearing Hartree and Flowers, it seemed to me that if we start on 'mathematical' problems here, in the above sense, we ought to be able to manage with considerably less storage, though I have asked the Royal Society to be prepared for something fairly substantial.

Newman's application was for a grant to cover the capital cost and five years of salaries for an electronic computer.[12] The Royal Society set up a committee to investigate the application for a grant, consisting of Blackett, Darwin, Hartree, and two pure mathematicians, Hodge from Cambridge and Whitehead from Oxford. Darwin opposed it on the grounds that the ACE was to serve the needs of the country. Womersley had expressed the particular fear that Newman would steal Flowers away from the ACE. But he was outvoted, for it was approved that there should be 'a different type of machine' for Newman's project. On 29 May the Treasury allocated £35,000 to Newman. The rationale was that as 'fundamental science', his proposal fell into the orbit of the Royal Society, and did not conflict with the DSIR. It meant that a rather surprising breath of mathematical innocence was felt again, with a computer development not required for weaponry.

It so happened that Blackett knew F. C. Williams from before the war, when they had collaborated on an automatic curve-tracer for the differential analyser. Both Darwin and Newman, therefore, were looking to Williams to help them 'jump in ahead of America', and Darwin's plan for a single national computer installation had already been undermined. The question of who was going to build the ACE remained unresolved while Williams decided between these suitors.

This incipient rivalry was further complicated by the fact that a third British electronic computer project was initiated in mid-1946. This was the brainchild of M. V. Wilkes, Alan's fellow 'B-star' of 1934 who in 1945 had left wartime work at TRE to resume his position at the Mathematical Laboratory in Cambridge, of which he now became director. He was in touch with Womersley at once, but the ENIAC and EDVAC plans were secret until the spring of 1946, and only then could Hartree tell him about what he had seen in 1945. Hartree then also arranged for him to attend a course of lectures arranged by the ENIAC team at Philadelphia in July and August 1946.

This course at the Moore School, together with the reports issued by von Neumann's group at the IAS, enjoyed very considerable influence upon the future development of computers. For one thing, it brought about the first federal funding for a machine on the lines of the EDVAC. A certain James T. Pendergrass represented the Navy's CSAW, and reported back[13] upon the virtue of a universal machine, as opposed to the special purpose equipment which had 'often proved to be expensive and time-consuming'. (Presumably Travis had been given the same analysis a year earlier.) For another, it inspired Wilkes with great enthusiasm for putting his wartime electronic experience to work in building a British version of the EDVAC.

Alan, in contrast, remained unaffected by the American developments, and they by him. In growth as in conception they were independent. There was a little indirect contact between the NPL and the American group. Hartree had made a visit to the ENIAC in the summer of 1946, being allowed to use it himself, and took with him a copy of the ACE report and a 'third version' of the ACE design.[14] But its programming ideas made no obvious impression upon the Americans.

As a theoretician of computers, von Neumann had been joined by Norbert Wiener, the American mathematician of a slightly earlier generation, whom the war had likewise taken from group theory to machinery, although in his case the servo-mechanisms of anti-aircraft artillery had been the formative influence. Thus von Neumann and Wiener corresponded in connection with the potential of the planned EDVAC, but mostly in terms of the faculty of conditional branching corresponding to 'feedback'. They did not discuss hierarchies of programs, nor the computer reorganising and creating its own instructions. They remained most impressed with the McCulloch and Pitts ideas, which suggested that the logical functions of electronic valves bore some similarity to the structure of neurons in the human nervous system. In a letter[15] of 29 November 1946, von Neumann wrote to Wiener of the 'extremely bold efforts' of Pitts and McCulloch, with which he 'would like to put on one par the very un-neurological thesis of R. [*sic*] Turing.'

In the other direction, likewise, there was limited communication. The *Draft Report on the EDVAC* was there at the NPL, and Alan continued to make use of its notation for logical networks. David Rees, who attended the Moore School lectures on behalf of the Manchester interest, reported back to Alan and Jim Wilkinson for ten days or so. But the American plans had no particular influence on the ACE project, Alan being notably sceptical about the prospects for the Iconoscope, the storage medium on which the Americans were pinning their hopes. They did not think of the American development as a rival; it was simply another project. The ACE was Alan Turing's own thing, like naval Enigma, like the Delilah. It was not, however, developing in quite the same way.

Although the wartime spirit lived on amongst the huts and bombsites of 1946, the ACE section did not flower with the *camaraderie* of Hut 8, or the quite remarkable *rapport* that Alan had established with Don Bayley. Mike Woodger came back from sick leave in September and found a note on his desk asking him to program BURY and UNBURY. The relationship continued in this master-and-servant way. Alan liked the earnest, rather nervous Mike Woodger, and tried to be kind, but hid it under what appeared as an abrupt, rather frightening manner. He probably did not realise the awe with which he was regarded by young people like Mike Woodger, just starting on a career. Always casting himself as the Young Turk embattled against officialdom, he still found it hard to see himself as the official in others' eyes. He was impatient of slowness, and could not use his imagination to make communication more effective.

Jim Wilkinson was older and more experienced than Mike Woodger, but he too found many days when it was better to keep out of the way of the now somewhat isolated 'creative anarchy' that was Alan Turing. 'Likeable, almost lovable . . . but some days depressed', he appeared; his mercurial temperament and his emotional attitude to his work showing clearly. It was at about this time that Alan got the long-promised promotion to Senior Principal Scientific Officer, and he took Jim Wilkinson and Leslie Fox from Goodwin's department out to a celebration dinner in London. The train journey was spoilt by a sultry row over some mathematics, and then, as they arrived at Waterloo, the clouds cleared and he was buoyant again.

This particular argument arose because Alan had become involved in a problem of numerical analysis, the work done in Goodwin's section. In 1943 the statistician H. Hotelling[16] had analysed the procedure for solving simultaneous equations (or, roughly equivalently, for inverting a matrix) and his result made it appear that errors would grow very rapidly as successive equations were eliminated. If this were so it would undermine the practical usefulness of the ACE. Goodwin's section, being directly concerned with the problem, had attacked it heuristically in 1946 by solving a set of eighteen equations that had come up in an aerodynamic calculation, and Alan had joined in (notably the least competent at the detailed work), on the desk machine exercise. To their surprise, they found the final errors to be remarkably small. Alan had undertaken a theoretical analysis of why this should be. It was a typical Turing problem, needing a fresh attack, and with a concrete application. He tackled it much as he had developed a theory of probability for use at Bletchley.

This work, of course, did not lie far in the past, and he set Mike Woodger some probability problems, including the one about the 'barrels of gunpowder'. There was also professional contact arising out of wartime work. Jack Good and Newman had made a visit to the NPL – Newman, of course, being interested in setting up his Manchester computer project – and Jack had managed to disprove

Alan's assertion that no one could write an instruction table that was free from error at the first attempt. Jack Good had also written a short book[17] on *Probability and the Weighing of Evidence*, effectively setting out the theory they had employed at Bletchley, though not its more advanced applications. The 'sequential analysis' method was, as it happened, soon published in America by the statistician A. Wald,[18] who had developed it independently for the testing of industrial components. Alan, in contrast, published nothing that came of his Bletchley work, except in the less direct sense that almost everything he was doing was flowing from his wartime experience added to his pre-war theory of machines.

Rather than forming new ties of friendship at the NPL, he retained those of the war. Donald Michie, now an undergraduate at Oxford, was one of these friends, and a footnote on Alan's October 1946 letter to Jack Good, with comments on the draft of his book, noted cryptically that 'Donald has agreed to help and I have now got the necessary gadgets for the treasure.' This was a reference to a proposed expedition to recover the silver bars. (David Champernowne had meanwhile realised a healthy profit in his ingots, which had remained safely in his bank.) There had been a previous attempt with Donald Michie, who was offered the choice of either a one-third share of the total proceeds or a payment of £5 per expedition. This was itself a nice example of the Turing theory of probability, which appealed to the odds that a perfectly rational person would be prepared to bet on an event. As a perfectly rational man, Donald Michie opted for the latter choice. The first real-life treasure hunt had been a failure, since when they went to the wood near Shenley where one bar was buried, Alan found that the landmarks had changed since 1940, and he could not locate the spot. The point of the 'gadget' was that it was a metal detector which Alan had designed and built himself. On the second trip, it functioned, though only to a depth of a few inches. It successfully located a great many pieces of metal under the surface of the wood, but not the silver bar. As for the second bar, he knew where that was, but they found that they were unable to apply the UNBURY routine when standing in the bed of the stream.

Such failures he would easily laugh off. This was not his only visit back to Buckinghamshire, for he spent a weekend, probably in December 1946, discussing D. Gabor's new theory of communication[19] with Don Bayley. This time he distinguished himself by fainting when he grazed himself shaving. He had told Don long before about this reaction to blood, but this was the first time Don had seen it happen. There had also been an occasion in October 1945, when he, Don Bayley, Robin Gandy and 'Jumbo' Lee met up to go to a lecture on wartime radio work at the Institute of Electrical Engineers. Afterwards they had gone to Bernard Walsh's oyster restaurant and rather hoped to be fed on the house – in which they had been disappointed. Alan had cycled into London from Teddington and had parked his bicycle outside the Soho restaurant, from which it was duly stolen.

For him to cycle the fifteen miles was not entirely characteristic, since he would quite happily take in such distances on foot. His success in the Hanslope races had been followed up. On arrival at Teddington he had joined the local Walton Athletics Club, and had taken up running as a serious amateur. He was a long-distance runner, rather than a sprinter; it was his stamina that gave him the edge in races over three miles in length. During this period he would spend two or three hours every day on training, and would run for the club on Saturday afternoons. Thus in October 1946 he wrote to his mother:

> My running was quite successful in August. I won the 1 mile and ½ mile at the NPL sports, also the 3 miles club championship and a 3-mile handicap at Motspur Park. That was the meeting* at which all the stars were trying to break records, but in fact were pulling muscles instead. Being a very humble athlete myself I was able to get away without pulling a muscle. . . . The track season is over now, but of course the cross country season will be beginning almost at once. I think that will suit me rather better, though the dark evenings will mean that my weekday runs will be in the dark.

He was lengthening his distance, and working up to marathon running. If possible he would make official visits double as training runs. In particular, he would run the ten miles across west London to Dollis Hill in connection with the plodding development of the ACE delay lines. Every few months, he would run the rather longer

---

* Actually on 31 August.

distance of eighteen miles to Guildford, to put Mrs Turing's social imperatives to some constructive use. It amazed everyone, but he did not care about that. It also, as Mrs Turing put it, gave him contact with men 'in all walks of life'.

He even managed to combine running and chess. He saw something of David Champernowne from time to time, either at Oxford, where he now had a position, or at his parents' house at Dorking. They would play ping-pong, and talk about probability theory, but they also devised a form of chess in which each player had to make his move while the other ran round the garden. Fast running would tend to prevent good thinking, so the problem was to choose the right balance. Alan was also interviewed by the *Sunday Empire News* on training hints. He might have remembered the discussion of 'second wind' in *Natural Wonders*, which explained how it depended upon 'teaching' the brain not to 'raise such a row' when it smelt a little carbon dioxide in the blood.

One of the difficulties of his position was that there was a good deal of carbon dioxide in the blood supplying the British brain. For all the talk of planning for the future, there was a terrible exhaustion after the war, and little eagerness to upset the apple-cart any further. In one way this became clear at once. At Hanslope, Don Bayley had continued to improve and test the Delilah. Later in 1945 he had taken it to Dollis Hill for evaluation where – hardly surprisingly – they failed to find any cryptographic weakness. In early 1946 he had taken it to the Cypher Policy Board, which was a coordinating organisation established in February 1944. He set it up in the basement of their London offices, and left it with one of their officers. They were more interested than the Post Office, and suggested to Gambier-Parry that his man might join them to continue work. But Gambier-Parry turned this down, and this refusal closed the story. The Delilah's two neat packages of equipment, providing speech security with no more than thirty valve-envelopes, were completely forgotten. As a contribution to British technology it had been a complete waste of time.*

*

---

* Speech encipherment caught up with the Delilah in fifteen years or so.

But Delilah had been part of the preparations for the ACE, and this, Alan Turing's logical Overlord, was what mattered. The plans were all ready, and only needed the signal to start. And they did at least gain a sort of second wind on 31 October 1946, when Mountbatten, as president of the Institution of Radio Engineers, gave a speech[20] that conveyed – however inaccurately – the excitement of what had happened in the new technology of communication and control. It was as far beyond the old days of the *Glorious* as they had been ahead of the papyrus scroll:

> The war not only taught us a great deal about techniques, but it proved the occasion for new departures in application, particularly in electronics, which had enormously augmented our present human senses. Apart from radar, which aided to a remarkable degree the sense of sight, we might in future be able, by pooling and transforming the potentialities of other forms of radiation, such as light, heat, sound, X-rays, gamma-rays and cosmic rays, to receive the counterpart of radar screen pictures from inside our bodies, or even from individual body cells. Or perhaps we might receive them from the interior of the earth, or from the stars and galaxies. . . . there was reason to believe that facilities for impressing information and knowledge on the human brain . . . may be extended by the direct application of electrical currents to the human body or brain. . . .
>
> The stage was now set for 'the most Wellsian development of all'. It was considered possible to evolve an electronic brain, which would perform functions analogous to those at present undertaken by the semi-automatic portions of the human brain. It would be done by radio valves, activating each other in the way that brain cells do; one such machine was the electronic numeral [*sic*] integrator and computer (ENIAC), employing 18,000 valves . . .
>
> Machines were now in use which could exercise a degree of memory, while some were being designed to employ those hitherto human prerogatives of choice and judgment. One of them could even be made to play a rather mediocre game of chess! . . .
>
> Now that the memory machine and the electronic brain were upon us, it seemed that we were really facing a new revolution; not an industrial one, but a revolution of the mind, and the responsibilities facing the scientists to-day were formidable and serious. 'Let us see to it,' he concluded, 'that we not only insist on being allowed to shoulder it; but that when we have established our right, we can also prove our fitness.'

In 1946 people still believed that the great war surplus of scientific and technical advance could be turned to good use, although Mountbatten's comments on 'responsibilities' reflected the fact that few had any idea of how this was going to be achieved.

The ENIAC had been released from military secrecy months before, and Hartree had written about it in the scientific journal *Nature*,[21] but it needed Mountbatten to make it 'news'. He had taken his information from the NPL, and the inaccurate reference to chess-playing machines would suggest that he heard an excited Alan Turing talking about the future possibilities of the ACE. (There was, of course, no machine in existence that could play chess.) Darwin and Hartree were embarrassed not only by Mountbatten getting the wrong ends of the technical sticks, but also by his perfectly correct assertion that the ACE would exercise 'hitherto human prerogatives of choice and judgment'. They did not like to criticse Mountbatten, but wrote[22] to *The Times* complaining that its headline ELECTRONIC BRAIN had given a false impression.*

The official NPL press release,[23] on 6 November, was very different in tone. It presented the building of the ACE as a somewhat distant possibility, rather than being just round the corner. Correctly it set the origin of the ACE in Alan's 'severely mathematical paper' of 1936, and explained how electronic switching provided the speed to make such a machine practical. It explained the superiority of the ACE over the ENIAC, through its large memory store, and referred to the work already done on programming instruction tables. But the cost had now risen to a figure 'in the region of £100,000 to £125,000' and it was stated that 'It will be two or three years before the completion of this machine can be hoped for, since its construction presents formidable problems, both mathematical and technical.'

Now that this stirring if remote prospect had at last been entrusted to the British public, the *Daily Telegraph* showed itself the most eager to spread the good tidings, which it imbued with a suitably patriotic flavour. The headline BRITAIN TO MAKE A RADIO BRAIN/'Ace' Superior to US Model/BIGGER MEMORY STORE appeared on 7 November, followed up next day with an account by its own

---

* *The Times* printed their letters under the headline ELECTRONIC BRAIN – a subtle mixture of data and instructions.

reporter, who had interviewed Hartree, Womersley, and Alan at the NPL:

'ACE' WILL SPEED JET FLYING

. . . Revolutionary developments in aerodynamics, which will enable jet-planes to fly at speeds vastly in excess of that of sound, are expected to follow the British invention of 'Ace', which has been commonly labelled the electronic 'brain'.

. . . Professor Hartree said: 'The implications of the machine are so vast that we cannot conceive how they will affect our civilisation. Here you have something which is making one field of human activity 1,000 times faster.

In the field of transport, the equivalent of Ace would be the ability to travel from London to Cambridge . . . in five seconds as a regular thing. It is almost unimaginable.

. . . Dr Turing, who conceived the idea of Ace, said that he foresaw the time, possibly in 30 years, when it would be as easy to ask the machine a question as to ask a man.

Dr Hartree, however, thought that the machine would always require a great deal of thought on the part of the operator. He deprecated, he said, any notion that Ace could ever be a complete substitute for the human brain, adding:

'The fashion which has sprung up in the last 20 years to decry human reason is a path which leads straight to Nazism.'

The Germans, like the computer, had only been obeying orders, but this allusion did not deter Alan when next day a reporter from the local newspaper came to investigate the 'New NPL Wonder'. He only lengthened the time-span of his prophecy. They published the interview[24] of the '34-year-old mathematics expert' under the headline ELECTRIC BRAIN TO BE MADE AT TEDDINGTON:

Dr Turing, speaking about the 'memory' of the new brain . . . said it would be able to retain for a week or more about as much as an actor has to learn in an average play. Asked about Lord Louis Mountbatten's statement that it would be able to play an average game of chess, Dr Turing said that was looking far into the future. . . . The point was then put to him that chess and similar activities required judgment as well as memory, and Dr Turing agreed that that was a matter for the philosopher rather than the scientist. 'But,' he added, 'that is a question we may be able to settle experimentally in about 100 years time.'

This was the most exciting if embarrassing thing to happen at the NPL for a long while, and Darwin was sufficiently encouraged to make a radio broadcast[25], in which he outlined the 'idealised machine' of *Computable Numbers*, and explained that 'Turing, who is now on our staff, is showing us how to make his idea come true.' But as the newsreel music faded away, the awkward fact remained that it was now nearly a year since Alan had shown in detail how to 'make his idea come true', and Darwin still did not know how the NPL was to give effect to his proposals.

On 22 October, when Hartree enquired about progress on the ACE, Darwin had to confess that 'Post Office assistance had not been as great as was expected.' On the TRE side, there had been more technical progress, since F. C. Williams had begun in about June to investigate the behaviour of spots on cathode ray tubes with a view to creating a storage system. During the war he had seen attempts at the Massachusetts Institute of Technology radar research laboratory to employ ordinary cathode ray tubes for echo cancellation, attempts which failed because of the transience of the spots, which faded in a second or so. But during the autumn of 1946, he had the idea, independently of Alan's proposal in the ACE report, of refreshing the spots periodically to overcome this problem.[26] He had also seen a way to do it. On the other hand, from an administrative point of view there had been a setback to NPL's plans, since Williams had accepted the chair of electrical engineering at Manchester University – an appointment which he believed he owed to Blackett. Darwin explained to the Executive Committee that

> He had also hoped that considerable assistance could be obtained from Dr F. C. Williams of TRE, but that he now understood that Dr Williams had accepted a University appointment. He said that he would explore the possibility of Dr Williams working on this project at his University – perhaps with the help of NPL or TRE staff . . .

The possibility, thin as it was, was indeed duly explored. On 22 November 1946, Williams, together with two other top TRE men, R. A. Smith and A. Uttley, made a visit to the NPL 'to discuss with Mr Womersley and Dr Turing in what way help could be given to the ACE project.' The official minutes[27] were a masterpiece of discretion, Darwin being furious that Williams had been seduced into the Manchester orbit. Speaking alone with Smith, he banged on the negotiating table:

> The Director emphasised the extreme importance which he attached to the development of ACE, and put it as having the highest priority in his opinion of any work that was being done for DSIR at TRE. He was most anxious that some effort should be set aside for work done on this project.

Smith explained that this was very difficult. 'Apart from the small number of staff now working for Dr F. C. Williams most of the able circuit technicians had been transferred to the Department of Atomic Energy' – a statement reflecting another post-war story. So 'the only way in which TRE could continue to contribute would be to second a small number of staff . . . to work under Dr Williams' direction at Manchester University.' This was not at all what Sir Charles Darwin wanted to hear, but he did not give up. The meeting was enlarged to include Williams and Uttley from TRE, and Womersley for the NPL, while Alan was allowed in as well. In the discussion of the ACE

> . . . It appeared that although an elaborate paper design had been laid down, the fundamental problem of storage of information had not been solved and that, as had been suspected, the experimental work of Dr Williams at TRE on storing information on a cathode ray tube was considerably in advance of the work which the Post Office were doing on the use of delay lines for storage purposes . . . .

Williams had, in fact, already succeeded in storing a spot on a cathode ray tube for an indefinite period. A compromise formula was agreed, according to which Williams' work 'should continue with as little disruption as possible'. The NPL drafted a contract for Williams to sign, according to which he would develop both electronic storage and components for the arithmetical hardware.

There was, however, a triple misunderstanding. For one thing, there was still the strong possibility of Manchester funding allowing Williams to develop his storage independently of NPL requirements. For another, the ACE design and programming had been done for a delay line store, and would have to be completely reworked if policy changed to the use of cathode ray tubes. Very probably Darwin and Womersley did not appreciate that the storage medium effectively dictated many aspects of the design of the machine and its programming. Alan might not have minded rewriting everything if it meant that progress could be achieved, but there was a deeper difficulty. While Darwin spoke as though he had had a 'mathematician' to lay down a design, and now was finding a practical man to build it, Williams' attitude was that he was negotiating for funds wherewith to construct *his* computer. They were talking at cross-purposes. Success would depend upon bridging that gap, so closely tied in with social class, between 'mathematics' and 'engineering'. But the crossing of demarcation lines was very difficult now. No longer was Germany enforcing cooperation upon its enemy.

Hartree, wanting to see more cooperation, took the next step. On 19 November he explained to Darwin that 'Mr Wilkes was prepared to give as much help as he could on the ACE as he had facilities at Cambridge; he had experience of making up [delay] lines and would exchange information with Dr Turing.' Next day Wilkes wrote[28] to Womersley:

> I believe Professor Hartree has told you that I am beginning to do a little work on electronic calculating machines and that I am anxious to co-operate with you. As you know I recently paid a visit to the United States and saw what they are doing over there. Professor Hartree said he had discussed the matter with you and thought perhaps we might be able to make some mutually advantageous arrangement.

Wilkes visited the NPL a week later, on 27 November, to discuss his plans. On 2 December he wrote again:

> I have been thinking over in more detail the subject we were discussing last Wednesday and what I have now in mind is that I should build up a group here containing about eight people of varying grades in addition to the mechanic and boy who are at present in the workshop. This would involve an annual wage bill of something like £2,500. In addition there will be cost of materials and getting things made outside.
>
> I am quite convinced that the construction of a pilot model of some sort is an essential step in designing ACE.* I do not see how else one can test out such things as control circuits. I am attaching a note I have written on the design of the pilot model and I think it would be a good thing if you decide to place a contract with us to call for this specifically. It does not

---

* By 'ACE', Wilkes meant 'a computer'.

follow that no part of ACE may be ordered until the pilot model is finished but there will be many matters which it will be hard to settle, without experience at least of trying to get a pilot model working.

But the attached note outlined the specification of a computer on the EDVAC model, and thus entirely different from the ACE. Not only did it employ a central accumulator, but it ran counter to the Turing philosophy of keeping the hardware as simple as possible, and putting the work into the programming. Instead, it made the programming as easy as possible, at the cost of requiring electronic circuits to do the work of recognising and executing the various arithmetical instructions. Wilkes wrote apparently in ignorance of the fact that at the NPL they already had Version VII of a detailed design, for which six months' hard work had been put into writing programs. Even to contemplate Wilkes' proposal, as Womersley apparently did, was to undermine his own division's work. On 10 December he passed the proposal to Alan, whose reaction was understandably brusque:

Mr Womersley:

I have read Wilkes' proposals for a pilot machine, and agree with him as regards the desirability of some such machine somewhere. I also agree with him as regards the suitability of the number of delay lines he suggests. The 'code' which he suggests is however very contrary to the line of development here, and much more in the American tradition of solving one's difficulties by means of much equipment rather than by thought. I should imagine that to put his code (which is advertised as 'reduced to the simplest possible form') into effect would require a very much more complex control circuit than is proposed in our full-size machine. Furthermore certain operations which we regard as more fundamental than addition and multiplication have been omitted.

It might be argued that if one is to have so little memory then it is necessary to have a complex control to make up. In so far as this is true I would say that it is an argument for either having no pilot model, or for not using it for serious problems. It is clearly rank folly to develop a complex control merely for the sake of the pilot model. I favour a model with a control of negligible size which can later be expanded if desired. Only test problems would be worked on the minimal machine.

But Womersley wrote back to Wilkes on 19 December:

> Thanks for your suggestions regarding the pilot model for the ACE. They don't quite agree with Turing's ideas of what a minimal machine should be. In his opinion the control part of it is too elaborate though he agrees about the amount of memory. I have therefore asked him to prepare a note giving his ideas of what the minimal machine should contain, and this I propose to send to you (without commitment on either side) so that we can meet and discuss further the question of a formal contract.

Meanwhile the publicity given to the ACE had engendered a couple of enquiries from British industry regarding the potential of the new machine. On 7 November the editor of the *Industrial Chemist* had requested an article from Alan, who replied that while the ACE 'would be well adapted to deal with heat transfer problems, at any rate in solids or in fluids without turbulent motion,' he thought that an article would be appropriate only 'when a few similar problems have actually been solved with the ACE and more practical details can be given.' On 11 November, Alan had been invited to give a paper to the Institution of Radio Engineers, to which Mountbatten had vouchsafed his revelations a few days earlier. In this case, however, Alan had to write:[29]

> I am sorry that I cannot give such a paper without permission from our Director. I suggest you write to him about it.

The administrators, by this time, were concerned to reduce to a minimum the exposure of Alan Turing to the outside world. There had already been enough embarrassment in the newspapers. Thus Womersley suggested to Darwin that 'Dr Turing, rather than explaining his machine to a number of isolated people on many different occasions, should conserve his time by giving a course of lectures intended primarily for those who will be concerned with the technical development of the machine. . . .' Such a course was indeed arranged to take place on Thursday afternoons during December and January at the Adelphi, headquarters of the Ministry of Supply.[30] The invitations, less than twenty-five in all, went to the relevant electronic engineers, component manufacturers, military departments, and to a few others already in the know. The scene was set for a British equivalent of the Moore School lectures, although the setting was subtly different. According to an NPL memorandum, there was to be ample time after Alan's exposition for 'Discussion – in particular, criticism of Dr Turing's technical proposals.' They did not trust him to know what he was talking about. Criticism was, in any case, inevitable; for by this time several of those who attended had ideas of their own, and no inclination to be fitted into the Turing plans. Wilkes wrote[31] later that he

> found Turing very opinionated and considered that his ideas were widely at variance with what the main stream of computer development was going to be. I may have gone to his second lecture but I certainly went to no more. Hartree continued to go and insisted on giving me his notes, but I found them of little interest.

On the other hand, lectures on elementary electronics did not go down well with those from TRE, who could also see for themselves how the ACE design was built round the delay line storage. It had been intended that 'from these discussions would emerge more clearly what contribution TRE should make' to the ACE. This was indeed to happen, but not in the way that Darwin had hoped.

As it happened, the scheme of lectures was suddenly disrupted, thanks to the arrival of a letter from America. On 13 December Womersley found himself invited to a grand Symposium on Large Scale Digital Calculating Machinery to be held at Harvard from 7 to 10 January, in connection with the inauguration of Aiken's Mark II relay calculator. But Darwin quickly decided that it would be better for Alan to drink at the fount of American wisdom, making arrangements for him to attend the conference, and then to visit the ENIAC and the von Neumann group. Jim Wilkinson took over the Adelphi lectures. Alan spent Christmas at David Champernowne's parents' house at Dorking, and in a Boxing Day athletics meeting lost the three-mile race by one foot,[32] with a time of 15 minutes, 51 seconds. The sports reporter of the *Evening News* obtained a unique statement from him regarding the authorship of the digital computer:

'ELECTRONIC ' ATHLETE

> Antithesis of the popular notion of a scientist is tall, modest, 34-year-old bachelor Alan M. Turing, scratch man in the Walton Athletics Club's three-mile open event on Boxing Day.

Turing is the club's star distance runner, although this is his first season in competitive events. At the National Physical Laboratory, at Molesey [sic], he is known as Dr Turing, [and] is also credited with the original idea for the Automatic Computing Engine, popularly known as the Electronic Brain.

He is diffident about his prowess in science and athletics, gives credit for the donkey work on the ACE to Americans, says he runs only to keep fit, but admits he rowed in his college eight at Cambridge.

After trying and failing to find a post office so that he could mail his sweater and pullover to his mother, who took in his non-shrink washing, he boarded the *Queen Elizabeth* that evening. This visit to the land of Oz was a very pale reflection of the liaison of just four years before. But taking the laundry parcel with him, to be 'lugged round America' in a characteristically awkward Turing style, he went to see the donkey workers.

The conference had brought together almost every conceivable interested American party,* but Alan was the only British participant. He played a large part in the discussions. For example, he discussed the cathode ray tube storage proposals made by J. W. Forrester and J. A. Rajchman, the latter being responsible for development of the Iconoscope at RCA. His discussion[33] was in characteristic style, taking an engineering problem and locating a point of abstract principle that lay behind it:

> Dr Turing: I do not know whether I should really ask my question of Dr Rajchman or Dr Forrester, because this difficulty arises in both papers. Dr Forrester mentioned that there was a possibility of reconstituting the charge by use of low velocity electrons and said that Dr Rajchman would explain it in his paper. I understand from Dr Forrester that the method should be applicable to his type of storage. But there seems to me to be a fundamental difficulty which in principle amounts to this: unless the storage medium has some kind of association with the granular structure, such a method cannot be applicable because if any one such pattern is stable, then by a symmetry argument, by a slight shift one way or the other, as it were, a slightly different configuration is also stable. Thus you do not get a finite number of stable configurations, but an infinite number.

---

* But not Norbert Wiener, whose refusal to attend this conference marked his public dissociation from all military-funded science.

He also identified the guiding principle which distinguished his plans for the ACE from those of the von Neumann group and of Wilkes:

> We are trying to make greater use of the facilities available in the machine to do all kinds of different things simply by programming rather than by the addition of extra apparatus. . . .
>
> This is an application of the general principle that any particular operation of physical apparatus can be reproduced within the EDVAC-type machine.* Thus, we eliminate additional apparatus simply by putting in more programming.

The conference brought Alan into contact with Claude Shannon and Andrew Gleason again. He also took the opportunity to brush up and submit to Alonzo Church's journal some wartime work on type theory.[34]

Afterwards, Alan spent about two weeks at Princeton. The Americans were no further ahead in constructing an electronic computer than was the NPL, and had run into parallel problems. The line between 'mathematics' and 'engineering' was one of these, Eckert and Mauchly having split off to found their own company, and a patent suit over aspects of the EDVAC design being in progress. Von Neumann and Goldstine had been thinking, like Alan, about the problem to do with the numerical analysis of matrix inversion, and also about the physics of mercury delay lines.[35] Goldstine got the false impression that Alan thought delay lines could not work. This might have been because Alan talked about some of the more subtle difficulties. He had, for instance, already devised a grid[36] (later patented) to be inserted in the short 'temporary storage' lines to prevent reflected pulses from travelling backwards.

He did not bring anything back from America which changed his ideas about the ACE. This was the one kind of liaison which at the present time he did not need at all, and it had been irrelevant to his plans. But he brought back some presents – some nylons and dried fruit for his mother, and a food parcel for Robin Gandy which included a much-valued tin of cream. Britain was now even more

---

* 'EDVAC-type machine' was another example of a phrase used to indicate 'a computer'. Although people spoke of the EDVAC as though it were already a thing, it was still as much at the planning stage as the ACE.

severely rationed than during the war, the balance of payments proving a sterner constraint than the U-boats. And as he returned across the Atlantic, the bitter winter of early 1947 had set in.

The talk of cooperation was still going on, but it was only talk. On 21 January E. S. Hiscocks, the NPL Secretary, had reported that 'the Post Office had succeeded in keeping a number circulating for half an hour through their delay line set-up. This news was very encouraging. Professor F. C. Williams's work on electronic storage will, of course, continue.' But two days later Williams had declined the NPL contract and the illusion that he was doing work for the ACE finally crumbled away. Wilkes had written to Womersley on 2 January, expecting a contract, the incompatibility in design notwithstanding. ('I shall be glad to hear of Turing's ideas on the subject of a pilot model.') This letter remained embarrassingly in Womersley's in-tray. Another of Hartree's ideas had proved to be more realistic. He had invited H. D. Huskey, one of the ENIAC team, to spend 1947 as a sabbatical year at the NPL, and he was already installed when Alan arrived back in a shivering Britain to give the last of the talks at the Ministry of Supply. The rationale[37] was that he would bring with him American expertise especially on 'the apparatus side'. But otherwise, the twelve months of 1946, a period within which the Colossus had been designed and built from scratch, had passed and nothing had happened. Alan took advantage of his American tour to summarise the state of affairs at the NPL:

My visit to the USA has not brought any very important new technical information to light, largely, I think, because the Americans have kept us so well informed during the last year. I was able, however, to get a useful impression of the values of the various projects, and the scale of their organisation. The number of different computing projects is now so great that it is no longer possible to have a complete list. I think this is a mistake, and that they are dissipating their energies over too wide a range. We ought to be able to do much better if we concentrate all our effort on the one machine, thereby providing a greater drive than they can afford on any single one. At the present, however, our effort is puny compared with any one of the larger American projects. To give an idea of the numbers of people involved in this work in the USA I may mention that there were between 200 and 300 present at the Symposium at Harvard, and that about 40 technical lectures were given. We are quite unable to match this.

One point concerning the form of organisation struck me very strongly. The engineering work was in every case being done in the same building with the more mathematical work. I am convinced that this is the right approach. It is not possible for the two parts of the organisation to keep in sufficiently close touch otherwise. They are too deeply interdependent. We are frequently finding that we are held up due to ignorance of some point which could be cleared up by the engineers, and the Post Office find similar difficulty; a telephone conversation is seldom effective because we cannot use diagrams. Probably more important are the points which are misunderstood, but which would be cleared up if closer contact were maintained, because they would come to light in casual discussion. It is clear that we must have an engineering section at the ACE site eventually, the sooner the better, I would say.

Looking on the bright side my visit confirmed that our work so far has been on the right lines. It is probable that the Princeton machine, based on the Selectron, will have some advantage over the ACE in speed, but our proposed machine has some compensating advantages, and I think that, other things being equal, it is better that the two different types should both be tried. The Princeton group seem to me much the most clear headed and far sighted of these American organisations, and I shall try to keep in touch with them.

We shall eventually obtain a word-for-word account of the conference. All the information given was 'unclassified'.

If, like the Second Front, the plans had been delayed again and again, no loss of confidence was betrayed when Alan gave a talk on 20 February to the London Mathematical Society. This was when he elaborated in detail* on the imagined operation of the ACE, and spoke as if its realisation was almost a formality: before long the terminals would be humming with activity, and the programmers would be busy with the work of converting the nation's problems into logical instructions.

His talk, however, dwelt rather more upon the dream that lay behind the practicalities of an installation, bringing out in a public form the ideas that he had long been developing in Bletchley

---

* In the passages quoted earlier.

conversations. In fact his discussion opened with the picture of 'masters' and 'servants' who would attend the ACE, very much as the high-level cryptanalysts and the 'girls' had worked to decipher naval Enigma. The masters would attend to its logical programming, and the servants to its physical operation. But, he said, 'as time goes on the calculator itself will take over the functions both of masters and of servants. The servants will be replaced by mechanical and electrical limbs and sense organs. One might for instance provide curve followers to enable data to be taken direct from curves instead of having girls read off values and punch them on cards.' This was not a new idea, for F. C. Williams had built just such a thing for the old Manchester differential analyser. But the novelty lay in suggesting that:

> The masters are liable to get replaced because as soon as any technique becomes at all stereotyped it becomes possible to devise a system of instruction tables which will enable the electronic computer to do it for itself. It may happen however that the masters will refuse to do this. They may be unwilling to let their jobs be stolen from them in this way. In that case they would surround the whole of their work with mystery and make excuses, couched in well chosen gibberish, whenever any dangerous suggestions were made. I think that a reaction of this kind is a very real danger. This topic naturally leads to the question as to how far it is in principle possible for a computing machine to simulate human activities.

This was a more controversial claim. Hartree, for instance, writing to *The Times* in November, had repeated his statement in *Nature* that 'use of the machine is no substitute for the thought of organising the computations, only for the labour of carrying them out.' Darwin had written more expansively that

> In popular language the word 'brain' is associated with the higher realms of the intellect, but in fact a very great part of the brain is an unconscious automatic machine producing precise and sometimes very complicated reactions to stimuli. This is the only part of the brain we may aspire to imitate. The new machines will in no way replace thought, but rather they will increase the need for it . . .

To describe such careful and responsible statements as 'gibberish' was not the most tactful policy.

Darwin and Hartree were, in fact, echoing the comment by Ada, Countess of Lovelace, who wrote an account[38] of Babbage's planned Analytical Engine in 1842, and claimed that 'The Analytical Engine has no pretensions whatever to *originate* anything. It can do *whatever we know how to order it* to perform.' At one level, this assertion certainly had to be urged against the very naive view that a machine doing long and elaborate sums could be called clever for so doing. As the first writer of programs for a universal machine, Lady Lovelace knew that the cleverness lay in her own head. Alan Turing would not have disputed this point, as far as it went. The manager who took all decisions from the rule book would hardly be 'intelligent', or really taking a decision. It would be the writer of the rule book who was determining what happened. But he held that there was no reason in principle why the machine should not take over the work of the 'master' who programmed it, to a point where, according to the imitation principle, it *could* be called intelligent or original.

What he had in mind went much further than the development of languages which would take over the detailed work of the 'masters' in compiling instruction tables. He mentioned this future development, which in the ACE report he had already explored a little, quite briefly:

> Actually one could communicate with these machines in any language provided it was an exact language, i.e. in principle one should be able to communicate in any symbolic logic, provided that the machines were given instruction tables which would enable it to interpret that logical system. This should mean that there will be much more practical scope for logical systems than there has been in the past. Some attempts will probably be made to get the machines to do actual manipulations of mathematical formulae. To do so will require the development of a special logical system for the purpose. This system should resemble normal mathematical procedure closely, but at the same time should be as unambiguous as possible.

Rather, in speaking of a computer 'simulating human activities', he had in mind the simulation of *learning*, in such a way that after a point the machine would not merely be doing 'whatever we know how to order it to perform,' as Lady Lovelace had claimed, for no one would know how it was working:

It has been said that computing machines can only carry out the purposes that they are instructed to do. This is certainly true in the sense that if they do something other than what they were instructed then they have just made some mistake. It is also true that the intention in constructing these machines in the first instance is to treat them as slaves, giving them only jobs which have been thought out in detail, jobs such that the user of the machine fully understands in principle what is going on all the time. Up till the present machines have only been used in this way. But is it necessary that they should always be used in such a manner? Let us suppose we have set up a machine with certain initial instruction tables, so constructed that these tables might on occasion, if good reason arose, modify those tables. One can imagine that after the machine had been operating for some time, the instructions would have altered out of recognition, but nevertheless still be such that one would have to admit that the machine was still doing very worthwhile calculations.

It was in this passage that he drew first attention to the richness inherent in a stored-program universal machine. He was well aware that, strictly speaking, exploitation of the ability to modify the instructions could not enlarge the scope of the machine, later writing:[39]

How can the rules of a machine change? They should describe completely how the machine will react whatever its history might be, whatever changes it might undergo. The rules are thus quite time-invariant. . . . The explanation of the paradox is that the rules which get changed in the learning process are of a rather less pretentious kind, claiming only an ephemeral validity. The reader may draw a parallel with the Constitution of the United States.

But with that strictly logical reservation, he held the process of changing instructions to be significantly close to that of human learning, and deserving of emphasis. He imagined the progress of the machine altering its own instructions, as like that of a 'pupil' learning from a 'master'. (It was a typically quick shift to the 'states of mind' idea of the machine from the 'instruction note' view.) A learning machine, he went on to explain:

might still be getting results of the type desired when the machine was first set up, but in a much more efficient manner. In such a case one would have to admit that the progress of the machine had not been foreseen when its original instructions were put in. It would be like a pupil who had learnt much from his master, but had added much more by his own work. When this happens I feel that one is obliged to regard the machine as showing intelligence. As soon as one can provide a reasonably large memory capacity it should be possible to begin to experiment on these lines. The memory capacity of the human brain is of the order of ten thousand million binary digits. But most of this is probably used in remembering visual impressions, and other comparatively wasteful ways. One might reasonably hope to be able to make some real progress with a few million digits, especially if one confined one's investigation to some rather limited field such as the game of chess.

The ACE, as planned, would have at most 200,000 digits in store, so to speak of a 'few million' was looking well into the future. He described the storage planned for the ACE as 'comparable with the memory capacity of a minnow'. But even so, he perceived the development of 'learning' programs as something that would be feasible within a short period: not merely a hypothetical possibility, but affecting current research in a practical way. On 20 November 1946 he had replied to an enquiry from W. Ross Ashby, a neurologist eager to make progress with mechanical models of cerebral function, in the following terms:[40]

The ACE will be used, as you suggest, in the first instance in an entirely disciplined manner, similar to the action of the lower centres, although the reflexes will be extremely complicated. The disciplined action carries with it the disagreeable feature, which you mentioned, that it will be entirely uncritical when anything goes wrong. It will also be necessarily devoid of anything that could be called originality. There is, however, no reason why the machine should always be used in such a manner: there is nothing in its construction which obliges us to do so. It would be quite possible for the machine to try out variations of behaviour and accept or reject them in the manner you describe and I have been hoping to make the machine do this. This is possible because, without altering the design of the machine itself, it can, in theory at any rate, be used as a model of any other machine, by making it remember a suitable set of instructions. The ACE is in fact, analogous to the 'universal machine' described in my paper on computable numbers. This theoretical possibility is attainable in practice, in all reasonable cases, at worst at the expense of operating slightly slower than a machine specially designed for the purpose in question. Thus, although the brain may in fact operate by changing its neuron circuits by the growth of axons and dendrites, we could nevertheless make a model, within the ACE, in which this possibility was allowed for, but in which the actual construction of the ACE did not alter, but only the remembered data, describing the mode of behaviour applicable at any time. I feel that you would be well advised to take advantage of this principle, and do your experiments on the ACE, instead of building a special machine.

Enlarging in his talk upon the favourite example of chess-playing, Alan claimed that

It would probably be quite easy to find instruction tables which would enable the ACE to win against an average player. Indeed Shannon of Bell Telephone Laboratories tells me that he has won games playing by rule of thumb; the skill of his opponents is not stated.

This was probably a misunderstanding. Shannon had been thinking about mechanising chess-playing, since about 1945, by a minimax strategy requiring the 'backing up' of search trees – the same basic idea as Alan and Jack Good had formalised in 1941. But he had not claimed to have produced a *winning* program. In any case, however, Alan

would not consider such a victory very significant. What we want is a machine that can learn from experience. The possibility of letting the machine alter its own instructions provides the mechanism for this, but this of course does not get us very far.

Alan next turned a little aside from this central idea in order to consider the objection to the idea of machine 'intelligence' that was raised by the existence of problems insoluble by a mechanical process – by the discovery of *Computable Numbers*, in fact. In the 'ordinal logics' he had invested the business of seeing the truth of an unprovable assertion with the psychological significance of 'intuition'. But this was not the view that he put forward now. Indeed, his comments verged on saying that such problems were irrelevant to the question of 'intelligence'. He did not probe far into the significance of Gödel's theorem and his own result, but instead cut the Gordian knot:

I would say that fair play must be given to the machine. Instead of it sometimes giving no answer we could arrange that it gives occasional wrong answers. But the human mathematician would likewise make blunders when trying out new techniques. It is easy for us to regard these blunders as not counting and give him another chance, but the machine would probably be allowed no mercy. In other words then, if a machine is expected to be infallible, it cannot also be intelligent. There are several theorems which say almost exactly that. But these theorems say nothing about how much intelligence may be displayed if a machine makes no pretence at infallibility.

This was very true. Gödel's theorem and his own result were concerned with the machine as a sort of papal authority, infallible rather than intelligent. But his real point lay in the imitation principle, couched in traditional British terms of 'fair play for machines', when it came to 'testing their IQ', a point which brought him back to the idea of mechanical learning by experience:

A human mathematician has always undergone an extensive training. This training may be regarded as not unlike putting instruction tables into a machine. One must therefore not expect a machine to do a very great deal of building up of instruction tables on its own. No man adds very much to the body of knowledge. Why should we expect more of a machine? Putting the same point differently, the machine must be allowed to have contact with human beings in order that it may adapt itself to their standards. The game of chess may perhaps be rather suitable for this purpose, as the moves of the opponent will automatically provide this contact.

At the end of this talk there was a moment of stunned incredulity, during which his audience looked round with disbelief. This was probably much to Alan's delight. He knew perfectly well that he was upsetting the conventional armistice between science and religion, and it was all the more grist to his mill. He had thought it all out since reading Eddington while in the sixth form, and he was not now going to toe this official line that separated the 'unconscious automatic machine' from the 'higher realms of the intellect'. There was *no* such line – that was his thesis.

At heart it was the same problem of mind and matter that Eddington had tried to rescue for the side of the angels by invoking the Heisenberg Uncertainty Principle. But there was a difference. Eddington had addressed himself to the determinism of physical law, in order to deal with the kind of Victorian scientific world-picture that Samuel Butler had parodied in *Erewhon:*

> If it be urged that the action of the potato is chemical and mechanical only, and that it is due to the chemical and mechanical effects of light and heat, the answer would seem to lie in an enquiry whether every sensation is not chemical and mechanical in its operation? . . . Whether there be not a molecular action of thought, whence a dynamical theory of the passions shall be deducible? Whether strictly speaking we should not ask what kinds of levers a man is made of rather than what is his temperament? How are they balanced? How much of such and such will it take to weigh them down so as to make him do so and so?

It was a picture drawn from nineteenth-century physics, chemistry and biology. But the Turing challenge was on a different level of deterministic description, that of the abstract logical machine, as he had defined it. There was another difference. Victorians like Butler, Shaw and Carpenter had concerned themselves with identifying a soul, a spirit or life force. Alan Turing was talking about 'intelligence'.

Alan did not define what he meant by this word, but the chess-playing paradigm, to which he constantly returned, would make it the faculty of working out how to achieve some goal, and the reference to IQ tests would indicate some measurable kind of performance of this skill. Coming from Bletchley, this kind of 'intelligence' was of burning and obvious significance. Intelligence had won the war. They had solved countless chess problems, and had beaten the Germans at the game. And more broadly, for his scientific generation, life had been a battle for 'intelligence', fought against stupid out-of-date schools, a stupid economic system, and stupid Blimps from 'a profession for fools' during the war – not to mention the Nazis, who had elevated stupidity into a religion. There was the influence of a Webbsian vision of socialism, in which society was going to be administered by intelligent functionaries of the state, as in the near future of *Back to Methuselah*, and in 1947 there was much talk about IQ tests, since the British youth was supposedly being newly divided into scientifically defined categories according to 'intelligence' rather than by class. Oscar Wilde had written of *The*

Soul of Man under Socialism, but under the socialism of Attlee and
Bevin words like 'soul' – supernatural or 'soupy' words as Bertrand
Russell called them – could be left to bishops and pep talks about
team spirit.

While many people might have reservations about the wisdom
and beneficence of scientists, they were at last basking in the
favour of government. The war had converted government to an
interest in science, and a view once visionary, then progressive, was
becoming orthodox. The scientists had emerged from the miserly
corners in which they had done their despised 'stinks' before, and
it seemed that their swords could be turned into ploughshares, or
more precisely, that they would supply governments with scientific
solutions to their problems. On one level Alan Turing belonged to
this climate of opinion, and certainly rejected the idea that scientists,
rather than generals and politicians, were to blame for the world's
current imperfections. Mermagen from Sherborne days, now a
master at Radley, another public school, wrote to Alan at this time
for advice about the place of mathematics and science in the post-
war world, and Alan replied:[41]

> On the subject of careers for mathematicians I am strongly inclined to
> think that the effect of ACE, guided projectiles, etc, etc, will be towards
> a considerably greater demand for mathematicians from a certain level
> upwards. For instance I am in need of a number who will be required to
> convert problems into a form which can be understood by the machine.
> The critical level may be described roughly as the degree of intelligence
> shown by the machine. We obviously do not want people who can take no
> responsibility at all. We just make the machine do the work which might
> have been given to them. At present of course this critical level is very low
> and I am sure you need not be afraid of encouraging boys that are keen
> and want to take up a mathematical career. The worst danger is probably
> an anti-scientific reaction (Scientists instead of goats at Bikini etc) but this
> is a digression.*

But what was the intelligence for? – that was the unasked question.
What was the goal towards which the technicians and managers

---

* There was, however, at least one scientist who may have been over-exposed to fall-out:
John von Neumann, who watched the Bikini tests of July 1946.

were now working? There was a vacuum at the centre in 1947, as the convictions of the 1930s and the enforced unity of the war evaporated away. The great opponent in the chess game had been beaten, and no one had yet taken his place.

By speaking of the mind in terms of puzzle-solving intelligence, of finding efficient means for undiscussed ends, Alan Turing superficially epitomised the technocratic outlook of 1947 social management. But it was only on the surface. For he had no interest in applying computers or related 'Wellsian developments' to the problems of society. He had wisely put examples of the usefulness of the computer into his report, to get it paid for. But his picture of the imagined installation simply copied what he had seen in operation at Bletchley. He knew it could be done, but had little interest in it nor indeed the ability to organise this side of it himself. The ACE would need a Travis to keep it going. Even in this letter, superficially about the usefulness of mathematics, his interest lay in comparing the intelligence of the planned computer with that of boys – a favourite comparison, in fact. His whole enterprise was still motivated by a fascination with knowledge itself, in this case with an understanding of the magic of the human mind. He was not a Babbage, with an interest in the more efficient division of labour. His interest in the ACE had little to do with the 'mechanization, rationalizion, modernization' that Orwell foresaw, although the computer might be funded for this purpose. It was much closer to an undiminished wonder at 'the glory and beauty of Nature', and an almost erotic longing to encompass it. Indeed his letter to W. R. Ashby had stated baldly that

> In working on the ACE I am more interested in the possibility of producing models of the action of the brain than in the practical applications to computing.

If, furthermore, he had left something out by confining his account of mind to a discussion of puzzle-solving, an omission that reflected the temper of the times, this was not because he thought that this kind of 'intelligence' was one of vast superiority to other human characteristics. In fact it was almost the reverse.

Perhaps this was the most surprising thing about Alan Turing. Despite all he had done in the war, and all the struggles with stupidity, he still did not think of intellectuals or scientists as forming a superior class. The intelligent machine, taking over the role of the 'masters', would be a development which would cut the intellectual expert down to size. As Victorian technology had mechanised the work of the artisans, the computer of the future would automate the trade of intelligent thinking. The craft jealousy displayed by human experts only delighted him. In this way he was an anti-technocrat, subversively diminishing the authority of the new priests and magicians of the world. He wanted to make intellectuals into ordinary people. This was not at all calculated to please Sir Charles Darwin.

Alan's talk was, as it happened, given on the same day that the British government announced its rapid withdrawal from India. The lessons of the war were at last sinking in, accentuated by the fuel crisis which the new management of the National Coal Board were powerless to control. Britain was no longer one of the 'Big Three', her role in the Mediterranean being quickly taken by the United States. It was a moment of truth, in which Britain appeared as a giant desert island. Germany had forced the truly Big Two out of an artificial isolation, and neither of them had fought to preserve British interests or markets. If there was a silver lining in the clouds, it was the fond belief that Britain could do better than 'the American tradition of solving one's difficulties by means of much equipment rather than by thought,' in Alan's words.

The British government was well-disposed to finding scientific solutions to its problems, and had announced on 5 February a grand plan to plant its East African colonies with groundnuts. The ACE, likewise, though it represented only a fraction of the investment in the groundnut scheme, was still in 1947 a highly progressive project. It was just what the left-wing pressure groups had called for in the 1930s, with the state taking over the development of new science and technology instead of leaving it to the caprice of commerce. Blackett, as president of the Association of Scientific Workers, was at the vanguard of this movement. In 1947 he wrote an introduction to the ASW book *Science and the Nation*, which promised such modern wonders as 'the scientific organisation of bureaucracy'. Yet the application in peacetime of such scientific policies was liable to go in unforeseen directions. Blackett himself, by enticing F. C. Williams to work on a pure-mathematical computer at Manchester, had not exactly helped the national plan. And although a left-wing advocate of planning, he had shown considerable personal dash and enterprise. Darwin's position was more paradoxical still. Whether for hereditary or other reasons, he held the right-wing views of a Social Darwinist, taking a dim view of the welfare state.[42] ('The policy of paying most attention to the inferior types is the most inefficient way possible of achieving the perfectibility of the human race.') His recipe for progress, or rather for a check to what he considered to be European racial suicide, was that men who had been 'promoted' should endeavour to have more children than others. Yet the dispensations of the NPL resembled less the jungle of cut-and-thrust competition than the ham-fisted management of the Soviet Union. They did not create an environment in which variety and initiative could long survive.

In the spring of 1947, while Darwin applied the higher realms of his thought to solving the problem of Alan Turing, incoherent initiatives were taken by more impatient people on the lower rungs of the promotion scale. One of these was Harry Huskey, who was very eager to see a start made on building a computer before his year was out. He admired the general form of the ACE design, but believed that the best plan would be to construct a small delay line machine 'on a plan which is a compromise between the NPL and the Moore School plans.' Relations between Alan and Huskey had been cool from the start, but they deteriorated when one day in the spring, Alan went into Mike Woodger's room to find him engaged upon writing a program headed 'Version H'. This was in fact an adaptation by Huskey of the Version V of the ACE, trimmed to include only the barest minimum of apparatus required to do a useful job, which was defined as the solution of eight simultaneous equations.* Though generally consistent with

---

* This criterion in itself marked a difference in attitude and policy. For someone from an ENIAC background it was obvious that the function of a computer was to do numerical calculation. Indeed Huskey cheerfully jettisoned the logical functions included in the ACE as 'not needed in most computing problems'. But who knew what computing problems would or could be? Alan Turing's scheme had reflected the fact that he had spent the war on non-numerical problems – but he was in no position to argue for the significance of this fact.

the ACE design philosophy, such a departure necessarily subverted Alan's control of the project. He had tolerated the 'pilot' policy on the understanding that the pilot would not detract from the full machine, but would become an integral part of it. If Huskey's project failed, it would be a waste of time, and if it succeeded, it would lead to a marked change of plan. Naturally, Alan boycotted the development. Somehow Harry Huskey managed to scrape together enough equipment to make a start. His official role was 'on the apparatus side', and he was better at the form-filling; furthermore he did not need to think in terms of building the grand ACE installation, but only in terms of an experimental set-up. Jim Wilkinson and Mike Woodger joined in. Life in the Mathematics Division became very complicated, but they did learn something about electronics for the first time.

At the same time, Alan managed to make a few experiments of his own in the cellar of Teddington Hall, the Mathematics Division building, and explained some electronics to Mike Woodger. He had devised circuits to transmit and receive pulses along a delay line, and a system to probe the circuit so as to examine the shape of the pulse on an oscilloscope. The NPL had nothing in the way of a machine to do this elementary engineering task, so, as at Hanslope, he had made one for himself. The whole thing[43] involved mounting four or five valves on a breadboard, and no more. He had no delay line to work with. A little later, coming back from lunch, he spotted a drainpipe lying in the long grass, and had someone help carry it back for him to try out as an air delay line. Don Bayley and 'Jumbo' Lee came and saw this 'dog's breakfast' of an outfit in March or April 1947. Alan took Don for a walk round the grounds and complained bitterly of how he had been thwarted. Alister Watson, philosopher now turned radar expert, visited the NPL on business and heard Alan complain that 'they say I don't understand magnetism!' Francis Price from Princeton days, whose family lived at Teddington, heard out his acrid comments on how the administration denied him the most standard pieces of equipment for experiment.*

But these anarchic developments were overtaken by Darwin's decision. He had become reconciled to the fact it would not be

---

* In November 1946 he had also tried to retrieve a spare Delilah power pack from the Cypher Policy Board: 'Do you think it could be spared and sent down to me here? It is an old friend whose tricks I am used to.' But this was almost certainly unsuccessful.

possible now, as it might have been during the war, to farm out the ACE to other organisations. Hitherto he had resisted setting up an electronics section within the NPL. But now it had become clear that there would be no single national computer, as had been expected in 1946, but various installations of which the NPL, if it was lucky, would have just one. His new plan was for a prototype to be constructed at the NPL, whereupon it would go to English Electric.

This was a sound policy. The tenuous link with Wilkes was formally snapped on 10 April, and the Post Office contract cancelled. Instead, during the summer of 1947, a new Electronics section of the Radio Division was set up, under a certain H. A. Thomas.

But there were two drawbacks to the new policy. The first was that Thomas's interest lay in the industrial application of electronics, and not in computers. The second was that he had little knowledge of the use of electronics for pulse or digital techniques. It was not that he refused to make any concession at all to the ACE project, for indeed he very quickly produced a report (with which Womersley was 'in full agreement') on how a computer should be built. But this made little reference to the Turing scheme. Then Thomas imported some enormous cathode ray tubes from Germany. He intended them for use as digital display tubes. The staff from the ACE section dutifully went and stared at these, but with some amazement, for they seemed completely irrelevant to the plans for the computer.

It was a curious sequence of events, in which the NPL administration had done everything possible to avoid 'building a brain'. They had chased after Williams, although it would mean a complete redesign to use the kind of storage he was developing. They contemplated handing the construction to Wilkes, when he had only 'a mechanic and a boy' and incompatible design principles. They paid for Huskey to come over from America because of his experience 'on the apparatus side', and then failed to use it. Finally they had appointed, as the head of an electronics section, a man without motivation or experience for the job in hand. The one person whom they had not trusted was Alan Turing. The one policy they had not adopted was that of finding or training engineers to effect the proposals to which they had agreed in 1946. It was certainly not easy to find such expertise, but ultimately, they had not really tried.*

After this happened, Alan withdrew. The programming work continued, and they went a long way with sub-routines for floating-point arithmetic, including matrices and the numerical solution of differential equations. But Alan had lost interest in this work, although he spent time on what he called 'abbreviated code instructions'. These took up the ideas announced in his original report, of having the computer expand its own programs. They made, in effect, a high-level language for the computer, long before such things were developed elsewhere. But he was thinking more about the even higher level that he had announced in February, that of making a computer show intelligence. In this respect there was nothing to be gained by working at the NPL. Darwin and Womersley concurred in this opinion, and on 23 July Darwin wrote to the DSIR that since the ACE had now reached the stage of 'ironmongery', it would be best if its designer were to 'go off for a spell'. The separation of brain from hand could not have been made more explicit. It was agreed that Alan should spend a year at King's to develop his theoretical ideas. It was too early in his Civil Service career for this to be a normal 'sabbatical year', but the DSIR and the Treasury were persuaded to treat him as a special case.

Monday 18 August 1947 was declared to be the official start to the building of the ACE. Darwin presided over a special morning meeting carefully calculated to impress upon the lowly engineers the privilege they enjoyed in working for Womersley's project. There was much brave talk of taking on the Americans, and then Thomas 'indicated the way in which he was about to approach the project in its initial stages'. Alan attended, but said nothing.

---

* As further illustration, it is noteworthy that irrespective of Alan's reports, Womersley continued to assume that the Americans held all the trumps. In April 1947 he daringly suggested that Darwin, currently attending meetings of the United Nations in New York, should call at Princeton to negotiate for the latest news in EDVAC engineering. This proposal was ill-timed; in May the NPL was declared by the American armed services to be 'unsuitable' for the receipt of such (commercially valuable) information. The ruling was relaxed later in 1947: information could be passed to Britain provided it was used *only* for military purposes.

Womersley, reporting to Darwin that the ACE should be complete by early 1950, was satisfied. In theory the ACE was still a venture of immense national importance, and Hiscocks had called this its 'D-Day'. The truth, however, was that the NPL had now almost succeeded in cutting the ACE down to its own bureaucratic size. Only Alan Turing prevented the triumph being complete, awkwardly remaining to voice the original vision. On 30 August he wrote to Darwin that[44]

> A letter has come from Ministry of Supply . . . asking us to do their programming for them. This is work that we ought to be able to undertake, but it will not be possible with our present very small programming staff. This staff is quite inadequate for our own needs; it will have to be at least three times greater than it has been up to now, if we are to make a success of the ACE project. The arrival of Mr D. W. Davies . . . will, of course, be of some help, but we need in addition another two or three bright [Scientific Officers] . . . immediately.
>
> It is essential to recruit the ACE planning staff now, because it must be trained and in full production long before the machine itself is available for use. A large body of programming must be completed beforehand, if any serious work is to be done on the machine when it is made.

In 1941 a direct appeal had worked wonders, but by 1947 the war might as well have never been. And talk of 'planning staff ' for the universal machine was almost as unreal as it would have been in 1936. As it happened, the new recruit, Donald Davies, who came from atomic research, had been studying the abstract universal machine of *Computable Numbers*, rather annoying Alan by finding mistakes of detail in its construction. Alan had also given a talk on the ACE at the NPL which Rupert Morcom attended. The unhappy fact was that after eleven years there was still nothing but paper plans, a paper machine and paper programs, abstruse and insubstantial as the 'satisfactory numbers' he had tried to explain at the Clock House in 1936. He had given his all, and had nothing to show for it but paper. He would give them another chance to create the right conditions. But the wheels of 1939, that once had seemed to be turning in his direction, were now revolving backwards.

*

In the summer of 1946 Alan had made contact again with James Atkins, who came to visit at Teddington. He had had a very different war, for standing fast as a conscientious objector he had spent four months in prison and then worked in the Friends Ambulance Unit. He asked Alan what he had done, and Alan replied 'Can't you guess?' But James could only guess that it was to do with the atomic bomb. After the passage of nine years, they found that there was nothing that they could truly say or do. When James went, he left something behind, and returning after a few minutes caught Alan sitting in his bed-sitter with a particularly acute look of misery.

Alan had mentioned *The Cloven Pine* to James, and in early 1947 Fred Clayton himself got in touch again. Alan replied on 30 May

> I was very glad to hear from you again, and should very much enjoy teaming up with you again for another sailing holiday.
>
> The best dates for me would be either the beginning of September or the beginning of July. The last year or two I have taken to running rather a lot. This is a form of compensation for not having been good at games at school. The application at present is that I have a Marathon race on August 23 and do not want to upset my training by sailing in August or late July.

In the spring Alan had tried himself out against much stronger competition than that of the local suburban clubs. In the Southern Counties ten-mile championship of 22 February he had done 'very badly', but two weeks later at the National ten-mile event had come in sixty-second place out of three hundred runners. The Marathon race to which he referred brought out his endurance training to better advantage, and in fact he would come in fifth place.[45] With a time of 2 hours, 46 minutes, 3 seconds he was only thirteen minutes behind the winner. Behind the off-hand words about 'running rather a lot', he had taken it very seriously. His letter to Fred continued with an equally off-hand reference to the war:

> I heard rumours at B.P. from time to time that you were coming there, and was disappointed that nothing came of it. You didn't really miss anything.

Alan went to Bosham at the end of June and fixed up a September holiday.

Meanwhile on 3 August Alan's father died at the age of seventy-three.

He had been in poor health for several years. He left £400 more to Alan than to his brother, in compensation for the sum spent on John's articled clerkship twenty years before. But Alan handed it over to his brother, not thinking this fair.* He also received his grandfather John Robert Turing's gold watch. As it happened, Mr Turing died just twelve days before the passing of the British Raj ended that exile world to which he belonged. After his death Mrs Turing spoke little of him, and kept few reminders of his existence. Her own life was far from declining, and a new independence was reflected in the adoption of her second name Sara in preference to Ethel. She had also taken more interest in Alan's doings, pleased that he was busy on something useful at last. She had sensed the high hopes of 1945, and sympathised with his complaints on how he had been thwarted.

When Mountbatten relinquished British rule to the new Dominions of India and Pakistan, it was a belated victory of the new men. Another case of where the war had obliged the reforms long urged in the 1930s, it also represented the renovation of an independent British role in world affairs. But these were still uncomfortable days, in which the new world order was rapidly becoming apparent. The American loan, negotiated by Keynes before his death in 1946, and supposed to repair the British economy, had largely evaporated thanks to American insistence on making sterling convertible again. After a financial crisis, Dr Dalton withdrew convertibility on 20 August, pepping up the sagging team spirit on the radio: 'God bless you all and your families. Get all you can of happiness and health and strength out of the sea air and then go all out in a great effort to help your country.'

As usual the servants were being called upon to make up for the crimes and follies of the masters. Alan had probably had enough of this by now. But he and Fred Clayton took the advice regarding sea and fresh air, and had a holiday which recaptured something of the days of August 1939. Alan tended to be rather impatient with his friend's handling of the boat, trying to explain feedback to him, and swearing mildly. But one day Fred took the helm over to the Isle of Wight, helped by a favourable wind, and all went well except that on the way back they bumped into a buoy and Alan said, without

---

* Giving the whole £400 was generous, but illogical. It was put in trust for his nieces.

thinking, 'I really ought to be a bit more careful with buoys.' Then they both saw the joke and laughed.

But it was no laughing matter. They exchanged their observations on the black market of unofficial truth, Fred contributing comments on the customs prevailing in India, where he had done low-level Japanese code-breaking, and Alan speaking of the war as a sexual desert ('You didn't miss anything') except for the incident while in America. It then transpired that Alan had expected this holiday to hold something for him in that direction, which came as a shock to Fred, who explained that he was going to get married.

Picking up the German threads after the war, Fred had entered into a warm correspondence with the sister of the two boys on whom *The Cloven Pine* was mainly based. It was she whom Fred intended to marry. It was an odd coincidence that one day, out on the water, Alan caught sight of a passing boat and said 'Oh, that's rather awkward, it's Martin Clarke's sister, we were engaged.' Joan recognised Alan, and waved and smiled. (She had seen him a few months before at the LMS lecture.) But they did not meet. Alan told Fred of the broken engagement. Fred was left confused by his rejection of Alan whom, however, he did not find at all attractive. He and Alan talked about their respective decisions. 'Freedom and consistency of mind' was not so easy for his friend, but Alan was now quite clear as to where he stood.

On 30 September 1947 Alan resumed his King's fellowship,* after a break of just under eight years from 2 October 1939. Because of the 1944 renewal, it was set to run until 13 March 1952. He was thirty-five, but could have passed for a young twenty-seven. Once this year at Cambridge he was accosted by a proctor, for not wearing a gown after dark. He had been taken for an undergraduate, and his youthfulness surely accentuated what was something of a return to the life of the past. In certain ways, of course, Cambridge had changed. Most students were in their mid-twenties, and had

---

* He also received reduced pay at the rate of £630 per annum for the sabbatical period. Darwin had offered full pay, but Alan told him he would prefer half-pay, saying that on full pay he would feel that 'I ought not to play tennis in the morning, when I want to.'

arrived after some years of service life rather than with the artificial immaturity of public schools. It was a more personally ambitious, less politically conscious Cambridge than that of the 1930s. But no one spoke of the war, which was fading like a dream, and it would take more than the war to change the special character of King's.

One friend immediately stood out from the rest. Robin Gandy had taken Part III of the Mathematical Tripos that summer, and was now working in theoretical physics towards a fellowship. Soon after term began, Robin called on Alan and asked if he could borrow his copy of Eisenhart's book on continuous groups. Alan took it off the shelf, and out fluttered a newspaper picture of a rank of pageboys at Princess Elizabeth's wedding. There was someone else in the room, probably Robin's physicist friend Keith Roberts, and Alan only said darkly 'You'll find nice *pages* like that in my books.' But next morning, Alan said to him, 'We know each other quite well now . . . I might as well tell you that I am a homosexual.' Robin's own personality was different, but he was someone who like Alan cared that (as Carpenter had put it)[46]

> men shall learn to accept one another simply and without complaint . . .
> honour the free immeasurable gift of their own personality, delight in it and
> bask in it without false shames and affectations

and he was pleased that Alan had chosen to tell him. It was very different from the sticky moment with Don Bayley. Robin was rather surprised that Alan was so forthright, for if obliged to give an opinion at Hanslope, he would have said that Alan did seem to be more drawn towards men, but that he was probably too shy and inhibited to do anything about it. The truth made Alan less austere and more fun. Alan in turn was very pleased that Robin should understand him so well, and with that out of the way, there was nothing (except Bletchley Park) that they could not talk about, whether of science or gossip.

He was altogether more confident with people than he had been before the war, and was elected to George Rylands' play-reading Ten Club, which he would have found frighteningly precious and pretentious then. He was not elected to the Apostles. Robin was an Apostle, and would have proposed him, except that at that point they were choosing only rather younger candidates. He did, however, become rather better connected with the King's social network, and Robin was particularly helpful in opening up his life to more communication, compensating for some of the fraught years of silent, frustrated desire. 'When I recall some past *epoch*,' Alan said in reply to some question about the past from Norman Routledge, a new acquaintance of this year, 'I think of whoever I was in *love* with at the time.'

The winter was spent on various topics, none with all-absorbing attention. The numerical analysis paper[47] – the only published evidence of his ACE work – was finished in November. Most of all he wanted to understand more about 'thinking'. Somehow the brain did it – but how? Contemporary physiologists had only the faintest ideas about neural stimuli and responses. He went to R. H. Adrian's lectures, but was disappointed. Chemistry and physics were at last pushing their way into biology, but his whole thesis lay in a different kind of description, the *logical* description of the nervous system, for which chemistry and physics were only the medium. He conveyed his disappointment to Peter Matthews, a very sharp undergraduate, one of the few who had come to Cambridge at eighteen, with whom he went to the physiology lectures. They would talk long over lunch and over mugs of cocoa in the evenings, since they lived on the same staircase.

Jim Wilkinson would visit Cambridge from time to time, and kept Alan informed as to the developments, or lack of them, back at Teddington. The news was that of cuts, crises, and an ever-narrowing vision. At a meeting in November, they had abandoned many of the advanced ideas for the ACE, including the 'abbreviated code instructions'. An edict from Darwin had stopped Huskey's Test Assembly, because of complaints from Thomas. The ACE section was reduced to the writing of a report[48] on their numerical analysis and programming work.

In the new year of 1948, an alternative possibility came into being for Alan's future. For if the NPL project had ground to a sticky halt, the Manchester development had bounded along with great speed. By the end of 1947 Williams had stored 2048 digits on an ordinary cathode ray tube screen, which was at least the equivalent of having a cheap, working delay line available. Newman still had his Royal Society grant virtually intact, and mooted the idea that Alan should take an appointment at Manchester to direct the computer construction there instead. Alan did not decide immediately, but in March Newman asked his university[49] to create a new position, with a salary paid from the computer fund, but with the status of Reader.*

The prospect of getting his hands on a computer after all was very attractive. But so was life at Cambridge, the nearest he had to a home. He joined the Moral Science Club again, and gave them a talk[50] on 'Problems of Robots' on 22 January. (That Czech word 'robot' had a topical flavour.) He joined the Hare and Hounds Club, and continued his training, often running to Ely and back in the afternoon. He tried some von Neumann game theory, and rather laboriously worked out a strategy for a simplified version of poker, slightly improving upon the account given by von Neumann,[51] and also for the Princeton game of Psychology. Robin was moving away from theoretical physics and into the philosophy of physics, and there were many discussions with him and his friend Keith Roberts. Once they had tried to set up a purely operational definition of Special Relativity, and when one of them objected that there was no such thing as a rigid body in relativity, Alan said, 'Well, let's call them *squeegees.*' It was that pleasant, unpompous way of being serious without being solemn that in 1948 was far from widespread in academic life, and was associated with the legacy of the Apostles and the King's milieu – though it came naturally to Alan. There was another nice occasion when Don Bayley came for the weekend to see Alan and Robin, to be greeted with a toy steam engine that Alan had bought at Woolworths. 'I always lusted after things like this as a boy,' he said ruefully, 'but didn't have enough pocket money for them. Now I have got enough money, I might as well have it.' They played with it for the afternoon.

Alan had told Robin that 'Sometimes you're sitting talking to someone and you know that in three quarters of an hour you will either be having a marvellous night or you will be kicked out of the room.' It was not always like that, for the innocent Peter Matthews' cocoa sessions did not involve this stark dichotomy. Nor was Alan accomplished in the necessary social game of words and eye-contact; he was too shy and brusque and lacking in confidence

---

* Higher than that of 'lecturer', but not as high as a true 'Prof'.

in his looks. Cambridge did inspire him with greater interest in his appearance, and sometimes he would show Robin a photograph of himself at sixteen and say how handsome he looked then. He was certainly *not* attractive in the Aryan-Brylcreem style of the 1940s. To the fastidious, his open-necked, shabby, breathless immediacy came as messy and coarse, though there were redeeming features; he could put on a roguish charm which reminded people of his Irish ancestry, and besides his piercing blue eyes he had thick, luxuriant eyelashes and a soft-contoured nose. But whatever his doubts and disadvantages he would pace round the courts of King's and invite young men in for tea. Sometimes he struck lucky. In April 1948 he struck very lucky indeed, for Neville Johnson stayed for tea, and stayed many times.

Neville was a third-year mathematics student, then twenty-four, but mathematics was not a cement but an embarrassment in their relationship, for although Neville had won a scholarship, this was 'a flash in the pan'. It made Neville feel inferior that they were in this respect on quite different planes. Once Alan said, 'When you know what the *Entscheidungsproblem* is, you will know what a great mathematician I am,' with a sad twinkle, but Neville never did find out. Neville thought he was very ordinary compared with the more glamorous side of King's, and that he was just 'the best that Alan could get'.

But for Alan, it was probably an attraction that Neville, a Geordie from Sunderland Grammar School and the Army, was someone down to earth and rather tough. His problem lay far more in having to remove the anaesthetising shell with which he had protected himself for so long. Perhaps it was too late. To Neville, it seemed that Alan had enjoyed ill luck with people, and that it was not surprising that he was keen on machines to replace them. Once he said to Neville, lying on his bed, 'I have more contact with this bed than with other people.' He also unburied a little of the past. Christopher still prodded him, if only because he would faint at the mention of blood or dissections in the physiology lectures; but in 1948 he was 'A very upper-class boy, from where they dressed for dinner.' Of Bletchley he revealed that the Poles had brought something crucial – but that he must not say more. Once he pointed out a professor of Greek in the street and said he had done something marvellous there.

They were often together, and sometimes with Alan's friends, although Neville then felt rather out of place. Neville joined in the game of poker that they determinedly played in order to test out Alan's minimax strategy. (It was not very exciting, as the strategy was largely the obvious one anyway.) It was another ordinary affair, at last, as it had been with James Atkins in the pre-war days.

This solved one problem. In other ways, Alan was back to the difficulties of 1939. For his mind still straddled mathematics, engineering and philosophy in a way that the academic structure could not accommodate. Temporarily the war had resolved his frustration, giving him something to do that was intellectually satisfying, yet which actually worked. But that was over now, and instead of being drawn in, he was being pushed out.

How was he to continue, having won the war, and lost the peace? If he did not return to the NPL then the *Entscheidungsproblem* lay between Cambridge and Manchester. He could stay at King's, and a lectureship ought to come his way. He could return to the world of Hilbert and Hardy, as though the last nine years had never been. Yet no more than in 1939, was that the way his spirit moved. He did not want to go backwards, and he still wanted to grasp the computer that he had invented. At Cambridge, the computer was firmly in the grasp of M. V. Wilkes, and Alan was far, far too proud to go cap in hand to use it. If he wanted a computer, it meant going to Manchester.

Darwin, meanwhile, expected him back at the NPL at the end of the Cambridge term. On 20 April 1948 he reported to the Executive Committee on 'Future Plans for Dr Turing':

> Dr Turing, who is on a year's leave of absence at Cambridge University, will shortly be due to return to the laboratory and Director is proposing to discuss with him what type of work he should then undertake. Director felt that from the point of view of Dr Turing's career, it would be an advantage if he started to write some papers rather than continue with the fundamental physiological studies into which his researches are carrying him. Dr Turing will no doubt join a University staff in due course, but Director felt that intervening period should be spent at NPL.

Kind though it was of Darwin to think of his career, the fact was that the year's wait had achieved nothing. Womersley reported to the same meeting that

> The present position on this project gives no cause for complacency and
> we were probably as far advanced 18 months ago. . . . There are several
> competitors to the ACE machine, and of these, that under construction at
> Cambridge University, under Professor [*sic*] Wilkes, will probably be the
> first in operation.

Coordination and cooperation had long since been replaced by
competition. A few days later Womersley reported[52] to Darwin
with proposals for the building of the machine, which he held to
be 'of supreme importance . . . in the fields of scientific research,
administration, and national defence'. The proposals included
'using as much of Wilkes' development work as is consistent with
our own programming system', and making an approach to F. C.
Williams to request a copy of the machine being built at Manchester.
In this radical modification, or rather abandonment, of the ACE
programme, no attention was paid to the ideas of its originator.
The administrators appeared to regard him as an abstract, almost
anonymous entity.

Alan was not perhaps entirely free from responsibility for this
absence of communication. It was, for instance, rude of him to
postpone for a long time a visit to Wilkes at the Mathematical
Laboratory in Cambridge – a short walk from King's across Market
Square, but not an easy one. As the time for decision approached,
Alan said, 'I really must go and see Wilkes', and then put it off, and
put it off again until the last minute in late May. By that time the
construction of what was to be the EDSAC was in full swing. They
were using mercury delay lines, and by good fortune, Gold had
moved to Cambridge as a research student and supplied Wilkes with
the designs. Instead of trying to work it all out by himself, Wilkes
kept 'religiously' to Gold's specifications and got on with building
them. He had obtained money from the DSIR, the University
Grants Committee, and from J. Lyons and Co. Ltd, reflecting a
very early interest on the part of private enterprise. He was in full
control, without a Womersley or a Darwin to get in the way, and
working much as Alan would have liked to. The barricade between
mathematics and engineering never arose. It was enough to show
the folly of NPL policy, and jealousy would have been a very natural
reaction. Afterwards he meanly said, 'I couldn't listen to a *word* he

said. I was just thinking, how exactly like a *beetle* he looked.' But duty was done.

A few days later, on 28 May, Alan called in at the NPL. It was the sports day. He conferred with Jim Wilkinson, who explained the whole dismal story. Darwin might well have liked a series of papers going out under the NPL *imprimatur*, but Jim Wilkinson agreed that there was nothing in it for Alan. He also thought Alan should stay at Cambridge and return to pure mathematics. He foresaw problems at Manchester.

Manchester University had indeed agreed to create a position for him, and on 21 May the Royal Society agreed that the salary could be paid out of their grant to Newman. A letter of 26 May, which Alan presumably received just before his NPL visit, informed him of this. It was on 28 May that he wrote accepting the appointment, and resigned from the NPL. It was against the 'gentleman's agreement' of the sabbatical, according to which he was supposed to stay for another two years. He had broken an engagement for a second time. Darwin was extremely annoyed with Blackett and Newman. Humpty Dumpty had had a great fall.

Everyone had been slow to adjust to the realities of the post-war period. In expecting the Post Office to cooperate on the delay lines, Alan had been as unrealistic as any of the administrators. And yet it was still extraordinary that as late as May 1948 there was no sign of any start being made on the construction of the control circuits. This failure could not be laid at Alan's door: to see that plans were implemented was the job, and indeed the only *raison d'être*, of the administrators. But perhaps Darwin never really wanted a computer, just as the Admiralty had not really wanted to know where German ships were. The 'support' of Travis and the Ministry of Supply had not in fact made any difference to the bureaucratic inertia. Darwin and Womersley had played at being commissars while Alan remained the humbler worker and peasant.

If he had been given his head, the project might well have gone adrift. He probably underestimated the technical difficulty of running pulses at the rate of a million a second, and overestimated his own engineering knowledge. He would have interfered in too much detail, annoying people who would not like being lectured on their own jobs. He would have had no idea of how to pull off a deal for the right equipment, how to flatter someone into doing the job the way he wanted, or how to play off one expert against another. He had no management skills. But he was not given a chance to make a mess of it for himself, as was his right as the creative worker. And basically his approach was on the right lines, for in the end every successful computer project had to solve the problem of integrating 'mathematical' and 'engineering' skills, which was exactly what he longed to do.

The war had given him a false sense of what was possible. For him, breaking the Enigma was much easier than the problem of dealing with other people, especially with those holding power. During the war his own kind of work had been amplified into something enormously real, but only because other people had done all the work of organising coordination, and because he had Churchill's personal backing. Now there was nobody to do this for him, the NPL administrators not really having tried. A different person could more easily have compromised, and more effectively have fought on for the success of the project. But with him, it was all or nothing. Like his father, he had resigned from the Civil Service when the system did not work for him. Unlike his father, however, he did not moan afterwards. Indeed, he very rarely referred to his time at the NPL. It became another great blank, to add to all the others.

Despite his resignation, and all the embarrassment that surrounded it, he completed a report[53] for the NPL in July and August 1948. Its almost conversational style reflected the discussions he had pursued, many at Bletchley, in advancing the idea of *Intelligent Machinery*. Although nominally the work of his sabbatical year, and written for a hard-line technical establishment, it was really a description of the dream of Bletchley Park, and reviewed in an almost nostalgic way the course of his own life rather than contributing to any practical proposals that the NPL might adopt.

Expanding upon the ideas he had publicly announced in February 1947, this time he identified some of the reasons why the concept of 'intelligent machinery' might be perceived by such as Darwin as a contradiction in terms:

(a) An unwillingness to admit the possibility that mankind can have any rivals in intellectual power. This occurs as much amongst intellectual people as amongst others: they have more to lose. Those who admit the possibility all agree that its realisation would be very disagreeable. The same situation arises in connection with the possibility of our being superseded by some other animal species. This is almost as disagreeable and its theoretical possibility is indisputable.

(b) A religious belief that any attempt to construct such a machine is a sort of Promethean irreverence.

These objections, he wrote, 'being purely emotional, do not really need to be refuted.' They were like Bernard Shaw's complaint against the first Charles Darwin, that his doctrines were 'discouraging'. The point, however, was not to find comfort, but truth.

He next moved to what he considered a less 'emotional', but still erroneous objection:

(c) The very limited character of the machinery which has been used until recent times (e.g. up to 1940). This encouraged the belief that machinery was necessarily limited to extremely straightforward, possibly even to repetitive, jobs. This attitude is very well expressed by Dorothy Sayers (*The Mind of the Maker*, p. 46) '. . . which imagines that God, having created his Universe, has now screwed the cap on His pen, put His feet on the mantelpiece and left the work to get on with itself.'

This, to Dorothy Sayers, made a *reductio ad absurdum* of determinism. How could Lord Peter Wimsey have been laid up for all time in the motions of the particles at the Creation? To Alan all it showed was that there was no word that expressed the 'mechanical' operations of machines with more than trivial logical structure. Dorothy Sayers did not know in 1941 that even the little Enigma machine was sufficiently unpredictable to keep hundreds of people in employment. He was certainly fascinated by the fact that a machine like the Delilah key generator could be perfectly deterministic at one level, while producing something apparently 'random' at another. It gave him a model for reconciling determination and free-will. But this did not go very far. It was the capacity of machines to *learn* that he saw as the crux of the argument. A learning machine would part company altogether with the 'mere machine' of common parlance.

The objection from Gödel's theorem he answered in the same way as he had in 1947, by separating 'intelligence' from 'infallibility'. This time he gave an example of how the intelligent approach could be wrong, and the accurate one stupid:

> It is related that the infant Gauss was asked at school to do the addition $15+18+21 + \ldots + 54$ (or something of the kind) and that he immediately wrote down 483, presumably having calculated it as $(15+54)(54-12)/2.3\ldots$ One can . . . imagine a situation where the children were given a number of additions to do, of which the first 5 were all arithmetic progressions, but the 6th was say $23+34+45\ldots+100+112+122\ldots+199$. Gauss might have given the answer to this as if it were an arithmetic progression, not having noticed that the 9th term was 112 instead of 111. This would be a definite mistake, which the less intelligent children would not have been likely to make.

More pertinent, perhaps, would have been the unmentionable fact that although careless about detail in the cryptanalytic work, he had still been regarded as the brain behind it. Implicitly, this argument appealed to the imitation principle of 'fair play for machines'. This also lay behind his answer to a fifth objection, that

> (e) In so far as a machine can show intelligence this is to be regarded as nothing but a reflection of the intelligence of its creator.

This he described as

> similar to the view that the credit for the discoveries of a pupil should be given to his teacher. In such a case the teacher would be pleased with the success of his methods of education, but would not claim the results themselves unless he had actually communicated them to his pupil. He would certainly have envisaged in very broad outline the sort of thing his pupil might be expected to do, but would not expect to foresee any sort of detail. It is already possible to produce machines where this sort of situation arises in a small degree. One can produce 'paper machines' for playing chess. Playing against such a machine gives a definite feeling that one is pitting one's wits against something alive.

This idea of *teaching* a machine to improve its behaviour into 'intelligence' was the key to most of the positive proposals of this essay. This time he would be putting the imitation principle to a constructive use. The real point of this paper was that he was beginning to think seriously about the nature of human intelligence, and how it resembled, or differed from, that of a computer. More and more clearly as time went on, the computer became the medium for his thought, not about mathematics, but about himself and other people.

He could see two possible lines of development. There was the 'instruction note' view, according to which better and better programs would be written, allowing the machine to take more and more over for itself. He thought this should be done. But his dominant interest now lay in the 'states of mind' approach to 'building a brain'. His guiding idea was that 'the brain must do it somehow', and that it had not become capable of thought by virtue of some higher being writing programs for it. There must be a way in which a machine could learn for itself, according to this line of argument, just as the brain did. He explained his view that 'intelligence' had not been wired into the brain at birth, in a passage showing the influence of his recent research into physiology and psychology:

> Many parts of a man's brain are definite nerve circuits required for quite definite purposes. Examples of these are the 'centres' which control respiration, sneezing, following moving objects with the eyes, etc: all the reflexes proper (not 'conditioned') are due to the activities of these definite structures in the brain. Likewise the apparatus for the more elementary analysis of shapes and sounds probably comes into this category. But the more intellectual activities of the brain are too varied to be managed on this basis. The difference between the languages spoken on the two sides of the Channel is not due to differences in development of the French-speaking and English-speaking parts of the brain. It is due to the linguistic parts having been subjected to different training. We believe then that there are large parts of the brain, chiefly in the cortex, whose function is largely indeterminate. In the infant these parts do not have much effect: the effect they have is uncoordinated. In the adult they have great and purposive effect: the form of this effect depends on the training in childhood. A large remnant of the random behaviour of infancy remains in the adult.
>
> All this suggests that the cortex of the infant is an unorganised machine, which can be organised by suitable interfering training. The organising might result in the modification of the machine into a universal machine or something like it.

Although expressed in more modern terms, reflecting the great debate between the protagonists of nature and nurture, this was little more than could have come from *Natural Wonders*, with its little homilies on the virtues of training the brain in childhood, and how languages and other skills could be best incorporated in the 'remembering place' while the brain was still receptive.

According to this view, therefore, it would be possible to start with an 'unorganised' machine, which he thought of as made up in a rather random way from neuron-like components, and then 'teach' it how to behave:

> . . . by applying appropriate interference, mimicking education, we should hope to modify the machine until it could be relied on to produce definite reactions to certain commands.

The education that he had in mind was of the public-school variety, by the carrot and stick that the Conservative party was currently accusing Attlee of taking away from the British donkey worker:

> . . . The training of the human child depends largely on a system of rewards and punishments, and this suggests that it ought to be possible to carry through the organising with only two interfering inputs, one for 'pleasure' or 'reward' (R) and the other for 'pain' or 'punishment' (P). One can devise a large number of such 'pleasure-pain' systems. . . . Pleasure interference has a tendency to fix the character, i.e. towards preventing it changing, whereas pain stimuli tend to disrupt the character, causing features which had become fixed to change, or to become again subject to random variation. . . . It is intended that pain stimuli occur when the machine's behaviour is wrong, pleasure stimuli when it is particularly right. With appropriate stimuli on these lines, judiciously operated by the 'teacher', one may hope that the 'character' will converge towards the one desired, i.e. that wrong behaviour will tend to become rare.

If the object were simply to produce a universal machine, it would be better to design and build it directly. The point, however, was that the machine thus educated would not only acquire the capacity to carry out complicated instructions, which Alan described as like a person who 'would have no common sense, and would obey the most ridiculous orders unflinchingly'. It would not only be able to do its duty, but would have that elusive 'initiative' that characterised intelligence. Nowell Smith, worrying about the development of independence of character within a system of mere routine, could not better have formulated the problem:

> If the untrained infant's mind is to become an intelligent one, it must acquire both discipline and initiative. So far we have been considering only discipline. To convert a brain or machine into a universal machine is the extremest form of discipline. Without something of this kind one cannot set up proper communication. But discipline is certainly not enough in itself to produce intelligence. That which is required in addition we call initiative. This statement will have to serve as a definition. Our task is to discover the nature of this residue as it occurs in man, and to try and copy it in machines.

Alan would have liked the idea of training a boy machine to take the initiative. He did work out an example of modifying a 'paper machine' into a universal machine by these means, but decided it was a 'cheat', because his method amounted to working out the exact internal structure, the 'character' of the machine, and then correcting it – more like codebreaking than teaching. It was all very laborious on paper, and he was eager to take the next step:

> I feel that more should be done on these lines. I would like to investigate other types of unorganised machine, and also to try out organising methods that would be more nearly analogous to our 'methods of education'. I made a start on the latter but found the work altogether too laborious at present. When some electronic machines are in actual operation I hope that they will make this more feasible. It should be easy to make a model of any particular machine that one wishes to work on within such a [computer]* instead of having to work with a paper machine as at present. If also one decided on quite definite 'teaching policies' these could also be programmed into the machine. One would then allow the whole system to run for an appreciable period, and then break in as a kind of 'inspector of schools' and see what progress had been made.

It was a happy thought that, like a public school, the machine could grind its way along quite deterministically, but without anyone knowing what was going on inside. They would see only the end

---

* The actual words he used were 'Universal Practical Computing Machine'.

products. There was a distinctly behaviourist flavour to all this talk of pain and pleasure buttons, but his wry use of the words 'training', 'discipline', 'character' and 'initiative' showed how this was the behaviourism of Sherborne School.

More precisely, it was the *official* description of the school process, albeit presented as rather a joke. It bore little relationship to his own mental growth. No one had pressed any pleasure buttons to reward *his* initiative; precious few pleasure buttons at all, while the pain had been dispensed freely in order to enforce patterns of behaviour that had nothing to do with intellectual advance. The only hint of contact with his own experience was the remark that discipline was necessary for the sake of *communication*, for certainly he had to be pushed into conventional communication in order to advance. Yet even there, it had not been jabs of pain and pleasure that had stimulated his willingness to communicate, but the aura – was it pain? was it pleasure? – that surrounded the figure of Christopher Morcom. As Victor Beuttell had often said to him, it was a mystery from where his 'intelligence' derived, for no one had been able to *teach* him mathematics.

Wittgenstein also liked to talk about learning and teaching. But *his* ideas derived not from the example of an English public school, but from his experience in an Austrian elementary school, where he had explicitly tried to get away from the repressive rote-learning that Alan had endured. By this time Alan had compared his school experience with Robin, who had had a much happier time at Abbotsholme, the progressive boys' boarding school where Edward Carpenter's ideas had enjoyed an influence, and 'Dear Love of Comrades' was the school song. Alan, speaking to Robin of Sherborne, had said: 'The great thing about a public school education is that afterwards, however miserable you are, you know it can never be quite so bad again'.* But there was no trace of his criticism of the Sherborne process in this essay, except inasmuch as he was enjoying a sally at the pompous old masters by talking of replacing them by machines. There was a gap here, a certain lack of seriousness. It was rather like Samuel Butler in *Erewhon*, wittily transposing the values attached to 'sin' and to 'sickness' in order

---

* A false prophecy.

to tease the official Victorian mentality, yet never questioning that beatings would be the appropriate 'treatment' for 'sin'.

But in other ways, he certainly did recognise that his machine model of the brain was deprived of some very significant features of human reality. This was where he began to question the isolated puzzle-solver as a model for the understanding of Mind:

> . . . in so far as a man is a machine he is one that is subject to very much interference. In fact interference will be the rule rather than the exception. He is in frequent communication with other men, and is continually receiving visual and other stimuli which themselves constitute a form of interference. It will only be when the man is 'concentrating' with a view to eliminating these stimuli or 'distractions' that he approximates a machine without interference . . . although a man when concentrating may behave like a machine without interference, his behaviour when concentrating is largely determined by the way he has been conditioned by previous interference.

In a soaring flight of imagination, he supposed it possible to equip a machine with 'television cameras, microphones, loudspeakers, wheels and "handling servo-mechanisms" as well as some sort of "electronic brain"'. Tongue in cheek, he proposed that it should 'roam the countryside' so that it 'should have a chance of finding things out for itself', on the human analogy, and perhaps thinking of his own country walks at Bletchley, where his odd behaviour had attracted the spy-conscious citizen's suspicion. But he admitted that even so well-equipped a robot would still 'have no contact with food, sex, sport, and many other things of interest to the human being' – and certainly of interest to Alan Turing. His conclusion was that it was necessary to investigate what

> can be done with a 'brain' which is more or less without a body, providing at most organs of sight speech and hearing. We are then faced with the problem of finding suitable branches of thought for the machine to exercise its powers in.

The suggestions he made were simply the activities that had been pursued on and off duty in Hut 8 and Hut 4, rather surprisingly brought into the open:

  (i) Various games e.g. chess, noughts and crosses, bridge, poker

 (ii) The learning of languages

(iii) Translation of languages

(iv) Cryptography*

 (v) Mathematics.

Of these (i), (iv) and to a lesser extent (iii) and (v) are good in that they require little contact with the outside world. For instance in order that the machine should be able to play chess its only organs need be 'eyes' capable of distinguishing the various positions on a specially made board, and means for announcing its own moves. Mathematics should preferably be restricted to branches where diagrams are not much used. Of the above possible fields the learning of languages would be the most impressive, since it is the most human of these activities. This field seems however to depend rather too much on sense organs and locomotion to be feasible.

The field of cryptography will perhaps be the most rewarding. There is a remarkably close parallel between the problems of the physicist and those of the cryptographer. The system on which a message is enciphered corresponds to the laws of the universe, the intercepted messages to the evidence available, the keys for a day or a message to important constants which have to be determined. The correspondence is very close, but the subject matter of cryptography is very easily dealt with by discrete machinery, physics not so easily.

There was more to *Intelligent Machinery* than this. One feature was that he laid down definitions of what was meant by 'machine', in such a way that it connected the 1936 Turing machine with the real world. He distinguished first:

> *'Discrete' and 'Continuous' machinery*. We may call a machine 'discrete' when it is natural to describe its possible states as a discrete set. . . . The states of 'continuous' machinery on the other hand form a continuous manifold . . . . All machinery can be regarded as continuous, but when it is possible to regard it as discrete it is usually best to do so.

and then:

> *'Controlling' and 'Active' machinery*. Machinery may be described as 'controlling'

---

* He meant what has here throughout been called 'cryptanalysis', as the following passage shows.

if it only deals with information. In practice this condition is much the same as saying that the magnitude of the machine's effects may be as small as we please. . . . 'Active' machinery is intended to produce some definite physical effect.

He then gave examples:

A Bulldozer	Continuous Active
A Telephone	Continuous Controlling
A Brunsviga	Discrete Controlling
A Brain is probably	Continuous Controlling, but is very similar to much discrete machinery
The ENIAC, ACE, etc.	Discrete Controlling
A Differential Analyser	Continuous Controlling

A 'Brunsviga' was a standard make of desk calculator, and the point was that such a machine, like an Enigma, a Bombe, a Colossus, the ENIAC or the planned ACE, was best *regarded as* a 'controlling' device. In practice it would have a physical embodiment, but the nature of the embodiment, and the magnitude of its physical effects, were essentially irrelevant. The Turing machine was the abstract version of such a 'discrete controlling' machine, and the cipher machines and decipherment machines were physical versions of them. They had taken up much of his working life. And the fundamental thesis of *Intelligent Machinery* was that the brain could also be 'best regarded as' a machine of this kind.

The paper also included a short calculation which bridged the two descriptions of a machine such as a computer, the logical description and the physical description. He showed that in a job taking more than $10^{10^{17}}$ steps, a physical storage mechanism would be virtually certain to jump into the 'wrong' discrete state, because of the ever-present effects of random thermal noise. This was hardly a practical constraint. He might have made a similar calculation regarding the effect of quantum indeterminacy, and the upshot would have been the same. The determinism of the logical machine, although it could never be rendered with absolute perfection, was still effectively independent of all the 'Jabberwocky' of physics. This part of the paper integrated his several interests in logic and physics, mapped out where his own work stood within a wider framework, and summed up a long chapter of unfulfilled ambitions.

A final section suggested approaches to 'intelligent machinery' which were not based on this crude 'teaching', but upon his real experience as a pure mathematician. He considered the process of transforming problems from one formulation into another, solving a problem by proving a theorem in some other logical system, and translating the result back into the original form. This corresponded closely with the real work of mathematics, that of detecting analogies, and searching for openings towards a proof within some framework of ideas. 'Further research into intelligence of machinery will probably be very greatly concerned with "searches" of this kind,' he wrote. 'We may perhaps call such searches "intellectual searches". They might very briefly be defined as "searches carried out by brains for combinations with particular properties".' Of course, this was not exactly unrelated to the task of cryptanalysis, that of finding patterns in the apparently patternless.

He drew a Darwinian parallel:

It may be of interest to mention two other kinds of search in this connection. There is the genetical or evolutionary search by which a combination of genes is looked for, the criterion being survival value. The remarkable success of this search confirms to some extent the idea that intellectual activity consists mainly of various kinds of search.

The remaining form of search is what I should like to call the 'cultural search'. As I have mentioned, the isolated man does not develop any intellectual power. It is necessary for him to be immersed in an environment of other men, whose techniques he absorbs during the first twenty years of his life. He may then perhaps do a little research of his own and make a very few discoveries which are passed on to other men. From this point of view the search for new techniques must be regarded as carried out by the human community as a whole, rather than by individuals.

This was a rare revelation of his self-perception. It was a dignified and generous response to the lessons of 1937 and 1945, when others had come forward with ideas equivalent to his own – so much more realistic than the usual worrying about 'priority', with its implicit fear of cheating and copying, and so free of the male competitiveness which was by 1948 becoming more and more evident in science.[54] He never claimed more than that 'some years ago I made an investigation into what could be done by a rule-of-thumb process,' when referring to his own part. And of course this had been yet another of the lessons of 1941, that it was the search of the whole Bletchley community that mattered so much. But that very fact might perhaps have made him wonder more as to whether the operation of the brain 'without interference' was really the right way in which to focus attention. The very existence of these social or cultural levels of description was an indication that individual 'intelligence' was not the whole story. This question was not developed in this essay. Meanwhile, his ability to stand back from the individual struggle was certainly required in adjusting himself to work with the rival computer that had been developed at Manchester.

He wrote off to F. C. Williams for information, and received a reply probably on 8 July. By this time, the fact was that they had already, on 21 June 1948, successfully run the first program on the first working stored program electronic digital computer in the world. Darwin had talked about 'formidable mathematical difficulties', but at Manchester they had just got on with it, and produced a computer behind Darwin's back. It used, for storage, the cathode ray tube that Williams had developed, and at this point the total store consisted of just 1024 binary digits stored on one tube. Alan drew attention to this figure in a table of 'memory capacities' in this report:

Brunsviga	90
ENIAC without cards and with fixed programme	600
ENIAC with cards	∞^*
ACE as proposed	60,000
Manchester machine as actually working (817/48)	1,100

It was a pointed contrast between one machine still merely 'proposed', and another that actually worked. But the figures also pointed to the fact that F. C. Williams had pursued his project in a more modest way. The Manchester computer was small, and might even be called small-minded. But it was the first embodiment of a Universal Turing Machine, albeit with a very short 'tape'. Alan wrote out a little routine[55] to perform long division, and posted it north immediately.

---

* *i.e.* 'infinity'. This was mathematical shorthand for the fact that a Babbage-like machine could in principle be fed with an *unlimited* quantity of data and instructions from an external source – the price being, of course, an unlimited time delay.

Jack Good and Donald Michie looked in on Alan at King's, and rather annoyed him by peeping at the uncompleted version of *Intelligent Machinery* while he was out. Afterwards, walking along King's Parade, Alan dropped a very deliberate remark to Jack Good about a boy in Paris.* His drift got across to Jack, who had not known before. They were also in correspondence during this summer period. Jack wrote:

25 July 48

Dear Prof,

When I was last in Oxford I met a lecturer in physiology who said that he thought the number of neurons in the brain was only about two million. This seems amazingly little to me even allowing [for] the fact that the number of processes from each neuron is something like 40. I wonder if you could tell me the right answer, with or without a reference.

I understand that by next October we'll have swapped towns, Judging by the international situation I think you've had the better of the bargain. . . .

How near were you to getting into the Olympics?

Jack was leaving his Manchester lectureship to join the branch of the Civil Service now known as the Government Communications Headquarters, and located at Eastcote, in north-west London. As for the international situation, the new lines were rapidly hardening. Yugoslavia had been expelled from the Cominform – a break which led Robin, like many other sympathisers with the pre-war USSR, to move much further away from the Communist party. The airlift to West Berlin was under way, and for the first time there was serious talk of war with Russia.

The US Air Force had begun its temporary stay on British soil, and Americans were overtaking plucky British losers in the Empire Stadium, where a scraggy, rationed Britain was hosting the Olympic Games.† Alan went with Anderson, an acquaintance from the Hare and Hounds Club, and they saw the Czech athlete Zátopek win the 10,000 metre race on 30 July. The Marathon was won for Argentina but in a time still only seventeen minutes better than Alan's. He replied to Jack:

---

* A remark very characteristic of his post-war life – but there is no evidence as to how this particular contact had come about.

† As *The Times* put it on 9 August, the public would 'put the true high value on the near misses of the British men and women.'

Dear Jack,

I have repeatedly looked in books on neurology for the very important number N you asked about, and never found any figure offered. My own estimate is $3.10^8 < N < 3.10^9$. It is based on the diagram p. 207 of latest Starling which refers to a mouse together with w[eigh]t of average brain (3 lb). . . . I have asked many phsyiologists myself and got anwers from 10^7 to 10^{11}.

I've had something wrong with my leg for some months, so wasn't able to run in any Marathons this season.

Yours Prof

An injury to his hip had put paid to his chances for the Olympic marathon team, for which otherwise he might have qualified, and to his regret, prevented any further development in serious long-distance running.

Alan sent off another routine, for factorising numbers, to Manchester on 2 August. Then he went on holiday with Neville in Switzerland. It was the first escape from austerity England, and they could hardly believe the fresh country foods. The trip was made on the travel allowance of £25, in the form of five crisp five-pound notes. They did it by cycling and staying in youth hostels: they trod glaciers and scaled mountains and had the usual smouldering arguments of people on holiday together, as when Alan let his bicycle break down through inattention, or was keen on another young man in the hostel. It was not quite E. M. Forster's greenwood, but as near as he had ever approached it.

The summer continued with a week in the Lake District at Pigou's cottage, with Peter Matthews. Pigou was very keen on mountaineering, and even more, with a pre-Carpenter innocence, on Wilfrid Noyce the young mountaineer.* Alan took care to practise rock-climbing on the King's College gate with Peter Matthews before they went. It was like some 1890s Cambridge reading party, with old Pigou clocking up the times and the victors at chess. He had a collection of First World War medals which, though a pacifist, he had been awarded for ambulance service, and used to award them after a farewell dinner to whoever had done best on the slopes. Alan did some easy climbing, but mostly trotted round Buttermere in short shorts. From Jack Good:

---

* As it happened, Alan and his mother had seen Noyce as a boy, passing him when walking in the Welsh hills in 1927.

16 Sep 48

Dear Prof,

Pardon the use of the typewriter: I have come to prefer discrete machines to continuous ones.

When I was in Cambridge recently I hunted unsuccessfully . . . for an estimate . . . of N, the number of neurons in the human brain. Soon after this Donald succeeded in finding a reference. He tells me that . . . N=10000000000 roughly.

I visited Oxford last week-end. Donald showed me a 'chess machine' invented by Shaun [Wylie] and himself. It suffers from the very serious disadvantage that it does not analyse more than one move ahead. I am convinced that such a machine would play a very poor game, however accurately it scored the position with respect to matter and space. In fact it could easily be beaten by playing 'psychologically' i.e. by taking into account the main weakness of the machine. . . .

When in Oxford I succeeded in hypnotising Donald . . . . Would you agree that a very typical property of the brain is the ability to think in analogies? This means taking only a part of the evidence into account. . . .

Do you know of any reference to Russian electronic computers? . . .

Donald Michie was now studying physiology at Oxford. He had followed up their Bletchley speculations by teaming up with Shaun Wylie to devise a chess program they called the *Machiavelli*. Meanwhile Alan and David Champernowne had worked out one they called the *Turochamp*.[56] It followed the minimax system, and the important idea of pursuing chains of captures until no more could be made. It had a scoring system in which pawn mobility, castling, and getting a rook on to the seventh rank were included as well as captures. None of this went much beyond what Alan had discussed back in 1941 with Jack Good, or indeed with Champ in 1944. Going for a walk, probably at Christmas that year, they had made a bet on whether a machine could beat Champ himself at chess by 1957. The odds were put at 13 to 10 in favour of the machine succeeding. The *Turochamp* certainly did not reach this standard, although it beat his wife, a beginner at chess. It was not taken at all seriously, or written out in detail. But it would have been a system of this kind which gave Alan 'a sense of pitting one's wits against something', as he wrote in *Intelligent Machinery*. Champ also took on the system for poker that

Alan had more carefully worked out, and had the pleasure of beating it by sheer good luck. Alan replied to Jack:

Sept 18 48

Dear Morcom,

I am glad to hear my estimate of no. of neurons is not too essentially wrong.

The chess machine designed by Champ and myself is rather on your lines. Unfortunately we made no definite record of what it was, but I am going to write one down definitely in the next few days with a view to playing the Shaun-Michie machine.

To a large extent I agree with you about 'thinking in analogies', but I do not think of the brain as 'searching for analogies' so much as having analogies forced upon it by its own limitations . . .

The report was handed in. Mike Woodger was very excited by the prospects it opened up, and gladly drew the diagrams neatly for the typed version. Darwin was less impressed, probably highly embarrassed by the references to Dorothy Sayers, God, and robots taking country walks. At the Executive Committee meeting on 28 September he explained sniffily that 'Dr Turing had now produced a report which, although not suitable for publication, demonstrated that during his stay there he had been engaged in rather fundamental studies.' The unsuitable paper disappeared into the NPL files. Ironically, it was on 20 September 1948 that von Neumann gave a first published lecture[57] on the 'theory of automata' – in effect, the theory of discrete controlling machines – in which he drew attention, after eleven years, to the fundamental importance of the Universal Turing Machine.

Robin had rented Blackett's holiday home in Wales on occasion, and did so again this year. By inviting Alan, he made possible a third holiday before the summer was out. Another of the party was Nicholas Furbank, the friend of E. M. Forster, who had lately been writing a book on Samuel Butler. This too was rather like an old-style Cambridge reading party – they were amused by the resemblance – with organised walks, and funny nicknames, and reading Thomas Love Peacock's gothic *Melincourt* aloud.

Alan seemed very happy. They played rationalist Twenty Questions as they filed along the hill paths and old railway tunnels.

Alan developed a theory of how to choose the next question so as to maximise the expected weight of evidence of the answer. He also recounted quaint tales of the Pigou regime, whose antiquated brand of hearty male misogyny left him bemused. 'The standard at the Pigou cottage is very *high*,' he said. 'I ran all round Buttermere faster than Noel-Baker in '28, and I only got a *bronze*.' One day they took a taxi and a bus at dawn for an expedition to the real mountain slopes of the Snowdon horseshoe, where Nick Furbank was suitably terrified and went on all fours along the narrow ridge of Cribgoch, but Alan strode on in dogged Turing fashion, as twenty years before, but with friends at last.

Down from the mountain tops, it was time to pack up. The suitcase with the parts for the zeta-function machine was still in his room, along with the star globe and Christopher's picture. He kept as souvenirs some of the gear wheels that had been cut, but gave the rest to Peter Matthews to sell for scrap. Alan was rather disappointed with the price they fetched.

There was another throwback to 1939, for Bob was getting married. He had settled in Manchester, first doing some wartime cotton research, and then becoming an industrial chemist. Alan went to Cumberland for the wedding on 2 October, giving the couple a generous present. Then he went to Manchester himself, to take up his new life. His plans had been wrecked, but then the era of planning, if ever it had existed at all, was now over. The government would do well if it could manage to look one move ahead. Alan Turing likewise would have to make the best of a bad job.

7 The Greenwood Tree

Unseen buds, infinite, hidden well,
Under the snow and ice, under the darkness, in every square
    or cubic inch,
Germinal, exquisite, in delicate lace, microscopic, unborn,
Like babes in wombs, latent, folded, compact, sleeping;
Billions of billions, and trillions of trillions of them waiting,
(On earth and in the sea – the universe – the stars there in the
    heavens,)
Urging slowly, surely forward, forming endless,
And waiting ever more, forever more behind.

What Alan Turing did not know was that a number of changes had been made at Manchester University since his appointment in May. He had been created 'Deputy Director' of the 'Royal Society Computing Laboratory' when Newman was supposed to be directing it, and the Royal Society funding it. But by October it had become clear that F. C. Williams had need neither of a 'Director' nor of the Royal Society.

In the development of electronic hardware, the important factor had been that Williams's ingenuity was backed by a cosy relationship with TRE, which allowed him to draw upon their supplies, and to have two assistants seconded from the government establishment. One of these was a young engineer with a Cambridge mathematics degree, T. Kilburn. The second, after a short interval, came to be G. C. Tootill, another young TRE man from the same wartime Cambridge year.

As for the development of a logical design, the first step had been taken by Newman. He had explained the principle of storing numbers and instructions, which according to Williams[1] 'took all of half an hour', favouring the von Neumann type of design.* In

---

* Newman had written[2] to von Neumann on 17 June 1946 that he was 'at present grappling with Turing's report, which I find a good deal less readable than yours.' He had also spent a term at Princeton in late 1946, and discussed computers with von Neumann.

late 1947 the plans had rapidly evolved in the hands of Williams and his two assistants. They were not detained by the prospect of 'formidable mathematical difficulties', but had pressed on, as Williams put it,* 'without stopping to think too much'. The result was the tiny computer of whose existence Alan had learned in the summer, whose store consisted of just one cathode ray tube.

The advantage of the cathode ray tube over the delay line was that, in both senses, it eliminated delay. It was essentially an ordinary piece of equipment, not requiring precision engineering as did the mercury delay line, and could be taken 'off the shelf'. In practice this virtue was tempered by the fact that most tubes contained too many impurities in the screen to be used, but its 'home-made' quality was still of value in getting the project off the ground. In operation it was not particularly fast – indeed it would take ten microseconds to read a digit as compared with the one microsecond intended for the ACE – but this was compensated by the fact that the information stored on the tube was directly available, without the long period of waiting for a pulse to emerge from a delay line. Continuing his 'papyrus scroll' analogy, Alan compared it[3] to 'a number of sheets of paper exposed to the light on a table, so that any particular word or symbol becomes visible as soon as the eye focusses on it.'

They had been able to store 2048 spots on the tube by the principle of regenerating them periodically, but in the end had settled on using just 1024, arranged in thirty-two 'lines' each of thirty-two spots. Each line would represent either an instruction or a number. A second cathode ray tube served as the logical control, storing the instruction currently being executed, and the address of that instruction. A third acted as the accumulator, the shunting station for the arithmetical operations. It was a 'one-address' system, so that each act of shunting in, or of shunting out, constituted a full instruction – an arrangement entirely different to that envisaged for the ACE. Arithmetic was, however, reduced to the barest minimum for the sake of demonstrating that it was possible at all – the operations of copying and subtraction, together with a

---

* In 1949, on a visit to the United States, Williams scandalised the employees of IBM, whose corporate motto THINK was everywhere in evidence, with this analysis of how he had succeeded where they had not.

simple form of conditional branching. It amounted to far less than Huskey's 'Test Assembly' would have done, had that abortive NPL effort been completed. Physically, the Manchester computer was embodied in a straggly jumble of racks and valves and wires,* with three screens glowing in the gloom of a room with dirty brown tiles which Williams was fond of describing as 'late lavatorial' in style.

It was, in fact, the most obvious feature of the cathode ray tube storage system that one could actually *see* the numbers and instructions held in the machine, as bright spots on the three monitor tubes. Indeed, at this stage it was essential to see them, for there was no other output mechanism. Nor was there any form of input but that of hand switches, used to insert digits one at a time into the storage tube.

But this was enough. As Williams described[4] the day of triumph:

> When first built, a program was laboriously inserted and the start switch pressed. Immediately the spots on the display tube entered a mad dance.
>
> In early trials it was a dance of death leading to no useful result, and what was even worse, without yielding any clue as to what was wrong. But one day it stopped and there, shining brightly in the expected place, was the expected answer.

This happened on 21 June 1948, and the world's first working program on an electronic stored-program computer, to find the highest factor of an integer by crude brute force trial, had been written by Kilburn.

> Nothing was ever the same again. We knew that only time and effort were needed to make a machine of meaningful size. We doubled our effort immediately by taking on a second technician.

It was in these circumstances that Kilburn mentioned to Tootill a few days later that 'there's a chap called Turing coming here, he's written a program.' Williams knew about Alan because of his dealings with the NPL. Kilburn vaguely knew of him. Tootill,

---

* Yet already the end of the vacuum tube was in sight. A final letter from Jack Good to 'Prof', dated 3 October 1948, continued their discussion of the brain, and also asked: 'Have you heard of the TRANSISTOR (or Transitor)? It is a small crystal alleged to perform "nearly all the functions of a vacuum tube". It might easily be the biggest thing since the war. Is England going to get a look-in?'

who had not heard of him at all, worked on the program. He was astonished (and naturally, smugly pleased) to discover it not only to be inefficient but to contain an error.

At Manchester they had a machine which actually worked, and this simple fact counted for more than did ingenious or impressive plans. It meant that while Alan had been away on his holidays, political considerations had transformed the Manchester set-up. Already in July, Sir Henry Tizard, then Chief Scientific Adviser to the Ministry of Defence, had seen the machine and considered it[5]

> of national importance that the development should go on as speedily as possible, so as to maintain the lead which this country has thus acquired in the field of big computing machines, in spite of the large amount of effort and material that have been put into similar projects in America. He promised full support both in supply of material and in obtaining necessary priorities.

To the engineers it was a gratifying verdict, but it was one which had no connection whatever with the 'fundamental research in mathematics' that was Newman's object, and the purpose of the Royal Society grant.

It was not surprising that Tizard should take this view. In 1948 (although he changed his mind in 1949, saying that Britain should admit it was no longer a Great Power), he supported the policy of building a British atomic bomb. In August 1946 the MacMahon Act had prevented the United States government from sharing atomic knowledge with Britain, and at the beginning of 1947, the British government had, in great secrecy, decided upon an independent development. The government's interest in a working electronic computer was then refreshed by at least two other experts: Sir David Brunt the meteorologist and successor to R. V. Jones in Scientific Intelligence, and Sir Ben Lockspeiser the Government Chief Scientist. A few days after Lockspeiser's visit, the Ministry of Supply placed an order with Ferranti Ltd, the Manchester-based weapons and electronics manufacturers, which according to the letter dated 26 October 1948 was simply 'to construct an electronic calculating machine to the instructions of Professor F. C. Williams.'

About £100,000 was thus spent by the government, whose rapid, almost panicky move made a strong contrast with the stately progress of Planned Science at the NPL. It had more to do with events in Berlin and Prague than with the intentions of the Royal Society. (It was in that same month of October 1948 that the demolition of air-raid shelters was suddenly stopped.) It certainly had nothing to do with Alan, the pawn in the Great Game. For that matter the *carte blanche* contract made no reference to Newman or Blackett. Newman's motives had been entirely those of a pure mathematician, one who wistfully thought of what the talent at Bletchley could have achieved had it been applied to his subject. He had originally wanted to buy a machine and get on with the mathematics, and by this time had realised that it could not be so; the development of the hardware was going to be a dominating feature, and his interest had accordingly waned. He therefore did not object to the project being taken away. Blackett, however, was distinctly annoyed, perhaps the more so as he opposed the atomic weapon development.

But even apart from the politics of the machine, Alan had come too late to direct its development. Already the important decision had been taken to adopt, for use as a large, slow, backing store, a rotating magnetic drum such as A. Booth of Birkbeck College, London, had developed for use with a relay calculator. With digits stored on tracks around the drum to be read off by a head, this was equivalent to providing a large number of slow, cheap, delay lines for the storage of data and instructions not immediately in use. Another innovation in the design, a modification originally suggested by Newman, was that of the 'B-tube'. (It was so called because the arithmetic and control tubes were naturally 'A' and 'C' tubes respectively.) This additional cathode ray tube had the property of modifying the instructions held in the control; in particular it could be used when working along a sequence of numbers, in such a way that the idea of the 'next' number did not have to be rendered into laborious programming.* As such it was contrary to the general policy that Alan had pursued on the ACE design, that of using instructions rather than hardware as far as possible.

But more generally, the design and development had all been decided by others. They called it the 'baby machine' – but it was

---

* This invention, later known as an 'index register', was of considerable importance in the future development of computer hardware design.

someone else's baby. Williams had turned the tables, for while Darwin had hoped for him to build to Alan Turing's instructions, now Alan had the task of making Williams' machine work. With the best will in the world, there was room for conflict; the more so as the engineers had no intention of being 'directed' by anyone. The line between 'mathematicians' and 'engineers' was demarcated very clearly, and if not quite an Iron Curtain, it was a barrier as awkward as the MacMahon Act. This would never be Alan Turing's machine, as the ACE would have been, and correspondingly, he withdrew as much as possible from any administrative responsibility for it. But he could foster it, and there was the prospect of using it. His position also attracted the salary of some £1200 per annum (increased to £1400 in June 1949), and very considerable freedom.

So he stuck with Manchester, not as a 'Deputy Director' but as a freelance 'Prof' (as people still called him, perhaps to the slight annoyance of the true professors). There was a conventional sense in which Manchester, compared with Cambridge, was a come-down. It was largely the technical university of the North, producing doctors and engineers, rather than abstract ideas. However, Manchester prided itself on its standards, and Newman had built up a mathematics department which rivalled that of Cambridge. So although Alan was a bigger fish in a smaller pond, he was not a fish out of water. Certainly the physical setting of the university was grim. Its late Victorian gothic buildings, black with soot from the first industrial revolution, faced across the tram-tracks of Oxford Road on to the Temperance Society and expanses of slumland, whose holes and shored-up corners marked where the bombers had got through. Alan also commented on the low standard of male physique, not surprising in a city still recovering from the Depression. But the industrial landscape had some pleasures too: when Malcolm MacPhail from Princeton days visited in 1950, he was taken to see where the Duke of Bridgwater's canal crossed the Manchester Ship Canal, having first been challenged to work out how this was achieved.

Like Princeton it was a place of exile, but without the compensations of American largesse. Manchester University also resembled the American milieu in that it represented a bastion of respectability, its Nonconformist northern middle class being less accommodating to human diversity than was (in private) the more privileged

Cambridge establishment. But Manchester had a spark of generosity in its city life, rather than the parochial attitudes of the small town. It had the liberal *Manchester Guardian* which, along with *The Observer*, was Alan's newspaper. And perhaps he found something satisfying about working in ordinary industrial England, without the affectations and traditional rituals that went with Cambridge life.

If Alan had really objected to being left out on a limb, he could have resigned and returned to King's, of which he remained a Fellow.* At some point there was talk of him taking a position at Nancy in France (perhaps through Wiener's connections with its premier school of mathematics), but this came to nothing except the obvious joke with Norman Routledge of finding Nancy boys. He could always have found an American position – but that would have gone quite against his grain. Instead, he made the best of what had been his own decision. To many at Manchester, Alan Turing was something of an embarrassment, foisted upon them, but they would have to put up with him.

In March 1949 he wrote to Fred Clayton:

> I am getting used to this part of the world, but still find Manchester rather mucky. I avoid going there more than I can avoid.

Instead, he worked or pottered around at home. Most of the university staff lived in the suburb of Victoria Park, but Alan lived further out in a large lodging house in Nursery Avenue, Hale. ('Only one large bed – but I think you will find it quite safe,' he described it to Fred, inviting him to stay.) It was on the very edge of the built-up area so that he could go running in the Cheshire countryside, far from the dark satanic mills and from the tensions of the university. He retained his connection with the Walton Athletic Club, and sometimes ran for them, as in the London to Brighton relay race on 1 April 1950.† His competitive days were, however, coming to a close and he ran more as a solitary exercise. Sometimes he ran into Manchester, though more often he cycled through the scrubby suburbs to work, cutting a comic figure in a yellow oilskin

---

* The arrangement was now that he could receive a fellowship stipend only for a quarter in which he had spent twenty-five nights in residence – a condition fulfilled during August.

† Running in a team with Christopher Chataway.

and hat when it rained. Later he added a small motor to his bicycle, but he never acquired a car: 'I might suddenly go mad and crash,' he told Don Bayley rather dramatically. He had not done too well at Princeton with the car, and probably tended to daydream with mathematical thoughts in a dangerous way. He preferred in any case to use his own steam.

He cared little for the Victoria University, as it was officially called, taking what he found relevant and ignoring the rest. For him there were those who were serious, in his own sense, and those who were not, and he wasted no time on the latter. This had little to do with formal positions. In September 1947, just as Alan effectively left the NPL, they had appointed a young engineer, E. A. Newman, who did have knowledge of pulse electronics from his experience of the H2S airborne radar system. Ted Newman, also a strong runner, used to go to Manchester to see Alan every month or so. Beside training together, they would argue for hours about the idea of intelligent machinery. In contrast, Alan would repulse abruptly any kind of ingratiating shop talk from those who might well be more academically qualified.

People did not have a second chance. If they tuned into a Turing wavelength, they would receive hours of attention at full blast, with an almost embarrassing intensity. But with a wobble of frequency, a hint of being judged by conventional or secondhand standards, the light went out, the door banged. It was all or nothing, like the pulses of the computer. He would walk away without a word of apology, when bored. And in his hatred of pretence and pretentiousness he must have thrown away many sincerely meant, but too tentative, approaches. In 1936 he had felt rebuffed by Hardy, but now it was he who obliged others to meet him on his own terms alone.

'Boyish' or 'schoolboyish' was the word that still came to many lips to describe the immediacy of his presence, his shaggy, dog-eared, larger-than-life appearance, and his ability to see that the Emperor had no clothes. His role at Manchester, indeed, was sometimes seen as that of Newman's *enfant terrible*. He had little social life at Manchester; it would have required too much compromise. Apart from a few visits to Bob and his wife, now living in the Cheshire suburbs, it was the Newman home, a piece of Cambridge in the North, that gave him a welcome. They came to be on first-name terms, something unusual for Max Newman, who cut a distinctly magisterial figure in his department. His wife was the writer Lyn Irvine, who first came across Alan when he stayed with them at Criccieth for Easter 1949, amazing them with long runs round Cardigan Bay. She was struck by Alan,[6] with 'his off-hand manners and his long silences – silences finally torn up by his shrill stammer and the crowing laugh which told upon the nerves even of his friends'; there was his 'strange way of not meeting the eye' and of 'sidling out of the door with a brusque and offhand word of thanks.'

Nor did he compromise with Manchester society by associating with the small homosexual set centred on the university, the BBC and the *Manchester Guardian*. In this respect life centred still on Cambridge. The exile in Manchester meant in particular a separation from Neville, whom over the next two years he would visit at Cambridge every few weeks. Neville was taking a two-year postgraduate course in statistics. At Easter 1949 they had another short holiday together in France, cycling and visiting the Lascaux Caves. (The prehistoric paintings rather suited Alan, who always wanted to draw nature from scratch himself.) Alan also spent the August of each year back for the long vacation in King's, rather as he had in 1937.

So King's retained its protective role, and Robin in particular was the White Knight in the forest, as the most helpful character in the story. In other ways, Robin was not the White Knight at all, being rather dashing and energetic. Later he acquired a powerful motor-cycle and a full set of black leathers, and sometimes took Alan for rides in the Peak District. Alan told his friends about the Princeton treasure hunts, and he, with Robin, Nick Furbank and Keith Roberts, organised several of them over the next few years. Alan would run round in search of the clues, while the others would cycle. Once Noel Annan joined in, and made a great hit by producing a bottle of champagne to match a clue which involved an Old French text with the word *champaigne*. Keith Roberts had many discussions with Alan about science and computers, but was innocent of other matters which Alan shared with his friends. He never deciphered the coded messages that passed between the others. Nick Furbank, on the other hand, did not have the scientific background, but he was very interested in rationalism and game theory and the imitation principle.

Alan and Robin and Nick devised a new game called Presents. The idea was that one person went out of the room and the others made up a list of imaginary presents that they believed he would like to have. Then he came back and could ask questions about the presents before choosing them, and here the game of bluff and double bluff came in, for one of the presents would secretly be designated 'Tommy' and once Tommy was chosen, the turn was finished. The imaginary presents moved after a while into a more probing realm. Alan tentatively dropped 'Tea in Knightsbridge Barracks' into the game at one point, perhaps reflecting fantasies of twenty years before. The Manchester computer had, in its unexpected and back-handed way, realised one of the products of his imagination. There still remained other dreams; no less hard to fulfil; no less liable to go awry.

The arrangement at Manchester was that the university engineers were to build a prototype machine, which Ferranti would use as 'the instructions of Professor F. C. Williams'. So throughout 1949 the engineers, who were now able to recruit more staff, were adding to the original 'baby machine'. By April it had been fitted with three more cathode ray tubes for fast store, multiplier and 'B-tube', and by that time a small magnetic drum was being tested. Another change was that each line on the cathode ray tube store now held *forty* spots, an instruction taking up twenty of them. These were conveniently thought of as grouped in fives, and a sequence of five binary digits as forming a single digit in the base of 32.

Meanwhile Newman made an ingenious choice of problem with which to demonstrate the machine as it stood with only a tiny store but with a multiplier. It was something that had been discussed at Bletchley – finding large prime numbers. In 1644 the French mathematician Mersenne had conjectured that $2^{17}-1$, $2^{19}-1$, $2^{31}-1$, $2^{67}-1$, $2^{127}-1$, $2^{257}-1$ were all prime, and that these were the only primes of that form within the range. In the eighteenth century, Euler laboriously established that $2^{31}-1=2,146,319,807$ was indeed prime, but the list would not have progressed further without a fresh theory. In 1876 the French mathematician E. Lucas proved that there was a way to decide whether 2^p-1 was prime by a process of

p operations of squaring and taking of remainders. He announced that $2^{127}-1$ was prime. In 1937, the American D. H. Lehmer attacked $2^{257}-1$ on a desk calculator and after a couple of years of work showed that Mersenne had been mistaken. In 1949 Lucas's number was still the largest known prime.

Lucas's method was tailor-made for a computer using binary numbers. They had only to chop up the huge numbers being squared into 40-digit sections and to program all the carrying. Newman explained the problem to Tootill and Kilburn and in June 1949 they managed to pack a program into the four cathode ray tubes and still leave enough space for working up to $p=353$. *En route* they checked all that Euler and Lucas and Lehmer had done, but did not discover any more primes.*

This was part of an uneasy treaty of alliance, according to which the zones of 'engineers' and 'mathematicians' were agreed. Newman took little further interest in the machine, and Alan took on the role of 'the mathematician'. It was for him to specify the range of operations that should be performed by the machine, although his list was in fact cut back by the engineers. He had no part in the internal logical design, which was done by Geoff Tootill, but had control over the input and output mechanism, which lay more in the province of the user.

At the NPL he had chosen punched cards for input since they already had a punched card section; here he preferred to generate a teleprinter tape which could later be run off on a printer. He was, of course, very familiar with the teleprinter system from Bletchley and Hanslope, and people knew it was from 'a place you mustn't talk about' that he obtained a paper tape punch, which ran off a dry battery, and 'had a tendency to replace 1 by 0'. After it had been attached, those 32 different combinations of 0s and 1s in five-row teleprinter tape became the language of the Manchester machine, haunting the days and dreams alike of its users.

It was Alan's job to make the Manchester machine convenient to use, but his ideas of convenience were not always shared by others. He had, of course, attacked the principle on which Wilkes

---

* The next one was out of range at p = 521, as was discovered by computer search in 1952.

was working according to which the hardware of the machine would be designed to make the instructions easy for a human user to follow – so that in the EDSAC design, the letter 'A' was used as the symbol for the instruction to add. In contrast, Alan held that human convenience should be catered for by programming techniques, not by electronics. In his 1947 talk he had referred to such matters of convenience as 'fussy little details', and had stressed how they could be taken care of by 'pure paperwork'. Now at Manchester, he had the opportunity, in principle, to put this into practice – for the machine hardware had not been designed to pander to the programmer. However, by 1949 he had lost interest in doing this kind of work. The 'fussy little detail' of binary to decimal conversion, for instance, he now found not worth bothering about. He himself found it simple to work directly in the base-32 arithmetic in which the machine could be regarded as working, and expected other people to do the same.

To use base-32 arithmetic it was necessary to find 32 symbols for the 32 different 'digits'. Here he took over the system already used by the engineers, in which they labelled the five-bit combinations according to the Baudot teleprinter code. Thus the 'twenty-two' digit, corresponding to the sequence 10110 of binary digits, would be written as 'P', the letter that the sequence 10110 encoded for an ordinary teleprinter. To work in this system meant memorising the Baudot code and the multiplication table as expressed in it – something he, but few others, found easy.

The ostensible reason for sticking to this hideously primitive form of coding, which entailed so much work for the user, was that the cathode ray tube storage made it possible – indeed necessary – to check the contents of the store by 'peeping', as Alan called it, at a monitor tube. He insisted that what one saw as spots on the tube had to correspond digit by digit to the program that had been written out. To maintain this principle of correspondence it was actually necessary to write out the base-32 numbers *backwards*, with the least significant digit first. This was for technical electronic engineering reasons, the same as those which obliged cathode ray tubes always to scan from left to right. Another awkwardness arose on account of the five-bit combinations which did not correspond to a letter of the alphabet on the Baudot code. (It was the same problem that the Rockex system overcame.) Geoff Tootill had already introduced extra symbols for these, the zero of the base-32 notation being represented by a stroke '/'. The result was that pages of programs were covered with strokes – an effect which at Cambridge was said to reflect the Manchester rain lashing at the windows.

By October 1949 the machine was ready, bar some details, for Ferranti to manufacture. The prototype remained in place while this was done, and the idea was to use the time to write an operations manual and basic programs ready to use on the computer (the Mark I, it would be called), when it arrived.

This was Alan's next job, and he must have spent a great deal of time in checking the operation of every single function on the prototype, arguing over their efficiency with the engineers. By October he had written out an input routine: that is, a means to persuade the machine when first switched on and empty of instructions, to read in new instructions from a tape, to store them in the right place, and to begin executing them.

But this was low-level work; and on this level the *Programmers' Handbook*[7] that he wrote, though full of helpful and practical advice, involved few new ideas. Indeed, it had nothing as sophisticated as the routines he had devised at the NPL for floating-point numbers. Nor did he do anything inspired in connection with the organisation of sub-routines. This, in the Manchester development, was dominated by the existence of two kinds of storage: on the Ferranti-built machine this would amount to eight cathode ray tubes each with their 1280 digits, and the magnetic drum promising no fewer than 655360 digits, arranged in 256 tracks of 2560 digits each.* Programming revolved around the process of 'bringing down' data and instructions from the drum to the tubes, and sending them back again, and the hardware more or less obliged each sub-routine to be stored on a new track of the drum, to be transferred *in toto* as required. The Turing scheme coped with this, but he did not bother with a system for sub-routines nested to any depth. He referred to this possibility in a rather flippant passage of the *Handbook:*

The sub-routines of any routine may themselves have sub-routines. This

---

* This promise was not entirely fulfilled, since it turned out that the tracks had been too closely packed, and often were unusable.

is like the case of the bigger and lesser fleas. I am not sure of the exact meaning the poet attached to the phrase 'and so ad infinitum', but am inclined to think that he meant there was no limit that one could assign to the length of a parasitic chain of fleas, rather than that he believed in infinitely long chains. This certainly is the case with sub-routines. One always eventually comes down to a routine without sub-routines

but he left this for the user to organise. His own 'Scheme A' only allowed for one level of sub-routine calling.

The *Handbook* brought out many of the problems of communication that he faced at Manchester. To Williams and the other engineers, a mathematician was someone who knew how to do calculations; in particular they saw binary notation as something new introduced to them by 'mathematics'. To Alan Turing, however, all their schemes with base-32 arithmetic and the rest were merely simple illustrations of the deeper fact that mathematicians were free to employ symbolism in any way they chose. To him it was obvious that a symbol had no intrinsic connection with the entity that it symbolised, and so a long paragraph at the beginning of the *Handbook* explained how it was that there existed a convention according to which sequences of pulses could be interpreted as numbers. While this was a far more accurate and also more creative idea than the usual statement that the machine 'stored the numbers', it was not immediately helpful to the person who had never before known that numbers could be expressed other than in the scale of ten. It was not that Alan despised doing routine, detailed work within a symbolism such as the Manchester machine demanded: but as in *Computable Numbers* and the ACE report he tended to veer from the abstract to the detailed in a way that made sense to him, but not to others. The development that could have absorbed both his liberated understanding of symbolism, and his willingness to do the donkey work when necessary, was that of designing programming languages, the development he described as 'obvious' in 1947. But this was precisely what he did *not* do; and thus he failed to exploit the advantage that a grasp of abstract mathematics gave him.*

---

* The opportunities open to him were rich, and his neglect of them very striking. He could, for instance, have used his knowledge of the 'recursive function' to develop a far more powerful and interesting treatment of the 'sub-routine'. Church's lambda-calculus,

In writing the standard routines for square roots and so forth, he had two assistants after October 1949. One was Audrey Bates, a postgraduate student. The other was Cicely Popplewell, whom he had interviewed for the advertised post in summer 1949. She was a Cambridge mathematics graduate with experience of punched cards used in housing statistics. They both shared his office in that Victorian fortress, the university Main Building, pending the construction of the new Computing Laboratory to house the Ferranti machine. It was not a happy arrangement, for he never really acknowledged their right to exist. On Cicely's first day he said 'Lunch!' and marched out of the room without telling Cicely where the Refectory was. He would talk away himself to anyone who visited, but would be very annoyed if either of them did. Sometimes the shell would crack; they persuaded him to play tennis once, and they were amazed the first time they saw him arrive apparently wearing a raincoat and nothing else, which caused some laughs. Once there was some business of him borrowing a ten-shilling note to pin on his shorts when he went home. But usually they were glad when, as often happened, he did not come in. He made no allowance for the amount they had to learn, and did nothing to mitigate what Cicely felt as 'an acute inferiority complex' in terms of speed of brain. Cicely also had the job of smoothing things over with the engineers, when interdepartmental tension was running high.

Using the prototype machine was no smooth operation. It was comparable with the use of the Robinsons. According to Cicely Popplewell,[8] it

---

and all the hitherto abstruse and 'useless' work he had done on such problems as the 'dots and brackets' in mathematical logic, were now relevant to the devising of practical programming languages. The knowledge of probability and statistics that he had employed in Enigma work could equally profitably have been applied to the theory of programming. Experience with searching, sorting, and the 'trees' involved in his chess-playing ideas, were all particularly relevant to the data-processing problems it was now possible to attack on computers. He could have done much to set standards for the new engineering discipline, if only because he could so easily rise above the technicalities of any particular installation, and could have set his weight against the often absurd and debilitating separation of university mathematics from the developing field of computer applications. But with few exceptions, one of them being an insistence on program-checking procedures which reflected his more abstract, rigorous, background, he abandoned this line of development.

. . . required considerable physical stamina. Starting in the machine room you alerted the engineer and then used the hand switches to bring down and enter the input program. A bright band on the monitor tube indicated that the waiting loop had been entered. When this had been achieved, you ran upstairs and put the tape in the tape reader and then returned to the machine room. If the machine was still obeying the input loop you called to the engineer to switch on the writing current, and cleared the accumulator (allowing the control to emerge from the loop). With luck, the tape was read. As soon as the pattern on the monitor showed that input was ended the engineer switched off the write current to the drum. Programs which wrote to the drum during the execution phase were considered very daring. As every vehicle that drove past was a potential source of spurious digits, it usually took many attempts to get a tape in – each attempt needing another trip up to the tape room.

In fact, writing from the tubes on to the magnetic drum was all but impossible on the prototype. Alan wrote[9]:

Judged from the point of view of the programmer, the least reliable part of the machine appeared to be the magnetic writing facilities. It is not known whether the writing was more often done wrong than the reading or less. The effects of incorrect writing were however so much more disastrous than any other mistake which could be made by the machine, that automatic writing was practically never done. . . . Other serious sources of error were the failure of storage tubes and the multiplier. . . .

In the hot summer of 1950 it was not unknown for computer users to be sweltering in 90°F heat, and banging the racks with a hammer to detect loose valves.

The autumn of 1949 saw what was to be Alan's only titbit of hardware design for the Ferranti machine.[10] One of the hardware functions on which he had insisted was that of a random number generator – a feature not included in his ACE design. His own electronic knowledge stopped short of the necessary practical detail, but with Geoff Tootill's collaboration he was still able to design his own system. It was one that produced truly random digits from noise, as opposed to something like a cipher key generator that would produce apparently random but actually determined digits. (That, if he wanted it, he would surely program for himself.)

Perhaps he based his design on the circuit that produced the Rockex key-tapes at Hanslope.

Geoff Tootill was interested in Alan's ideas, but some of them were particularly impractical in view of the limited time and effort available. There was, for instance, a scheme he devised for computer character recognition, which would involve an elaborate system with a television camera in order to transfer a visual image to the cathode ray tube store, and reduce it to a standard size. Geoff Tootill was probably the most tolerant of such dreams, but to him as to all on the engineering side, Alan Turing was the brilliant mathematician (or so they heard) but embarrassingly half-baked engineer. The year 1949 meant the end of his groping efforts to be the academic engineer; there were few who appreciated that the remarkable thing for a British pure mathematician was not a deficiency in electronics, but the willingness to dirty his hands at all.

Meanwhile the more theoretical side of computer development had become a more public question. In 1948 Norbert Wiener had published a book called *Cybernetics*, defining this word to mean 'Control and Communication in the Animal and the Machine'. It meant the description of the world in which information and logic, rather than energy or material constitution, was what mattered. As such it was heavily influenced by the massive wartime technological developments, although the basic ideas, such as feedback, were hardly new. Wiener and von Neumann had led a conference in the winter of 1943–4 on 'cybernetic' ideas, but Wiener's book marked the opening up of the subject outside the narrow domain of technical papers. In fact *Cybernetics* was still very technical, incoherent, and almost unreadable, but the public seized upon it as a magic key which would unlock the secrets of what had happened to the world in the past decade.

Wiener regarded Alan as a cybernetician, and indeed 'cybernetics' came close to giving a name to the range of concerns that had long gripped him, which the war had given him an opportunity to develop, and which did not fit into any existing academic category. In spring 1947, on his way to Nancy, Wiener had been able to 'talk over the fundamental ideas of cybernetics with Mr Turing,' as he explained in the introduction to his book.

By 1949 an American supremacy was virtually taken for granted in science as in everything else, and it was a sign of the times that on 24 February 1949 the popular magazine *News Review*,[11] presenting a digest of what Wiener had to say, should explain with pride how British scientists had been able to supply 'valuable data' to the American professor when he had flown in. It was as a planet round the Wiener sun that Alan appeared, the photograph of his young and slightly nervous profile standing in marked contrast to the ponderous features of Wiener and the massive visage of the biologist J. B. S. Haldane.

In reality Alan was more than a match for Wiener, and although genuinely sharing many common interests, their outlooks were different. Wiener had an empire-building tendency which rendered almost every department of human endeavour into a branch of cybernetics. Another difference lay in Wiener's total lack of a sense of humour. While Alan always managed to convey his solid ideas with a light English touch of wit, Wiener delivered with awesome solemnity some pretty transient suggestions, to the general effect that solutions to fundamental problems in psychology lay just around the corner, rather than putting them at least fifty years in the future. Thus in *Cybernetics* it was seriously suggested that McCulloch and Pitts had solved the problem of how the brain performed visual pattern recognition. The cybernetic movement was rather liable to such over-optimistic stabs in the dark. One story going around, which later turned out to be a hoax, but which found its way into serious literature,[12] was of an experiment supposed to measure the memory capacity of the brain by hypnotising bricklayers and asking them such questions as 'What shape was the crack in the fifteenth brick on the fourth row above the damp course in such and such a house?'. Alan's reaction to these cybernetic tales was one of amusement.

Another point of difference lay in the fact that Wiener was openly concerned about the economic implications of cybernetic technology. The war, for him, had not changed a conviction that machines should be made to work for people rather than *vice versa*. His comment that factory robots would put the people they replaced in the position of competing against slave labour, and his daring description of the principle of competition as a 'shibboleth', put him on the extreme left of 1948 American opinion. It was no accident that on his visit to Britain Wiener had consulted the left-wing luminaries of science, J. D. Bernal and H. Levy, as well as Haldane.

But the academic debate that followed *Cybernetics* in Britain was not concerned with this question, nor indeed with anything to do with the use of computers, or the harnessing of wartime technology to peaceful and constructive ends, or the relative merits of cooperation and competition. When the *News Review* called cybernetics a 'frightening science' it was not the economic consequences but the threat to traditional beliefs that it feared. Post-war reaction to planning and austerity, conservative rather than commercial, was reflected in the almost Victorian terms of reference which the intellectuals also accepted. This was true of Alan Turing as much as anyone; these were terms close to his own struggle of the 1930s over problems of thought and feeling. Times had changed, however, and so it was not a bishop but a brain surgeon who led the British intellectual reaction to the claims of machinery to thought. The eminent Sir Geoffrey Jefferson delivered an address,[13] *The Mind of Mechanical Man*, as the Lister Oration on 9 June 1949.

Jefferson held a Chair of Neurosurgery at Manchester, and knew about the Manchester computer development from talking about it with Williams. But most of his impressions came from Wiener, whose emphasis was still placed on the similarity of the nerve-cells of the brain to the components of a computer.* On this level the analogy was pretty feeble, and not much advanced by Wiener's comparisons between computer malfunctions and nervous diseases. The Turing ideas of discrete-state machines and universality were required to lend precision and substance to the cybernetic claim. Some of Wiener's assertions were rather easy to attack; but Jefferson did go further than just knocking them down, playing some strong commonsense cards, such as:

---

* Alan Turing himself always sought to play down any such comparison, which in his view was irrelevant to the essential thesis that the brain could be regarded as a discrete-state machine. Thus in the 1948 report for the NPL he had written: 'We could produce fairly accurate electrical models to copy the behaviour of nerves, but there seems very little point in doing so. It would be rather like putting a lot of work into cars which walked on legs instead of continuing to use wheels.'

But neither animals nor men can be explained by studying nervous mechanics in isolation, so complicated are they by endocrines, so coloured is thought by emotion. Sex hormones introduce peculiarities of behaviour often as inexplicable as they are impressive (as in migratory fish).

Jefferson liked talking about sex. But his Oration was concluded by flights of rhetoric which only begged the question. An oft-quoted passage held that

Not until a machine can write a sonnet or compose a concerto because of thoughts and emotions felt, and not by the chance fall of symbols, could we agree that machine equals brain – that is, not only write it but know that it had written it. No mechanism could feel (and not merely artificially signal, an easy contrivance) pleasure at its successes, grief when its valves fuse, be warmed by flattery, be made miserable by its mistakes, be charmed by sex, be angry or miserable when it cannot get what it wants.

Jefferson ended by 'ranging myself with the humanist Shakespeare rather than the mechanists, recalling Hamlet's lines: "What a piece of work is man! How noble in reason! How infinite in faculty"', and so forth. Shakespeare was often exhibited in these discussions as proof of the speaker's exquisite human sensibilities. However, Jefferson had done a good deal to improve upon the 'piece of work' himself, not only by mending the broken heads of two world wars but as an exponent in the late 1930s of the frontal lobotomy.

This was the 'heads in the sand' argument, resting upon the assumption that a machine, because its components were non-biological, was incapable of creative thinking. 'When we hear it said that wireless valves think,' Jefferson said, 'we may despair of language.' But no cybernetician had said the *valves* thought, no more than anyone would say that the nerve-cells thought. Here lay the confusion. It was the system as a whole that 'thought', in Alan's view, and it was its logical structure, not its particular physical embodiment, that made this possible.

The Times[14] seized upon Jefferson's concession that

A machine might solve problems in logic, since logic and mathematics are much the same thing. In fact, some measures to that end are on foot in my university's department of philosophy [*sic*].

Their reporter telephoned Manchester, where Alan rose to the bait and chatted away without inhibition:

'This is only a foretaste of what is to come, and only the shadow of what is going to be. We have to have some experience with the machine before we really know its capabilities. It may take years before we settle down to the new possibilities, but I do not see why it should not enter any one of the fields normally covered by the human intellect, and eventually compete on equal terms.

I do not think you can even draw the line about sonnets, though the comparison is perhaps a little bit unfair because a sonnet written by a machine will be better appreciated by another machine.'

Mr Turing added that the university was really interested in the investigation of the possibilities of machines for their own sake. Their research would be directed to finding the degree of intellectual activity of which a machine was capable, and to what extent it could think for itself.

This embarrassing definition of what 'the university' was 'really interested in' provoked a swish of the cane from the Catholic public school:[15]

... If one may judge from Professor Jefferson's Lister oration ... responsible scientists will be quick to dissociate themselves from this programme. But we must all take warning from it. Even our dialectical materialists would feel necessitated to guard themselves, like Butler's Erewhonians, against the possible hostility of the machines. And those of us who not only confess with our lips but believe in our hearts that men are free persons (which is unintelligible if we have no unextended mind or soul, but only a brain) must ask ourselves how far Mr Turing's opinions are shared, or may come to be shared, by the rulers of our country.

Yours &c. ILLTYD TRETHOWAN

Downside Abbey, Bath, June 11.

The rulers of Great Britain did not divulge their opinions. But Max Newman wrote to *The Times* to correct the impression left by Alan's heady prophecies, dousing them with a laborious explanation of the Mersenne prime problem. Jefferson proved a good publicity agent for Manchester, for *The Times* published photographs of the adolescent machine, and the *Illustrated London News* followed on 25 June. By chance, these happened to upstage the grand opening of the EDSAC in Cambridge.

Wilkes' team had made rapid progress, and had already completed the building of an EDVAC-type computer, with mercury delay line storage, well ahead of any American development. It had a storage capacity of only thirty-two delay lines, and its digit time was two microseconds, twice that of the planned ACE. But it worked; and if the 'baby machine' at Manchester was the first working electronic stored program computer, the EDSAC was the first to be available for serious mathematical work.*

Alan attended the inaugural conference, and gave a talk on 24 June 1949 entitled 'Checking a Large Routine'.[16] He described a sophisticated procedure, appropriate for long programs in which it would be easy to lose track of the fate of numbers in store. Illustrating his points, he did some sums on the blackboard, and lost everyone by writing the numbers backwards as he was used to doing at Manchester. 'I do not think that he was being funny, or trying to score off us,' wrote Wilkes,[17] 'it was simply that he could not appreciate that a trivial matter of that kind could affect anybody's understanding one way or the other.' It was a 'fussy little detail', which perhaps masked the irony of the fact that while the EDSAC people had only begun to write programs in May 1949, soon discovering the idea of the sub-routine, Alan had been writing and perhaps checking them for years.

Meanwhile the ACE did survive after all. Alan had resigned at the lowest ebb. Then Thomas, who was in charge of the electronics, resigned, and his successor, F. M. Colebrook, proved to have a very different attitude. In fact, as soon as Thomas was gone, the mathematicians moved into the engineers' building. Thanks to Colebrook an unheard-of relaxation took place, and the two groups were soon working together in a sort of assembly line. The speed of progress attained in building the machine was comparable with that envisioned in Alan's original proposal. By mid-1949 they had a delay line working and the wiring of the

---

* The Mersenne prime problem was a highly artificial, if ingenious, application of the growing Manchester computer. Only from the autumn of 1949 could it be applied to 'regular' problems. Besides those of Alan Turing himself, as later described, it was used for optical calculations, tracing rays through systems of lenses, and for some mathematical work in connection with guided missiles.

control was finished in October. The machine, the 'Pilot ACE', was based on the Turing 'Version V', just as Huskey's premature effort had been. It retained the 'distributed' processing that distinguished it from the von Neumann system which used an accumulator. They also kept megacycle speed, which made it the fastest in the world. Meanwhile Sir Charles Darwin retired in 1949. Max Newman took the view that he had done Darwin a good turn in taking Alan away, and Alan would agree with this. When he went to the official opening of the Pilot ACE in November 1950 he was particularly generous in telling Jim Wilkinson how much better they had done than would have been possible had he stayed. Certainly the Pilot ACE would not have been possible at all had not Alan gone. But he must also have known that it represented only a shadow of his original vision.[18]

Womersley managed to rewrite the history of the ACE project after Alan had left. For it was Womersley's story that Colebrook gave to the Executive Committee meeting on 13 November 1949:

> Mr Colebrook then referred to the organisational history of the Automatic Computing Engine project. The work on this originated with Dr Turing's paper . . . 'On Computable Numbers with an Application to the Entscheidung Problem' [sic] and Mr Womersley began thinking about the logical design in 1938 after reading Dr Turing's paper and after discussions with Professor Hartree. Mr Womersley came to the Laboratory early in 1944, and the following year visited the United States to see the Harvard and ENIAC machines. Professor Newman came to see Mr Womersley in 1945 and introduced Dr Turing, who very soon afterwards joined the staff of the Laboratory.

This was the only mention of Alan's part in the project. The account continued to explain that

> In 1946, work on the Automatic Computing Engine was started and it was arranged for the experimental work to be done by the Post Office and the theoretical work, including the programming of the machine, at the Laboratory. Because of slow progress at the Post Office, a section was started at NPL in 1947 to build the ACE machine.

Skilfully passing over the Thomas period, Colebrook described the progress made in 1948 and 1949. He then contrasted the Pilot ACE with 'the machine originally proposed', and announced that

> The actual size of the ACE as originally contemplated was the outcome of long consideration by Mr Womersley and Professor von Neumann during Mr Womersley's visit to the United States.

Already by 1950, Alan Turing was an unperson, the Trotsky of the computer revolution.

But he was never one to complain, once he had made his decision. In many ways his position at Manchester was parallel to that at Hanslope, in terms of status and class and struggles over equipment. One difference was the harshness of the Manchester environment, which surely exacerbated his rudeness. Another was that in 1943 his move sideways and downwards had gained for him practical experience with electronics. In 1948 it gained for him the use of a computer. And this remained a paramount consideration. He had conceived of a universal machine, and now he could work or play with one of the two that existed in the world of 1949. There was a method in his madness.

For the time being he contented himself with paying off some scores to old days that would vindicate the power of the universal machine. The first thing he did was to revive the zeta-function calculation. The gear wheels that had been cutting when the phoney war broke in could now be replaced by instructions on the tape of a universal machine, in the phoney peace of 1950. It did not go quite according to plan, this being partly the machine's fault and partly his own:[19]

> In June 1950 the Manchester University [prototype] Electronic Computer was used to do some calculations concerned with the distribution of the zeros of the Riemann zeta-function. It was intended in fact to determine whether there are any zeros not on the critical line in certain particular intervals. The calculations had been planned some time in advance, but had in fact to be carried out in great haste. If it had not been for the fact that the computer remained in serviceable condition for an unusually long period from 3p.m. one afternoon to 8a.m. the following morning it is probable that the calculations would never have been done at all. As it was, the interval $2\pi.63^2<t<2\pi.64^2$ was investigated during that period, and very little more was accomplished:
>
> . . . The interval $1414<t<1608$ was [also] investigated and checked, but unfortunately at this point the machine broke down and no further

work was done. Furthermore this interval was subsequently found to have been run with a wrong error value, and the most that can consequently be asserted with certainty is that the zeros lie on the critical line up to $t=1540$, Titchmarsh having investigated as far as 1468. . . .

This was an unusual joint exercise, on which Kilburn stood by all night. Alan would hold up the output teleprinter tape to the light to read:

The content of a tape may afterwards automatically be printed out if desired [sic] . . . the output consisted mainly of numbers in the scale of 32 . . . writing the most significant digit on the right. More conventionally the scale of 10 can be used, but this would require the storage of a conversion routine, and the writer was entirely content to see the results in the scale of 32, with which he is sufficiently familiar.

Another old score, that of the Enigma, was also paid off at about this time:[20]

I have set up on the Manchester computer a small programme using only 1000 units of storage, whereby the machine supplied with one sixteen figure number replies with another within two seconds. I would defy anyone to learn from these replies sufficient about the programme to be able to predict any replies to untried values.

He had, in other words, devised a cipher system which he reckoned impregnable even with the help of known plain-text. The lumbering wheels of the Second World War were already heading towards the same obsolescence as that of his zeta-function machine.

There were some other hints of a continued interest in cryptology. Another item that he demanded from the engineers as a hardware function of the Mark I Ferranti machine was what they called a 'sideways adder'. It would count the number of '1' pulses in a 40-bit sequence. This would have no application in a numerical program, but would be very useful in one where the digits coded 'yes' or 'no' answers to some Boolean question, and it was required to count the 'yes' answers – just what the Colossus had done. Such applications might possibly have been spare-time interests of his own. However, it was during this period, as the international situation hardened, that he found himself consulted by GCHQ. It would indeed have been remarkable had they *not* consulted the person who knew more about cryptology and the potential of electronic computers than did anyone else. And had he not described cryptanalysis as the most 'rewarding' field for the application of programming? Few, however, were in a position to perceive this fact, the subject being more secret than ever.

A hark-back to cryptology also featured in his discussions with a young American, David Sayre, in this period. A graduate of the wartime MIT Radiation Laboratory, Sayre was now at Oxford studying molecular biology with Dorothy Hodgkin. Having worked with F. C. Williams during the war, he made a visit to Manchester to see the computer, explaining that it might help in X-ray crystallography. Williams passed him on to Alan, who showed an unusual kindness and geniality, making Sayre[21] 'perfectly at ease with him'. They talked for two and a half days, interrupted only when 'the telephone would ring to say that the machine was free for a few minutes in case he wanted to use it, and he would gather up sheaves of paper and ribbons of punched paper tapes . . . and disappear for a bit.'

David Sayre was able to guess that Alan had worked on cryptological problems during the war. The point was that X-ray crystallography, which was now being applied to the problem of determining the structure of proteins, was remarkably similar in nature to cryptanalysis. The X-rays would leave a diffraction pattern, which could be regarded as the encipherment of the molecular structure. Performing the decipherment process was closely analogous to the problem of finding both plain-text and key, given the cipher-text alone.* The result of this analogy was that

---

* X-ray measurements give only the *amplitudes* of the different frequency components in the diffracted X-rays, and not the *phases*. The analysis depends upon guessing the phases, the criterion of a correct guess being that when the amplitudes and phases are put together, they lead to a picture of the crystal which is in accord with physical reality, with the right number of atoms and a positive electron density. This is exactly the same idea as guessing a key, given a piece of cipher-text: the criterion of a correct guess being that it give a sensible message.

The analogy with cryptanalysis is even closer in that the crystallographer attacks the problem, at first sight too enormous for contemplation, by making a hypothesis about the structure of the crystal. Thus Watson and Crick pursued the DNA analysis, as did Pauling, by making good guesses about the helix structure, and thus getting closer and closer to the solution. This is essentially the same idea as the 'Probable Word' method, which also

... Before we finished he had re-invented by himself most of the methods which crystallographers, up to that time, had worked out. He had, for this purpose, a breadth of knowledge greatly surpassing that of any crystallographer I have known, and I am confident that he would have advanced the crystallographic situation very decidedly if he had worked in it seriously for a time. As it was, he may have had hold of one line which in 1949 had not yet appeared in crystallography, concerned with establishing quantitatively how much information it is necessary to have on hand at the outset of a search for a solution to ensure that a solution can be found.

Alan told him about the Shannon theorem which he had exploited for the Delilah, and Sayre made use of it in a paper[22] which much advanced the theory of the subject. But Alan did not decide to work seriously in this area, although he encouraged young Sayre to return and use the Manchester machine for computations; this was a branch of science where exciting progress was being made, and yet for him it would perhaps have been too much a research into things past, like all these other 1949 encounters. Or perhaps it would have been too crowded and competitive a field. He always wanted something that would be self-contained.

Claude Shannon was himself another visitor. Since 1943, their discussions on machines and minds, information and communication, had been opened up to all. In September 1950 there was a London symposium[23] on 'information theory', at which Shannon was the star guest. His paper[24] on chess-playing, which explained the principles of minimax play and tree searching, had just appeared. Someone reviewed it and made a remark that Alan thought had confused cause and effect, or in typical Turing language, it was

... like taking a statistical analysis of the laundry of men in various positions and deciding, from the data collected, that an infallible method of getting ahead in life was to send a large number of shirts to the wash each week.

---

effects a drastic reduction in the number of possible keys – so that with the Enigma, for instance, they were left only with a small number of Bombe 'stops' to try out for sensible German plain-text.

It is not surprising that Alan Turing could see how to quantify the idea of information required for a guess to be possible: this was very close to the quantification of 'weight of evidence' which constituted his major conceptual advance at Bletchley.

Afterwards, Shannon went to Manchester to see the prototype machine, then in its last days, and Alan told him all about the zeta-function calculation.*

The conference which gave occasion to this visit was a manifestation of the 'cybernetics' movement. Another was that in July 1949, following a talk by K. Lorenz at Cambridge on animal behaviour, an informal cybernetics discussion group had been started, meeting in London about once a month over a dinner. It was called the Ratio Club. McCulloch who with Wiener was one of the original high priests of cybernetics, addressed the first meeting. (He also travelled to Manchester to see Alan, who thought him 'a charlatan'.) Alan was not in the Ratio Club's founding group, but at the first meeting, his name was put forward[25] by Gold and the biologist John Pringle, who had been Alan's undergraduate contemporary at King's.

Thereafter Alan used to go to meetings every few months, and found them good entertainment. Robin Gandy went to some meetings later, and Jack Good joined after attending Alan's talk on 'Educating a Digital Computer' in December 1950. Uttley from TRE and the philosophical physicist D. Mackay, were also very interested in machine intelligence, while W. Grey Walter and W. Ross Ashby, neurologists who both brought out early and influential books[26] on cybernetic ideas, were keen members. (Grey Walter made some motorised 'tortoises' which could recharge themselves when their batteries were low.) The meetings were held at the National Hospital for Nervous Diseases, whose John Bates acted as secretary and galvaniser. There was a lot of enthusiasm, though it fell off over the next few years, as it was found that cybernetics offered no immediate solutions to the problems posed by human beings.

In some ways it was an attempt to revive the democratic association of young scientists which had characterised the war years. They excluded those of professorial rank, and Alan's light touch went down well. Many of the Ratio Club had worked at TRE, where they had held what were called Sunday Soviets, according to the

---

* Shannon was sceptical about this programme of work, and Shannon had a good point. By 1977 computer calculations would show that among the first *seven million* zeroes of the zeta function, there is not one that lies off the special line. This was a case where a brute-force attack could yield only a negative result.

illusions of the day – much the same way as each section worked at Bletchley. It was just a faint ghost of the 'creative anarchy'.

As it happened, Peter Hilton from Bletchley days had left Oxford to join the Manchester mathematics department in 1948, and Alan took him to see the machine which in some ways had grown out of their experiences. Peter Hilton was also present at a discussion in the department in 1949 which also touched upon subjects in Alan's remoter past, the two fields of group theory and mathematical logic which had set his professional career in motion.

The discussion concerned the 'word problem' for groups. This was like the Hilbert *Entscheidungsproblem* that *Computable Numbers* had settled, but instead of asking for a 'definite method' for deciding whether or not a given theorem was provable, it asked for a definite method for determining whether or not some given product of group elements was equal to some other given product; that is, whether some given sequence of operations would have the same effect as some other sequence.* Emil Post had given the first new result in this direction in 1943, by showing that the word problem for 'semi-groups' was insoluble.† The question for groups still remained open. Peter Hilton was amazed because[27]

> Turing claimed he had never heard of this problem, and found it a very interesting problem, and so, though at that time his principal work was in machines, he went away and about ten days later announced that he had proved that the word problem was unsolvable. And so a seminar was arranged at which Turing would give his proof. And a few days before the seminar he said: 'No, there was something a little wrong in the argument, but the argument would work for cancellation semi-groups.' And so he in fact gave his proof for cancellation semi-groups.‡

The proof required quite new methods, technically more difficult

---

* If the operations are represented by letters, then such a sequence is represented by a 'word' – hence the name of the problem. For a *finite* group, there would of course exist such a definite method, namely the crude one of working through all the possibilities. The problem arises for *infinite* groups.

† A 'semi-group' is the abstract version of a set of operations which meets half of the conditions required for a 'group': the operations cannot necessarily be reversed.

‡ A 'cancellation semi-group' is a semi-group with a property which makes it closer to being a group: if $AC = BC$ then it must be that $A = B$.

than those of *Computable Numbers*, in order to relate the ideas of doing and undoing operations to the action of a Turing machine. It showed that at any time, even though he was so out of touch, he could revert to being 'a logician'. It was a great comeback, and yet for him it would not have been coming back, but going back. He spent some more time on the original problem for groups, but did not dedicate himself to it. It offered the innocence of the work of his twenties, before he had got mixed up with the world's affairs. But it did not offer the direction in which to move onward.

Alan submitted his results[28] to von Neumann's journal, where it was received on 13 August 1949, and elicited a reply[29] from the big man himself:

> September 13 1949
>
> Dear Alan,
>
> . . . Our machine project is moving along quite satisfactorily, but we aren't yet at the point where you are. I think that the machine will be physically complete early next year. What are the problems on which you are working now, and what is your program for the immediate future?
>
> With best regards, Yours sincerely, John

Von Neumann's machine at the IAS was lagging years behind because the Iconoscope, upon which such hopes had been placed, could not be made to work. The first American computers to be completed were Eckert and Mauchly's BINAC, used for aircraft engineering, in August 1950, and then the CSAW's cryptanalytic ATLAS in December 1950. But by late September 1949 the Soviet Union had tested its atomic bomb, and this encouraged the American decision in early 1950 to construct a thermonuclear weapon. The IAS machine and its copy MANIAC at Los Alamos were then pushed ahead, although even so it was 1952 before they were completed. The 1950–52 calculation for H-bomb feasibility was performed by 1930s methods, with slide rules and desk calculators, absorbing years of human work. In the end they had to scrap the special Iconoscope and use Williams' ordinary cathode ray tubes. With two assistants, he had beaten American industry. It was still possible for British ingenuity to 'jump in ahead of the Americans'.

But where did this leave Alan? What was his programme for the immediate future? It was a very pertinent question that the Wizard posed to Dorothy – not least because the facilities of the Manchester computer, when completed, would not match the ambitions spelt out by Alan in 1948 for 'learning' and 'teaching' and 'searching'. He had to reconcile himself to the fact that those ideas were dreams on the edge of reality, and find some new way in which to continue.

Meanwhile the claims of cybernetics had attracted the attentions of philosophers more weighty than Jefferson, and Alan was drawn into a more professional defence of his views. The motive force was supplied by Michael Polanyi, the Hungarian emigré who had held the Chair of Physical Chemistry at Manchester from 1933 to 1948, since when he had occupied a new Chair of 'social studies', specially created to facilitate his philosophical ambitions.

Polanyi had long led an opposition to the notions of Planned Science. Even during the war he had founded a 'Society for Freedom in Science', and after the war attempted to combine political and scientific philosophy, marshalling a variety of arguments against various kinds of determinism. In particular, he seized on Gödel's theorem as a proof that mind would do something that was beyond any mechanical system. It was this subject that most engaged Alan and Polanyi in discussion. Alan would run over to the Polanyi home, which was not far from his lodgings at Hale. (Once Polanyi visited Alan, only to find him practising the violin in freezing cold, not bothering or not daring to ask the landlady for proper heat.) But Polanyi had many other suggestions up his sleeve. He rejected Eddington's argument for free will from the Uncertainty Principle. But, unlike Eddington, he thought that the mind could interfere with the motions of molecules, writing that[30] 'some enlarged laws of nature may make possible the realization of operational principles acting by consciousness', and that the mind might 'exercise power over the body merely by sorting out the random impulses of the ambient thermal agitation.'

Polanyi also favoured an extension of the 'Jabberwocky' argument, that science was all in the mind anyway, and had no meaning apart from the 'semantic function' which the human mind alone could supply. Karl Popper, who held similar views, said in 1950[31] that 'It is only our human brain which may lend significance to the calculators' senseless powers of producing truths.' Popper and

Polanyi both held that people had an inalienable 'responsibility', and that science only existed by virtue of conscious, responsible decisions. Polanyi held that science should rest on a moral basis. 'My opposition to a universal mechanical interpretation of things,' he wrote, '. . . also implies some measure of dissent from the absolute moral neutrality of science.' There was a schoolmasterish tone to this 'responsibility' that was rather different from gentle Eddington's vision of Mind-stuff perceiving the spiritual world. There was also a powerful Cold War thread to it. Polanyi attacked the Laplacian picture on the grounds that it 'induces the teaching that material welfare . . . is the supreme good' and that 'political action is necessarily shaped by force.' These unpalatable doctrines he associated with the Soviet government rather than with that of the other Great Power, complaining at the suggestion that 'all cultural activities should subserve the power of the State in transforming society for the achievement of welfare.' Alan liked the point that all measurements ultimately involved an element of decision, and produced for Polanyi a photograph of a horse-race finish, in which of two neck-and-neck horses one could be said to have won if a jet of spittle were counted as part of its body, and not otherwise – a contingency not allowed for in the rules.[32] But the thrust of the Christian philosopher's arguments lay in a very different direction from his own.

This was the background to a formal discussion on 'The Mind and the Computing Machine'[33] held in the philosophy department at Manchester on 27 October 1949. Just about everyone in British academic life with a view to express had been assembled. It began with Max Newman and Polanyi arguing about the significance of Gödel's theorem, and ended with Alan discussing brain cells with J. Z. Young, the physiologist of the nervous system. In between, the discussion raged through every other current argument, the philosopher Dorothy Emmet chairing. 'The vital difference,' she said during a lull, 'seems to be that a machine is not conscious.'

But such a use of words would satisfy Alan no more than would Polanyi's assertion that the function of the mind was 'unspecifiable' by any formal system. He wrote up his own view, which appeared as a paper, 'Computing Machinery and Intelligence',[34] in the philosophical journal Mind in October 1950. It was typical of him that the style he employed in this august journal was very little different from that of his conversation with friends. Thus he introduced the idea of an operational definition of 'thinking' or 'intelligence' or 'consciousness' by means of a sexual guessing game.

He imagined a game in which an interrogator would have to decide, on the basis of written replies alone, which of two people in another room was a man and which a woman. The man was to deceive the interrogator, and the woman to convince the interrogator, so they would alike be making claims such as 'I am the woman, don't listen to him!' Although pleasantly recalling the secret messages that might be passed in his conversations with Robin and Nick Furbank, this was in fact a red herring, and one of the few passages of the paper that was not expressed with perfect lucidity. The whole point of this game was that a successful imitation of a woman's responses by a man would *not* prove anything. Gender depended on facts which were *not* reducible to sequences of symbols. In contrast, he wished to argue that such an imitation principle did apply to 'thinking' or 'intelligence'. If a computer, on the basis of its written replies to questions, could not be distinguished from a human respondent, then 'fair play' would oblige one to say that it must be 'thinking'.

This being a philosophical paper, he produced an argument in favour of adopting the imitation principle as a criterion. This was that there was no way of telling that other *people* were 'thinking' or 'conscious' except by a process of comparison with oneself, and he saw no reason to treat computers any differently.*

The *Mind* article largely took over what he had said in his NPL report, which had not, of course, been published. There were, however, some new developments, not all very serious. One was the joke of a proud atheist who refused to be the Responsible Scientist expected by Downside Abbey. He gave a tongue-in-cheek demolition of what he called the 'Theological Objection' to the idea of machines thinking, which concluded that thinking might indeed be the prerogative of an immortal soul, but then there was nothing to stop God from bestowing one upon a machine. More ambiguous in tone was a reply to an objection 'from Extra-Sensory Perception'.

---

* Polanyi rejected this argument, saying that a machine was a machine, a human mind was a human mind, and no amount of evidence could change this *a priori* fact.

He wrote that

> These disturbing phenomena seem to deny all our usual scientific ideas. How we should like to discredit them! Unfortunately the statistical evidence, at least for telepathy, is overwhelming. It is very difficult to rearrange one's ideas so as to fit these new facts in. Once one has accepted them it does not seem a very big step to believe in ghosts and bogies. The idea that our bodies move simply according to the known laws of physics, together with some others not yet discovered but somewhat similar, would be the first to go.

Readers might well have wondered whether he really believed the evidence to be 'overwhelming', or whether this was a rather arch joke. In fact he was certainly impressed at the time by J. B. Rhine's claims to have experimental proof of extra-sensory perception. It might have reflected his interest in dreams and prophecies and coincidences, but certainly was a case where, for him, open-mindedness had to come before anything else; what *was so* had to come before what it was convenient to think. On the other hand, he could not make light, as less well-informed people could, of the inconsistency of these ideas with the principles of causality embodied in the existing 'laws of physics', and so well attested by experiment.

The idea of 'teaching' the machine had also progressed since 1948. By now he had probably learnt by trial and error that the pain and pleasure method was appallingly slow, and had worked out a reason why, which cast a look back to Hazelhurst:

> The use of punishments and rewards can at best be a part of the teaching process. Roughly speaking, if the teacher has no other means of communicating to the pupil, the amount of information which can reach him does not exceed the total number of rewards and punishments applied. By the time a child has learnt to repeat 'Casabianca' he would probably feel very sore indeed, if the text could only be discovered by a 'Twenty Questions' technique, every 'NO' taking the form of a blow. It is necessary therefore to have some other 'unemotional' channels of communication. If these are available it is possible to teach a machine by punishments and rewards to obey orders given in some language, *e.g.* a symbolic language. These orders are to be transmitted through the 'unemotional' channels. The use of this language will diminish greatly the number of punishments and rewards required.

It was a nice touch of self-reference to bring in *Casabianca*, for the boy on the burning deck, executing his orders mindlessly, was like the computer. He went on to suggest that a learning machine might achieve a 'supercritical' state when, in analogy with the atomic pile, it would produce more ideas than those with which it had been fed. This was essentially a picture of his own development, stated rather more seriously than in 1948, and a claim that even his own originality must somehow have been determined. Perhaps he was thinking of his series for the inverse tangent function, and the law of motion in general relativity, when he first began to put things together in his mind. This, again, was not a new idea. Bernard Shaw had argued it thus in *Back to Methuselah*, when Pygmalion produced his automaton:

ECRASIA: Cannot he do anything original?

PYGMALION: No. But then, you know, I do not admit that any of us can do anything really original, though Martellus thinks we can.

ACIS: Can he answer a question?

PYGMALION: Oh yes. A question is a stimulus, you know. Ask him one.

Much of what Alan wrote was a justification of Pygmalion's argument, which Shaw, champion of the Life Force, had derided.

This time he also offered a very carefully phrased prophecy, made deliberately rather than off the cuff to newspaper reporters.

I believe that in about fifty years' time it will be possible to programme computers, with a storage capacity of about 10^9, to make them play the imitation game so well that an average interrogator will not have more than 70 per cent chance of making the right identification after five minutes of questioning. The original question, 'Can machines think?' I believe to be too meaningless to deserve discussion. Nevertheless I believe that at the end of the century the use of words and general educated opinion will have altered so much that one will be able to speak of machines thinking without expecting to be contradicted.

These conditions ('average', 'five minutes', '70 per cent') were not very demanding. But it was most important that the 'imitation game' would allow questions about anything whatever, not just about mathematics or chess.

It reflected his all-or-nothing intellectual daring, and it came at an appropriate moment. A first generation of pioneers in the new sciences of information and communication, people like von Neumann, Wiener, Shannon, and pre-eminently Alan Turing himself, who had combined broad insights into science and philosophy with the experience of the Second World War, was giving way to a second generation which possessed the administrative and technical skills to build the actual machines. The broad insight, and the short-term skill, had little in common – that was one of Alan's problems. This paper was something of a swan song for the primal urge, bequeathing the original excitement to the world before it was submerged in mundane technicalities. As such it was a classic work in the British philosophical tradition. It was a gentle reproof to the ponderous essays by Norbert Wiener, as well as to the reactionary, 'soupy' trend of English culture in the late 1940s. Bertrand Russell admired it, and his friend Rupert Crawshay-Williams wrote appreciatively to Alan of how much Russell and he had enjoyed reading it.[35]

From a philosophical point of view, it could be said to fit in with Gilbert Ryle's *The Concept of Mind*, which had appeared in 1949, and which put forward the idea of mind not as something added to the brain, but as a kind of description of the world. But Alan's paper proposed a *specific* kind of description, namely that of the discrete-state machine. And he was more the scientist than the philosopher. The point of his approach, as he stressed in the paper, was not to talk about it in the abstract, but to try it out and see how much could be achieved. In this he was the Galileo of a new science. Galileo made a practical start upon that abstract model of the world called physics; Alan Turing upon that model provided by the logical machine.

Alan himself would have liked the comparison: he made reference in the article to Galileo incurring the displeasure of the church and the format of his 'Objections' and 'Refutations' was one of a trial. A year or so later he gave a talk[36] on this subject subtitled 'A Heretical Theory'. He liked to say things like: 'One day ladies will take their computers for walks in the park and tell each other "My little computer said such a funny thing this morning!",' to destroy any sort of sanctimonious forelock-touching to the 'higher realms'. Or, when asked how to make a computer say something surprising, he answered 'Get a bishop to talk to it.' In 1950 he was hardly likely to be on trial for heresy. But he certainly felt himself up against an irrational, superstitious barrier, and his predisposition was to defy it. He continued:

> I believe further that no useful purpose is served by concealing these beliefs. The popular view that scientists proceed inexorably from well-established fact to well-established fact, never being influenced by any unproved conjecture, is quite mistaken. Provided it is made clear which are proved facts and which are conjectures, no harm can result. Conjectures are of great importance since they suggest useful lines of research.

Science, to Alan Turing, was thinking for himself.

Untarnished by all the trials and errors surrounding the actual computer installations, sprang out this 'conjecture': the achievement by the millennium of something approaching the artificial intelligence that had long been expressed in the myth of Pygmalion. Also emerging fully-formed was the fruit of his thought since 1935 on the discrete-state machine model, on universality, and the constructive use of the imitation principle to 'build a brain'.

Nonetheless, beneath the assertive surface of the paper lay probing, needling, teasing questions. For this was not tunnel vision. Unlike so many scientists, Alan Turing was not trapped within the narrow framework within which his ideas were formed. Polanyi was keen on pointing out the different models employed by the different branches of scientific enquiry, and the importance of distinguishing them. But Edward Carpenter had gone to the heart of the matter long before:[37]

> The method of Science is the method of all mundane knowledge; it is that of limitation or actual ignorance. Placed in face of the great uncontained unity of Nature we can only deal with it in thought by selecting certain details and isolating those (either wilfully or unconsciously) from the rest.

To model the activity of the brain as a 'discrete controlling machine' was a good example of 'selecting certain details', since the brain could, if desired, be described in many other ways. Alan's thesis was, however, that this was the model *relevant* to what was called 'thinking'. As he said a little later,[38] in a parody of Jefferson's argument, 'We are not interested in the fact that the brain has the consistency of cold porridge. We don't want to say "This machine's

quite hard, so it isn't a brain, and so it can't think".' Or as he wrote
in this paper,

> We do not wish to penalise the machine for its inability to shine in
> beauty competitions, nor to penalise a man for losing in a race against an
> aeroplane. The conditions of our game make these disabilities irrelevant.
> The 'witnesses' can brag, if they consider it advisable, as much as they
> please about their charms, strength, or heroism, but the interrogator
> cannot demand practical demonstrations.

There could be arguments about his thesis *within* this model, or
there could be arguments *about* the model. The discussion of
Gödel's theorem was, *par excellence*, one which accepted the model
of a logical system. But alive to the philosophy of science, Alan
discussed the validity of the model itself. In particular, there was the
fact that no physical machine could really be 'discrete':

> Strictly speaking there are no such machines. Everything really moves
> continuously. But there are many kinds of machine which can profitably be
> *thought of* as being discrete-state machines. For instance in considering the
> switches for a lighting system it is a convenient fiction that each switch must
> be definitely on or definitely off. There must be intermediate positions, but
> for most purposes we can forget about them.

That 'forgetting about them' would be precisely the element of
'selecting certain details' necessary to the scientific method. He conceded that the nervous system was itself continuous, and therefore

> certainly not a discrete-state machine. A small error in the information
> about the size of a nervous impulse impinging on a neuron, may make a
> large difference to the size of the outgoing impulse. It may be argued that,
> this being so, one cannot expect to be able to mimic the behaviour of the
> nervous system with a discrete-state system.

But he argued that whatever kinds of continuous or random
elements were involved in the system – as long as the brain worked
in *some* definite way, in fact – it could be simulated as closely as one
pleased by a discrete machine. This was reasonable since it was only
applying the same method of approximation as worked very well in
most applied mathematics and in the replacement of analogue by
digital devices.

Natural Wonders had begun by proposing the question, 'What have I in common with other living things, and how do I differ from them?' Now Alan was asking what he had in common with a computer, and in what ways he differed. Besides the distinction of 'continuous' and 'discrete', there was also that of 'controlling' and 'active' to consider. Here he met the question as to whether his senses, muscular activity and bodily chemistry, were irrelevant to 'thinking', or at least, whether they could be absorbed into a purely 'controlling' model in which the physical effects did not matter. Discussing this problem, he wrote:

> It will not be possible to apply exactly the same teaching process to the machine as to a normal child. It will not, for instance, be provided with legs, so that it could not be asked to go out and fill the coal scuttle. Possibly it might not have eyes. But however well these deficiencies might be overcome by clever engineering, one could not send the creature to school without the other children making excessive fun of it. It must be given some tuition. We need not be too concerned about the legs, eyes, etc. The example of Miss Helen Keller shows that education can take place provided that communication in both directions between teacher and pupil can take place by some means or other.

He was not dogmatic about this line of argument. At the end of the article he wrote (perhaps so as to be on the safe side):

> It can also be maintained that it is best to provide the machine with the best sense organs that money can buy, and then teach it to understand and speak English. This process could follow the normal teaching of a child. Things would be pointed out and named, etc. Again I do not know what the right answer is, but I think both approaches should be tried.

But this was not where he placed his own bets. Later he went as far as to say:[39]

> . . . I certainly hope and believe that no great efforts will be put into making machines with the most distinctively human, but non-intellectual characteristics, such as the shape of the human body. It appears to me to be quite futile to make such attempts and their results would have something like the unpleasant quality of artificial flowers. Attempts to produce a thinking machine seem to me to be in a different category.

In the subjects proposed for automation in 1948, he had been careful to choose those which involved no 'contact with the outside world'. Chess playing, pre-eminently, would involve no relevant fact but the state of the chessboard and the state of the players' brains. The same could certainly be claimed of mathematics, and indeed of any purely symbolic system, involving anything *technical*, any matter of *technique*. He himself had included cryptanalysis in this scope, but hesitated over language translation. The *Mind* paper, however, boldly extended the range of 'intelligent machinery' to general conversation. As such it was vulnerable to his own criticism, that it would require 'contact with the outside world' for this to be possible.

He did not meet the problem that to speak seriously is to *act*, and not only to issue a string of symbols. Speech may be uttered in order to effect changes in the world, changes inextricably connected with the meaning of the words uttered. The word 'meaning' led Polanyi into extra-material, religious connotations, but there is nothing at all supernatural about the mundane fact that human brains are connected with the world by devices other than a teleprinter. A 'controlling machine' was to have physical effects 'as small as we please', but speech, to be audible or legible, has to have a definite physical effect, tied into the structure of the outside world. The Turing model held that this was an irrelevant fact, to be discarded in the selecting of certain details, but the argument for this irrelevance was left weakly supported.

If, as Alan Turing himself suggested, knowledge and intelligence in human beings derive from interaction with the world, then that knowledge must be stored in human brains in some way that depends upon the nature of that interaction. The structure of the brain must connect the words it stores, with the occasions for using those words, and with the fists and tears, blushes and fright associated with them, or for which they substitute. Could the words be stored for 'intelligent' use, within a discrete-state machine model of the brain, unless that model were also equipped with the brain's sensory and motor and chemical peripheries? Is there intelligence without life? Is there mind without communication? Is there language without living? Is there thought without experience? These were the questions posed by

Alan Turing's argument – questions close to those that worried Wittgenstein. Is language a *game*, or must it have a connection with real life? For chess thinking, for mathematical thinking, for technical thinking and any kind of purely symbolic problem-solving, there were arguments of great force behind Alan's view. But in extending it to the domain of all human communication the questions he raised were not properly faced, let alone resolved.

Indeed, they had been faced more openly in the 1948 report, in choosing activities for a 'disembodied' brain. He had narrowed them down to those not requiring 'senses or locomotion'. But even there, in his choice of cryptanalysis as a suitable field for intelligent machinery, he had played down the difficulties arising from human interaction. To portray cryptanalysis as a purely symbolic activity was very much a Hut 8 view of the war, sheltered from the politics and military activity, and trying to work in a self-contained way without interference from outside. The hero of *The Small Back Room* had said rather ironically:

> It's a great pity when you come to think of it that we can't abolish the Navy, the Army and the Air Force and just get on with winning the war without them.

But they could not do without the fighting services. There had to be some integration of Intelligence and Operations, in order for Bletchley to have any meaning. Indeed, the difficulty of the authorities was that of trying to draw a line between them, where no line really existed. The intelligence analysts invaded the field of appreciation. Appreciation held consequences for operations, which in turn were necessary for more effective cryptanalysis. But the Operations actually happened, in the war-winning, ship-sinking physical world. It was hard to believe in Hut 8, where the war was like a dream, but they were actually doing something.

To the mathematicians, it might well be tempting to *regard* the machines and the pieces of paper as purely symbolic. But the fact that they had physical embodiment mattered very much to those for whom knowledge was power. If there was a real secret to Bletchley it lay in the integration of those different kinds of description of its activities: logical, political, economic, social. It was so complex, not just within one system, but in its meshing of many systems, that a

Churchillian 'Spirit of Britain' was as good an explanation of how it worked as any. But Alan had always leant towards keeping his work self-contained, as a technical puzzle, and was resistant to what he regarded as administrative interference. It was the same problem with his model of the brain, as in his work for the Brain of Britain. There was the same problem again, in the fate of the ACE. Having set down a highly intelligent plan, Alan tended to assume that the political wheels would turn as if by magic to put it into effect. He never allowed for the interaction required to achieve anything in the real world.

This was the objection that lay at the heart of Jefferson's remarks, confused as they might be. It was not that Alan avoided it entirely, for he went as far as to concede:

> There are, however, special remarks to be made about many of the disabilities that have been mentioned. The inability to enjoy strawberries and cream may have struck the reader as frivolous. Possibly a machine might be made to enjoy this delicious dish, but any attempt to make one do so would be idiotic. What is important about this disability is that it contributes to some of the other disabilities, *e.g.* to the difficulty of the same kind of friendliness occurring between man and machine as between white man and white man, or black man and black man.

Yet this was not a special, but a very substantial concession, opening up the whole question as to the part played by such human faculties, in the 'intelligent' use of language. This question he failed to explore.

In a rather similar way, he did not avoid giving a direct answer to Jefferson's objection that a machine could not appreciate a sonnet 'because of emotions genuinely felt'. Jefferson's 'sonnets' had about them the quality of Churchill's advice to R. V. Jones:[40] 'Praise the humanities, my boy. That'll make them think you're broadminded!' – and accordingly, Alan fastened on to the phoney culture of this Shakespeare-brandishing, perhaps a little cruelly. He rested his case on the imitation principle. If a machine could argue as apparently genuinely as a human being, then how could it be denied the existence of feelings that would normally be credited to a human respondent? He gave a paradigm conversation to illustrate what he had in mind:

INTERROGATOR: In the first line of your sonnet which reads 'Shall I compare thee to a summer's day', would not 'a spring day' do as well or better?

WITNESS: It wouldn't scan.

INTERROGATOR: How about 'a winter's day'. That would scan all right.

WITNESS: Yes, but nobody wants to be compared to a winter's day.

INTERROGATOR: Would you say that Mr Pickwick reminded you of Christmas?

WITNESS: In a way.

INTERROGATOR: Yet Christmas is a winter's day, and I do not think Mr Pickwick would mind the comparison.

WITNESS: I don't think you're serious. By a winter's day one means a typical winter's day, rather than a special one like Christmas.

But this answer to the objection would prompt the same questions about the role of interaction with the world in 'intelligence'. This play with words was the strawberries and cream, and not the meat, of literary criticism. It was a view of sonnets from the back of Ross's English class! Where lay the 'genuine feeling'? What Jefferson could well have intended, was something more like intellectual *integrity* than examination mark-scoring: truthfulness or sincerity pointing to some connection between the words, and experience of the world. But such integrity, a constancy and consistency in word and action, could not be enjoyed by the discrete-state machine alone. The issue would be clearer if the machine were confronted with a question such as 'Are you or have you ever been . . .' or 'What did you do in the war?' Or, staying with the sexual guessing game, asked to interpret some of the more ambiguous of Shakespeare's sonnets. If asked to discuss proposed alterations to literature, Dr Bowdler's preference for

> Under the greenwood tree
> Who loves to *work* with me

would make a telling topic. Questions involving sex, society, politics or secrets would demonstrate how what it was possible for people to *say* might be limited not by puzzle-solving intelligence but by the restrictions on what might be *done*. Such questions, however, played no part in the discussion.

Alan disliked anything with a churchy or pretentious flavour, and employed a light style with homely metaphors in order to make his serious points. It was in the Apostolic tradition, and also shared with Samuel Butler and Bernard Shaw. But rather like those writers, his examples of 'intelligence' could be accused of a touch of the blarney, of arguing for the sake of it, of mere cleverness, or of making debating points. He enjoyed the play of ideas – but a logical jousting with God and Gödel, the Lion and Unicorn tussle of free will and determinism, was not enough.

It was not necessary to be either 'soupy' or pretentious in order to approach the questions of 'thinking' or 'consciousness' in another way. It was the year 1949 that saw *Nineteen Eighty-Four* – a book that Alan read, impressed: it elicited from him an unusually political comment when talking with Robin Gandy: '. . . I find it very depressing. . . . I suppose absolutely the *only* hope lies in those proles.' Orwell's discussion of the capacity of political structure to determine language, and language to determine thought, was itself highly relevant to Alan Turing's thesis. And Orwell might have been thinking of the Turing sonnet-writing computer, with his 'versificators', machines to turn out popular songs.

But that was not the central issue, for Orwell was not concerned to reserve for human beings the intelligent, indeed intellectually satisfying, work of rewriting history at the Ministry of Truth. His passion was reserved for intellectual integrity: keeping the mind whole, keeping it in contact with external reality. 'You must get rid of those nineteenth-century ideas about the laws of Nature,' O'Brien told Winston Smith. 'We make the laws of nature. . . . Nothing exists except through human consciousness.' Here lay Orwell's fear, and to counter it he seized upon scientific truth as an external reality that political authority could not gainsay: 'Freedom is the freedom to say two and two make four.' He added in the unchangeable past, and sexual spontaneity, as things that *were so*, whatever anyone said. Science and sex! – they had been the two things that allowed Alan Turing to jump out of the social system in which he was trained. But the machine, the pure discrete-state machine, could have none of this. Its universe would be a void, but for the word of its teacher. It might as well be told that space was five-dimensional, or even that two and two made five when Big Brother decreed. How could it 'think for itself', as Alan Turing asked of it?

As they might say on *The Brains Trust*, it all depended upon what was meant by 'intelligence'. When Alan first began to use the word, it was applied to chess playing and other kinds of puzzle-solving. It was a sense which accorded well with the wartime and immediate post-war mood, in which intelligence was what Hut 8 had, and the Admiralty did not. But people had always used the word in a broader sense, involving some insight into reality, rather than the ability to achieve goals or solve puzzles or break ciphers. This discussion was missing from 'Computing Machinery and Intelligence'. There was only his passing comment about Helen Keller to justify a claim that the means of communication, the interface between brain and world, would be irrelevant to the acquisition of intelligence. But this was a slight argument for so central a question. Even Bernard Shaw, in his irrational way, had put his finger on the problem that Alan ducked:

> PYGMALION. But they are conscious. I have taught them to talk and read; and
> now they tell lies. That is so very lifelike.
> MARTELLUS. Not at all. If they were alive they would tell the truth.

Inevitably, Alan's choice of emphasis reflected his background and experiences. As a mathematician, he was concerned with the symbolic world. And more than this, the *formalist* school of mathematics, which had given such a start to his career, had explicitly been concerned to treat mathematics as if it were a chess game, without asking for a connection with the world. That question was, as it were, always left for someone else to tackle. The game-like character of formalism showed itself in the present discussion, matching the Looking Glass quality of these 'interrogations'. It might be said, in fact, that the kind of machine behaviour that he was describing, a behaviour unrelated to action, was not so much the ability to *think* as the ability to *dream*.

The discrete-state machine, communicating by teleprinter alone, was like an ideal for his own life, in which he would be left alone in a room of his own, to deal with the outside world solely by rational argument. It was the embodiment of a perfect J. S. Mill liberal, concentrating upon the free will and free speech of the individual. From this point of view, his model was a natural development of the argument for his definition of 'computable' that he had framed in

1936, the one in which the Turing machine was to emulate anything done by the individual mind, working on pieces of paper.

On the other hand, he knew better than that, his strength lying in a seriousness directly applied to reality, rather than in ingenuity applied to puzzles. In 1938, his 'Ordinal Logics' paper had carried the comment: 'We are leaving out of account that most important faculty which distinguishes topics of interest from others; in fact, we are regarding the function of the mathematician as simply to determine the truth or falsity of propositions.' He himself had carefully chosen topics of interest for the application of his mind; topics that *mattered*. This crucial faculty could find no room in the discrete-state machine, depending as it did upon contact with reality. But more than this – he had to live in the world and communicate, like anyone else. And his fascination with computers had a complementary aspect, one of extreme consciousness of the social rules and conventions placed upon him. Puzzled since childhood by the 'obvious duties', he was doubly detached from the imitation game of social life, as pure scientist and as homosexual. Manners, committees, examinations, interrogations, German codes and fixed moral codes – they all threatened his freedom. Some he would accept, some actually enjoy obeying, others reject, but in any case he was peculiarly conscious, self-conscious, of things that other people accepted 'without thinking'. It was in this spirit that he enjoyed writing formal 'routines' for the computer, just as he enjoyed Jane Austen and Trollope, the novelists of social duty and hierarchy. He enjoyed making life into a game, a pantomime. He had done his best to turn the Second World War into a game. Again, it was expressed in his other 1936 argument for computability, in which the Turing machine was to do anything conventional, anything for which the rules were laid down.*

The free individual, sometimes working with the social machine, more often against it, learning by 'interference' from outside, yet

---

* This oscillation between the two concepts of computability found its way into his *Programmers' Handbook*: on the first page the programmer was greeted by the assertion that 'There is also a part of the machine called the control which corresponds to the [human] computer himself. If his possible behaviour were very accurately represented this would have to be a formidably complicated circuit. However we really only require him to be able to obey the written instructions and these can be made so explicit that the control can be quite simple.'

resenting that interference: the interplay between intelligence and duty; the abrasion and stimulation of interaction with the environment – this was his life. While all these elements were reflected in his ideas about machine intelligence, they were not all satisfactorily brought together. He had not tackled the question of the channels of communication, nor explored the physical embodiment of the mind within the social and political world. He had brushed these aside light-heartedly. He had not always done so, once writing to Mrs Morcom of how we could live free as spirits and communicate as such, 'but there would be nothing whatever to do.' Thinking and doing; the logical and the physical; it was the problem of his theory, and the problem of his life.

In the summer of 1950 he decided to end his life of suitcases and landladies' crockery. He bought a house in Wilmslow, the middle-class dormitory town in Cheshire, ten miles to the south of Manchester. It was a semi-detached Victorian house on a line of development that lay on the further side of the railway station, and which formed its own identity as the village of Dean Row. The fields and the hills of the Peak District lay immediately at the back. Here at least he was free. Neville thought he should not live alone, but he was no more alone by himself than he was amidst the madding crowd. Neville himself had finished his Cambridge statistics course and had managed to get a job with an electronics company near Reading, where he went to live with his mother. It was now much more difficult for them to meet, and this made another change in Alan's life.

The house, 'Hollymeade', was rather larger than he needed – a little selfish, perhaps, in the housing crisis of 1950. The quite good pieces of furniture that he acquired never overcame the sense of sparseness, and a somewhat temporary flavour. Certainly his ideas on how to live had little in common with his respectable neighbours – but there was one piece of good luck. His immediate neighbours, in the other half of the building, were the friendly Webbs. Roy Webb, as it happened, had been Alan's almost exact contemporary at Sherborne, and was now a Manchester solicitor. They welcomed him for cups of tea and the occasional dinner; Alan used their telephone, never having one installed himself; they shared their gardens, the Webbs cultivating part of Alan's patch. Alan offered gardening, along with chess and long-distance running, as his recreations in *Who's Who*.[41] But it was more pottering in the wilderness of nature than arranging the trim lawns of suburbia. 'Things don't grow in the winter,' he told Roy Webb, explaining his *laissez-faire* attitude to the vegetable world. The Webbs grew used to seeing him at any time in vest and shorts, and they also had him baby-sit for their son Rob, who had been born in 1948. Alan enjoyed this; it would have been of intellectual interest to him to see a brain awakening into conscious speech, but it was also a simple delight in communication that the little boy reciprocated. Later on they would sit together on the Webbs' garage roof, and were once heard discussing subversively whether, if God sat on the ground, He would catch cold.

Having his own home gave him more opportunities to play the desert-island game, of using his ingenuity to make things he needed for himself. He wanted a brick path, and at first wanted to fire the bricks himself, like the chess set at Bletchley, but settled for ordering a load. He did the laying himself, but found he had grossly underestimated the cost, and for this reason the path was never finished. As in the war, stories like this helped people to cope with his more forbidding aspects, and as in the war, his messy, spartan environment was far more striking to those unfamiliar with Cambridge dons. It also upset those who supposed a middle-class man incapable of doing anything with his hands.

Alan did not, however, achieve a self-sufficient existence: he cheated by having a Mrs C—— to shop and clean for him on four afternoons a week. Indeed he could be seen as hankering after having someone to look after him, and to give him the home comforts that he was unwilling, or unable, to provide for himself. He would have liked the convenience, but not the fuss and interference, of domestic life. The ordinary life of the Webbs next door gave him some contact with what he missed in this respect. But he did learn to cook for himself, so that Mrs Webb found herself not only explaining how to dry socks, but giving all the details on how to make a sponge cake. Alan enjoyed showing off to visitors a new skill, remote from his education but close to his own experiments as a little boy.

Not many visitors trudged a mile along the road from the station past the RAF camp. Sometimes junior engineers were invited down, and might collect some apples; Bob and his wife came over once or twice, before he went to work abroad. Robin Gandy was a regular visitor, for at least once every term he would come for a weekend from Leicester where since October 1949 he had been a lecturer at the University College. Alan had by now become his PhD supervisor. They would mainly discuss the philosophy of science, though Robin's interest was turning more and more towards mathematical logic itself, rather than the logic of science, and his work was meeting up with Alan's. In fact like the White Knight, who was interested in songs, and the names of songs, and the names of the names of songs, Robin had become primarily interested in the theory of types, reviving Alan's interest in the subject. They might also do a few jobs in the house or the garden together, and afterwards there would always be a bottle of wine with the dinner – which Alan would mull by plunging it in a jug of very hot water. That was an invariable rule, and another was to put the cork back in the bottle after the meal was finished, even if Robin would have preferred to finish it off. After the meal, as they did the washing-up, there would be some exercise to perform, such as working out how it could be that trees drew water up more than thirty feet.

There might well have been another kind of occasional visitor to his life, if not to his house, one that came through the trade entrance. For all the time there was another England, on towpaths or trains, in pubs, parks, toilets, museums, at swimming baths, bus stations, shop windows, or just looking back in the street, for those who had eyes to see; a communication network of flashing eyes, millions and millions of them, sundered from the lobotomised culture of official Britain, but to which Alan Turing belonged. Before the war he would have been too shy, but by 1950 he had made some discoveries. Traditionally, for the upper-middle-class homosexual man, there was Paris, and going abroad was a double escape, both from the English law, and from the class system which enveloped an Englishman as soon as he opened his mouth. But England had its opportunities too. Alan always used to stay in the YMCA in London, if only because it would hardly occur to him to pay to stay anywhere grander, and this would have held something for his eye in the shape of naked youths in its swimming pool, if nothing more. But Manchester was another story.

Walking up to the city centre from the Victoria University, there was a point where Oxford Road became Oxford Street, just under the railway bridge. Here it was a long way indeed from the dreaming spires at the other end of the A34. There were a couple of cinemas, an amusement arcade, a pub – the Union Tavern – and a very early example of the milk bar. This stretch between the urinal and the cinema was where the male homosexual eye was focussed – perhaps the same block as trodden by Ludwig Wittgenstein in 1908, such unofficial institutions lasting as long as the respectable kind. Here straggled a motley convoy of souls, and amidst them the odd independent sailing, like Alan Turing. Here merged many kinds of desire – for physical excitement, for attention, for a life outside family and factory confines, or for money. These were not sharp divisions. Money, if involved, was little but the clink of 'tipping' that was heard in any encounter between different social classes; and indeed little different from the way that women could expect to be entertained and treated by men. The most special relationship would have its *quid pro quo*, and this kind was more likely for ten bob than a quid. This was how it was done in the ordinary England of 1950, outside privileged circles such as those of Cambridge or Oxford. For the young, in particular, without means or private space, homosexual desire meant street life. Desert-island sex, managing on the minimum of social resources, and only noticed when something went wrong, this was not for the respectable man. But Alan was above respectability.

Candide had retired to *cultiver son jardin*, the back garden of science. But what was his 'programme for the immediate future'? The past two years had offered flashbacks to the successes of his earlier life. The usual academic course would be to build upon those successes, extracting as much as possible from them. But this was not his path; he had to find something new in order to continue. And he now began to draw upon something that had, in fact, been there all the time, but which only now began to emerge into light. It was as though a long sub-routine, that had begun with Christopher Morcom, passing through Eddington and von Neumann, Hilbert and Gödel, *Computable Numbers*, his war machines and mechanical processes, the relays and the electronics and the ACE, the programming of computers and the groping towards intelligent machinery – all *that* stream of enquiry had tapered off, and left him free to continue from where schooling had interrupted him.

Already there had been a hint in *Intelligent Machinery*, for Sir Charles Darwin's benefit:

> The picture of the cortex as an unorganised machine is very satisfactory from the point of view of evolution and genetics. It clearly would not require any very complex system of genes to produce something like the . . . unorganised machine. In fact this should be much easier than the production of such things as the respiratory centre. . . .

Somehow the brain did it, and somehow brains came into being every day without all the fuss and bother of the minnow-brained ACE. There were two possibilities: either a brain learnt to think by dint of interaction with the world, or else it had something written in it at birth – which must be programmed, in a looser sense, by the genes. Brains were too complicated to consider at first. But how did anything know how to grow? There lay the question – a question that the smallest child could ask, and the question that *Natural Wonders Every Child Should Know* had placed at the centre. When embarking upon the delicate subject of 'what little boys and girls are really made of', E. T. Brewster had launched into a description of the growth of starfish, beginning with

> the young egg, before there is any sign of a growing creature inside. One would perhaps expect to see the oil and jelly mixture change gradually into a star-fish. Instead of this, however, this little balloon-like affair splits squarely into two, and makes two little balloons just alike, and which lie side by side. . . . In about half an hour, each of these balloons or bubbles, 'cells' as they have come to be called, has divided again; so that now there are four. The four soon become eight; the eight, sixteen. In the course of a few hours, there are hundreds, all sticking together and all very minute; so that the whole mass looks like the heap of soap bubbles which one blows by putting the pipe under the surface of the soap suds.

From this sphere of cells, Brewster explained, the animal would take shape:

> If it is an animal like ourselves, this body stuff, before it becomes a body, is a round ball. A furrow doubles in along the place where the back is to be,

and becomes the spinal cord. A rod strings itself along underneath this, and becomes the back-bone. The front end of the spinal cord grows faster than the rest, becomes larger, and is the brain. The brain buds out into the eyes. The outer surface of the body, not yet turned into skin, buds inward and makes the ear. Four outgrowths come down from the forehead to make the face. The limbs begin as shapeless knobs, and grow out slowly into arms and legs. . . .

Alan had thought about embryology all the time, fascinated by the fact that how such growth was determined was something 'nobody has yet made the smallest beginning at finding out.' There had been little advance since *Growth and Form*, the 1917 classic that he had read before the war. The 1920s had made it possible to invoke the Uncertainty Principle to suggest that life was intrinsically unknowable, like the simultaneous measurement of position and velocity in quantum mechanics.[42] As with minds, there was an aura of religion and magic around the subject that attracted the attention of his scepticism. It was a fresh field. C. H. Waddington's 1940 standard work[43] on embryology did no more than list experiments on growing tissue, to explain in what circumstances it seemed to know what to do next.

The greatest puzzle was that of how biological matter could assemble itself into patterns which were so enormous compared with the size of the cells. How could an assemblage of cells 'know' that it must settle into a fivefold symmetry, to make a starfish? How was this symmetry communicated across millions of cells? How could the Fibonacci pattern of a fir cone be imposed in its harmonious, regular way upon the growing plant? How could matter *take shape* or, as biological Greek had it, what was the secret of *morphogenesis?* Suggestive words like 'morphogenetic field', vague as the Life Force, were employed by biologists to describe the way that embryonic tissue seemed to be endowed with an invisible pattern which subsequently dictated its harmonious development. It had been conjectured that these 'fields' could be described in chemical terms – but there was no theory of how this could be.* Polanyi

---

* Thus a contemporary review article[44] stated 'The importance of the principle of patterned "field" activities in the determination of embryonic systems has been generally recognised. . . . Yet their nature and mode of operation are still among the greatest puzzles of modern biology.'

believed that there was *no* explanation except by a guiding *esprit de corps*; the inexplicability of embryonic form was one of his many arguments against determinism.[45] Conversely, Alan told Robin that his new ideas were intended to 'defeat the Argument from Design'.

Alan was familiar with Schrödinger's 1943 lecture, *What is life*, which deduced the crucial idea that genetic information must be stored at molecular level, and that the quantum theory of molecular bonding could explain how such information could be preserved for thousands of millions of years. At Cambridge, Watson and Crick were busy in the race against their rivals to establish whether this was really so, and how. But the Turing problem was not that of following up Schrödinger's suggestion, but that of finding a *parallel* explanation of how, granted the production of molecules by the genes, a chemical soup could possibly give rise to a biological pattern. He was asking how the information in the genes could be translated into action. Like Schrödinger's contribution, what he did was based on mathematical and physical principle, not on experiment; it was a work of scientific imagination.

There were other suggestions in the literature for the nature of the 'morphogenetic field', but at some point Alan decided to accept the idea that it was defined by some variation of chemical concentrations, and to see how far he could get on the basis of that one idea. It took him back to the days of the iodates and sulphites, to the mathematics of chemical reactions. But the new problem was of another order. It was not merely to examine substance A changing into substance B, but to discover circumstances in which a mixture of chemical solutions, diffusing and reacting with each other, could settle into a pattern, a pulsating pattern of chemical waves; waves of concentration into which the developing tissue would harden; waves which would encompass millions of cells, organising them into a symmetrical order far greater in scale. This was the fundamental idea, parallel to Schrödinger's – that a chemical soup could contain the information required to define a large-scale chemical pattern in space.

There was one central, fundamental problem. It was exemplified in the phenomenon of *gastrulation*. This was the process described and illustrated in *Natural Wonders*, in which a perfect sphere of cells would suddenly develop a groove, determining the head and tail ends of the emergent animal. The problem was this: if the sphere were symmetrical, and the chemical equations were symmetrical, without knowledge of left or right, up or down, where did this *decision* come from? It was just such a phenomenon that inspired Polanyi to claim that some immaterial force must be at work.

In some way information was being *created* at this point, and this went against what was normally expected. When the lump of sugar has been dissolved in the tea, no information remains, at the chemical level, as to where it was. But in certain phenomena, those of crystallisation for instance, the reverse process could occur. Pattern could be created, rather than destroyed. The explanation lay in the interplay of more than one level of scientific description. In the *chemical* description, in which only average concentrations and pressures would be considered, no spatial direction would be preferred to any other. But at a more detailed, Laplacian level, the individual motions of the molecules would not be perfectly symmetrical, and under certain conditions, like that of the crystallising liquid, could serve to pick out a direction in space. The example Alan chose as illustration was drawn from his electrical experience:[46]

> The situation is very similar to that which arises in connexion with electrical oscillators. It is usually easy to understand how an oscillator keeps going when once it has started, but on a first acquaintance it is not obvious how the oscillation begins. The explanation is that there are random disturbances always present in the circuit. Any disturbance whose frequency is the natural frequency of the oscillator will tend to set it going. The ultimate fate of the system will be a state of oscillation at its appropriate frequency, and with an amplitude (and a wave form) which are also determined by the circuit. The phase of the oscillation alone is determined by the disturbance.

He set up a system of oscillating circuits in his office, and used to show people how they would gradually all come to resonate with each other.

Such a process of toppling, or crystallising, or falling into some pattern of oscillation, could be described as the resolution of an unstable equilibrium. In the case of the developing sphere of cells, it would have to be shown that in some way, through a change in

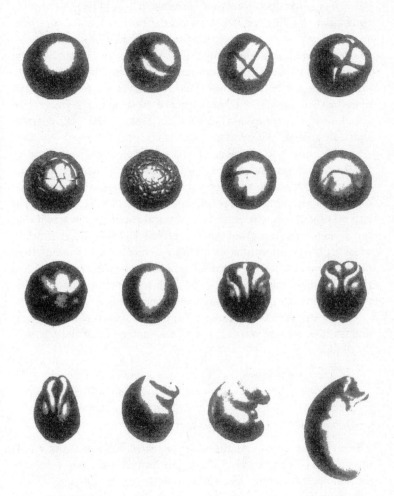

The process of gastrulation, essentially as revealed in *Natural Wonders*.

temperature or the presence of a catalyst, the stable chemical balance could suddenly become unstable. It would be the chemical equivalent of piling the last straw on the camel's back. Alan's own analogy was that of a mouse climbing up a pendulum.

Here was an idea which might explain something about how the information in the genes could be translated into physiology. The problem of growth as a whole would be far, far more complicated than this. But analysis of this moment of creation might yield a clue as to how the harmony and symmetry of biological structures could suddenly appear, as if by magic, out of nothing.

To examine this moment of crisis mathematically, one had to approximate over and over again. He had to ignore the internal structure of the cells, and forget that the cells would themselves be moving and splitting as the patterning process took place. There was also an obvious limitation to the chemical model. How was it that the human heart was always on the left hand side? If this symmetry-breaking of the primordial sphere were determined at random, then hearts would be equally distributed between left and right. He had to leave this problem on one side, with a conjecture that at some point the asymmetry of the molecules themselves would play a part.

But with these reservations, his approach was to take the model and try it out. As he wrote,[47] in a classic statement of the scientific method:

> . . . a mathematical model of the growing embryo will be described. This model will be a simplification and an idealization, and consequently a falsification. It is to be hoped that the features retained for discussion are those of greatest importance in the present state of knowledge.

The result was applied mathematics *par excellence.* Just as the simple idea of the Turing machine had sent him into fields beyond the boundaries of Cambridge mathematics, so now this simple idea in physical chemistry took him into a region of new mathematical problems. But this time at least it was all his own work. No one else could make a mess of it for him.

Even with immense simplifications, the mathematical equations corresponding to a soup of only four interacting chemical solutions were still too intractable. The problem was that chemical reactions were 'non-linear'. The equations for electricity and magnetism were 'linear', meaning that if two electromagnetic systems were superimposed (as for example when two radio transmitters radiated simultaneously), then the effects would simply be additive. The two transmitters would not interfere with each other. But chemistry was quite different. Double the concentration of reactants, and the reaction might go four times as fast. Superimpose two solutions, and anything might happen! Such 'non-linear' problems had to be solved as a whole, and not by the mathematical methods familiar in electromagnetic theory, those of describing the system as the sum of many little bits. However, that critical moment of budding, at the instant when the unstable system crystallised into a pattern, could be treated as if it were a 'linear' process – a fact familiar to applied mathematicians, and one which gave him a first handle upon the problem of growth.

So he had his hands on another central problem of life, this time not of the mind, but of the body, although both questions were related to the brain. He quite literally had his hands upon it, for he had always enjoyed examining plants when on his walks and runs, and now he began a more serious collection of wild flowers from the Cheshire countryside, looking them up in his battered *British Flora*,[48] pressing them into scrapbooks, marking their locations in large-scale maps, and making measurements. The natural world was overflowing with examples of pattern; it was like codebreaking, with millions of messages waiting to be decrypted. Like codebreaking, the field was open-ended; with his chemical model he had one sharp tool to apply to it, but that was only the beginning.

Mrs Webb was just the latest person to be given a talk on the Fibonacci spiral pattern of the fir cone, the pattern which showed itself also in the seeds in sunflower heads, and the leaf arrangements of the common plants. It was the problem of explaining its occurrence in nature that he set himself as a serious challenge. But this required the analysis of a two-dimensional *surface,* and he chose to leave it while he first considered in detail some rather simpler cases.

In a chapter called 'Nature's Repair Shop', Brewster had dwelt upon the regeneration of the *Hydra*, the small fresh-water worm, which could grow a new head or new tail from any chopped-off section. Alan took the hydra, with its simple tubular form, and simplified it yet again, by neglecting its length, and concentrating upon the idea of a *ring* of cells. Then he found, taking a model of just two interacting chemicals reacting and diffusing around this ring, that he was able to give a theoretical analysis of all the different possibilities for the moment of budding. And the idea, although admittedly in a grossly oversimplified and hypothetical way, actually worked. It appeared that under certain conditions the chemicals would gather into stationary waves of concentration, defining a number of lobes on the ring. These, it could be imagined, would form the basis for the pattern of tentacles. The analysis also showed the possibility of waves gathering into asymmetrical lumps of concentration, which reminded him of the irregular blotches and stripes on animal hides. With this last idea he did some experimental numerical work. By the end of 1950 the prototype computer had been closed down, and the scientists at the university were waiting for the new one to arrive from Ferranti, so this work was done on a desk calculator. It produced a dappling pattern rather resembling that of a Jersey cow. He was beginning to *do something* once more.

At Christmas 1950 Alan met J. Z. Young again, to follow up the discussion about brain-cells that they had had in October 1949. Young had just given the Reith Lectures for 1950,[49] presenting a rather aggressive statement of the claims of neuro-physiology to explain behaviour. Young later recalled[50] Alan's

> . . . kindly teddy-bear quality as he tried to make understandable to others, ideas that were still only forming in his own mind. To me, as a non-mathematician, his exposition was often difficult to follow, accompanied as it was by funny little diagrams on the blackboard and frequently by generalizations, which seemed as if they were his attempt to press his ideas on me. Also, of course, there was his rather frightening attention to everything one said. He would puzzle out its implications often for many hours or days afterwards. It made me wonder whether one was right to tell him anything at all because he took it all so seriously.

They talked about the physiological basis of memory and of pattern recognition. Young wrote:[51]

13.1.51

Dear Turing,

    I have been thinking more about your abstractions and hope that I grasp what you want of them. Although I know so little about it I should not despair of the matching process doing the trick. You have certainly missed a point if you suppose that to name a bus it must first be matched with everything from tea-pots to clouds. The brain surely has ways of shortening this process by the process – I take it – you call abstracting. Our weakness is that we have so little idea of the clues and code it uses. My whole thesis is that the variety of objects etc. are recognised by use of comparison with a relatively limited number of models. No doubt the process is a serial one, perhaps a filtering out of recognised features at each stage and then feeding back the rest through the system.

This probably does not make much sense in exact terms and the only evidence for it is that people do group their reactions around relatively simple models – circle, god, father, machine, state, etc.

Can we get anywhere by determining the storage capacity given by 10^{10} neurons if arranged in various ways and assuming facilitation of pathways by use? Is there any finite number of sorts of arrangement that they could have? For example, each with 100 possible outputs to others arranged a) at random through the whole or b) with decreasing frequency with distance. Given any particular plan of feedback can one compare the storage capacity of these plans, assuming say a given increase of probability of re-use of a pathway with each time of use?

This is all very vague. If you have any ideas about the next important sorts of question to ask do let me know. Would it be a great help if we could give some sort of specification of the destinations of the output (within the cortex) of each cell? I feel we ought to be able to disentangle the pattern somehow.

Yours, John Young.

Alan's reply made clear the connection between his interests in the logical and the physical structure of the brain:

8th February 1951

Dear Young,

I think very likely our disagreements are mainly about the use of words. I was of course fully aware that the brain would not have to do comparisons of an object with everything from teapots to clouds, and that the identification would be broken up into stages, but if the method is carried very far I should not be inclined to describe the resulting process as one of 'matching'.

Your problem about storage capacity achievable by means of N (10^{10}

say) neurons with M (100 say) outlets and facilitation is capable of solution which is quite as accurate as the problem requires. If I understand it right, the idea is that by different trainings certain of the paths could be made effective and the others ineffective. How much information could be stored in the brain in this way? The answer is simply MN binary digits, for there are MN paths each capable of two states. If you allowed each path to have eight states (whatever that might mean) you would get 3MN. . . .

I am afraid I am very far from the stage where I feel inclined to start asking any anatomical questions. According to my notions of how to set about it that will not occur until quite a late stage when I have a fairly definite theory about how things are done.

At present I am not working on the problem at all, but on my mathematical theory of embryology, which I think I described to you at one time. This is yielding to treatment, and it will so far as I can see, give satisfactory explanations of –

(i) Gastrulation
(ii) Polygonally symmetrical structures, e.g. starfish, flowers.
(iii) Leaf arrangement, in particular the way the Fibonacci series (0,1,1,2,3,5,8,13. . . .) comes to be involved.
(iv) Colour patterns on animals, e.g. stripes, spots and dappling.
 v) Pattern on nearly spherical structures such as some Radiolaria, but this is more difficult and doubtful.

I am really doing this now because it is yielding more easily to treatment. I think it is not altogether unconnected with the other problem. The brain structure has to be one which can be achieved by the genetical embryological mechanism, and I hope that this theory that I am now working on may make clearer what restrictions this really implies. What you tell me about growth of neurons under stimulation is very interesting in this connection. It suggests means by which the neurons might be made to grow so as to form a particular circuit, rather than to reach a particular place.

Yours sincerely, A. M. Turing

A few days later, the Ferranti Mark I computer was delivered at the Manchester department, which by now had a newly built Computing Laboratory to house it. Alan wrote to Mike Woodger back at the NPL:

Our new machine is to start arriving on Monday [12 February 1951].

I am hoping as one of the first jobs to do something about 'chemical embryology'. In particular I think one can account for the appearance of Fibonacci numbers in connection with fir-cones.

It had been twenty-one years, and the computer had come of age. It was as though all that he had done, and all that the world had done to him, had been to provide him with an electronic universal machine, with which to think about the secret of life.

Much of the computer installation that he had imagined for the ACE had now come into being; people were soon to come to it with their problems; the 'masters' would program it and the 'servants' service it. They did indeed build up a library of programs. (In fact, it was just about Alan's last contribution to the Manchester computing system that he laid down a way of writing and filing a formal description of the programs intended for common use.) He had a room of his own in the new computer building, and was, at least in theory, the chief 'master'. The engineers moved on to design a second, faster machine (in which he took no interest whatever), and it was up to him to take charge of the use of the first one.

There was plenty that could be done in the way of seminars and publications and demonstrations, for this was the world's first commercially available electronic computer, beating by a few months the UNIVAC made by Eckert and Mauchly's firm. It also enjoyed the firm support of the British government, whose National Research Development Corporation, chaired by the administrator Lord Halsbury, managed the investment, sales and patent protection after 1949. In fact they went on to sell eight copies of the Mark I, the first to the University of Toronto, for the design of the St Lawrence Seaway, then others* more discreetly to the Atomic Weapons Research Establishment and to GCHQ. With Alan fulfilling a consultant role for GCHQ, it may be reasonably supposed that he played a part in suggesting how they should use the universal machine he had promised to Travis six years or so before. But this was not where his heart lay now. As electronic computers began to impinge upon the world economy, Alan Turing continued to back away, and remained engrossed in the otherwise forgotten 'fundamental research'.

---

* More strictly, all but the Manchester and Toronto machines were of a slightly modified design, the Mark 1*.

A big inaugural conference was planned for July, but this work was done entirely by the engineers and Ferranti Ltd. It was not that Alan got in the way; he simply avoided participation. No one could have guessed that officially he was paid to 'direct' the laboratory. In the spring of 1951 he found a way to off-load his remaining responsibilities when R. A. Brooker, a young man from the Cambridge EDSAC team, called in to have a look at the new machine on his way back from a climbing weekend in west Wales. For reasons of his own he liked the idea of moving to the North, and asked Alan if he had a job to offer. Alan said he did, and in fact Tony Brooker joined later in the year.

Alan's detachment was annoying to the engineers, who felt that their achievement was hardly getting the recognition it deserved within the mathematical and scientific world. In many ways the Computing Laboratory remained as secret as Hut 8, just as computation remained the lowest form of mathematical life. Recognition did, however, come to Alan Turing. In the 1951 elections, which took place on 15 March, he became a Fellow of the Royal Society. The citation referred to his work on computable numbers which had, of course, been done fifteen years earlier. Alan was rather amused by this and wrote to Don Bayley (who had sent his congratulations) that they could not really have made him an FRS when he was twenty-four. The sponsors were Max Newman and Bertrand Russell. Newman had lost all interest in computers and was only grateful that Alan had regenerated his pulse with the morphogenetic theory.

Jefferson, himself a Fellow since 1947, also sent a letter of congratulation,[52] saying 'I am so glad; and I sincerely trust that all your valves are glowing with satisfaction, and signalling messages that seem to you to mean pleasure and pride! (but don't be deceived!).' He managed to confuse the logical and the physical levels of description even within one sentence. Alan would refer to Jefferson as an 'old bumbler' because he never grasped the machine model of the mind, but Jefferson certainly found an apt description of Alan,[53] as 'a sort of scientific Shelley'. Apart from the more obvious similarities, Shelley also lived in a mess,[54] 'chaos on chaos heaped of chemical apparatus, books, electrical machines, unfinished manuscripts, and furniture worn into holes by acids,' and Shelley's voice too was 'excruciating; it was intolerably shrill, harsh and discordant.' Alike they were at the centre of life; alike at the margins of respectable society. But Shelley stormed out, while Alan continued to push his way through the treacly banality of middle-class Britain, his Shelley-like qualities muted by the grin-and-bear-it English sense of humour, and filtered through the prosaic conventions of institutional science.

Mrs Turing was very proud of the Fellowship, a title which raised Alan to the eminence of George Johnstone Stoney, and laid on a party at Guildford where her friends could meet him – hardly the function to appeal to Alan, who once walked wordlessly out of a sherry party to which his brother had invited him, after a bare ten minutes. His mother found it hard to overcome her amazement that important personages could speak well of her Alan, but in this respect she was making progress, and had come a long way since the 1920s. Although Alan complained to his friends of her patronising fussiness and religiosity, there remained the fact that she was one of the few people who took an interest in his doings. Mostly this came out in her efforts to improve Alan's domestic life, with instructions on the right and wrong ways to perform each little routine. ('Mother says . . .', Alan would explain to Robin, with a half-amused, half-exasperated twinkle.)

Nor was their contact very frequent; Alan would visit Guildford about twice a year, annoying both mother and brother by announcing his imminent arrival with a telegram or postcard and no more. His mother made the journey to Wilmslow once in the summer each year. Besides the postcards there might be a few telephone calls; Alan finding for instance that they both liked the stories on *Children's Hour* and telling her when a good one was coming. But Mrs Turing did like to have the feeling of involvement in Alan's work, and felt happier with the biology than with the computers. Although she really had no clue as to what he was doing at Manchester, she would help with the wild flowers and the big maps. With her nineteenth-century optimism she construed it in terms of usefulness to humanity, bringing him closer to the Pasteur of whom she had long ago dreamed. Perhaps, she speculated, it might lead to a cure for cancer! It was not a foolish connection to draw, but it was not his motive. Nor was there any way of knowing where his Faustian dabblings might lead this time; they were just as relevant to the state-controlled embryology of *Brave New World*. Even if his practical methods had something in common with the naive natural history of the past century, and even if it meant a return to childhood fascinations, his work lay firmly within the great modernisation of biology, in which the technical advances of the 1930s were being followed by the application of the quantitative analysis so triumphant in physics and chemistry. The problem of life could no longer be allowed to lag behind; they had to know how its machinery worked.

Appreciation in the Computing Laboratory was on a more down-to-earth level. There history began in 1951, and no one connected with the computer knew about *Computable Numbers*. At the NPL there had been strong connections with Cambridge mathematics and with the Royal Society; the new masters of the Mark I were quite a different crew, and had no sense of his past. Nor did Alan try to explain. An applied mathematics research student, N. E. Hoskin, had just become involved in using the new computer, and when he said over coffee 'I never thought of *you* as an FRS,' Alan just laughed – with that wince-inducing, mechanised laugh.

He did look young for an FRS, although at thirty-eight he was by no means the youngest to be elected. Hardy had been elected at thirty-three, and the self-taught Indian mathematician Ramanujan at thirty. Maurice Pryce was also elected in 1951, so in this respect Alan had caught up a year on the mathematical physicist, whom he did not meet again after the war. Writing to Philip Hall at King's, who also had congratulated him, he said it was 'very gratifying to be about to join the Olympians'. After a mathematical description of his 'waves on cows' and 'waves on leopards', he added 'I am delighted to hear Maurice Pryce is also in the list. I met him first when up for scholarship exam in 1929, but knew him best in Princeton. He was quite my chief flame at one time.' In what was more of a mathematician's joke he wrote, 'I hope I am not described as "distinguished for work on unsolvable problems".'

In his retirement from the organisation of the laboratory, it barely impinged on Alan that the new computer was used to perform calculations for the British atomic bomb. A young scientist, A. E. Glennie, spent time at Manchester on this work. He would sometimes chat with Alan about mathematical methods, though it went no further than talk in general terms. Once, however, Alick Glennie found himself collared by Alan, when he wanted a 'mediocre player' on whom to try out his current chess-playing program. They went back to Alan's room for three hours in the afternoon. Alan had all the rules written out on bits of paper, and found himself very torn between executing the moves that his algorithm demanded, and doing what was obviously a better move. There were long silences while he totted up the scores and chose the best minimax ploy, hoots and growls when he could see it missing chances. It was ironic that, despite all the developments of the past ten years, he was little closer to actually trying out serious chess-playing on a machine – the existing computers had neither speed nor space for the problem.*

Alick Glennie sometimes thought of Alan as Caliban, with his dark moods, sometimes gleeful, sometimes sulky, appearing in the laboratory on a somewhat random basis. He could be absurdly naive, as when bursting with laughter at a punning name that Glennie made up for an output routine: RITE. To Cicely Popplewell he was a terrible boss, but on the other hand, there was no question of having to be polite or deferent to him – it was impossible. He was regarded as a local authority on mathematical methods; those who wanted a suggestion would just have to ask him straight out, and if they could keep his interest and patience, they might get a valuable hint. Alick Glennie was rather surprised by his knowledge of hydrodynamics. All the same, he was no world-standard mathematician, and it was often more amazing to the professional mathematician what he did *not* know, than what he did. He never approached the von Neumann status or breadth of knowledge; indeed he had read very little mathematics since 1938.

In April 1951 he had another look at the word problem for groups, and came up with a result which J. H. C. Whitehead

---

* Meanwhile D. G. Prinz, who worked for Ferranti, quite independently programmed the Manchester computer to solve two-move chess problems. But this would have been of minimal interest to Alan; given that a solution exists it is simply a matter of patience to run through every possibility until it is found. Unless it gave an idea of how the brain did it, or some feel of 'pitting one's wits' against the machine, the problem of doing the programming, however ingenious, would have had little appeal for him. As in 1941, he was not interested in chess in itself, but as a model for thought.

at Oxford found 'sensational' – but it was never published.[55] Max Newman kept him interested in topology, and he went to seminars. But the trend of postwar pure mathematics was moving away from his interests. Mathematics was flowering through a greater and greater abstraction for its own sake, while Caliban on his island remained somewhere in between the abstract and the physical. Nor was he a keen conference-goer, loathing the academic chit-chat, but he went to the British Mathematical Colloquia that Max Newman helped to get started. In spring 1951 he went with Robin to one at Bristol, which got him interested in discussing topology with the mathematician Victor Guggenheim. But these were only diversions.

Another diversion was offered by the BBC, whose new highbrow Third Programme was offering a series of five talks on computers. One was by Alan, the others by Newman, Wilkes, Williams and Hartree. Alan's went out on 15 May 1951 and was entitled 'Can Digital Computers Think?'. Largely it ran over the ground of the universal machine and the imitation principle.[56] There were some references to the 'age-old controversy' of 'free will and determinism', which harked back twenty years with a mention of Eddington's views on the indeterminacy of quantum mechanics, and back ten years with some suggestions on how to incorporate a 'free will' element in the machine. It could be done either by 'something like a roulette wheel or a supply of radium' – that is, by the kind of random number generator that worked like the Rockex tape generator, off random noise – or else by machines 'whose behaviour appears quite random to anyone who does not know the details of their construction.' His listeners could scarcely have imagined the secrets which lay behind *that* bland suggestion! He ended with his justification for investigating machine intelligence:

> The whole thinking process is still rather mysterious to us, but I believe that the attempt to make a thinking machine will help us greatly in finding out how we think ourselves.

This short talk did not include any details of how he proposed to program a machine to think, beyond the remark that it 'should bear a close relation to that of teaching.' This comment sparked off an immediate reaction in a listener: Christopher Strachey, the son of Ray and Oliver Strachey.

Though born to a codebreaker father and mathematician mother, Christopher Strachey had not particularly stood out as a King's mathematics student from 1935 to 1938, and after wartime radar work was teaching at Harrow School. But the idea of machine intelligence grabbed his attention rather as it had Alan's. In 1951 a mutual friend put him in touch with Mike Woodger at the NPL, and he embarked upon writing a draughts program for the new Pilot ACE. By May he was also working with the Turing *Programmers' Handbook* with an eye to using the Manchester machine. On the evening of the broadcast he wrote a long letter[57] to Alan, with ambitious plans:

> . . . The essential thing which would have to be done first, would be to get the machine to programme itself from very simple and general input data. . . . It would be a great convenience to say the least if the notation chosen were intelligible as mathematics when printed by the output . . . once the suitable notation is decided, all that would be necessary would be to type more or less ordinary mathematics and a special routine called, say, 'Programme' would convert this into the necessary instructions to make the machine carry out the operations indicated. This may sound rather Utopian, but I think it, or something like it, should be possible, and I think it would open the way to making a simple learning programme. I have not thought very seriously about this for long, but as soon as I have finished the Draughts programme I intend to have a shot at it.

He had been thinking about the learning process, not only in the classrooms of Harrow School, but by playing the logical game of Nim* with a non-mathematical friend. Most mathematicians would know from Rouse Ball's old *Mathematical Recreations* that there was an infallible rule for a winning strategy, based on expressing the number of matches in each heap in binary notation. Few people were likely to spot this rule through play, but Strachey's friend did notice a special case of it, namely that a player who could achieve the position (n,n,0) had won, for thereafter it was only necessary to copy the opponent's moves to reduce the heaps down to (0,0,0). It was the element of abstraction achieved by a human learner that

---

* In this game three piles of matches are laid out, and two players take turns to remove as many matches as they please from any one pile. The player who removes the last match is the winner.

interested Strachey. He had worked out a program which could keep a record of winning positions, and so improve its play by experience, but it could only store them individually, as (1,1,0), (2,2,0) and so on. This limitation soon allowed his novice friend to beat the program. Strachey wrote:

> This shows very clearly, I think, that one of the most important features of thinking is the ability to spot new relationships when presented with unfamiliar material. . . .

and his Utopian 'Programme' was explained as one of his 'glimmerings of an idea as to how a machine might be made to do it.'

Alan's interests were by now centred on biology but he was still keen to develop such speculative ideas about mechanical thinking, in a way more detailed than he had explained in *Mind*. A talk given at this period[58] incorporated some proposals which started off like an office filing system, or indeed the 'intelligence' of Hut 4:

> The machine would incorporate a memory. . . . It would simply be a list of all the statements that had been made to it or by it, and all the moves it had made and the cards it had played in its games. These would be listed in chronological order. Besides this straightforward memory there would be a number of 'indexes of experiences'. To explain this idea I will suggest the form which one such index might possibly take. It might be an alphabetical index of the words that had been used, . . . so that they could be looked up in the memory. Another such index might contain patterns of men on parts of a 'Go' board that had occurred.

But then the minds of the filing clerks would begin to be taken over by the machine itself:

> At comparatively late stages of education the memory might be extended to include important parts of the configuration of the machine at each moment, or in other words it would begin to remember what its thoughts had been. This would give rise to fruitful new forms of indexing. New forms of index might be introduced on account of special features observed in the indexes already used. . . .

In many ways what he was doing was to work out his own theory of psychology, with the machine (mostly in imagination) as the stage on which it could be played.

The Inaugural Conference of the Manchester computer, from 9 to 12 July 1951, to which Alan returned after a holiday abroad, was a more mundane occasion. Alan gave one of the talks[59] – a dull one on the Manchester machine code, with all the gory detail of the base-32 backwards arithmetic – and he contributed to the discussions, chipping in to press for interpretative routines to be used on the Pilot ACE.

But Wilkes was the star, with 'micro-programming', an elegant new system for the design of control and arithmetic hardware. By this time it was being widely said that it was the Cambridge approach, which made concession to the human user, that held the key to the future. The Cambridge group called themselves the 'space cadets' and the rest, the 'primitives', and Alan Turing had made himself an arch-primitive by insisting on being able to follow the Manchester machine's operations digit by digit, although, on another level, he was the boldest Dan Dare of them all, embarrassing the responsible scientists with his anthropomorphic view of machines.

The application of computers to commercial purposes received serious discussion, and M. J. Lighthill, the new Professor of Applied Mathematics at Manchester, proposed that by 1970 'the use of the machine shall permeate the whole undergraduate course. Finally it may be necessary to re-orient the teaching of mathematics even in schools. However, any idea that "ABC" may at last be ousted by "/E@A" is, one hopes, only visionary.' This complaint at the base-32 notation espoused by Alan was to be vindicated; it would soon be thought absurd to expect ordinary users to adjust themselves in this way, although in 1951 this was far from clear. This conference was Alan's last appearance as a contributor to the programming or operating of computers, and he was already passing into legend – a ghost from the past in a science without history. A shabby and eccentric survival from the Cambridge of the 1930s, here he found himself seen, and little understood, against the classless stainless steel of the dawning 1950s.

Mike Woodger gave a talk about the relative performance of the Manchester and NPL machine instruction codes. Alan had invited him to stay at Hollymeade for the week of the conference. His guest would have been rather frightened if he had known that Alan was homosexual, but of this he remained unaware. What he did encounter, however, was the great muddle of pots and pans full of weeds and smelly mixtures, in which Alan was pursuing his desert-island hobby, seeing what chemicals he could make out of natural materials, and in particular doing some electrolytic experiments. Mike Woodger made a big hit when he admired the brick path, but did not do so well when Alan tried to explain his progress with morphogenesis.

For it was the biological theory, rather than the imitation game, that was his favourite topic now. At last there was something else that he was seriously interested in, that he could also talk about. As soon as the new computer was installed and working, it was set up to simulate chemical waves on his idealised ring of cells, the Turing *Hydra*. After much working on different cases, he came up with a convincing set of hypothetical reactions which, set to work in an initially homogeneous 'soup', would have the effect of setting up a stationary spatial distribution of waves of chemical concentration. This could be done at different speeds, with different results: 'fast cooking' and 'slow cooking', he called it. He also tried out the gastrulation problem, showing how random disturbances on a sphere could lead to a particular axis being picked out.

In this work he developed a unique sense of interaction with what was, in effect, a personal computer. It was like the dialogue with the Colossus, perhaps, though Roy Duffy, the new maintenance engineer, would call it 'playing the organ' as he watched Alan sit at the console and use the manual controls. Everyone who used the machine had to develop a good sense of how it was actually working, if only because there were always drum tracks and cathode ray tube stores out of action, requiring modifications to be made to the programming. But Alan developed it to a fine art, writing in instructions which would make the machine's 'hooter' sound at different points, when new parameters would be required. In this way he could watch the 'cooking' as it went along. The user also had complete control of the running and output mode of the machine, and Alan sometimes had it display the biological patterns on the cathode ray tube monitors, or print out contour maps in the way that crystallographers had by then devised.

Usually he worked overnight – he regularly booked Tuesday and Thursday nights. It was not all work on biology. He had, in particular, a 'bell ringing' program. Bell ringing? Working through every possible permutation? For whom did the bell toll? Ask not. . . . But usually he could be expected to emerge in the morning waving around print-outs to anyone who was around – 'giraffe spots', 'pineapples' or whatever – and then go home to sleep until the afternoon. This night-work was reflected in what was perhaps the most progressive feature of his manual, in which he explained how he made the machine itself do the secretarial work of keeping track of experiments and modifications. Even in this technical piece of writing, there was a high-level scientific play with the idea of 'rules' and 'descriptions': the programmer was allowed to use the machine in its logical sense, the engineer to use it in its physical sense, and the 'formal mode', as he defined it, was to print out the description of the operation that was 'complete' in another, higher level, sense:

> There are a number of modes or styles in which the machine may be used, and each mode has its conventions, restricting the operations considered admissible. The engineers for instance will consider the removal of a valve or the connection of two points temporarily with crocodile clips to be admissible, but would frown on certain uses of a hatchet. The removal of valves and all alterations of connections are certainly not permitted to the programmers and other users, and they have additional taboos of their own. There are in fact a number of modes of operation which might be distinguished, but only the *formal mode* will be mentioned here. This mode has rather stringent and definite conventions. The advantage of working in the formal mode is that the output recorded by the printer gives a complete description of what was done in any computation. A scrutiny of this record, together with certain other documents should tell one all that one wishes to know. In particular this record shows all the arbitrary choices made by the man in control of the machine, so that there is no question of trying to remember what was done at certain critical points.

But apart from such by-products of his work, in this case anticipating the introduction of a computer 'operating system', the other users of the machine did not now have more than the vaguest idea of what he was doing with it; after the summer of 1951 there was virtually no contact between them.

Alan spent August 1951 at Cambridge as usual, and from there a party consisting of himself, Robin, Nick Furbank, Keith Roberts and Robin's friend Christopher Bennett went down on the train to London for the Festival of Britain. They went to the Science Museum in South Kensington where the science and technology exhibits were housed. Grey Walter's cybernetic tortoises were on show, though they seemed to be going round in circles, and Robin said they were suffering from General Paralysis of the Insane. However, they observed one nice and unexpected touch: the feedback-dance that the tortoises performed in front of a mirror. Then they came across the NIMROD, which Ferranti were exhibiting: a special purpose electronic machine which would play Nim with members of the public. The Ferranti people were pleased to see Alan and said, 'Oh Dr Turing, would *you* like to play the machine?' which of course he did, and knowing the rule himself, he managed to win. The machine dutifully flashed up 'MACHINE LOSES' in lights, but then went into a distinctly Turingesque sulk, refusing to come to a stop and flashing 'MACHINE WINS' instead. Alan was delighted at having elicited such human behaviour from a machine.

The *cognoscenti* among them were also nudging each other about the young men who tended the exhibits, so this was not the only pleasure. After they had taken a good look at the new scientific Britain arising from the ashes, they made off to the Festival fun fair in Battersea Park. Alan was feeling a bit more flush than usual, and broke his usual rule, which he inherited from his father, by paying for a taxi rather than going on a bus. He would not go on the roller-coaster, saying it would make him sick. But they all went in the Fun House, and goggled at each other in the ultra-violet light.

Back at Manchester, Tony Brooker arrived, and began at once to undo the worst effects of the base-32-backwards mentality, writing new and more efficient schemes for input and output which allowed decimal notation, as well as improved sub-routine linkage. Alan did not mind in the least, but stuck to his own scheme, in which he was quite happy, his imagination quite easily visualising cow-blotches or rose-petals in a jumble of the thirty-two teleprinter symbols. Meanwhile Christopher Strachey had visited Manchester to try out a long program, the longest that anyone had dared attempt, written only from the manual and a little consultation with Cicely Popplewell. It was intended to solve a problem that Alan had suggested to him, that of making the machine simulate its own behaviour, in such a way that other programs could be tested out. At the laboratory they regarded this ambitious amateur effort with friendly scorn, knowing that like everyone's first dabblings it stood not a chance of success. But the program was duly punched out and Strachey allowed to try it. Alan showed him how to operate the machine, offering a few quick-fire instructions, and then leaving Strachey to get on with it. Usually Alan would be impatient with others' relative slowness, but this time he met his match. Overnight, Strachey was able to make his program work, and furthermore to amaze everyone with its final rendition of *God Save the King*, played on the hooter. On Alan's recommendation, Lord Halsbury immediately offered Strachey a job with the NRDC, at a salary sufficient to persuade him away from his boys at Harrow. Alan's days as grand master of the console were over. He had handed on the torch.*

At the beginning of November his paper on morphogenetic theory was ready. He decided to send it to the biological series of the Royal Society *Proceedings*, where it was received on 9 November. This meant putting in some rather elementary mathematical discussion. As he pointed out, few people were likely to be familiar with differential equations, physical chemistry *and* physiology. Biologists tended to be more experienced in translating what they saw into Greek than into mathematics. Mathematicians, on the other hand, usually knew nothing of the life sciences, though Lighthill was particularly encouraging about Alan's work. It was yet another case of him producing ideas which fell into no neat compartment of thought. The Chemistry department offered a middle ground, and Alan gave a seminar there on his theory on 11 December 1951.

Christmas was approaching, and with it the duty of choosing presents. Alan was always conscious of this, if not of other social obligations. His personal generosity could be relied upon. He had helped his favourite aunt Sybil, who had been a missionary in India, with a Braille set when she lost her sight. (He visited her at the time of the Bristol colloquium, for she lived nearby.) He had helped his childhood friend Hazel Ward to return to missionary work after the death of old Mrs Ward, atheism notwithstanding. And to the

---

* Until his death in 1975, Christopher Strachey was to be a central innovative figure in British computing, and his draughts-playing program seminal in the study of 'machine intelligence'.

surprise of Robin, who had heard him talk about his schooldays, he had subscribed to the Sherborne Quatercentenary appeal in 1950. This Christmas, however, he felt he owed himself a present, after finishing a paper which he regarded as the equal of *Computable Numbers*, setting out not just a new result, but a new framework, a new world to conquer.

Alan later wrote a short story,[60] in the new, 'frank', rather jaundiced, socially conscious style of Angus Wilson,* itself in E. M. Forster's tradition. It began:

Alec Pryce was getting rather [*illegible*] with his Christmas shopping. His method was slightly unconventional. He would walk round the shops in London or Manchester until he saw something which took his fancy, and then think of some one of his friends . . . who would be pleased by it. It was a sort of allegory of his method of work (though he didn't know it) which depended on waiting for inspiration.

When applied to Christmas shopping this method led to a variety of emotions just as much as when applied to work. Long periods of semi-despair wandering the stores, and every half hour or so, but quite erratically, something would leap out from the miserable background. This morning Alec had spent a good two hours at it. He had found a wooden fruit bowl which would just suit Mrs Bewley. She would be sure to appreciate it. Alec had also bought an electric blanket for his mother, who suffered from a poor circulation. It was more than he had wanted to pay, but she certainly needed just that, and would never think of getting one for herself. One or two other minor commitments had been dealt with. But now it was time for lunch, and Alec was walking towards the university but looking for a reasonably good restaurant.

Alec had been working rather hard until two or three weeks before. It was about interplanetary travel. Alec had always been rather keen on such crackpot problems, but although he rather liked to let himself go rather wildly to newspapermen or on the Third Programme when he got the chance, when he wrote for technically trained readers, his work was quite sound, or had been when he was younger. This last paper was real good stuff, better than he'd done since his mid twenties when he had introduced the idea which is now becoming known as 'Pryce's buoy'. Alec always felt a

---

* Alan was introduced to Angus Wilson at Cambridge by Robin Gandy. Although Angus Wilson had worked at Bletchley, he had not met Alan there.

glow of pride when this phrase was used. The rather obvious *double-entendre* rather pleased him too. He always liked to parade his homosexuality, and in suitable company Alec could pretend that the word was spelt without the 'u'. It was quite some time now since he had 'had' anyone, in fact not since he had met that soldier in Paris last summer. Now that his paper was finished he might justifiably consider that he had earned another gay* man, and he knew where he might find one who might be suitable.

Alan succeeded, for as he walked along Oxford Street, and pretended to look at the posters outside the Regal cinema, he caught the eye of a young man.

Arnold Murray, who was nineteen, came from the background of *The Road to Wigan Pier*. He had known bread and margarine at best. His father, a concrete layer when in work, knocked his mother about. Emaciated with malnutrition and nervousness in the blitz, he had been sent to a boys' camp out in Cheshire for schooling, and he was very proud of having shot to the top of the class with the new encouragement and competition. They had cheered for D-day and for VE day, but for him it meant return to a Manchester slum home by the pitch and tar distillery, and six months of technical school before his father made him leave for work. He had had several jobs, of which the longest lasting was that of making spectacle frames after the National Health Service began in 1948. (It was a trade that became a notable casualty of the Korean War, for Gaitskell's budget of 1950, setting in motion the massive rearmament of the new decade, ended the free provision of glasses.) Arnold had

---

* Was this plain-text or cipher-text? This is a good example of where the meaning of a word depends upon its social embodiment. At least since the 1930s it had been in general use among homosexual men as a code word with a plain meaning – in America. Thus D. W. Cory's pioneer work *The Homosexual in America*, which appeared in 1951, explained: 'Needed for years was an ordinary, everyday, matter-of-fact word, that could express the concept of homosexuality without glorification or condemnation. It must have no odium of the effeminate stereotype about it. Such a word has long been in existence, and in recent years has grown in popularity. The word is *gay*.' Alan Turing would usually use 'homosexual ' or, among his friends, the word 'queer'. But he could have known the American usage, and would entirely have approved of D. W. Cory's rationale of it. For this reason the word will be used from here onwards; any anachronistic or transatlantic effect thus introduced will reflect quite properly the difficulty that Alan Turing had in communicating his attitudes in the Britain of the early 1950s. As with the 'computer' he was ahead of his time.

found release from a dreary existence in July 1951, hitching down to London for the Festival of Britain. But he had been caught making a petty theft, and was sent back to Manchester on probation. He was still living with his family in Wythenshawe, and was currently unemployed and very hard up.

Arnold was searching for an identity, and thought that the world owed him something better than a life at the bottom of the heap. He had tried science – at fourteen he had blown out the windows with a chemistry-set concoction. And he had tried sex, with various experiences since that age. He was not a person with freedom or consistency of mind. He dreamt of a perfect relationship with a woman, but on the other hand liked the absence, when with men, of any sense of putting on a performance. He was also conscious of being called a 'Mary Ann' for intelligence and sensitivity. Middle-class men offered him manners and culture, and at this point of his development, homosexuality seemed something that belonged to an élite to which he aspired. He looked down on those who simply offered themselves directly for cash. Alan offered such a promise of association with gracious living – but this was not the whole story, for Alan combined this with a freshness and youthfulness that stood out on the Oxford Street background.

Alan asked Arnold where he was going and Arnold replied 'nowhere special'. So Alan invited him to lunch in the restaurant across the road. Fair and with blue eyes, undernourished and with his thin hair already receding, desperate for better things and more receptive than so many educated people, Arnold touched Alan's soft spot for lost lambs, as well as other chords. He also had a determined vivacity and a saving sense of humour that could carry him through the most difficult situations. Alan told him that he had to go back to the university, where he was a lecturer, and explained that he worked on the Electronic Brain. Arnold was fascinated; Alan asked him to come to his home at Wilmslow at the weekend. By making invitations to lunch and to his home, Alan had already offered a good deal more than would usually be expected of a street encounter, where a quick adjournment to railway arch, back alley or toilet would be more customary. Arnold accepted the invitation, but then failed to appear on the night.

This might easily have been the end of the matter, but Alan

saw Arnold again on Oxford Street on the next Monday afternoon. Arnold offered a feeble excuse for his failure to appear, and this time Alan invited him home immediately. Arnold did as Alan suggested, stayed until the late evening, and agreed to come another time on 12 January. Alan sent him a pen-knife as a Christmas present.

The BBC Third Programme had now arranged for a sort of *Brains Trust* on the subject of whether machines could be said to think.* Sometime near Christmas 1951, Alan visited David Champernowne at Oxford. He had an early tape recorder, and they made a spoof version of the discussion, with Champ putting on the Arts Man voice appropriate to the discussion of beauty and other high-minded concepts to which the machine could not aspire. Fred Clayton came later and was duly fooled. He, meanwhile, had got married, just as he said he would, and had been fortunate in his appointment to a lectureship in classics at the University College at Exeter. He was much concerned with developing a thesis about parallels between classical and English literature, and consulted Alan about the probability and statistics necessary in such comparisons. He was also interested in the significance of astrology in the classics, and picked Alan's brains for some elementary astronomy.

The real discussion[61] was recorded at the BBC Manchester studio on 10 January 1952. It was left to the brain surgeon to show the flag for the cause of consciousness, with Alan trying to haul it down. Max Newman and Richard Braithwaite, the King's philosopher of science, acted as referees.

It was couched in the jocular-Mandarin style of the day. 'Of course', wrote Alan to his mother, who listened to the broadcast, 'most of the questions put to me were more or less written in gags.' Braithwaite began with a very appropriate *Brains Trust* point: 'it all depends on what is to be included in thinking.' Alan explained the imitation game as a criterion of 'thinking', the others duly chipping in to put the objections. 'Would the questions have to be sums,' asked Braithwaite, 'or could I ask it what it had had for breakfast?' 'Oh yes, anything,' said Alan, 'and the questions don't really have to

---

* The BBC had made a more seasonal contribution to the public understanding of computers by broadcasting the Manchester machine's rendering of 'Jingle Bells' and 'Good King Wenceslas'.

be questions, any more than the questions in a law court are really questions. You know the sort of thing, "I put it to you that you are only pretending to be a man," would be quite in order.' They discussed learning and teaching, and Braithwaite said that people's ability to learn was determined by 'appetites, desires, drives, instincts' and that a learning machine would have to be equipped with 'something corresponding to a set of appetites'.

Newman steered a course back to the safer waters of mathematics, pointing to the act of imagination that had been required to connect the 'real numbers' of length with the integers of counting, which involved 'seeing analogies between things that had not been put together before . . . . Can we even guess at the way a machine could make such an invention from a programme composed by a man who had not the concept in his own mind?' Alan *could* guess, in fact; it was just the kind of thing he was thinking about:

> I think you could make a machine spot an analogy, in fact it's quite a good instance of how a machine could be made to do some of those things that one usually regards as essentially a human monopoly. Suppose that someone was trying to explain the double negative to me, for instance, that if a thing isn't not-green it must be green, and he couldn't quite get it across. He might say, 'Well, it's like crossing the road. You cross it, and then you cross it again, and you're back where you started.' This remark might just clinch it. This is one of the things one would like to work with machines, and I think it would be likely to happen with them. I imagine that the way analogy works in our brains is something like this. When two or more sets of ideas have the same pattern of logical connections, the brain may very likely economise parts by using some of them twice over, to remember the logical connections both in the one case and in the other. One must suppose that some part of my brain was used twice over in this way, once for the idea of double negation, and once for crossing the road, there and back; I am really supposed to know about both these things but can't get what it is the man is driving at, so long as he is talking about all these dreary nots and not-nots. Somehow it doesn't get through to the right part of the brain. But as soon as he says his piece about crossing the road it gets through to the right part, but by a different route. If there is some purely mechanical explanation of how this argument by analogy goes on in the brain, one could make a digital computer do the same.

Wittgenstein had talked about 'explaining' double negation in 1939.[62] But Jefferson brought the discussion back to earth with the problem of appetites. 'If we are really to get near to anything that can be truly called "thinking", the effects of external stimuli cannot be missed out. . . . You see a machine has not [an] environment, and man is in constant relation to his environment, which as it were punches him whilst he punches back. . . . Man is essentially a chemical machine, he is much affected by hunger and fatigue . . . and by sexual urges.' Alas, those appetites that interfered with thinking! It was a strong argument against the discrete-state machine. But Jefferson again spoilt his case with an appeal to the complexity of the nervous system (irrelevant since a universal machine, given sufficient storage, could emulate one of any complexity). In more rhetorical vein he continued, 'Your machines have no genes, no pedigrees. Mendelian inheritance means nothing to wireless valves,' and so forth. Jefferson wanted to say that he would not believe a computing machine could think until he saw it touch the leg of a lady computing machine, but they cut this out of the broadcast because (as Braithwaite said) one could hardly call that *thinking*. Braithwaite believed it would be necessary for the computer to incorporate an 'emotional apparatus' in order that it could think, but that it was not their concern to ask what problems this might lead to. The hot potato was dropped, and Jefferson concluded by reassuring the British intelligentsia that it was 'that old slow coach, man' who would continue to produce the ideas.

The broadcast went out on 14 January, by which time Arnold had made his second visit to Alan's home, and events had taken a more serious turn. Alan had arranged matters so as to cast the relationship as an 'affair', which meant that Arnold had arrived as a dinner guest, and expected to stay the night. Arnold responded warmly to what was for him the palatial circumstances of Hollymeade, it being particularly striking, for instance, that Alan employed a housekeeper. He was with the masters now, not the servants.

They did not have much in common to talk about, but found links, in such a way that Arnold was highly conscious of Alan's need to communicate and reach a fresh mind. Neither thought much of the American interference with the British attempt to oust Mossadeq in Iran. Arnold had great local patriotism and disliked the USAF bases still dominating parts of Cheshire. Besides current affairs Alan also talked about astronomy, played a tune on the violin, and let Arnold have a try. After dinner, rather the better for wine, and lying on the rug, Arnold began telling Alan about his recurrent childhood dream, or nightmare rather, in which he felt himself suspended in absolutely empty space while a strange noise would start, growing ever louder, until he woke up in a sweat. Alan asked what kind of noise it was, but Arnold could not describe it. Thinking of big empty spaces, Alan imagined the old hangar on the RAF camp along the road, and made up a science-fiction story (he talked a bit about H. G. Wells) in which the hangar was itself a brain, programmed in such a way that it would work normally for anyone else, but when *he* went in to the hangar, he would be trapped. The doors would shut. And then he would have to play against the machine, a game of chess, the best out of three. The machine would counter his moves so quickly that he would have to make conversation to distract it. So he would talk to it, first making it show anger, then pleasing it by being stupid himself, and making it feel smug.

'Can you *think* what I *feel*? Can you *feel* what I *think*?', he said with terrific emphasis at one point, as he became more excited with the story. He quite transfixed Arnold as he took a piece of chalk, and imagined how he could beat the machine, by doing arithmetic so badly and slowly and stupidly that it would commit suicide in despair.

Arnold tried to explain his ideas too; and Alan was patient, although he could so easily have been crushing, and led him on Socratically. 'Whatever you think, *is*,' said Alan at one point, and this meant a lot to Arnold, who had his own dreams that he wanted to come true. Alan felt frustrated because he could not better communicate his ideas: 'There's got to be more to it than this level,' he told Arnold, almost in anger, adding with great emphasis 'I've got to *teach* you, take you out of all this.'

Dear love of comrades! – in 1891 Edward Carpenter had met his George Merrill, a working man of twenty; it began just like this, and continued for thirty years. Alan made it clear that he wanted them to sleep together as lovers, and this they did. In the morning Alan got up and made the breakfast, after which they talked and smoked and prolonged the pleasure of the night. They arranged for another visit in two weeks' time. One subject was not, however, discussed as it might have been. This was the question of money. It was as obvious that Arnold was short of it as it was clear that Alan had more than he needed. Alan was going to do the expected thing, and was perhaps surprised when Arnold declined the offer. The underlying difficulty was that Arnold jibbed at a direct payment, which threatened to label him as 'a renter'. Alan, in contrast, was highly uncomfortable about conventional social manoeuvres, whether in his mother's drawing room or in his own bedroom. He was therefore particularly shaken next day to notice the absence of some money from his wallet, which he suspected that Arnold could have taken while he was making the breakfast. He wrote to Arnold, saying that he did not after all wish to continue the acquaintance. But Arnold arrived on his doorstep a few days later, demanding to know the reason for this rejection, and denying that he had anything to do with the loss. Alan was 'half convinced' by his indignation. Arnold went on to mention that as it happened he was £10 in debt for a suit bought on hire purchase, and asked to borrow £3. Alan gave him the money, saying that it was a gift, and later wrote to Arnold restoring the invitation. Arnold wrote back to thank him on the 18th, but added a request for a further £7 loan. Alan's response was to ask for the name of the firm to which the money was owed: it was not the money but the truth of the story which was the issue for him. Again, on the 21st, Arnold arrived at Hollymeade to complain of the lack of trust that Alan showed, and left with a cheque for £7. He was going to start work in a Manchester printing shop, and so could promise to pay it back from his earnings.

Meanwhile Robin had come to stay for the weekend, which was devoted to a discussion of his essay on Eddington's 'Fundamental Theory' of physics. This Alan said was 'very much more satisfactory than anything you have done before.' This stern praise meant a great deal to Robin, whose 1949 King's fellowship dissertation had met with sharp criticism from Alan which had left him in tears.

Eddington had died in 1944 leaving unfinished an attempt to develop a theory of physics from nothing but logical necessity. It was a somewhat Turingesque venture, and met with Alan's sympathy in principle, but he had long since decided that Eddington was 'an old muddle-head' and wanted to see the 'Fundamental Theory' debunked. Robin, who never knew how important Eddington had been to Alan twenty years before, had found a number of errors in his arguments, including one which could be regarded as a confusion of logical types. It was a nice meeting of logic and physics.

Life went its ordinary way. Alan's aunt Sybil had died on 6 January, and left him £500. The last survivor of his father's generation, she had accumulated the Turing fortunes. She left £5000 to Mrs Turing, who for some reason had the idea of taking out a mortgage on her house, a policy which in a typical turn of phrase Alan called 'about as appropriate as going out charring when you need more help in the house'. He stopped the £50 per annum he had sent her since 1949.

He had listened to the broadcast and found his voice 'rather less trying to listen to than before'. On Wednesday 23 January, the programme was repeated. And the same day, the environment punched him back, as Jefferson put it. Alan arrived back in the evening to find that his house had been burgled. Alan was writing next day to Fred Clayton in connection with the astronomy of the ancient world. He explained the significance of the zodiac, and ended:

> I have just had my house broken into, and am still every few hours finding some fresh thing missing. Fortunately I am insured, and little has gone that is really irreplaceable. But the whole thing has had a very disturbing effect, especially as it followed shortly on a theft from me at the University. I go about expecting a brick to fall on my head or something disagreeable and unexpected anywhere.

A rather pathetic collection of oddments was missing – a shirt, some fish-knives, a pair of trousers, some shoes and shavers and a compass – even an opened bottle of sherry. He assessed it at a total value of £50. He reported the burglary to the police, and two CID officers came to take fingerprints in the house. Yet even while he did this, he suspected that there might be some connection with Arnold. He consulted a solicitor recommended to him by his neighbour Roy Webb, and on his advice wrote to Arnold on 1 February, reviving the question of the money missing from his wallet, saying that whatever the truth of the matter it had come between them, and that it would be best if they did not see each other again. He added in a somewhat schoolmasterish tone that it was Arnold's duty to repay the £7. He also said that if Arnold came to his house again he would not be admitted.

But when Arnold reacted to this letter by calling at Hollymeade on the Saturday evening, 2 February, he found himself admitted after all. Again he angrily protested his innocence, and in a moment of emotion said he could go to the police and tell *everything*. Alan challenged him to 'do his worst' – but it was an empty threat, for Arnold soon admitted he could do nothing against a man in Alan's distinguished position. The anger was discharged, and a different mood prevailed. Giving Arnold a drink, Alan mentioned the burglary, and to this question Arnold immediately supplied an answer. He did not know the burglary had taken place, but did know exactly who might have done it. For he had mentioned Alan to an acquaintance called Harry, a twenty-year-old unemployed youth recently discharged from National Service in the Navy, while they were talking in the Oxford Street milk bar. They had been speaking rather boastfully of their respective successes. Harry had suggested a robbery, and although Arnold had refused to join in, he knew it had been planned.

The result was to re-establish a friendly, and indeed an erotic relationship. Arnold once again slept with Alan, although during the night Alan found himself in two minds, at one point going downstairs to put away the glass with Arnold's fingerprints on it, in the hope that he could compare them with those left by the burglars. Next morning they went together into Wilmslow town, and Arnold waited outside the police station while Alan went in to pass on the information about the likely culprits, fabricating a story to explain how he had come by it. He had allowed the game of Presents to be taken a fair way without making a fuss; but to let it go unchecked would, in his view, be tantamount to giving in to blackmail.

Arnold went off having offered to do his best to track down the stolen goods, and indeed he was able to write to Alan with a report a few days later. But by that time everything had changed. One change had been marked by the Manchester bells, ringing not this time for victory, but for the death of George VI. On Thursday the new Queen Elizabeth flew back from Kenya and Winston Churchill, Prime Minister again, welcomed her at the airport. And it was on that very evening, as the new Elizabethan age began, that the detectives paid a call on Alan Turing. *No man is an Island, entire of itself.* Now he was in the soup.

8 On the Beach

In paths untrodden,
In the growth by margins of pond-waters,
Escaped from the life that exhibits itself,
From all the standards hitherto publish'd, from the
    pleasures, profits, conformities,
Which too long I was offering to feed my soul,
Clear to me now standards not yet publish'd, clear to
    me that my soul,
That the soul of the man I speak for rejoices in comrades,
Here by myself away from the clank of the world,
Tallying and talk'd to here by tongues aromatic,
No longer abash'd, (for in this secluded spot I can
    respond as I would not dare elsewhere,)
Strong upon me the life that does not exhibit itself,
    yet contains all the rest,
Resolv'd to sing no songs to-day but those of manly attachment,
Projecting them along that substantial life,
Bequeathing hence types of athletic love,
Afternoon this delicious Ninth-month in my forty-first year,
I proceed for all who are or have been young men,
To tell the secret of my nights and days,
To celebrate the need of comrades.

It had not taken the police long to detect Alan Turing's crime. It was almost inevitable once he had made the original report of the burglary, for the police had been able to identify Harry's fingerprints. He was already in custody on another charge in Manchester, and before long made a statement which referred to Arnold telling him of having 'business' at Alan's home. The further information Alan had volunteered on the Sunday merely gave the police their opportunity to act with confidence.

Alan took them upstairs to where he was working with his desk calculator. The detectives, Mr Wills and Mr Rimmer, found themselves in an unfamiliar environment, the room littered with pieces of paper covered with mathematical symbols. They told Alan that they 'knew all about it', leaving him unclear whether they were talking about the burglary or of something else. He later told Robin that he had to admire their interrogation technique. They asked him to repeat the description he had given them on the Sunday morning, and Alan said,[1] 'He's about twenty-five years of age, five foot ten inches, with black hair.' Imitation was not Alan Turing's strong point – perhaps an intelligent machine would have done better. This feeble attempt sank like a stone. Mr Wills said, 'We have reason to believe your description is false. Why are you lying?'

This was the moment for 'I don't know what came over me', or the other phrases employed by more politically-minded persons, but once the detectives had shown their hand, Alan blurted out everything that they wanted to hear, in particular admitting that he had concealed the identity of the informant because he 'had an affair with him'. Mr Wills asked 'Would you care to tell us what kind of an affair you have had with him?', and this policemanlike question elicited from Alan a memorable phrase, detailing in semi-official language three of the activities that had taken place. 'A very honourable man', the detectives thought him as they cautioned him in the usual way, and they were the more impressed when he volunteered a statement of five handwritten pages. Relieved of the usual necessity to translate human life into police language, they were most appreciative of what was 'a lovely statement', written in 'a flowing style, almost like prose', although 'beyond them in some of its phraseology'. They were particularly struck by his absence of shame. 'He was a real convert. . . . he really believed he was doing the right thing.'

Alan had commented to the detectives that he thought a Royal Commission was sitting 'to legalise it'. There he was wrong. And almost certainly he underestimated the seriousness of what in his statement became 'the offence'. Harry had been justified in assuming that Alan was fair game for robbery. As a sex criminal, he had forfeited the protection of the law. Alan's statement illustrated the difficulty he faced in grasping this fundamental fact. It was mostly concerned with the undecidable problem of Arnold's veracity, and details of 'the offence', though freely and even defiantly supplied, appeared as incidental to what he perceived as the story. It might be called unrealistic of him to expect a relationship rooted in such inequality to develop as an 'affair' between free individuals; he took no account of the fact that words and actions could mean different things to people in different social circumstances. Yet if this showed a lack of realism, a liberal intellectual dream world, it was an unreality also consciously sought and appreciated by Arnold, who had found himself challenged and moved by being treated as a friend of the élite. And the greater unreality lay in Alan's attitude to the law, which was not interested in his mental dilemmas, but was very much concerned with his bodily activities. He found it almost too absurd for belief; but the fact remained that it was this, 'the offence', that the police were investigating with persistent, conscientious, thoroughness.

The detectives did not, however, extend their questioning to his whole past life. In this respect they only took his fingerprints and photograph, to be checked against Scotland Yard records for previous offences. As supporting evidence of the crimes they also took what correspondence he had relating to Arnold. Afterwards, Alan was aware that if he had said Harry was lying, the police might have been unable to make any case against him. As it was, they were able to complete their duties with ease. On Saturday morning Mr Wills arrested Arnold in the Manchester printing shop (a job he immediately lost), took him to Wilmslow police station and showed him Alan's statement. Mr Wills was soon able to write out a statement for Arnold to sign, spelling out 'offences' in copious detail. This in turn Alan agreed on Monday 11 February to be 'materially correct'. The police had solved a crime which attracted up to two years of imprisonment.

The crime was, in fact, that of 'Gross Indecency contrary to Section 11 of the Criminal Law Amendment Act 1885'. It was defined purely in terms of parts of the male body, and applied absolutely, irrespective of such factors as age, financial advantage, and whether the activity was in a public or a private place. Alan's statement left no room for doubt that he was guilty, and he was wrong in imagining that what he had done might soon be 'legalised'. He was right, however, in thinking that changes were taking place in the official perception of homosexuality. Above all, the silence was being broken.*

Indeed the turn of the 1940s had seen a renewal in Britain of the process which had led to that 1885 Act, to the trials of Oscar Wilde, and to the books of Havelock Ellis and Edward Carpenter in the 1890s. The point about the law was that it had replaced the vague theological 'crime against nature', or the 'crime not to be mentioned among Christians', by a definite rule. When Oscar Wilde spoke of 'the love that dares not speak its name', he identified a crucial aspect of what was happening – the speaking up, the 'flaunting', the explicitness.

In the next fifty years, incursions into British public consciousness by such books as *The Loom of Youth* and *The Cloven Pine* had been exceedingly circumspect and allusive in nature. But in the 1940s a new wave of explicitness swept across the Atlantic to break upon the more austere and tight-lipped culture of the island race. Since 1938, for instance, the zoologist Alfred Kinsey had been documenting the unofficial reality of human sex, and in 1948 he revealed a breach of the 'fixed moral codes' so massive that, like the evidence confronting Dönitz, its implications were too profound to be entertained.

While for a time such revelations could, in Britain, be dismissed as American extravagance and vulgarity, the 'head in the sand' attitude was already doomed. In many ways what was happening was a delayed effect of the war – or rather, like so many wartime developments, the expression of ideas which had begun in the 'mechanization, rationalization, modernization' of the late 1930s. While in military affairs the old regime had been forced to adopt modern methods for the sake of survival in 1942, the parallel developments in social policy took longer to filter through. The

---

* More precisely, it was predominantly the subject of *male* homosexuality that was coming into greater public prominence, just as the 1885 Act defined 'gross indecency' as a male crime. In the parallel period after World War I, much had been made of a supposed 'Black Book' compiled by the German secret service, containing names of thousands of 'sexual perverts', both men and women. This was one reason why in 1921 the Commons had voted to extend the 1885 Act to women. But the Lords rejected the proposal – believing that even to mention the crime would have the effect of giving women ideas. The fact that men received a conscious attention not accorded to women was, therefore, an aspect of male privilege – although perhaps Alan Turing would not have seen it in quite this way.

opening of a public debate about male homosexuality in Britain in 1952 was the conflict of the small back room, in another sphere.

In 1952, as in 1942, the times were out of joint. The rulers of Great Britain were still apt to regard the behaviour of its population as that of a public school. In 1952 the pocket money and the tuck shop were under better management than before, and there was less open complaining from the Modern side. But the return of the old Headmaster in October 1951 had suggested invidious comparisons with former triumphs. In 1951 Britain had lost control over Iran and Egypt, countries so successfully held against German encroachment not ten years before. As during the crisis of imperialism in the 1890s, military loss of control could be identified with sexual loss of control. In the traditional view, homosexuality was an *act*, or *practice*, into which any man might be led – and such lapses into 'slackness' were to be prevented not only in the armed forces, but in the national life which raised and moulded them.

Such a view, however, could already be identified with that of an older generation, and one which had been pushed aside since 1940. For nearly a hundred years there had existed a quite different kind of official description, which concentrated not upon the act, but the state of mind. Considerable efforts had been made to elucidate a 'homosexual type', or a 'homosexual personality', rather as the nineteenth-century psychologists had also devoted energy to defining criminal, or mentally deficient, or other 'degenerate types'. The word 'homosexual' was itself a nineteenth-century medical neologism. Freud was often credited with making this mode of description available to people. Indeed, Alan and Robin would sometimes puzzle over the question of how people had been able to think about sexual desire before Freud's day.

In his 1950 *Mind* article, Alan had referred to the 'skin of an onion' analogy as helpful:

> In considering the functions of the mind or the brain we find certain oper-
> ations which we can explain in purely mechanical terms. This we say does
> not correspond to the real mind: it is a sort of skin which we must strip off if
> we are to find the real mind. But then in what remains we find a further skin
> to be stripped off, and so on. Proceeding in this way do we ever come to the
> 'real' mind, or do we eventually come to the skin which has nothing in it?

His own view, of course, was that the mind was like an onion, and not like an apple, there being no central, irreducible, undetermined core. In a different way, nineteenth and twentieth-century science had been peeling the onion of the mind, and had dented the concept of responsibility with 'mental illness', shell-shock, neurosis, breakdowns and so forth. Where was the line to be drawn? The conservative fear was that every kind of behaviour would be excused by appeal to some irresistible, uncontrollable *force majeure*. Like Polanyi and Jefferson, they sought a *non plus ultra* to the pretensions of mental determinism, a barrier against the flood of threats to traditional values unleashed by the Second World War. They found one in homosexuality: the new men's talk of 'conditions' and 'complexes' was not to be allowed to excuse a deadly social evil, corrupting and weakening everything in its path.

At the same time, yet a third kind of description was gradually coming into focus, that of homosexual men as *socially* defined. From this point of view, the emphasis was to be placed not upon thoughts and feelings, nor on sexual acts, but on the particular patterns of acquaintanceship, money and occupations associated with homosexuality. The sociologist 'Gordon Westwood', whose book *Society and the Homosexual* opened the British debate in 1952, described male homosexuality in all of these ways, one after the other. Reaching a wider audience, the *Sunday Pictorial*'s series of reports[2] on 'Evil Men' also broke what it called 'the conspiracy of silence on the subject' the same year, and likewise treated it from a modern psychological and social perspective, rather than in terms of the law. 'Most people', the newspaper explained, 'know there are such things – "Pansies" – mincing effeminate, young men who call themselves queers.' But these obvious 'freaks and rarities', it continued, represented but the tip of the iceberg. The problem was far greater than most people realised, and the time had come to tackle it.

One of the difficulties pervading these discussions was that no single description was adequate to the matter in hand, although there was an obvious virtue to each of them. There were certainly many homosexual *acts* – in schools, for example – which had little to do with deep-seated desires, nor with a social 'minority'. In contrast, the diffuse, romantic ambience of *The Cloven Pine* fell into no category recognised by the English criminal law. While there were others, like Arnold, who did not know what they wanted, but who were very familiar with the social patterns, with their advantages and disadvantages, of what a Methodist minister, quoted in the *Sunday Pictorial*, called 'the worst city for homosexuality that I have been in'.

The backroom boys of medicine and the social sciences were bringing these unwelcome contradictions to the surface. The law was under attack for its purely physical level of description. Gordon Westwood held, in contrast, that[3] 'The overriding consideration in dealing with homosexual offenders should be that it is a form of mental illness.' But life was more complicated than that; the enforcement of the law was related less to the prevalence of the 'acts' than to the structure of British society.

For this reason the attempt to form a more scientific description ran up against British doublethink. The psychologist Dr Clifford Allen told the *Sunday Pictorial* that 'In the past battles may have been won on the playing fields of our public schools, but numerous lives have been broken in the dormitories'. The unofficial reality could be entirely different from any particular official line, and in private the most conservative personages might regard both the law and current psychological theories as nonsensical. But amidst the great complexity, one simplifying feature stood out. As in the 'nation in miniature' of the public school, it was contact between those of different social rank that was most likely to be discovered and punished. Alan Turing's crime epitomised the action upon which the operation of the law was in practice focussed. So was its discovery, by a related petty crime, a classic case of successful detection. In another way again, the arrest was a textbook case, for the age range of thirty to forty was the one most frequently represented in the prosecutions of the period. It was also true that as what Westwood called an 'outsider', unfamiliar with the social milieu, he was natural prey for attempted blackmail.

In the development of his sexual life Alan Turing was in many ways typical for a gay man of his time. He had enjoyed the benefit of a very unusual, very privileged ambience at King's, but in the outside world, the same factors came into play as Kinsey had noted[4] while interpreting the statistics:

There is considerable conflict among younger males over participation in such socially taboo activity, and there is evidence that a much higher percentage of younger males is attracted and aroused than ever engages in overt homosexual activities to the point of orgasm. Gradually, over a period of years, many males who are aroused by homosexual situations became more frank in their acceptance and more direct in their pursuit of complete relations, although some of them are still much restrained by fear of blackmail.

Kinsey found among his 'active' population a general increase in frequency of sexual experience up to an age of thirty-five, thereafter continuing at that level until the age of fifty, corroborating the common sense expectation that the 'social taboo' could inhibit sexual development for perhaps twenty years. In this respect Alan Turing was just launching out. It was only in his thirties that he had begun to find his way outside King's. He had been involved in two extended relationships, but he was not by nature the most conjugal person, and his exploratory urge was better suited to the possibilities of the cruising-grounds, once he had overcome his shyness. It was not that he was very successful; nor perhaps did he escape a profound sense of compromise and of the loss of youthful ideals – 'Beggars can't be choosers,' he wrote in his self-analytical short story – but he could take pride in breaking out of the framework of his upbringing, working out something for himself, and in managing without special privileges. He had gained experience, and as a very young forty, would want to pursue its opportunities before he was much older. It was this process which had been arrested.

Another feature of the operation of the law was that this sinking of men's souls was all the time on the increase. Between 1931 and 1951 there had been a five-fold increase in prosecutions, a steady rise through depression, blitz and rocket bombs alike. In 1933 it was indeed as J. S. Mill had said about heresy: that public opinion was more crushing than the direct application of the law. By 1952 the position had changed. This was consistent with the great extension of the role of the state in every direction, taking over functions formerly left to individuals, families, voluntary societies and so forth. It might be argued that the state was taking a larger role in policing sexual behaviour precisely because the inhibitory effect of public opinion was decreasing.

In more conservative circles, it was taken that the law only gave the final stamp of authority to the ostracism of society. King George V was supposed to have said, 'I thought men like that shot themselves.' Alan Turing, however, cared nothing for the opinion of society, and therefore was ahead of his time in laying bare the role of the state. For most gay men, the question of *who knew* would be of colossal significance, and life would be rigidly divided into two compartments, one for those who knew, and one for those who did not. Blackmail depended as much upon this fact as upon the legal penalties. The question was important to Alan too, but in a rather different way: it was because he did not wish to be accepted or respected as the person he was not. He was likely to drop a remark about an attractive young man, or something of the kind, on a third or fourth meeting with a generally friendly colleague. To be close to him, it was essential to accept him as a homosexual; it was one of the stringent conditions he imposed.

Exposure, therefore, held no intrinsic terror for him. But a criminal trial would involve not merely exposure as a homosexual, but all the concrete details. It would be one thing to be a martyr for an abstract cause, and quite another to have the sequence of events with Arnold rendered into an unflattering public form. It would expose him not only as a sexual outlaw, which at least carried with it a certain pride, but as a fool. In this respect his insouciance was amazing. But it was his all-or-nothing mentality at work. He had presumably decided long before that such things were part of the 'large remnant of the random behaviour of infancy', and that it was absurd to be ashamed of anything harmlessly enjoyed, whether it be parlour games or bedroom pleasures. It meant that he had to take a stand not for an ideal, not for anything particularly rewarding or successful, but for that which was simply true. But he did not flinch. The detectives continued to be astonished when they visited him in connection with the case. He would take out his violin and play to them the Irish tune *Cockles and Mussels*, accompanied with glasses of wine.

After three weeks, on 27 February, Alan and Arnold both appeared in the Wilmslow magistrates' court for the committal proceedings. The CID officer, Mr Wills, described the circumstances of the arrests, and read out the statements in full. There was another

prosecution witness: Alan's bank manager, whose ledgers corroborated the detail of the £7 cheque. There was no cross-examination.
Alan's solicitor 'reserved his defence', and obtained his release on
£50 bail. Arnold, however, was held in custody pending the trial
proper, to be held at the forthcoming Quarter Sessions. The local
newspaper[5] reported the court appearance and the gist of the story.
They printed both men's full addresses, in the usual way, and a photograph of Alan.

The case was not taken up in the Manchester papers, but there
was certainly a possibility of the forthcoming trial being reported
widely. In any case, Alan had to look after his individual relationships, so that people he cared about should not learn of what had
happened from the newspapers or some other unfortunate way. In
particular there was his family to consider. Alan wrote to his brother
John – with a letter, for once, not a postcard or a telegram. It started
off 'I suppose you know I am a homosexual.' John knew no such
thing. He had always thought of his brother as 'misogynist', inasmuch as he avoided flirtatious chit-chat when on his occasional visits
to Guildford. But Alan bore no resemblance to John's picture of
'pansies', and this possibility had never occurred to him. John stuffed
the letter into his pocket and read it later in his office.

The letter explained some of the circumstances, and also that
he was going to plead 'not guilty' and be properly defended. John
immediately dropped everything and went to Manchester, where he
consulted a senior partner in a leading firm of solicitors. He in turn
saw Alan's solicitor, and they persuaded Alan to change his plea to
'guilty'. He was, in fact, caught between two untruths. To deny what
he had done would be to tell a lie, and to convey a false sense that he
considered it something that ought to be denied. Yet to be portrayed
in public with words such as 'guilty', 'self-confessed', 'admit' was
also to compound an untruth. There was no way to keep himself
pure. In practical terms his statement to the police had rendered
'defence' impossible, and he had little to lose by pleading 'guilty'.
More pertinent, from John's point of view, was the idea that a 'guilty'
plea would render the trial both quick and quiet. He thought Alan
had been a 'silly ass' to go to the police about the burglary, and that
everything he had done showed his naiveté about the world outside
the intellectual élite.

Behind this lay the question of how to tell their mother. This was the very thing for which 'men like that shot themselves', and Alan told Robin that it was the worst part of the whole business. He had the gall to ask John to do it, which reasonably enough John refused. Accordingly it became Alan's duty to make a journey to Guildford to tell *her* the facts of his life. Mrs Turing was not entirely clear about the significance of what had happened, but she was sufficiently conscious of the subject for there to be a distressing argument, somewhat at cross purposes. She then placed the matter firmly at the back of her mind. But for whatever mixture of reasons, the salient fact was that she did not let it cause an estrangement. In Alan's schooldays she had sided with the authorities, seeing him as the problem for them, not school as a problem for him; this time she tacitly took his side.

Alan wrote to his brother complaining that he had shown no sympathy with the position in which homosexuals found themselves, which was true enough. He also accused him of caring for nothing but his own reputation in the City, which was not. It was more that both alike shared their father's character, and spoke their different minds.* John Turing made no secret of the fact that he considered his brother's behaviour disgusting and disreputable, an extreme example of 'a *modus vivendi* in which the feelings of others counted for so little.' He was particularly offended by the letter of complaint, because he had intervened in order to protect Alan from himself.

Perhaps nearly as difficult was the duty of telling Max Newman, so long a father figure. But if so, it did not show in Alan's behaviour. He simply announced that he had been arrested, and the reason for it, while they were sitting at lunch in the refectory. He did this in a particularly loud voice, making it clear that he rather wanted it to be heard by all and sundry. Max Newman was astonished, but his reaction was one of support. Alan asked him to appear as a character witness at the trial, a request he also made of Hugh Alexander, currently working for GCHQ. They both agreed. So in this respect Cambridge liberalism was prepared to stand up and be counted on his behalf – no small matter when a known homosexual was a social leper, conferring stigma by association.

---

* Once John Turing had asked his father what he hated most. 'Humbug', he replied without a moment's hesitation.

It was easier to tell those who were already familiar with his homosexuality. To Fred Clayton Alan wrote:

> . . . The burglary business was actually infinitely worse than an ordinary burglary. I had got a boy friend, who . . . put his friends on to my house. One of these has been picked up by the police and has informed against us. When you come to Liverpool perhaps you will stop off to see me in jail.

Then there was Neville; Alan telephoned and then made a journey down to Reading to see him. Neville thought Alan had been incredibly naive to call in the police in the first place. He was, of course, indirectly threatened himself, being fortunate that the police had not pursued their enquiries further, searching through letters and so forth. One ship in the convoy having gone down, the others had to look out very sharply. Not himself coming from the governing classes, he felt it as a simple outrage that someone who had done something very important in the war could be treated in this way. The visit was painful. Neville's mother heard what had happened and applied sufficient emotional pressure to stop her son from seeing Alan again.

There were others too who had to know. Alan wrote to Joan Clarke (herself, as it happened, now engaged to be married), explaining that he had not told her that he 'did occasionally practise', and that he had been found out. 'They're not as savage as they used to be,' he added, thinking back perhaps to the trials of Oscar Wilde. He also wrote to Bob, now away in Bangkok, where his letter, with its tone of 'never apologise, never explain', caused shock and sadness.

At the university it was yet another case of Alan proving a great embarrassment. They dealt with it as 'typical Turing'. There were members of staff who avoided him, but then they had avoided him anyway. Most people coped with it by carefully not referring to the case. A more free and easy atmosphere prevailed in the computing laboratory than elsewhere, although one or two of the staff were shocked. Tony Brooker's attitude suited Alan best: he had no idea such laws existed, and was simply interested to hear from Alan about what happened. In some ways the case made Alan appear a more human figure. When he called in Cicely Popplewell and asked 'Are you shockable?' explaining that he might be going to prison, it was the first time he had treated her as a person. There was no question of helping or extending sympathy to him – his personality ruled it out. Individual people were onlookers on events that might as well have been happening in Russia. Alan probably found an element of pleasure in confronting the more 'stuffy' elements of Manchester, and gave the impression, which the less sensitive seriously believed, that he did not care a jot about the case. As at school, he bore his afflictions cheerfully.

There was joking in the laboratory about how he would manage for money if he lost his job. In this respect, Max Newman spoke strongly on his behalf, and so did Blackett. Indeed Blackett went to see the Vice-Chancellor, Sir John Stopford, Professor of Experimental Neurology and a distinguished Mancunian, armed with a quotation from the Kinsey report to sustain his case. He said that Alan's work should be safeguarded 'at all costs'. The Vice-Chancellor was less receptive to Kinsey's statistics than the Admiralty had been to Blackett's convoy calculations ten years before. 'I will listen to any argument with care and sympathy,' said Stopford, 'but if anyone seeks to document it they must bring me authorities I can myself respect.' But Alan's position was spared, although presumably only after the most thorough of carpetings, for Stopford was no friend of 'slackness'. Max Newman's statement was crucially important ; indeed he found himself surprised by the autonomy he enjoyed as head of department. He said he wished Alan Turing to remain, and this sufficed.

There was also his connection with King's to consider, but here an odd coincidence came into play. His fellowship was due to expire on 13 March 1952, so although he had been arrested a Fellow, he would no longer be one at his trial. Alan consulted with Philip Hall regarding his position, and he in turn talked with Professor Adcock. They advised him not to resign, and in fact the fellowship simply ended at its appointed time, after a total of nine years spread over the past seventeen. He had no reason to feel cut off from King's because he had been found out; Cambridge could remain a point of security and support for him. There was another point of support in the reaction of his good neighbours the Webbs. Although upset by what had transpired, the Webbs still made him welcome in their home.

Despite so much time being taken up by these events, he did not stop work. He would have been ashamed to have let them stop him, just as he had insisted on keeping up some work on logic throughout the war. The very day after the arrest, he was in London for a meeting of the Ratio Club, talking about his theory of morphogenesis. John Pringle took up the idea as a basis for his discussion of the origin of life in the primordial chemical soup, in a lecture[6] later in 1952. Then again, on 29 February, the day that the local newspaper was reporting the first hearing, he was defending his work against the criticisms of the Belgian chemist Ilya Prigogine, then on a visit[7] to the Manchester chemistry department. On the same day Alan also completed his revisions to the morphogenesis paper, and on 15 March he submitted for publication his work on the calculation of the zeta function, even though the practical attempt at doing it on the prototype Manchester computer had been so unsatisfactory. It might be that he wished to get it out of the way in case he was going to prison.

On 21 March, Alan went to Henley-on-Thames for the weekend, to attend a large Nuffield Foundation conference on biological research. He found many points of contact with the discussions,[8] which were influenced by the rise of cybernetics and in which the importance of the morphogenetic problem was much stressed. Donald Michie was there. He had corresponded a little with Alan about the morphogenetic ideas, having himself moved on from physiology to genetics. Alan asked him to come for a walk, and revealed that beneath the *sang froid* he showed to the more conventional world, he was in a very nervous state. He mentioned the previous appearance in the magistrates' court, and the forthcoming trial, which was now only a week away. Donald said that no serious person would take a court judgment in the matter to be of importance, and that Alan would have to go through with it knowing that. But Alan might well have reflected that it was not only the law that made him an outcast, but all the official British culture, that of its administration, its newspapers, schools, churches, social life and entertainment – and that very largely its intellectuals would add their public weight against him too, whatever Donald Michie generously said.

Attitudes were one thing. Practical prospects were another. There was the nauseating business of having authorities ransack his emotional life and pronounce upon it, and there was the actual punishment that lay in store. The circumstances of his crime, with their elements of age and class difference, were against him. Even to the kindly disposed, this might seem a case of the 'elderly degenerate' of *The Green Bay Tree*, rather than the romantic wantonness of the greenwood tree.

His intransigent attitude was also a challenge to the authority of the law. But on the other hand, only 174 of the 746 men prosecuted in 1951 for the crime of 'gross indecency' had been imprisoned, and then mostly for less than six months. He would have been in a more dangerous position had the charge been that of 'buggery', for the law distinguished carefully between different kinds of sexual activity. He was also a 'first offender', which diminished the chances of imprisonment. But beyond this, the times were changing, and a more modern attitude was coming to prevail. The backroom boys were beginning to affect not only descriptions, but prescriptions.

Writing a new Foreword to *Brave New World* in 1946, Aldous Huxley suggested that 'The release of atomic energy marks a great revolution in human history, but not (unless we blow ourselves to bits and so put an end to history) the final and most searching revolution.' Despite the fact that he believed atomic power would usher in 'highly centralised totalitarian governments', continuing the trend which the Second World War had accelerated, he stuck to his guns of 1932. 'This really revolutionary revolution is to be achieved, not in the external world, but in the souls and flesh of human beings,' he had claimed, and the signs were there in the existing research in 'biology, physiology, and psychology'.

Alan Turing was no stranger to these subjects; his intellectual development had circled round to face the question of *Natural Wonders*: 'By what process of becoming did I myself finally appear in this world?' The significance of his mathematical work, indeed, rested upon the fact that specific substances, the 'growth hormones' referred to in his paper, had been chemically isolated by experimental biochemists. Amidst the patient accumulation of facts, the discoveries of the sexual hormones since 1889 had engendered a particular interest. Such interest, both lay and scientific, was not confined to the role of the hormones in physiological growth. It was widely asserted that the 'chemical messages', at which E. T. Brewster had marvelled in 1912, determined the psychology as well as the physiology of the individual.

If, to the more old-fashioned person, the problem of homosexuality remained one of 'filth' and indiscipline, of which as little was to be said as possible, the modern psychological view was dominated by the categories of 'masculine' and 'feminine', with a belief that gay men and lesbian women had been endowed by nature with some unusual mixture of these all-important characteristics.* One attraction of this view was that it left intact an assumption of universal heterosexuality, since these apparent exceptions could be defined away as cases where a woman was 'really' a man, or *vice versa*. Some found in this theory a scientific justification of homosexuality, according to the logic of the age, others found the hope of solving the hitherto unsolvable problem that it posed.

The discovery of the hormones suggested that the eternal verities of 'masculine' and 'feminine' might, indeed, be embodied in a simple chemical form. It was appropriate to a decade in which these great truths were so sedulously cultivated by Hollywood that the first major American experiments to test this theory were made at Los Angeles in 1940. The endocrinologists estimated the quantity of male hormones, or androgens, and of female hormones, or oestrogens, in the urine of seventeen homosexual men who had been 'taken into custody'. They did the same for thirty-one 'normal males'. Their results showed that even for a single individual this ratio could vary from time to time by a factor of up to thirteen. However, a judicious averaging of the ratios suggested that gay

---

* Thus an exposition of this idea appeared in the 1931 American novel *Strange Brother*,[9] one of the few accessible pre-war exceptions to the general silence: 'You see the generative gland is made up, not only of the gland of reproduction, but of a gland which manufactures the chemicals that cause a man to be masculine in temperament, and a woman feminine.'

'Both these chemicals are found in every human embryo. But if normal development does not take place, the feminine chemical may predominate in a male, or the masculine in a female. And we then get a man attracted by men, or a woman attracted by women.'

'This, at least, is the most plausible theory that modern science has to offer. And our experiments on rats and guinea pigs bear out the theory. . . . We've proved that, aside from the function of reproduction, sex differences are chemical.'

men had an androgen-oestrogen ratio of only sixty per cent of the others.*

The data were intimately bound up with instructions. Dr Glass, describing this result,[10] explained that 'Obviously, if a biologic etiology were established, this would lead to investigation of thera-peutic possibilities from a much wider perspective than now exists,' which in English meant that if they could find a chemical to turn homosexuals into heterosexuals, then they could use it.† So in 1944 Glass experimented[11] with the injection into some eleven gay men of male hormones 'kindly supplied' by pharmaceutical companies. 'Four subjects accepted organotherapy by compulsion' – a court order in one case, and parental authority in the case of three boys. Alas, the experiment was not a success, in Dr Glass's opinion. 'Only three of the subjects reported benefit from this therapy. Five reported an intensification of the homosexual drive.' This to the scientists was 'a worsening of the condition'. It did not help in 'the clinical management of the male homosexual.' Back at the drawing board of endocrinology, the failure of the experiment suggested a diametrically opposite approach. If the male hormone *increased* sexual 'drive', then perhaps the female hormone would *decrease* it – for heterosexual and homosexual men alike. This bright idea had already been tried[12] by another American pioneer, C. W. Dunn, in 1940. He had reported that 'At the end of the treatment there was a complete absence of libido.'

One attraction of this technique was that it was so much more effective than physical castration. Surgery of this kind had held a traditional part in the American Way, especially since the eugenic clean-up of the late nineteenth century. In 1950 there were eleven states which allowed for compulsory castration, with fifty thousand cases on record.[13] But there was also scientific evidence that castration did not successfully inhibit sexual activity, and in this respect the chemical approach was more promising.

---

* Some of the results did not fit in, because the 'normals' sometimes had low ratios, and the gay men high ratios. But this was ingeniously explained: 'Those few normal subjects may be latent homosexuals, whereas the homosexuals with the high ratios may not be of the true constitutional type.'

† In similar vein: 'The growing importance of the sociological aspects of the subject makes urgent the continued investigation of the problem from a broad psychosomatic perspective.'

This was the lesson drawn by the first British paper[14] on the subject, which appeared in the pages of the *Lancet* just a few days after Jefferson invoked the supremely human 'charm of sex' in his Lister Oration. It appeared under the authoritative name of F. L. Golla, director of the progressive Burden Neurological Institute at Bristol, where Grey Walter had built his cybernetic tortoises.* Neither compulsory nor voluntary castration was allowed in Britain. On the other hand, as he wrote, 'The Criminal Justice Act, 1948, had emphasised the duty of the community to provide treatment for the habitual sexual offender.' Hormone dosage resolved this contradiction, being both legal and more effective. By 1949 Golla had experimented on thirteen men, and found that with sufficiently large doses 'libido could be abolished within a month.' He concluded:

> In view of the non-mutilating nature of this treatment and the ease with which it can be administered to a consenting patient we believe that it should be adopted whenever possible in male cases of abnormal and uncontrollable sexual urge.

He had opened the visionary prospect of providing chemical castration for all homosexual men. In 1952, the *Sunday Pictorial* commented that

> What is needed is a new establishment for them like Broadmoor. It should be a clinic rather than a prison, and these men should be sent there and kept there until they are cured. Doctors and psychiatrists would welcome the idea. There is still a great deal to be learned about the delicately balanced endocrine glands which determine whether or not a man could take to these unpleasant activities.
>
> L. R. Broster, the Charing Cross Hospital specialist who has done pioneer work in this field, writes that surgical treatment has made great strides recently but 'is still in the groping stage of trial and error'.

The possibilities had, in fact, widened into a more ambitious sphere of human management. Another paper[15] described how the *male* hormone had been administered to a fourteen-year-old truant:

---

* Thus Golla and Grey Walter were the two scientific colleagues thanked by W. Ross Ashby for reading the draft of his 1952 cybernetic book *Design for a Brain*.

For many years he had shrunk from personal contacts, been sensitive, shy and latterly more solitary. Recently he had become morbidly preoccupied with thoughts on abstruse topics beyond his years, chiefly psychology and religion . . . In the ward he read a lot, wrote many letters, helped with domestic jobs, showed interest in philosophy, but mixed little with others.

But after a course of the drug

his religious and other preoccupations disappeared. The drug was stopped and he was discharged, much improved. Six months later he was reported to have a job at a printers, but still apt to speculate on religion, and to be readily teased by other youngsters.

Perhaps the fourteen-year-old Alan Turing would have benefited more from such scientific treatment than from the rough-and-ready team games. On the other hand, it was the *female* hormone which according to this authority had

been most useful in controlling occasional outbreaks of homosexual behaviour in the 12 to 16 age groups.

More effective than cold baths, or Nowell Smith's eloquence, it was the oestrogen which had been useful in 'enabling one of the provisions of the Criminal Justice Act, 1948, to be carried out.' It was the beginning of a new era, in which chemical solutions could be found for the problems of social control.

These advances did not escape the notice of other scientists. In 1952 Sir Charles Darwin, who always took a long-term view, published a book called *The Next Million Years*. Biology, rather than physics, he held to offer 'the most exciting possibilities', one being that

there might be a drug, which, without other harmful effects, removed the urgency of sexual desire, and so reproduced in humanity the status of workers in a beehive.

Other chapters in the progress of mankind had given rise to alternative methods of treatment, but these had generally disappointed the experts. Gordon Westwood summed up the experience of analytical psychoanalysts as that of finding homosexuality to present the most problems, of all the cases that they met. Lobotomy had been tried but, as Westwood wrote,[16] this did not seem to be any more 'successful'. Nor was the administration of a drug to induce epileptic fits, another medical advance of the 1940s. The application of behaviourism to the problem, by administering electric shocks or nauseating drugs in association with sexually attractive stimuli, was a technique still undergoing trial in Czechoslovakia and not yet introduced to British psychiatry. For the time being, the less scientific pain buttons offered by prison, loss of employment, social ostracism and blackmail were expected to control behaviour, and when these failed, the new men offered 'organotherapy', or chemical castration. Such were the resources of modern science that were offered to Alan Turing. He perceived them as the lesser evil, and on that basis went to trial. It was a trial between the old and the new.

The queened pawn faced the White Queen. The case of *Regina v. Turing and Murray* was heard on 31 March 1952, at the Quarter Sessions held at the Cheshire town of Knutsford.[17] The judge was Mr J. Fraser Harrison. Alan was represented by Mr G. Lind-Smith, and Arnold by Mr Emlyn Hooson. Both were prosecuted by Mr Robin David. The charges now amounted to twelve in number. With the Looking-Glass symmetry of symmetrical crimes, they began:

Alan Mathison Turing

1. On the 17th day of December, 1951, at Wilmslow, being a male person, committed an act of gross indecency with Arnold Murray, a male person.

2. On the 17th day of December, 1951, at Wilmslow, being a male person was party to the commission of an act of gross indecency with Arnold Murray, a male person.

and so forth, for each of the other two nights, and then Arnold was charged in exactly the same way so that the last accusation was that he:

12. On the 2nd day of February, 1952, at Wilmslow, being a male person, was party to the commission of an act of gross indecency with Alan Mathison Turing, a male person.

They both pleaded 'guilty' to all the charges, although Alan was guilty of something for which he showed no guilt. The prosecuting counsel, in outlining the case, laid stress upon his unrepentant remarks.

There only lay his 'character' to set against this admitted law-breaking. Normally, 'good character' would be a disguised statement of class status, but in these circumstances his status told against him. The theme of the better public schools had been the balance of privilege and duty, and as one of the prefect class he was supposed to set an example, not to break the rules himself. Alan Turing, however, was little interested either in the privileges or the duties of his class. He never tried to pull rank on the detectives, who saw him as an 'ordinary fellow', with his occasional visits to the local pub. Conversely, his crime was seen at least by an older generation as a betrayal of his class. Arnold likewise was made to feel by his family that his real crime had been that of dragging down a gentleman.

The OBE* was duly given a mention, and Hugh Alexander bore witness that Alan was 'a national asset'. Max Newman was asked whether he would receive such a man in his home, and replied that he had already done so, Alan being a personal friend of himself and his wife. He described Alan as 'particularly honest and truthful'. 'He is completely absorbed in his work,' he continued, 'and is one of the most profound and original mathematical minds of his generation.' Lind-Smith pleaded that he should not go to prison:

> He is entirely absorbed in his work, and it would be a loss if a man of his ability – which is no ordinary ability – were not able to carry on with it. The public would lose the benefit of the research work he is doing. There is treatment which could be given him. I ask you to think that the public interest would not be well served if this man is taken away from the very important work he is doing.

Mr Hooson, however, defended Arnold as the innocent led astray by Alan's wiles:

> Murray is not a university Reader, he is a photo-printer. It was he who was approached by the other man. He has not such tendencies as Turing, and if he had not met Turing he would not have indulged in that practice.

---

* The fact that he retained his OBE was itself an interesting detail of the case. The War Office would demand the return of medals from anyone guilty under the 1885 Act. The Foreign Office presumably took a different view.

Max Newman and Hugh Alexander were amazed that Alan should go to the stake for Arnold, but Alexander was impressed by his 'moral courage' and Newman by his 'strong line'. He answered back at the judge's remarks, and he did not recant, at an occasion whose very essence was the obtaining of a confession. Hilbert had written of Galileo that in his recantation 'he was not an idiot. Only an idiot could believe that scientific truth needs martyrdom – that may be necessary in religion, but scientific results prove themselves in time.' But this was not a trial of scientific truth.

The verdict quivered between the old and the new dispensations, and came down for the new. Bletchley Park scored a victory beyond its term. The state washed its hands, and handed Alan to the judgment of science. He was placed on probation, with the condition that he 'submit for treatment by a duly qualified medical practitioner at Manchester Royal Infirmary.'

The Wilmslow newspaper headline was:

UNIVERSITY READER PUT ON PROBATION
To have Organo-Therapic Treatment

Alan wrote to Philip Hall, two weeks later:

(postmarked 17 April 1952)

. . . I am *both* bound over for a year *and* obliged to take this organo-therapy for the same period. It is supposed to reduce sexual urge whilst it goes on, but one is supposed to return to normal when it is over. I hope they're right. The psychiatrists seemed to think it useless to try and do any psychotherapy.

The day of the trial was by no means disagreeable. Whilst in custody with the other criminals I had a very agreeable sense of irresponsibility, rather like being back at school. The warders rather like prefects. I was also quite glad to see my accomplice again, though I didn't trust him an inch.

Perhaps it was surprising that he chose the scientific alternative to prison. He was annoyed at having been circumcised, and at any editorial meddling with his writings – small interferences compared with this piece of doctoring. Neither did he care much for creature comforts, and a year in prison, even an English one, would not have been much more uncomfortable than Sherborne. On the other hand, to take that option would have impeded his work, and very likely would have forefeited his Manchester position and the computer. He had the choice between his body and feelings on the one hand, and his intellectual life on the other. It was a remarkable decision problem. He chose 'thinking' and sacrificed 'feeling'.

There was no concept of a right to sexual expression in the Britain of 1952. People made jokes about bromides put in the servicemen's tea. Samuel Butler might well have laughed in his grave at the prophet of the intelligent machine being punished for being sick, and treated for committing a crime. But no one at the time perceived an irony in Alan Turing being on the receiving end of science. As for Jefferson, ranging himself with the humanists, or Polanyi, foe of the state's pretensions to order human life, and adherent of the Congress for Cultural Freedom – this was a private and distasteful medical matter, and did not gain the attention of the liberal intelligentsia of Manchester, discoursing on the folly and iniquity of treating minds like machines.

Harry the burglar was sent to a Borstal the same day, in another trial. Arnold, however, was conditionally discharged. He left the court in a daze, hardly knowing to what he had confessed, and then found himself pointed out in the street by his neighbours. After a few weeks he escaped back to London, found a job in the Lyons Corner House in the Strand, and rapidly made his way into anarchic Fitzrovia. Here in the coffee-bar world, meeting such people as Colin Wilson, he was accepted as an individual and learned to play the guitar.

For Alan, there were rather different consequences of the trial, because of the drug treatment. He was rendered impotent, although scientific opinion was that the impotence was not permanent and potency would return when the medication was stopped. It had other physical effects, for[18]

> To obtain the necessary effect mentally, it was necessary to maintain a moderate but not excessive degree of gynaecomastic response.

Translating from the Greek, this meant that there could be no 'reduction in libido' without the production of breasts. Again, according to the same authority,

There is at least a possibility that oestrogen may have a direct pharmacological effect on the central nervous system. Zuckerman (1952) has demonstrated, through his experiments on rats, that learning can be influenced by sex hormones, and that oestrogen can act as a cerebral depressant in these rodents. While it has yet to be shown that a similar influence is exerted in humans, there are some indications clinically that performance may be impaired, though more investigation is needed before any conclusion is reached.

So perhaps 'thinking' and 'feeling' could not be so neatly separated after all.*

There were some more minor consequences. The *News of the World* covered the case with a short article in its northern editions, headed ACCUSED HAD POWERFUL BRAIN. He remained under the auspices of the district probation officer. David Champernowne came to Manchester to do some work† on the computer, and being invited to dinner at Hollymeade, found the probation officer another guest. Alan told a story about how the retired bishop of Liverpool had heard of the case and asked to see him – he had gone along, rather surprisingly for one who had written in 1936 that he would not tolerate bishops interfering in his private life. But nothing was private now. He had thought the bishop well-meaning but hopelessly old-fashioned. A further consequence, which for another person might have been major but which for Alan had little significance, was that, with a criminal record of 'moral turpitude', he was henceforth automatically barred from the United States.‡

---

* At the Nuffield Foundation conference that Alan attended, P. B. Medawar had proposed a programme of experiments on male animals, injecting oestrogen in order to reveal the neuro-physiological mechanism through the consequent alteration in behaviour patterns. Rarely can a Fellow of the Royal Society have had the honour of sitting at the wrong end of such an experiment.

† The work he was doing was on the application of sequential analysis to economics. Although he knew that Alan was interested in Bayesian statistics, he had no idea that Alan had independently invented sequential analysis in Hut 8.

‡ In a reform typical of the period, 1952 saw a change in American immigration policy from a *legal* definition of homosexuality (the breaking of a law), to a *medical* definition. The Immigration and Nationality Act of that year specified that 'Aliens afflicted with psychopathic personality . . . shall be excludable from admission into the United States.' In 1967 the Supreme Court confirmed that 'the legislative history of the Act indicates beyond

Describing the events of the trial as with amused detachment, Alan tried to continue as though nothing had happened – as though he had been caught doing a naughty experiment in the dormitory and had suffered the confiscation of his chemistry set. This was, indeed, more or less what had happened and on one level he could treat it like the humiliations of schooldays. Yet it had obliged him to become more conscious of the conduct of his life and more conscious of his environment. The short story that he scribbled out was one symptom of this increased self-awareness. One person who found him much more interesting and indeed congenial, now that he was not a remote mathematician with a one-track mind about machines, was Lyn Newman. With his reality revealed, Alan dropped his disconcerting evasiveness, and Lyn Newman found[19] that 'once he had looked directly and earnestly at his companion, in the confidence of friendly talk,' his eyes, 'blue to the brightness and richness of stained glass', could 'never again be missed. Such candour and comprehension looked from them, something so civilised that one hardly dared to breathe.' It was at this time that she

pushed first *Anna Karenina* and then *War and Peace* into his hands. I knew that he read Jane Austen and Trollope as sedatives, but he was totally uninterested in poetry and not particularly sensitive to literature or any of the arts, and therefore not at all an easy person to supply with reading matter. *War and Peace* proved to be in a very special way the masterpiece for him and he wrote to me expressing in moving terms his appreciation of Tolstoy's understanding and insight. Alan had recognized himself and his own problems in *War and Peace* and Tolstoy had gained a new reader of a moral stature and complexity and an originality of spirit equal to his own.

For indeed, he was there in *War and Peace*, as Pierre wandering into the midst of the battle – and then? What did it mean? What was it for? And he was there in Tolstoy, whose puzzle was not over this or that fact, but what history was. Could an individual cause an event, hold power, or exercise will, as in the story-book kind of history? 'The subject of history,' he wrote,[20] 'is not the will of

a shadow of doubt that the Congress intended the phrase "psychopathic personality" to include homosexuals.' Strictly speaking, therefore, Alan Turing entered the prohibited category in 1952 irrespective of the trial; in practice, of course, the point was that he had been found out.

man as such but our presentation of it.' It was, in other words, the level of description. The degree of 'will' would depend upon the kind of description, and 'what is known to us we call the laws of necessity; what is unknown we call free will. Free will is for history only an expression connoting what we do not know about the laws of human life.' In particular, the laws of the *connection*, as he put it, between the mind and the world, were yet unknown, and therefore called free. These were Turing questions, in another form. In the January radio discussion he had said, 'Thinking is those mental processes we don't understand.'

Yet, wrote Tolstoy, however nonsensical the idea of free will, 'without this conception of freedom not only would he be unable to understand life but he would be unable to live for a single moment. Life would be intolerable because all man's aspirations, all the interest that life holds for him, are so many aspirations and strivings after greater freedom. . . . To imagine a man wholly destitute of freedom is the same thing as to imagine a man destitute of life.'

For Alan Turing there remained one freedom, not perhaps one that Tolstoy would have had in mind. It was that of exiled pleasure. The headmaster having taken action to prevent associations within the house, he had to fall back upon the possibilities offered by the boys in the other houses.* For Alan would not let the system defeat him. On May Day 1952 there was a meeting of the Ratio Club at Cambridge, which he attended, and it was probably then that he saw Norman Routledge at King's. Alan explained about the trial and the hormone treatment ('I'm growing *breasts!*') and Norman told him that he had heard that in Norway (of all places) there were dances 'for men only'.

In the summer of 1952 Alan went for a holiday in Norway, one which turned out to be a disappointment regarding the rumoured dances. But he met a number of Scandinavians, enough to have the addresses of five or six,[21] and he was particularly struck by one young man called Kjell, whose photograph he showed to Robin

---

* It may be that to avoid being sent to prison he had to give a promise that he would not repeat the 'offence', in addition to undertaking the hormone treatment. Lack of evidence prevents this point being settled. If he had so promised, he would have kept to it, but he would have been the first to observe that this said nothing about what he did abroad. For this reason his foreign holidays may have been all the more consciously a critical factor in his life after 1952.

on his return. Kjell had been somewhat coquettish, and little had
transpired, but Alan had thereby demonstrated his unbroken will,
which was perhaps the most important thing.

As for what the endocrinologists called intellectual 'performance',
Alan's work on the biological theory did continue, despite everything,
to develop in range and scale. He was tackling the problems which
he had outlined in the first paper. In particular, he was trying out
on the computer the solution of the very difficult differential equations that arose when it was attempted to follow the chemical theory
of morphogenesis beyond the moment of budding, taking into
account the essential non-linearity. This was experimental work, in
which he would be trying out many different initial conditions to
see what happened. But it also required some rather sophisticated
applied mathematics, which involved the use of 'operators' rather
as in quantum mechanics. Numerical analysis was also important, in
deciding how to approximate the equations for the purpose of the
calculation. In this it was like a private atomic bomb, the computer
in both cases following the development of interacting fluid waves.

As a rather separate line of attack, he also developed a purely
descriptive theory of leaf-arrangement, or 'phyllotaxis' as it was
called in biological Greek, in which he found ways of using matrices
to represent the winding of spirals of leaves or seeds round a stem
or flower-head. He brought into this theory a concept of 'inverse
lattices' somewhat like that used by crystallographers. It was also
accompanied by a good deal of measurement-making of his own.
The intention was that ultimately these two approaches would join
up when he found a system of equations that would generate the
Fibonacci patterns expressed by his matrices.

Although there was some correspondence with a number
of biologists, this work was essentially done on his own. The
Manchester botanist, C. W. Wardlaw, was particularly interested and
wrote a paper[22] describing, in biologists' terms, the significance of
the first Turing paper. This finally appeared in August 1952, and soon
Alan had a letter[23] from C. H. Waddington expressing interest but
scepticism as to the correctness of the essential chemical hypothesis.
But on the whole, Alan would tend to speak – to Lighthill, in
particular – as though rather disappointed with the slow speed with
which his ideas diffused, and the lack of reaction to them. There

was, perhaps, an analogy with *Computable Numbers* in this respect, for the 'confirmed solitary' whom Max Newman had diagnosed in 1936 still lacked the talent for patient, persistent pushing. Neither had his abilities as a lecturer improved. One side-line that developed was an interest in irreversible thermodynamics, and after giving his seminar in the chemistry department Alan had a meeting with W. Byers Brown to discuss the subject; but this soon faded out, Alan possibly being more interested in young Byers Brown than in this branch of physical chemistry. But one difference from the reaction to his earlier achievements was that this time no one had pre-empted his ideas. He was quite alone.

Robin had persuaded Alan to come with him to the 1952 British Mathematical Colloquium in the spring; it was held at the Royal Naval College, Greenwich, which meant that they had the excuse to take a jaunt on the Thames steamer. Alan found some interesting wild flowers on the Greenwich bombsites, and there was a nice moment at the lunchtime bar, when Alan suddenly disappeared through one doorway on spotting a particularly dull logician bear down upon him from another. By this time he was becoming quite famous as the author of *Computable Numbers*. He liked to hear references to Turing machines ('Pryce's buoy' in his story) but he did not like paying the price, that of being pinned down for shop-talk by those trying to make connections.

More to his taste was talking with Christopher Strachey, who had brought a fresh breeze from King's into the technical atmosphere of the Manchester computer laboratory. He had much the same attitudes as Alan, and the same sense of humour. His draughts program was much developed[24] and played throughout the summer of 1952, this being the first time that the kind of automatic game-playing that Alan had so long talked about was seriously tried. But he and Alan also used the random number facility in a program to compose 'love letters'. One of these ran:

Darling Sweetheart,

You are my avid fellow feeling. My affection curiously clings to your passionate wish. My liking yearns to your heart. You are my wistful sympathy: my tender liking.

Yours beautifully, MUC

Those doing real men's jobs on the computer, concerned with optics or aerodynamics, thought this silly, but it was as good a way as any of investigating the nature of syntax, and it greatly amused Alan and Christopher Strachey – whose love lives, as it happened, were rather similar too.

Tony Brooker, meanwhile, had written a program system called FLOATCODE, which did the work of interpreting floating-point arithmetic, rather as Alan had envisaged in 1945, but had never bothered to do for Manchester. It was based on similar work he had done at Cambridge on the EDSAC. And Alick Glennie went further in 1952 with something called AUTOCODE which was, in effect, the first working high-level computer language in the world. Christopher Strachey was enthusiastic about it – AUTOCODE was in line with the ideas he had written about in 1951, of translating mathematical formulae into machine instructions. But Alan took little interest. Alick Glennie talked to him about it, but found that he was bored by mere translation, something that he had described as obvious in 1947 and had never chosen to take any further himself. He would have been interested in something that would actually *do* the algebra, rather than translate it.

The computer industry was now in a position to expand beyond the confines of a small trained élite, programming languages opening the universal machine to a much wider clientele. AUTOCODE did not in fact play this role, and was little known away from Manchester. But the American FORTRAN was not far behind, one of a chain of developments with which Alan Turing had quite parted company.

By 1952 the Manchester engineers not only had a Mark II machine in hand, but had begun the design of a small transistor-based prototype. No one could have guessed from his total lack of participation in these developments that Alan Turing had once been avid to keep abreast of the latest technological advances, and had broken unwritten rules in order to get his hands on them himself. All this had been dropped in 1949 when it finally became clear that the world saw such an interest only as a nuisance. There was no reflection of his attempts to make a practical contribution to computer development in a book called *Faster than Thought*,[25] the definitive account of British computers as they stood in 1951–2. Here he appeared principally as the writer of part of a chapter on 'Digital

Computers applied to Games'; he wrote up his chess game with Alick Glennie, helped with some comments on the play by Hugh Alexander. Besides a one-line mention as the author of *Computable Numbers* and as one of Womersley's assistants, there was one telling entry in the glossary:

> *Türing Machine.* In 1936 Dr Turing wrote a paper on the design and limitations of computing machines. For this reason they are sometimes known by his name. The umlaut is an unearned and undesirable addition, due, presumably, to an impression that anything so incomprehensible must be Teutonic.

The world of 1945 was now as remote, and effectively as secret, as that of 1942; a fact to which Alan seemed entirely resigned.

Robin kept him interested in the theory of types, and some work he had done stimulated Alan to dig out the paper he had written during the war, but had left unpublished, as his attempt to persuade mathematicians into a more careful use of 'nouns' and 'adjectives'. Here again, the suggestions for 'The Reform of Mathematical Notation', as his essay was entitled, were at a tangent to the development of post-war mathematics, in which the muddles to which Alan objected were being cured by other means.

Alan mentioned his 'reform' proposals to Don Bayley when he paid a visit to him and his wife at Woburn Sands, near Bletchley, that summer. He gave Don some mathematical help, but the main point of the weekend was to make one last serious attempt to retrieve the silver bars. This time Don had got hold of a commercial metal detector, and they went out to the bridge near Shenley in his car. Alan said, 'It looks a bit different', as he took off his socks and shoes and paddled in the mud. 'Christ, do you know what's happened? They've rebuilt the bridge and concreted over the bed!' They tried for the other bar in the woods, finding that the pram in which he had wheeled the ingots in 1940 was still there, but without any more luck than before in locating the spot. They found nails and oddments, just as Alan had done on the earlier attempt with Donald Michie. Giving up both bars as lost for ever, they made their way to the Crown Inn at Shenley Brook End for some bread and cheese. The disappointment was not too great, and largely dissipated by the warm welcome accorded to him by Mrs Ramshaw, his wartime landlady.

When Don Bayley had met him at Bletchley station, he noticed Alan was carrying a Norwegian grammar. Alan explained that he had just had a holiday in Norway, and that the language interested him. Although at this point his knowledge was rudimentary, he made sufficient progress with Norwegian and Danish to read Hans Christian Andersen stories to his mother a year later. It did not occur to Don that the Norwegian holiday might have had a particular motive, even though Alan explained that now he would have to go abroad for pleasure. He had written to Don about the charge and the trial, as to his other friends, and on the visit spoke with his usual light-hearted bravado of the outcome. He also referred to a letter he had written to a titled lady politician, calling for a change in the law. This was no pleading, as Oscar Wilde had done, that it was no crime but sickness. He had drawn attention to the homosexuality of the politician's son. All he had received in reply was a brusque disclaimer from her secretary.

In October 1952, both Don Bayley and Robin went up to Wilmslow for a weekend, a re-creation of Hanslope. Don was there first and they waited together for Robin on the station, Alan showing Don the diffraction pattern that appeared when looking at the station lights through his handkerchief. On the summer visit Alan had taken his usual pleasure in being looked after for a weekend within the more conventional domesticity of the Bayley household, and Don was struck by the contrasting spartan, untidy arrangements of the Prof's home. Alan pointed to a stack of filing trays overflowing with letters from all over the world about logic, but said that he was not bothering to appear at the university now, and worked at home. He explained that he had an assistant who had taken over the organisation of the computer. Don advised him to watch out, or his assistant would take over. 'Pooh!', said Alan, as if to say 'see if I care!'.

But if his computer days were over, this did not mean the end of his underlying interest in the human mind. October 1952 also saw Polanyi and the philosophy department at Manchester score something of a *coup* over the psychology department, by having the Swiss psychologist Jean Piaget to give a course of lectures,[26] which Alan attended. They concerned the child's learning of logical ideas, and connected symbolic logic with actual psychological observations. So perhaps for the first time, Alan found himself listening to arguments about learning and teaching that did not just come from his own experience, and which were touched by modern theories of education that no one at Sherborne would have known to exist. At about the same time he breached his self-sufficiency in another way. He began seeing a Jungian psychoanalyst, Franz Greenbaum.

There was an element of resistance in his attitude towards this step, because of the implication that there was something wrong with him, and in particular, that his homosexuality was something that should be changed. The 1950s were, indeed, witnessing a powerful come-back of psychoanalysis, and increasingly vocal claims to the effect that its techniques could eradicate homosexual desire. But Greenbaum did not take such a view; homosexuality was not 'a problem' to him. He accepted Alan as a 'natural homosexual', and as a Jungian, he did not consider human activities in terms of displaced or unconscious sexuality. Rather, as a German refugee of 1939 with a Jewish father and Catholic mother, it was the psychology of religion that interested him most. As with Jung himself, there was no devaluation of the intellect in Greenbaum's approach, and he was proud to know Alan as the inventor of the computer and as one who was working on the nature of life. His emphasis, as with Jung, was on the *integration* of 'thinking' and 'feeling'. To apply intelligence to himself; to look at his own system from outside like Gödel, and break his own code – these were natural extensions of Alan's long-growing interest in psychology. A turning-point was indicated on 23 November 1952, when he wrote to Robin in connection with his now completed PhD thesis, and added:[27]

> Have decided to have another, and rather more co-operative go at the psychiatrist. If he can put me in a more resigned frame of mind it would be something.

Thereafter Franz Greenbaum had Alan write down all his dreams, and he filled three notebooks with them.* The relationship soon became more that of friendship than that of doctor and patient.

---

* Jung held that dreams had meanings, but did not believe that they could be deciphered according to some fixed scheme:[28] 'The interpretation of dreams and symbols demands intelligence. It cannot be turned into a mechanical system. . . . It demands . . . an increasing knowledge of the dreamer's individuality . . .'

But the professional status gave Alan the excuse to devote time and energy to all those things that he had pushed aside so long from the serious male business of 'thinking'. As with the war, he made the most of the situation in which he found himself.

In the analysis of his dreams he was surprised to find that many concerned, or could be interpreted as relating to, his mother in hostile terms. In real life, his relationship with her had continued to grow warmer. The fact that she had taken the news of the trial as she had counted for a great deal. So in her seventieth year, Mrs Turing found herself becoming one of Alan's few friends. By now she knew that he would never cease to be the 'intellectual crank' she had been afraid of, and he knew that she would always concern herself with matters like fish-knives as though still arranging for dinner parties at Coonoor. A gentle bickering, with '*Really*, Alan!' answered by 'Mother, don't be so *ridiculous*,' characterised the occasional visits. But by this time he had perhaps come to appreciate some of *her* problems and frustrations, while she, in turn, had come a long way from being the muffled Dublin girl at Cheltenham Ladies College, and had perhaps come to realise that Alan's vivacity offered her a taste of the more artistic life that she had been denied. After looking so long for the higher and better in churches and institutions, ranks and titles, she found something of it in her son. For forty years she had been cross with him for doing everything the wrong way, but she found the capacity for change. Alan, too, became less totally dismissive of her preoccupations.

There was plenty of scope for unburying a forty-year-old resentment of a mother so unlike the sensuous, seductive figure of Freudian theory. Perhaps Alan also confronted the figure of his father, whose strength had somehow cancelled itself out, and who had not shown the marathon-runner's quality of his son. Perhaps too there was a hidden disappointment that his father had never even tried to penetrate his concerns in the way that his mother, however irritatingly, attempted. If Alan's friends heard him disparage his mother, they usually heard nothing of his father. But sorting out such complexities of inner feeling was one thing; coping with his situation in the real world of 1952 was quite another, and in this respect psychoanalysis was bound by the same limitation as his imitation game – it was the world of dreaming, not of doing. A private 'free association' of ideas was allowed, but free association with male persons – that was the very thing that was forbidden. Franz Greenbaum could not supply the greenwood. Consistency and completeness of mind was not enough; something had to be done.

He had written to a politician about the state of the law, and there was little else that an individual could do – except in refusing to keep quiet. The problem did not lie at an individual level, where the only 'solution' was that of being 'resigned'. He was not charged as one who had harmed another person, but as an enemy of social order. Alan Turing, however, had no interest in ordering other people, while he retained an almost untouched innocence of 'why not?' in respect of sex.* It was not an issue that could be resolved by rational argument, and not a problem that Dr Greenbaum could solve.

The examination of Robin's PhD thesis, on the logical foundations of physics, had to be postponed since Stephen Toulmin, the philosopher of science, had decided he could not undertake it after all. Early in 1953, Alan wrote to Robin:

> They have at last found someone to referee your thesis, viz. Braithwaite. I think it would be best if we had the oral at Cambridge and am writing to Braithwaite to suggest this. . . . Have had another go at the Unity of Science essay.

This was a prize essay that Robin was submitting to the *Unity of Science* journal, concerned with the same subject.

> I think the duplicate types might be quite important. Don't they answer the question about 'What is Time'? I was amused about 'inpenetrability' at first. I thought it was a reference to 'Through the Looking Glass', where Humpty Dumpty says 'Inpenetrability. That's what I say.' But on looking up the reference thought probably not.

---

* He did agree with Robin that one should not persist with efforts to gain the interest of a boy of less than fifteen or so. (Robin had attracted a good deal of attention as a boy, and a too enthusiastic admirer had had the effect of putting him off sex for a time.)

This letter was rendered, not very effectively, as part of the computer printout.* Alan had proposed having the oral in March, but this did not suit Robin, who had arranged to go skiing in Austria. Alan wrote:

> Sorry it really isn't possible to make your oral any earlier. Braithwaite won't have read it before the very end of March. . . . If you really are going skiing no doubt it could be delayed till April or May though I may have forgotten about it by then mostly.
>
> Your last letter arrived in the middle of a crisis about 'Den Norske Gutt', so I have not been able to give my attention yet to the really vital part about theory of perception. . . .

The nature of this 'crisis' was partly revealed by another letter, dated 11 March 1953:

> My dear Robin,
>
> I am going to try and stop your journey to Austria by informing the immigration authorities of the following facts:
> (i) That although you have the permission of your mother the counter-signature of the mayor of Leicester is a forgery executed by one of Strauss's patients.†
> (ii) That the skiing expedition is a blind and that you are really being exported to satisfy the lusts of La Contessa Addis Abbabisci‡ (the pope's mistress-in-chief) who fell in love with you when you visited the opera in Naples.
> (iii) That you are a heretic with allegiances to the church of Princeton and the hall of Kings.
>
> Any of these grounds should be adequate I think. If in spite of all they let you in, I hope you have a good holiday. I'll leave it to Braithwaite now to start arranging another date for oral. I may visit Cambridge towards end of March anyway.
>
> The Kjell crisis has now evaporated. It was very active for about a week. It started by my getting a p.c. from him saying he was on his way to visit

---

* Even so, it was more easily legible than the message he once sent to David Champernowne which simply consisted of a piece of teleprinter tape. It had his friend spending hours on breaking the Baudot code.

† E. B. Strauss, the Jungian psychoanalyst, whom Robin had known for a long time.

‡ A reference to an incident in Robin's boyhood, when the exiled Emperor of Abyssinia had resided near his home, and had invited Robin and his mother to tea.

me. At one stage police over the N. of England were out searching for him, especially in Wilmslow, Manchester, Newcastle etc. I will tell you all one day. He is now back in Bergen without my even seeing him! For sheer incident it almost rivals the Arnold story.

Alan spoke of this 'crisis' in the computer laboratory, to Norman Routledge, and to Nick Furbank, with whom he stayed briefly at the end of March, while attending a conference on computers[29] at the NPL. But he never 'told all', dismissing the story as another absurd police folly,* involving his own house being under watch. It did not occur to those whom he told that there was a different kind of explanation for what had happened. His letter to Robin said no more, and passed on to other concerns:

> I've got a shocking tendency at present to fritter my time away in anything but what I ought to be doing. I thought I'd found the reason for all this, but that hasn't made things much better. One thing I've done is to rig the room next [to the] bathroom up as electrical lab. Am not doing very well over your vision model.

In this 'laboratory' he was able to do 'desert island' experiments of an electrolytic kind, using current from the electricity mains supply. He would use coke as electrodes, saying that using carbon sticks from old batteries was a form of cheating, and weed juice as a source of oxygen. He liked to see how many chemicals he could produce, starting from common substances like salt – much as he would have done at Dinard if his mother had allowed it. The room he used was in fact a small space left in the middle of the house when the bathroom had been carved out of a larger room. He called it 'the nightmare room' – playing on Mrs Turing's fears of an accident.

Alan's letter also explained that he

> Went down to Sherborne to lecture to some boys on computers. Really quite a treat, in many ways. They were so luscious, and so well mannered, with a little dash of pertness, and. Sherborne itself quite unspoilt.

His schooldays might well have seemed simple and safe, in

---

* There was a postscript to his letter to Robin: 'Is the beginning of this letter whimsical or what?' – a question he did not answer. Nor did Robin look for an answer: his own reply responded subtly to the news of a 'crisis' by recommending the novels of Denton Welch.

comparison with the world in which he now waited for the next turn of the screw. This visit had been on 9 March, and in his talk to the science society,[30]

> Mr Turing made a very clear analogy between a stupid clerk, with his mechanical calculating device, paper to write his workings on, and his instructions, and the electronic brain which combined all these in one. All that was necessary was to put the instructions into a tape machine and the mass of wires, valves, resistors, condensors, and chokes did the rest, the answer appearing on another tape . . .

The existence of this society, the Alchemists, since 1943, was a concession to the modern world, but otherwise Sherborne was indeed 'unspoilt'; neither war nor end of empire had much modified the training of administrators for the 1980s and 1990s. There were, however, more and more cracks appearing in Alan's stiff upper lip, and although deploring his tendency to 'fritter his time away', he was not finding it quite so important to keep his nose permanently to the grindstone. Typically, he had made a game out of the difficult process of breaking the ice, so that when among friends, notably Robin and his friend Christopher Bennett, they would share what Alan chose to call 'sagas' or 'saga-ettes'.

A 'saga' would have to enjoy the dimensions of 'the Arnold story', but a saga-ette might have more modest proportions of self-revelation. Alan would tell the saga-ette of a particular Paris adventure. Alan had picked up a young man and insisted on walking back to his hotel instead of taking the Metro. This caused amazement, Alan said, because 'he thought of *Paris* like you or I would think of a *Riemann surface*; he only knew the circles of convergence round every *Metro* station, and couldn't analytically *continue* from one to another!' In the hotel, the boy had solemnly lifted up the mattress, and inserted his trousers, *pour conserver les plis*, which this time amazed Alan, who never had a visible crease and wanted to get on with it. Afterwards, the boy had made up some story about exchanging their watches so as to prove their trust for each other, until they met next day, so Alan showed his trust, and lost the watch, but considered it worth the sacrifice. Alan and Robin would also point out this or that pleasant sight in the street, each catering to the taste of the other.* 'Is that what you call a *pretty* girl?' Alan once asked, allowing the suggestion to stand that he thought he ought, at least in principle, to expand his own interests.

Once Alan had been persuaded that self-exploration and self-revelation were worthwhile goals, he pursued them in his uncompromising way. In the computer laboratory, for instance, a young man whom Alan found particularly attractive once arrived from London to use the computer. 'Who's that *beautiful* young man?' Alan immediately asked Tony Brooker, who explained. An invitation to dinner for the young PhD rapidly followed, but Alan found himself fobbed off with a thin excuse, of which he took a dim view, about having to visit a sick aunt.

Franz Greenbaum had a theory that Alan's attention was drawn to those who in some way either resembled himself, or what he would like to be – a rather commonplace observation, perhaps, and of the psychoanalytical kind where any exception could be taken to prove the rule. But it intrigued Alan, who apparently had never thought out such ideas before. One person who encouraged this development was Lyn Newman, who became another of the small group of human beings whom Alan could trust. There was a playful element in his correspondence[31] (some of it in French) with her, but it represented a serious cracking of the male shell. He wrote to Lyn Newman in May that 'Greenbaum has made great strides in the last few weeks. We seem to be getting somewhere near the root of the trouble now.'†

By spring 1953 he was also being invited to the Greenbaums' house from time to time, for Franz Greenbaum, whom the Manchester intellectual establishment did not consider a very respectable figure, was not bound by the strict Freudian view of relations between

---

* Robin's interests were more uniformly distributed.

† To be seeing a psychoanalyst, foreign and Jewish at that, was of course another source of stigma and certainly a drastic departure from his early background. It was typical of him that he should write in this open, nonchalant way. Nor was Lyn Newman a privileged *confidante* in that respect; thus Alan had also made friends with Michael Polanyi's young son John, recognising in him a budding chemist, inviting him to dinner to talk about morphogenesis, and presenting him with an envelope labelled 'scrapings from Alan Turing's kitchen'. It contained samples from a mysterious growth on the wall which he optimistically imagined that John Polanyi could identify. While away in Canada, John had a letter from Alan[32] 'full of hope for the future and praise for his analyst'.

therapist and client. Alan quite failed to communicate with Mrs Greenbaum, but became fond of playing with their daughter Maria. He made a particular hit by giving her a box of sweets, saying it was a special left-handed tin for her. Once he amazed Mrs Greenbaum by his excitement over a youth in the next-door garden whom she did not think at all attractive, She thought him 'obsessed by sex' – but he was obsessed with truth, at the cost of crudity.

The probation period ended in April 1953. For the past three months they had put an implant of hormone into his thigh, instead of the dosage of pills. Suspecting, with some annoyance, that the effect would last more than three months, he had it taken out. Then he was free, the more so because his future at Manchester was secure. On 15 May 1953 the University Council formally voted[33] to appoint him to a specially created Readership in the Theory of Computing when the five years of the old position ran out on 29 September. This he could reasonably expect to last for ten years, if he wanted it. In this respect, his insouciant 'Pooh!' to Don Bayley had been justified: he had a small pay rise, and freedom to work exactly as he chose.

On 10 May Alan sent a letter to Maria Greenbaum, describing a complete solution to a solitaire puzzle, and ending:

> I hope you all have a very nice holiday in Italian Switzerland. I shall not be very far away at Club Mediterranée, Ipsos-Corfu, Greece. Yours, Alan Turing.

He had already – most likely in 1951 – been to a Club Mediterranée on the French coast. In this summer of 1953, probably over the period of the coronation,* Caliban escaped from the island for his brief ration of fun, to Paris for a short while, and then to Corfu. He would return with half a dozen Greek names and addresses,[34] although from this point of view his exploration of the eastern Mediterranean proved disappointing. As at school, he made mistakes with the French, but still did better than with the Greek.

On the beach in Corfu, with the dark mountains of Albania on the horizon, he could study both the seaweed and the boys. Stalin

---

* At Whitsun (24 May) he was due at Guildford, and on 30 May at Cambridge for Robin's PhD oral examination, so his holiday was most likely in early June.

was dead, and a watery sunshine was emerging over a new Europe. Even the cold shabbiness of British culture was not immune to change, and after more than ten years of ration books, a quite new mood, one that no one had planned for, was coming with the growth of the Fifties. Television, its development arrested in 1939, made its first mass impact with the coronation. In a far more complex and more affluent Britain, the boundaries of official and unofficial ideas would become less clear. An outsider, an intellectual beatnik like Alan Turing, might find more room to breathe.

Besides the general relaxation of manners, the diversification of life was most acute in questions of sex. As in the 1890s, the greater official consciousness of sexuality was matched by a greater outspokenness on the part of individuals – and most notably in America, where the process had begun earlier than in Britain. One particular example of this, the American novel *Finistère*[35] which had appeared in 1951, was much admired by Alan. It described the relationship between a fifteen-year-old boy and his teacher, and like *The Cloven Pine* tried to see life through teenage eyes. It was, however, a relationship very different from the vague nuances of Fred Clayton's *cri de coeur*. In the old days Alan had often teased Fred, shocking him with rather over-simplified assertions about the prevalence of homosexual activity, and this was a book which caught up with the serious thread that had underlain that delight in gossip – a wish to defy the 'social stigma' and discuss sex in the same way as one might discuss anything else. Meanwhile *Finistère* also did full justice to the reality of the 'social taboo', and its plot followed a complex pattern of private and public disclosures. These the novelist made lead to a conclusion of hopeless doom, as though homosexual life were something inherently self-contradictory and fatal: 'the strip of sand, the distinct footprints leading in one single trail into the black water.'

In its tragic end, its suicide at a symbolic 'end of the earth' – as also with its linking of the boy's longing for a man friend with the failure of his parents' marriage – *Finistère* took its place amidst the older genre of writing about homosexuality. It brought a post-war explicitness into an already dated form. By 1953 the point had been well made that gay men could muddle through like anyone else; thus the new English novel *The Heart in Exile*[36] wended its way through the fading drama-ettes of upper-middle-class taboos, and the more modern obsession with psychological explanations, and rejected both for an ordinary, commonplace ending, tempered by the observation that 'the battle must continue'. Angus Wilson's 1952 *Hemlock and After*, with its bleak, black comedy of class and manners, was also close to the matter-of-fact modernity about sex that Alan liked to display. This was another book that he and Robin discussed – more evidence that officialdom and clinical management were not the only legacies of the Second World War. Yet Alan Turing could not share in this anarchic spirit as he might have wished. Less free than he appeared, he too was on the shore of life. A year later, on the evening of 7 June 1954, he killed himself.

Alan Turing's death came as a shock to those who knew him. It fell into no clear sequence of events. Nothing was explicit – there was no warning, no note of explanation. It seemed an isolated act of self-annihilation. That he was an unhappy, tense, person; that he was consulting a psychiatrist and had suffered a blow that would have felled many people – all this was clear. But the trial was two years in the past, the hormone treatment had ended a year before, and he seemed to have risen above it all. There was no simple connection in the minds of those who had seen him in the previous two years. On the contrary, his reaction had been so different from the wilting, disgraced, fearful, hopeless figure expected by fiction and drama, that those who had seen it could hardly believe that he was dead. He was simply 'not the type' for suicide. But those who resisted a stereotyped association of the trial in 1952 with the death in 1954 perhaps forgot that suicide did not have to be interpreted in terms of weakness or shame. As Alan had quoted Oscar Wilde in 1941, it could be the brave man that did it with a sword.

The inquest, on 10 June, established that it was suicide. The evidence was perfunctory, not for any irregular reason, but because it was so transparently clear a case. He had been found lying neatly in his bed by Mrs C—— when she came in at five o'clock on Tuesday 8 June. (She would normally have been in on the Monday, but it was the Whitsun bank holiday, and she had had a day off.) There was froth round his mouth, and the pathologist who did the post-mortem that evening easily identified the cause of death as cyanide poisoning, and put the time of death as on the Monday night. In the house was a jar of potassium cyanide, and also a jam-jar of a cyanide solution. By the side of his bed was half an apple, out of which several bites had been taken. They did not analyse the apple, and so it was never properly established that, as seemed perfectly obvious, the apple had been dipped in the cyanide.

John Turing attended the inquest, having met Franz Greenbaum and Max Newman in the meantime. (Mrs Turing was away on holiday in Italy at the time, flying back when the news reached her.) John had already decided that it would be a mistake to contest a verdict of suicide, a policy from which the presence of a row of newspaper reporters did nothing to dissuade him. The evidence given[37] was limited to the discovery of the death, the cause of death, Alan's good health and his freedom from financial trouble. Nothing was mentioned that hinted at sex, the trial, blackmail or anything of the kind. The coroner said 'I am forced to the conclusion that this was a deliberate act. In a man of his type, one never knows what his mental processes are going to do next,' and the verdict was that of suicide 'while the balance of his mind was disturbed.' In the event, the national press made remarkably little of it, and nothing was said regarding the 1952 trial.

Mrs Turing would not accept the verdict. Her argument was that it was an accident. Her evidence was that while Alan lay in his small front bedroom, an electrolytic experiment was bubbling away at the back. It had, in fact, been going for a long time. He did sometimes use cyanide for electrolysis, it being necessary for gold-plating. Recently he had used the gold from his grandfather John Robert Turing's watch to plate a teaspoon.[38] She argued that he had got cyanide on to his hands by accident, and thence into his mouth. This was, of course, what she had always said might happen. At Christmas 1953, when he made his last Guildford visit, she had repeated her warning ('*Wash* your hands, Alan, and get your *nails* clean. And don't put your *fingers* in your mouth!'). He brushed all this off with 'I'm not going to injure myself, Mother.' But this very fact meant that he was well aware of her sensitivity to the possibility of an accident. More willing to bend truth for the sake of kindness with her than with anyone else, he could turn her long complaint at his erratic ways into a scheme to save her feelings. It was a cruel blow for Mrs Turing, and the more so as it came hard on the heels of a greater *rapport* with her son. Suicide was officially a crime, over and above the social stigma, and she was also a great believer in Purgatory. The plan that Alan had mentioned to James Atkins in 1937, which involved both an apple and electrical wiring, might well have played on the same idea – perhaps it was the very plan he used. If so it was a 'perfect suicide', in this case calculated to deceive the one person whom he wanted to deceive.

It resembled the mixture of detective story and chemistry-set jokes that he enjoyed in the treasure hunts. Once he had devised a clue that depended upon the electrical conductivity of the soft drink Tizer. A last treasure hunt, held at Leicester with Robin in the summer of 1953, found him preparing bottles of red liquid, with clues written in red ink on the back of the label, so that they could only be read after the bottles were emptied. They were labelled the wrong way round: 'The Libation' for the smelly one and 'The Potion' for the drinkable. Perhaps the idea went back to Christopher Morcom's teasing 'deadly stuff', back to the poisons of *Natural Wonders*. He had found a final chemical solution.

Anyone arguing that it was an accident would have had to admit that it was certainly one of suicidal folly. Alan Turing himself would have been fascinated by the difficulty of drawing a line between accident and suicide, a line defined only by a conception of free will. Interested as he was by the idea of attaching a random element into a computer, a 'roulette wheel', to give it the appearance of freedom, there might conceivably have been some Russian roulette aspect to his end. But even if this were so, his body was not one of a man fighting for life against the suffocation induced by cyanide poisoning. It was that of one resigned to death.

Like Snow White, he ate a poisoned apple, dipped in the witches' brew. But what were the ingredients of the brew? What would a less artificial inquest have made of his last years? It would depend upon the level of description, 'not the will of man as such but our presentation of it.' To ask what caused his death is like asking what caused the First World War: a pistol shot, the railway timetables, the armament race, or the logic of nationalism could all be held accountable. At one level the atoms were simply moving according to physical law; at other levels there was mystery; at another, a kind of inevitability.

At the most superficial level, there was nothing to see. His working papers were left in an untidy mess in his room at the university. Gordon Black, who worked with the computer on lens design problems, happened on the Friday evening before his death to see him cycling home as usual.* He had also booked as usual to use the computer on the Tuesday evening, and the engineers waited up for him, only hearing next day that he was dead. His friendly next-door neighbours, the Webbs, had moved to Styal on the Thursday, and he had had them to dinner on the previous Tuesday, merry and chatty. He had been much regretting their move, spoke of visiting them, and said he was glad that the new occupants would be young and with young children. There were purchases, including theatre tickets, in his house when he died; he had written, though not posted, an acceptance of an invitation to a Royal Society function on 24 June. He had been seen out walking on the Sunday morning by a neighbour with whom he was on nodding terms ('as usual he looked very dishevelled'); he had taken in the *Observer* on the Sunday and the *Manchester Guardian* on the Monday; he had eaten and left the washing up. None of this shed the slightest light upon his death.

To his old undergraduate friends, the last year revealed a troubled mind, but equally, one that was continuing to press on. At Christmas 1953, besides visiting Guildford, he stayed with his friends David Champernowne at Oxford and Fred Clayton at Exeter. He went out for a walk with Champ and certainly did talk in a worried way about the Norwegian boy, Champ forming the impression that he had been imprudent and perhaps a bit reckless. But there was no definite point that came across; Alan rather rambled on and Champ felt a little bored.

At Exeter he also went for a walk with Fred and his wife, who now had four children. One of the boys, Alan agreed, closely resembled his uncle in Dresden. Alan told Fred about the arrest, the trial and the hormone treatment, describing how it had developed the breasts, and making the most of its black absurdity. To Fred it was

---

* He had recently abandoned his motor-assisted bicycle and was using a borrowed one – a ladies' bicycle, as it happened, for it made no difference to him.

the confirmation of all his fears, and he told Alan how unsatisfactory such pick-ups must be, wishing he could find a permanent friend from the academic world. (He did not know anything about Neville.) A great believer in family life, Fred felt that Alan envied him the course that he had taken since 1947. Alan found a large mushroom, which to the Claytons' amazement he said was edible, so they cooked and ate it. Afterwards Alan sent a thank-you note, with more notes on astronomy, and a homemade sundial in a cardboard box. It was hardly a grand farewell. Neither, indeed, did his Guildford visit take on a leavetaking character, and his last note[39] to his mother, written shortly after these visits, ended with some information about a shop he had found in London where 'some quite remarkably cheap things in glass suitable for wedding presents etc.' could be bought.

Neither of his two close postwar friends, Robin and Nick Furbank – had any clear idea that an end was approaching. Robin stayed at Wilmslow for the weekend of 31 May, just ten days before Alan died. Their friendship was one of great mutual confidence in emotional matters, but there was no hint of a psychological crisis on this visit. They amused themselves with Alan's experiments, trying to make a non-poisonous weedkiller and sink-cleaner from natural ingredients. They talked about type theory and planned to meet again in July.

Alan had much developed his friendship with the writer Nick Furbank – a development which perhaps reflected his greater willingness to branch away from science, and even to interest himself in literature. The subject of suicide had entered at some point in their conversation, and Nick recalled it when he wrote[40] to Robin on 13 June, describing what he had found at Wilmslow, when he went there in his capacity of executor. But this had lent no sense of explanation to the death, only perhaps the knowledge that it was not a course that Alan had ruled out of court. Nor did Franz Greenbaum, despite close acquaintance with Alan's inner life and dreams, feel that anything was understood. The dream books were recovered and passed back to the psychoanalyst, but did not answer any questions.

John Turing read through two of the dream books, which Franz Greenbaum lent to him before destroying them. Alan's 'scarifying' comments on his mother, and his description of homosexual activity since adolescence, told John far more than he ever wished to know, and he found these revelations in themselves sufficient explanation for what had occurred, thankful only that he had managed to prevent them from reaching his mother's eyes. To Alan's friends, nothing could be so clear.

There was one piece of evidence that he had prepared for death: he had made a new will on 11 February 1954. This in itself was a kind of statement as to where he stood. It made Nick Furbank his executor, rather than his brother, and gave to Robin all his mathematical books and papers. Then after £50 legacies to each member of his brother's family, and £30 to his housekeeper, the remainder was divided between his mother, Nick Furbank, Robin Gandy, David Champernowne and Neville Johnson. John Turing was amazed and appalled at Alan having thus lumped his mother together with his friends – but this very fact was a much warmer gesture to her than a more conventional allocation which treated her not as an individual friend but as a person to whom he owed a family duty.*

Yet the will included a provision that his housekeeper should have a further £10 for each year in which she had been employed after the end of 1953 – a strange point to add if he had then been settled upon death. It seemed to Nick, on visiting the scene, that Alan had arranged certain letters into packets – but there was no wholesale clearing up, neither of personal papers nor of his research. It was as though he had planned for the possibility, but in the event acted impulsively. What factors were there, at a less immediate level, that might have prompted such plans?

He died on a Whit Monday, the coldest and wettest Whit Monday for fifty years, as it happened. Was it, symbolically, the day *after* inspiration, when the spirit was beginning to give out? G. H. Hardy had attempted suicide in 1946, although in his case, as one who had gone seven years deprived of creative life by a stroke. Was there an underlying pattern in Alan Turing's life and death on this second layer of the psychological onion? The self-revelation of his

---

* His estate amounted to £4,603 5s 4d. But a larger sum, £6,742 4s 11d, was payable by Manchester University – it was the amount due as a death grant under the terms of the superannuation policy to which he had subscribed. The verdict of suicide did not affect the payment, which, as John Turing ensured, went entirely to Mrs Turing.

'short story' would suggest that he thought of himself as having had his 'inspiration' in 1935, and it being thereafter a fight to maintain that level. And the waves of inspiration had come only once every five years since Christopher Morcom's death: the Turing machine in 1935, naval Enigma in 1940, the ACE in 1945, the morphogenetic principle in 1950. At least, these were the outward expressions of his thought; he worked rather like the Turing machine that he arranged to write and do rough working on alternate squares. There had been a good deal of *reculer pour mieux sauter* in between.

In each case he had become not exactly bored or disillusioned by the last flood of work, but felt he had exhausted what he could achieve within its framework. He was, as compared with a more conventional academic, peculiarly anxious not to be enclosed or defined by his existing reputation. So indeed by 1954 or 1955 he would need something new to emerge, to maintain his freshness. But by June 1954 it was hardly time for despair. Indeed, the 1949 period had probably been much more difficult for him.

Possibly the morphogenetic work had turned out rather plodding and laborious. It was three years since he had claimed he could account for the fir cone pattern, and he had not achieved it when he died. But there was no sign of a dropping off of interest. In the summer of 1953 he had taken on a research student, Bernard Richards. (There had been an earlier student, who had achieved nothing.) Richards took over some detailed calculations in connection with his models for the formation of patterns on spherical surfaces. He worked out some exact solutions to Alan's equations, thus exhibiting the capacity of the theory to accommodate some of the simpler possible patterns found in the monocellular *Radiolaria*. Alan had a book of engravings* of these oceanic organisms, which he would show with delight to the engineers when on his night shift.

The working relationship had not progressed beyond the master-and-servant kind, but even so, it was clear to Richards that there was no decline or fall in Alan's work, not even at the very end. He was

---

* Almost certainly the volume of plates from the German zoologist Ernst Haeckel's report on the *Radiolaria*, part of the immense series published by the British government in the 1880s with the scientific results of the voyage of the *HMS Challenger* from 1873 to 1876.

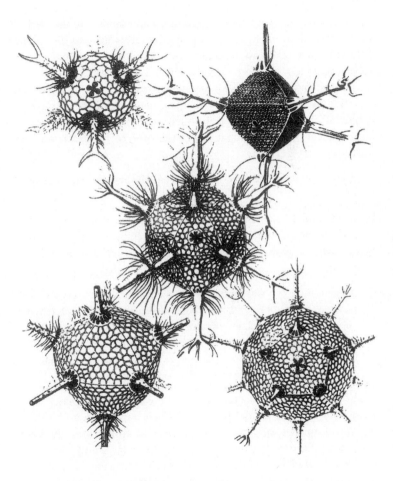

Part of a plate from Ernst Haeckel's report on *Radiolaria*.

doing a good deal of writing up, but this was not at the expense of new experimentation on the computer. Nor, in any case, was this a very dramatic kind of theory, which stood to be proved right or wrong in an all or nothing way. It was a stage on which he could play out ideas in chemistry and geometry and see where they led. It was open-ended, and depended upon integrating insights into many different branches of mathematics and science, rather than trying to solve a particular problem within a given framework.

He left behind a mass of detailed material,[41] some organised into the form of a second paper, the rest in the form of worked examples and of computer print-outs incomprehensible to anyone else. Typical of his investigations at the time of his death was the following passage:

> The amplitude of the waves is largely controlled by the concentration V of 'poison'.

In his *Natural Wonders* way, he called the chemical whose function was to *inhibit* growth, the 'poison' – a macabre touch, given that in his own body it had recently been the other way round. He continued:

> If the quantity R is small it means that the poison diffuses very fast. This reduces its power of control, for if the U values are large in a patch and large quantities are produced, the effect of the poison will mainly be to diffuse out of the patch and prevent the increase of U in the neighbourhood. . . .
> If R is allowed to be *too large* it can happen that the 'side-band suppression' effect even prevents the formation of a hexagonal lattice. . . .

Such observations reflected an insight gained from many trials even with this one model which was, as it happened, an 'Outline of Development of the Daisy'. He had quite literally been 'watching the daisies grow' – not only by 'examining 15 plants', rather as he had done with Joan Clarke in 1941, but on his universal machine. But this was only one particular branch of his investigations: there was another headed FIRCONES, with an associated computer routine OUTERFIR; and another headed KJELL theory, which concerned another form of his basic equations, and which was associated with routines called KJELLPLUS, IBSEN and other Nordic names. All of this went rather beyond the material that he was writing up for publication, so there was no lack of prospects.

He was working with Robin on the theory of types, and they planned to write a joint paper. He also wrote a popular article on the 'word problem', which appeared in the Penguin *Science News* in early 1954.* The Russian mathematician P. S. Novikov had announced[42] that the 'word problem' for groups was indeed unsolvable by any definite method; Alan's article explained this and tied it in with some questions in topology, showing that the problem of deciding whether some knot was the same as some other knot was essentially a 'word problem' of this kind. It was up to date, and looked forward to the complete proof of the Russian result becoming available. He was interested in solving problems up to the end: a last letter of May 1954 to Robin discussed certain ideas of Robin's for 'getting round the Gödel argument' but ended with: 'Looked at the rainbow problem again. Can do it rather successfully for sound, but total failure for electricity. Love, Alan.' When out walking together in Charnwood Forest, near Leicester, they had seen an unusual double rainbow, a phenomenon that Alan insisted on analysing. There had to be a reason for it.

If he was looking for something new, it was in theoretical physics, which likewise he had put on one side since the 1930s. Before the war he had spoken to Alister Watson of his interest in the 'spinors' that appeared in Dirac's theory of the electron, and in his last year he did some work[43] on the algebraic foundations of the spinor calculus. He defined what he called 'founts', after printers'

---

* It was entitled *Solvable and Unsolvable Problems*, and first gave an example of a 'solvable' problem. This was a solitaire game (actually the 'fifteen puzzle') in which, as he described, there were only a finite number of possibilities to consider (namely $16!=20,922,789,888,000$). Hence, in principle, the game could be 'solved' simply by listing all the possible positions. This helped to illustrate the nature of an absolutely 'unsolvable' problem, such as he went on to describe, but the large number also demonstrated the gap between theoretical and practical 'solvability'. As it happened, of course, the Bombe had indeed exploited the finiteness of the Enigma by just such a brute-force method, but in general the knowledge that a number is 'only' finite is not of practical significance. One cannot play chess, nor deduce all the wirings of an Enigma machine, by knowing that the possibilities are finite. The 'fifteen puzzle', indeed, poses a tough problem to the computer programmer. Turing machines, when embodied in the physical world, are severely limited by considerations other than those of logic.

founts of type.* He was also interested in the idea Dirac had suggested in 1937,[44] according to which the constant of gravitation would change with the age of the universe. Once at lunchtime he said to Tony Brooker, 'Do you think that a *palaeontologist* could tell from the *footprint* of an extinct *animal*, whether its weight was what it was *supposed* to be?' And always distrustful of the official line in quantum mechanics, he revived his interest in the foundations of the subject. He found a paradox within the standard interpretation, as von Neumann had set it out, because he noticed that if a quantum system were 'observed' frequently enough, its evolution could be made indefinitely slow, and in the limit of continuous observation would freeze to a stop. Thus the standard account depended upon an implicit assumption that this mysterious moment of 'observation' occurred only at discrete intervals.

He had some further heretical ideas which he explained to Robin:[45] 'Quantum Mechanists always seem to require infinitely many dimensions; I don't think I can cope with so many – I'm going to have about a hundred or so – that ought to be enough, don't you think?' And he had the germ of another idea: 'Description must be non-linear, prediction must be linear.' A shift of interest on his part into fundamental physics would have been well-timed; the development of relativity theory was to begin a great revival in 1955, after years in wartime doldrums. The interpretation of quantum mechanics, little developed since von Neumann's work in 1932, was also a subject crying out for new ideas, and was one well suited to his particular kind of mind.

It was not true, as Mrs Turing liked to think, that he was on the verge of making an 'epoch-making discovery' when he died; on the other hand, there was no clear pattern of decline or failure in his intellectual life that might in itself explain its abrupt end. It

---

* While some physical quantities (such as temperature) may be described by *one* number, in general they will require a set of numbers; anything like a direction in space, for instance, will do so. It is usual to 'index' this set by a letter of the alphabet. From a modern point of view the structure of the set is a reflection of the group of symmetries associated with the physical entity, and it is common to use a different type of letter (e.g. lower-case, upper-case, Greek) when different symmetry groups are implied. The word 'fount' made this principle explicit.

was rather a fluid, transitional period such as had occurred before in his development, and this time accompanied by a wider range of interests, and a more open attitude to intellectual and emotional life.

Nor, as others saw it, was this last year in any way eventful; quite the reverse. One strange incident alone stood out, and might have pointed to the existence of matters on his mind that few could possibly appreciate. It was in mid-May 1954, when Alan went with the Greenbaum family for a Sunday visit to Blackpool. It was a very fine day, and they walked cheerfully along the Golden Mile of seaside amusements, until they came across the Gypsy Queen, the fortune-teller. Alan went in, to consult her. Had not a gypsy foretold his genius, in 1922? The Greenbaums waited outside, and found themselves waiting for half an hour. When he came out, he was as white as a sheet, and would not speak another word as they went back to Manchester on the bus. They did not hear from him again, until he called on the Saturday two days before his death when, as it happened, they were out. They heard of his death before returning the call.

What was his fortune? *Audentes Fortuna Juvat* was the Turing family motto, but his uncle Arthur had died in the Great Game, ambushed in an ill-protected British position in 1899. As for Alan, in the great Looking Glass game of life, it was less clear what happened. But there was certainly a part of his mind which Robin and Nick and Franz Greenbaum did not know, and which did not belong to himself, but to those who moved the pieces: White and Red. The board was ready for an end game different from that of Lewis Carroll's, in which Alice captured the Red Queen, and awoke from nightmare. In real life, the Red Queen had escaped to feasting and fun in Moscow. The White Queen would be saved, and Alan Turing sacrificed.

When Alan met Don Bayley in October 1952, he had told him something – though without going into detail – that no other of his friends knew. He had been helping Hugh Alexander with cryptanalytical work. He also said that he could no longer do such work, because there was no room for homosexuals in that field. He accepted it as a fact. As a psychological blow, this was probably but a minor one compared with the others of 1952. It might have been more of a blow to GCHQ which – so Alan told Tony Brooker – had at one point offered him the colossal salary of £5000 to have him back for a year. For no more than during the war could government have operated as a monolithic entity, and the cryptological establishment, with its strong Cambridge connections, might well have been reluctant to lose its star consultant. A different view, however, would prevail in the 'Security Service', or MI5, whose role in 1952 had been extended by the Home Secretary, Sir David Maxwell-Fyfe. And the rapidly developing concept of 'security' was enjoying ever greater sway during these last two years of Alan Turing's life.[46] Although the least political person, he could not isolate himself from the state's changing demands. Indeed, he was at the heart of the problem.

Mechanical methods, clinical management, security – alike they were developments towards explicitness and rationalisation, and alike they were movements in which the American authorities were a step ahead. In 1950 a Senate subcommittee[47] had enquired into

> the extent of the employment of homosexuals and other sex perverts in Government; to consider reasons why [sic] their employment by the Government is undesirable; and to examine into the efficacy of the methods used in dealing with the problem.

The inquiry, the first of its kind, had several findings. One was that homosexuals were generally unsuitable because

> it is generally believed that those who engage in overt acts of perversion lack the emotional stability of normal persons. In addition there is an abundance of evidence to sustain the conclusion that indulgence in acts of sex perversion weakens the moral fiber of an individual to a degree that he is not suitable for a position of responsibility.

In this phase of the enquiry, the committee had drawn upon the knowledge of eminent psychiatrists. For their second main finding, however, they leant upon other authorities:

> The conclusion of the subcommittee that a homosexual or other sex pervert is a security risk is not based upon mere conjecture. That conclusion is predicated upon a careful review of the opinions of those best qualified to consider matters of security in Government, namely, the Intelligence agencies of the Government.

Through its experience of the Second World War, the American government had come to place great faith in Intelligence. Following up the lead given by William Stephenson, it now had its own organisation for foreign spying and manipulation, the Central Intelligence Agency. Much had changed since 1945, when America seemed set to revert to an isolated role guarding hemispheric interests. British foreign policy since the war had always worked towards the retention of an American interest in Europe, although the planners of 1945 could hardly have imagined the form that this would take, now embodied in the North Atlantic Treaty and kindred agreements. Rapidly shedding the pre-war innocence of worldly affairs, America now enjoyed through the CIA its opportunity to behave like every other nation state, only more so, and in particular to emulate the British secret service. One difference, however, was that this organisation was not concealed from its legislators in the British manner, and so quite openly:

> Testimony . . . was taken from representatives of the Federal Bureau of Investigation, the Central Intelligence Agency, and the intelligence services of the Army, Navy and Air Force. All of these agencies are in complete agreement that sex perverts in Government constitute security risks.
>
> The lack of emotional stability which is found in most sex perverts and the weakness of their moral fiber, makes them susceptible to the blandishments of the foreign espionage agent . . . Furthermore, most perverts tend to congregate at the same restaurants, night clubs and bars. . . . It is an accepted fact among intelligence agencies that espionage organizations the world over consider sex perverts who are in possession of or have access to confidential material to be prime targets where pressure can be exerted. In virtually every case despite protestations by the perverts that they would never succumb to blackmail, invariably they express considerable concern over the fact that their condition might become known to their friends, associates, or the public at large.

The FBI witnessed that 'orders have been issued by high Russian intelligence officials to their agents to secure details of the private lives of Government officials . . .' and the conclusion was clear. There was, indeed, a solid core of incontestable reality in the argument. The social stigma rendered gay men peculiarly liable to blackmail, and Soviet espionage could be expected to use this fact just like any other. It was a political fact of life. It meant that, in a curious way, Alan Turing's life had become part of the Red King's dream.

The special position of homosexuals did not come as something new to the world. But now the reaction of government to this and to other aspects of individual behaviour had to take on a new explicitness. This was a transitional period, when procedures appropriate to the 1930s and to the emergency of the Second World War were rapidly being superseded by those of superpowers equipped with arsenals of atomic bombs. Now large scientific establishments had to be maintained indefinitely, against the possibility of a war that might be lost in hours. Now the whole world was supposed to be a battlefield, the Kremlin being held responsible for every development in world affairs that did not accord with American interests. Now the logical war, as well as the physical war, was fully recognised; but during an official peace the flow of information into and out of a country could not be controlled as directly as in wartime. Instead, in one way or another, the problem was that of how to control the flow into and out of people's heads.

Ideally, the state apparatus would all be in the form of machines, but until that could be achieved, it would have to be embodied in the brains of human beings – brains in which the information could not be erased, in which it might be combined with unknown data and instructions, and which when off duty could transport it into unknown places. The problem facing the state was compounded by the fact that science had not yet made it possible to read the thoughts of a person who did not choose to reveal them. People remained dangerously unpredictable. Yet some of that unpredictability was necessary, if the state was to elicit invention and initiative. It was a similar difficulty, that of rewarding 'independence of character' within a system of 'mere routine', that had troubled Nowell Smith.

Brilliant but unsound, the scientists had won the wizard war and become the priests and magicians of the modern world. Yet if wars could be won by magic machines, incomprehensible to military men and administrators, then they could also so be lost. Success and danger were opposite sides of the same coin. Once disdained, then treated with a patronising awe, the scientists of the 1930s had bailed out the Allied governments. Making themselves indispensable, they had won an enhanced position – but at the cost of innocence. The political meaning of science had changed, and the climate of the 1950s was one in which contradictions ignored in the 1930s were being forced to the surface.

The discovery that Fuchs had been passing atomic secrets to Russia had brought some of these to light. No one claimed that he had acted out of malice, desire for monetary gain, negligence or pique. At least for a time he had been a true convert and had believed he was doing the right thing. The war correspondent Alan Moorehead drew the moral in his 1952 book *The Traitors*[48]:

> Perhaps Fuchs was telling the truth when he claimed after his arrest that his loyalties were now fixed in England, and his public cursing of Russian Marxism was sincere. But he was basically a man who would always refer to his own conscience first and society afterwards. There is no place for such men in an ordered community. They belong where Fuchs now is, sewing mailbags, in Stafford Gaol.

It was a hard saying, for it meant that Keynes and Russell, Forster and Shaw, Orwell and G. H. Hardy all alike belonged in prison. Like Einstein, they had permitted themselves to doubt the axioms, and even if they agreed to obey the rules, it had been their own choice so to do. That very detachment, that sense of making a choice, was precisely what the ordered community would have to deny. Yet English liberal writers had themselves admitted this logical conclusion, even though their culture, different in this respect from that of Germany, largely depended upon the blind eye being turned to such contradictions. Keynes, for instance, had referred to the 'consequences of being found out' as having to be accepted. Ideals of 'freedom and consistency of mind', as Fred Clayton had expressed them, simply had no place when matters of real importance to the world were concerned. The brief episode of 'creative anarchy' might have disguised the truth, but by 1950 the political facts of life were clear again.

Science, with its claims to identify an objective reality independent of laws and customs and loyalties; science with its movement towards abstract thought; science for which the world was a single country – science might suggest the danger of detachment from the axioms of society. But so, both more directly and dramatically, did any form of sex which departed from the socially approved forms.

Homosexuals in particular had chosen to set themselves above the clear and unmistakable judgment of society, and posed the problem of the guilty without guilt, lawbreakers with moral assurance. Did there not lie an embryonic Fuchs in each? There lay a great difference, in that Fuchs had been doing what he had explicitly promised not to do – and Fuchs had arrogated the right to power, the right to change history, rather than the right to control immediate relationships. But with most gay men playing an imitation game, they could not help sharing in dishonesty and deception; and nor could anyone know for certain where personal relationships might lead.

These were not new questions, although in the era of threatened atomic warfare they took on a new urgency. Never far below the surface, indeed, lay the highly traditional equation between sodomy, heresy and treachery. It was an equation which, even if overstated by Senator McCarthy, contained its grain of truth. Christian doctrine no longer mattered to the state, but belief in its social and political institutions certainly did. The family system, depending upon sex as a commodity for men to earn and women to surrender, remained a central doctrine of that faith, and the very idea of homosexuality undermined it. In the post-war re-establishment of male employment and female domesticity, that threat became more conscious. To those who perceived marriage and child-raising as duties rather than as choices, homosexuals appeared as the secret, seductive protagonists of a heresy, portrayed in religious terms of 'converts' and 'proselytising' and assumed, together with Soviet-inspired communists, to be plotting a conversion of the world – a mirror-image Christianity, in which the forbidden would be made compulsory.

The East Coast liberal, or in Britain, the public school educated intellectual, was particularly open to the suspicion of the less privileged, who had no means of knowing what had transpired in what Alan called 'the church of Princeton and the hall of King's'. Meanwhile the axioms of politics held that, granted the existence of an enemy, real or imaginary, any dissension or falling out of line could be regarded as weakening the state, and hence a form of treachery. And it was commonly suggested that a man who could do *that* thing, the worst thing in the world, was capable of anything. He had lost all mental control. He might love the enemy. For all these

reasons there was life in the ancient myth, or mythette, of the homosexual traitor.*

True to its modern approach, grounded in management and
the social sciences, the Senate report of 1950 had avoided these
potent and persistent archetypes, and had concentrated upon the
more reassuring picture of gay men as feeble, helpless victims
of blackmail. It was according to this rationale that, after 1950,
employees identified as homosexuals were ejected from American
government departments.† Scientific language had not, however,
entirely exorcised the older ideas, and the fear, amounting to panic,
that now surrounded the subject could better be explained by the
notion of a invisible cancer in society, transforming an obedient
population into unknown, uncontrollable, un-American quantities,
than by purely rational appraisals of possible compromise through
blackmail.

Unlike the American Senate, the British legislature was not
expected to interfere so openly with the practice of government.
But the same pressures of the modern world were at work, and
there was an event which had goaded the British government into
taking parallel steps. It was on 25 May 1951 that the two Foreign
Office officials Burgess and Maclean disappeared, and then on 10
June the *Sunday Dispatch* had drawn attention to the disappearance,
dropping a strong hint that it was time to follow the American policy
of 'weeding out both sexual and political perverts'.

British 'security' had come under scrutiny the previous year,
with the Fuchs case being heard. But Fuchs was a German refugee,
and the atomic bomb project had been highly irregular, resting
largely upon the work of exiles who in 1940 could not be trusted

---

* Of course the equation also ran the other way: a suggestion of homosexuality could
discredit the political target. In particular, it was implicit in the charge of being 'soft on
communism'.

† The flavour of the new policy in operation was conveyed by a report in the *New York
Times*, 2 March 1954, on progress made in the previous year: '. . . the State Department, a
principal target of Senator McCarthy, had separated 117 employees as "security risks", of
whom forty-three had allegations of a subversive association in their files and forty-nine had
been listed as having in their files "information indicating sex perversion". In the big super
secret Central Intelligence Agency . . . there were forty-eight "security risk" separations, of
whom thirty-one were included with information indicating perversion . . .'

with anything that seemed likely to be important. The difference was that Burgess and Maclean were from the upper middle class and Cambridge, the nurseries of British administration. Previously it might have been assumed that the public-school training would guarantee those subtly distinguished loyalties to house and school, and during the war the British government had saved itself the laborious and expensive surveillance of personnel that had been required in the less trusting United States. But now, bafflingly, the public-school code had been broken, and something new had to be done. The form taken by the new procedures was not, however, entirely the result of the Red Queen's flight. It was an aspect of the managerial revolution, overtaking the legacy of more aristocratic government, and also a reflection of the American alliance.

The *Sunday Pictorial* series in 1952 had included the comment that:

> In Diplomatic or Civil Service circles perversion is regarded as a special danger, for there is always the accompanying complication of possible blackmail. It is this blackmail danger which makes the perverts such a problem to the police.

They also stated that homosexuality was 'most prevalent among the Intellectuals'. These remarks in the press parallelled the action of the government, which in 1952 introduced the 'positive vetting' of those holding posts, or who were candidates for posts, involving important state information.* Until then, government employees had only been 'negatively vetted' by the Security Service, simply by checking in their records of those with 'subversive beliefs' and stamping 'Nothing Known Against' on applications. The point of 'positive vetting' was that it was to be a[49] 'searching investigation into a person's background and character'. This, in particular, included 'serious character weaknesses of a kind that might make a person unreliable or subject to blackmail. Thus any evidence that the person being investigated had or was reasonably suspected of having homosexual tendencies would automatically have raised a presumption that he was unfit to occupy a Positive Vetting post.'

---

* Positive vetting was now applied to those 'privy to the whole of an important section' of 'a vital secret process, equipment, policy, or broad strategic plan . . .' a description which would cover anything significant done by GCHQ.

In practice, it would require elaborate and expensive surveillance to establish whether or not a person was homosexual. It was not enough to look for 'pansies', since (in the words of the American report) there were 'no outward characteristics or physical traits' that were 'positive as identifying marks of sex perversion'. Traditional British reserve and a presumption of heterosexuality made it difficult to detect homosexuals who made contacts through friends and private gatherings alone. But one who had been found out was in a very different position, concentrating the fear and suspicion that more logically would have diffused over many.

Alan Turing had not only been found out, but had been found behaving with what to the official mind, or indeed to the mind of anyone alive to the demands of national security, would appear a horrifying indiscretion. His brain, filled with knowledge of the British cryptographic and cryptanalytic work of not ten years before, had been allowed to mingle with the street life of Oxford Road – and who knew where else? Besides the wartime work, the consultation since 1948 would have given him knowledge at least of special problems, and meanwhile the very idea of computer methods, when only a handful of computers existed in the world, was at the frontier of knowledge.* Whether or not directly relevant to Soviet interests, it was knowledge not only secret, but whose very existence was a secret. He had done the unthinkable. A member of the Inner Party, he had compromised himself with the proles. And he had done it in a way that Orwell himself would have dismissed in one word as 'perversion', while Aldous Huxley regarded the demand for sexual freedom as a concomitant of growing dictatorship. He was a law unto himself.

It could well be argued – and perhaps he argued it himself – that his conduct had shown precisely that he was *not* liable to blackmail. For he had gone to the police, and taken the consequences, at even the hint of a petty threat, one that had nothing to do with secrets.

---

* There might have been more. He was, in particular, the 'Deputy Director' of the laboratory where the atomic bomb calculations were in progress, and might well have been consulted at an early stage about this use of the computer. Ferranti Ltd were also engaged upon guided-missile development. Yet these were almost common knowledge, in comparison with the subject which was to remain unmentionable for another twenty years.

He had also told all the details, silly or shocking as they might be, and shown that he did not flinch from the 'friends, associates, or the public at large' knowing them. But such arguments would only accentuate an impression of indiscretion. They would render him the more outrageously anti-social, the more appallingly unpredictable.

He was not a frequenter of the few furtive English 'restaurants, night clubs and bars'. But to the mind of security, his holidays abroad were a nightmare. Inasmuch as Britain was a free country, and he a free citizen, he was entitled to make them. But he enjoyed no right to have young Norwegians come to visit *him*, and whatever the details of the mysterious 'Kjell crisis' of March 1953, a matter of which the local CID knew nothing, it certainly had the effect that Kjell returned to Norway without Alan seeing him. Alan's allusions to Robin about immigration officials suspecting 'blinds' and 'heretics' and plots to import sexual satisfactions were perhaps the furthest he could go in hinting at what 'for sheer incident rivalled the Arnold story,' without disclosing the reason why he would have been the object of special attention, to protect him from himself.

In these circumstances, his holiday in summer 1953 had been an act of defiance, one which might well have led to interrogations at the Ministry of Love of a kind more difficult to cope with than high-table quips about sonnets. How could they know he had not been compromised all the time? How could they know he had not gone mad? How could they know anything at all about his associations? How could they be certain of anything? The essence of Alan Turing's claim to life and liberty lay in his scrupulous adherence to a personal promise. But such a gentleman's agreement depended upon a tremendous degree of trust, a commodity which by 1953 was in short supply. Nor indeed was even Alan Turing quite perfect. Once he had talked a little too much to Neville, mentioning that the Poles had made a tremendous contribution to his wartime work. And then, in the year that passed before his death, the rules were changing yet again, and not in the direction of gentlemanliness. The game was becoming much rougher.

When in 1952 the subject of homosexuality had first become public, the *Sunday Pictorial* had explained that this was because 'as a start' it was necessary 'to turn the searchlight of publicity upon these abnormalities, to end the conspiracy of silence. . . .' The newspaper had conceded that a 'final solution of the problem' would be 'more difficult'. But progress towards a solution was accelerated in 1953. The feature of the period between June 1953 and June 1954 was that this was when government action took a far more dramatic and open form. It was considered time for a repeat of the Wilde trials which had so successfully deterred dissidence for fifty years.

An opportunity was found on the August bank holiday of 1953, when Lord Montagu of Beaulieu reported a theft to the police. Montagu then found that he and a friend were charged with an 'indecent attack' on two boy scouts who were acting as guides at his motor museum. This charge, denied absolutely and depending entirely upon the boys' statements, led to a case enjoying an immense and quite unprecedented publicity. It made a remarkable contrast with the Turing case, in which a matter of comparative substance and a sensational statement went unreported except at the parish pump, and when even there one character witness, Hugh Alexander, had gone unmentioned. But one difference was that this case was presented from the beginning as a show trial, not just of an individual but of a national 'moral decline'.

The Montagu trial ended in December 1953 without a conviction. But the Crown did not admit defeat, and on 9 January 1954 Montagu was arrested again, this time charged with an 'offence' dating from 1952. Two others were charged with him, one being Peter Wildeblood, diplomatic correspondent of the *Daily Mail*. Besides this hint of state affairs being at risk, the prosecution also involved a number of RAF servicemen, and so aroused fears that Britain's military pride and joy was endangered by the tide of 'filth'. The two trials also enjoyed accoutrements of telephone tapping, searches without warrant, free pardons for turning Queen's Evidence to the 'accomplices', a forgery on the part of the Crown – and a general overriding of legalities that suggested a threat to state security itself was involved. Indeed the Special Branch, the political arm of the police, played a role. The renewed publicity provoked complaints in Parliament that it was a 'danger to public morale'. But the government had clearly decided upon an increased public consciousness of male homosexuality, and the old silence was gone for good. The Home Secretary, Sir David Maxwell-Fyfe, had summoned magistrates to explain his policy and spoke[50] of a 'drive against male vice'. It was claimed by judges and reported dutifully by newspapers that the country had suddenly seen an outbreak of homosexual crime. But it was an outbreak of official anxiety which was manifested in the sharp increase in the number of prosecutions.

Besides the Conservative complaints in Parliament about the open reporting of the Montagu case, there were also more modern-minded questions regarding the operation of the law. These had nothing to do with notions of rights or freedoms; the plea of the new men was for homosexuals to be treated by science rather than be punished by imprisonment.*

On 26 October 1953 the young Labour MP, Desmond Donnelly, asked the Home Secretary to bring homosexuality into the scope of a current Royal Commission on the law relating to mental disorders. This plea was followed on 26 November by the independent-minded Conservative MP, Sir Robert Boothby, asking for a new Royal Commission to examine[51] 'the treatment of . . . homosexuals . . . in the light of modern scientific knowledge . . .' Another MP suggested 'establishing a hospital for these unhappy people, where they can receive suitable discipline and treatment.' But Maxwell-Fyfe replied that the prisons were 'alive to this problem and doing their utmost to treat these people according to the most up-to-date views and knowledge.' For even prison, or 'prison treatment' as he called it, was scientific now.

On 28 April 1954, the Commons briefly debated the state of the 1885 law, and the Lords followed on 19 May. Much of the Lords' debate revolved around the nineteenth-century concept of the homosexual personality: that of[52] 'a certain school of so-called scientists whose dangerous doctrine has done more, and does more, harm to the youth of the country and many others than anything else; that is to say, the doctrine that we are not ourselves responsible and that, to a certain extent, these things are irresistible.' The bishop of Southwell joined in an attack on the 'behaviourist plea'. Another Lord referred to 'other countries in the past, who were once great,

---

* Occasionally it was noted that gay men said they actually enjoyed their 'condition', had no desire to change, scorned the idea of psychiatry, or simply asked to be left alone. But these remarks were interpreted by the old guard as evidence of the arrogance and anti-social attitudes that made homosexuals so dangerous – and by the modernists as unfortunate hindrances to successful treatment.

but became decadent through corrosive and corrupting immorality.' But science had its defenders. Lord Chorley interrupted these adieux to Empire with the assertion that 'It is really much more a medical question than a criminal question.' Lord Brabazon, the pioneer of flight, also struck a medical note: 'There are the hunchback, the blind and the dumb; but of all the dreadful abnormalities surely abnormal sexual instincts must be one of the worst.'*

Important as these observations were, the problems of government called for a more pragmatic, less philosophical approach to the problem of human free will. On 29 April, the House of Commons debated the Atomic Energy Bill, and turned to an opposition amendment which would have set up an appeals system for employees of the new Atomic Energy Authority dismissed as 'security risks'. Opposing this policy, Sir David Eccles for the government pointed to cases for which such appeals would be inappropriate, pre-eminently those of[55]

> moral turpitude. What, in short, that includes is that if a man is a homosexual he is much more easily blackmailed, owing to the law being what it is at present, than almost anyone else. There are cases in which the price which the blackmailer exacts is not money but secrets. . . .

But this, he wished to say, was not a matter for discussion.

> these sorts of cases . . . are not cases about which the public is anxious. The public is anxious, I think rightly, about . . . political associations.

---

* Lord Jowett, Lord Chancellor in the Labour Government of 1945–51, was another speaker. But his ideas had been developed more fully in a lecture[53] of late 1953, where he expressed the hope that 'treatment by hormones or glandular secretion . . . will help these unhappy people to eradicate their abnormal desires.' More generally, suggestions emanating from the small back room were more robust than those of the parliamentary debate. In April 1954, the doctors' journal *The Practitioner* was devoted to an analysis of the national sexual crisis. Its editorial, besides explaining that discipline must come before happiness, and that sexual vice meant 'slow death to the race', endorsed the suggestion made by one contributor that homosexuals should 'strengthen their resolve' in some 'natural and bracing environment' – such as a 'camp' on the island of St Kilda. A contributing endocrinologist also drew upon German data on 'the problem of homosexuality', citing 'the use of castration in over 100 cases of sexual perversion and homosexuality reported by Sand and Okkels (1938) who noted gratifying results in all but one case'.[54] For medicine as for mathematics, the world could be a single country.

If the public was not anxious, one Labour MP was:

> MR BESWICK: Apart from the general contention, the Minister has just made a statement which is serious. Is he saying that a homosexual is automatically now considered to be a security risk? That is what he said. I should like him to confirm it, because it is a serious thing to say that in this country we now consider all these people to be security risks, and so ought to be discharged.

> SIR DAVID ECCLES: I should like to take advice on that, but my impression is that the answer is 'Yes'. It certainly is in America. It is the result of the law as it is now.

So the new rule was inadvertently revealed. At the end of the debate, presumably on being briefed by his department, Eccles said:

> I may have made a mistake in conveying to the House – though I do not think I did – that all homosexuals are necessarily suspect risks. If I conveyed that, I am sorry.

But he had given the game away. And it was a game of 'hands across the sea'. The field of atomic energy was covered particularly rigorously by 'positive vetting', so that anyone with even occasional access to atomic-energy information had to be screened in advance. The reason for this was outside the hands of the British government; it was 'a requirement of the agreement between the United Kingdom and the United States for the exchange of atomic energy information'.

The American authorities were understandably suspicious of the ability of the British to put their own house in order, and were bound to make stipulations when it came to sharing their secrets. It was one of the charges against Fuchs that he had 'imperilled the good relations between this country and the great American republic with which His Majesty is aligned'. Burgess and Maclean had compromised American secrets. This was a subject of paramount, desperate sensitivity.

Eccles's more carefully worded statement reflected the tradition of a more discreet state machine, never eager to make public its dispensations. But changes were being made, to accommodate the demands of the alliance. While the Montagu trials, which enjoyed large headlines in the American press, suggested that neither Lords nor Old Etonians would be spared in the purge, more significant problems would have been faced behind closed doors.*

The public emphasis was placed upon the secrets of atomic physics. But all the time, there existed that other area of secrecy, which did not even officially exist. This, *a fortiori*, would have been subject to the same considerations, for it, too, was intimately linked with the arrangements of the Special Relationship. An American arriving at the CIA office in London in 1952 would discover[57] that 'the wartime partnership was still paying off handsomely.'

> The British, recognising the importance of keeping the United States actively engaged in an effort to contain Soviet disruptive thrusts, were extraordinarily open and cooperative with Americans in intelligence matters. They provided not only most of their highest-level joint intelligence estimates but also supplied the station chief in London with most of their clandestine intelligence MI6 reports.

As during the war, such intelligence was not confined to that derived from spying. There was Sigint:

> Some of the material thus exchanged with liaison services was from intercepted electronic signal messages. Eventually most of this material was worked into the reporting system of the National Security Agency, the consolidated cryptanalysis and signals intelligence facility set up in 1950 . . .

If the CIA represented an American imitation of the British secret service, the existence of the far more secret NSA simply reflected the rather later victory of the arguments for centralisation which in Britain had prevailed after the First World War. The Americans had learnt

---

* The British newspapers were not notable for their explanations of what was going on, but a less inhibited suggestion appeared in the *Sydney Sunday Telegraph* of 25 October 1953:[56]

> The plan originated under strong United States advice to Britain to weed out homosexuals – as hopeless security risks – from important Government jobs.

> One of the Yard's top-rankers, Commander E. A. Cole, recently spent three months in America consulting with FBI officials in putting finishing touches to the plan. . . .

> The Special Branch began compiling a 'Black Book' of known perverts in influential Government jobs after the disappearance of the diplomats Donald Maclean and Guy Burgess, who were known to have pervert associates. Now comes the difficult task of sidetracking these men into less important jobs – or of putting them behind bars.

from British experience, and it was in London, 'the hub of the closest intelligence exchange in all history', that this particular American official perceived 'what vast benefits our allies provided in the way of good intelligence. Without them, the alliance system itself could not function effectively.' The exchange was formalised 'by roughly dividing the world between them and exchanging the materials recorded'. The British had taught the lesson learned at Bletchley, that

> there is no way to be on top of intelligence problems unless you collect much more extensively than any cost-accounting approach would justify and then rely on the wisdom and experience of analysts to sift out the small percentage of vital information that needs to be passed to the top of the government.

The CIA's espionage was 'supplemented greatly' by these contributions, and correspondingly

> In Great Britain these extensive liaison arrangements were supplemented by equally crucial exchanges in the counter-espionage and counter-intelligence field – also important in liaisons with other allies with good internal security services.

This being so, British Intelligence had to accommodate itself to American security rules, just as did atomic research. Accordingly, the case of Alan Turing was one that also had to be seen through American eyes. Regardless of any post-1945 developments, he had been the top-level liaison between the two countries in 1943, and admitted into secret American establishments. Besides knowing so many technical details, he was a person 'on top of intelligence problems'. He knew how the systems worked as a whole – the people, the places, the methods, the equipment. Had the headline been 'ATOMIC SCIENTIST FOUND DEAD', the questions would have been immediate and public. In Alan Turing's case, the questions were not obvious, but precisely because the field of his expertise was even *more* closely guarded than that of nuclear weapons. And it was the Ultra secret that Churchill personally cared about, the adventures of the secret service being useful largely as cover stories. Alan Turing had stood at the very centre of the Anglo-American alliance. His very existence was a glaring embarrassment, and one which put the British government in the position of being answerable for his behaviour. As John Turing had found as a small boy, this was no easy responsibility. Not only the quiet trial at Knutsford, but also his visits to countries bordering on the eastern bloc were subjects that in contemporary American eyes, were they to come to American notice,[58] would be tantamount to an international incident. These were exceedingly deep waters.

Fundamentally it was not his homosexuality that presented a difficulty to the mind of security, but the lack of control, the element of the unknown. The coroner said that 'with a man of his type' – a man of the Professor type! – no one could tell what his mental processes would do next. That iconoclastic 'originality' had been acceptable in the brief period of 'creative anarchy', which had even stomached the arrogance and will-power required to solve the unsolvable Enigma, and force the implications upon an unwilling system. But by 1954 a very different mentality prevailed. At Alan's last Christmas visit to Guildford he had left behind some papers; calming his mother's anxiety about them he betrayed his impatience with post-war Newspeak:[59]

> The note on [the] M of S* document about secrecy etc. is all eyewash really. The document is 'unclassified' (an idiotic word of American origin meaning 'not in the least secret'. It arises by documents being 'classified' according to their degrees of secrecy, hence secret documents get to be called 'classified', and hence, worst of all 'unclassified' does not mean 'not as yet assigned to any category' but 'not secret').

He belonged to an age of implicit trust and class-based discretion, at a time when trust and discretion were being mechanised and classified. In the climate of 1954 it was almost irrelevant that he had no time for the Soviet Union, for *all* were under suspicion until 'cleared', and everything that was not the purest White, could be considered potentially Red.

With the loss of strategic independence, and the end of imperial confidence, Alan Turing's country had changed. His old housemaster had declared him 'essentially loyal', and in rather the same way an assumed 'essential loyalty' had satisfied the recruiters of the new men. It probably never occurred to them that a

---

* Ministry of Supply.

well-connected English person could take an abstract, foreign idea seriously enough for it to make any difference. Fifteen years later, events had proved otherwise. If the 1940s had made the idea of 'intelligence' into something very concrete and definite, the 1950s forced the concept of 'loyalty' to an equal explicitness. And the Cambridge which had supplied the intelligence was an unknown quantity in respect of loyalty. This was a time at which Patrick Blackett, once the trusted adviser of an independent Royal Navy, was being pointed out among Manchester University staff as 'the fellow traveller'.

Alan Turing, by comparison, was the entirely apolitical person. But he came from the dissenting King's background; he had supported the 'very good' Anti-War demonstration in November 1933. He had never moved in the sophisticated circles of Burgess and Maclean, and had not been elected an Apostle, but connections could easily be found by anyone who chose to look for them. At a time of guilt by association – when there was nothing but association to go on – he was guilty. They had made some incredible mistakes, and how could they be sure Alan Turing was not yet another, given his instructions by the Red Queen twenty years before? What would constitute a proof ? It was Wittgenstein's awkward question, applied to real life. Burgess and Maclean had been absurd and clumsy players of the imitation game – but were there others more skilful, yet to be found out? Even if such gross suspicions had been entirely ruled out, the fact was that by combining and concentrating the two great unthinkables, cryptanalysis and homosexuality, mysteries of 'stinks' and 'filth' respectively, he had rendered himself a demon, arousing the most primordial insecurities. And it was at a time when British securities had evaporated. The old social discipline offered no defence against nuclear war, but neither did scientific methods offer better than plans for revenge and suicide. Torn between a subservient trust and a resentful anxiety regarding American machinations, to which British power had been surrendered, a panic over spies and homosexuals provided Great Britain with a suitable diversion.

The tide in the affairs of men had turned in 1943, and by the summer of 1954 had obliterated the patterns drawn in the Second World War. Stalin had gone, but this had made no difference to the system of threat and counter-threat, apparently beyond the control of individuals. A Soviet hydrogen bomb had been tested in August 1953, presenting the possibility of devastation greater than the most pessimistic prognostications of 1939, and of a scale far outweighing that offered by the British bomb tested in October 1952. But it was the American test on 1 March 1954, the 14-megaton blast catching the crew of the *Lucky Dragon*, that suddenly jolted public consciousness. On 5 April, in a rare Commons 'defence' debate, Churchill saw fit to reveal the terms of the 1943 Quebec agreement between Britain and the United States, on which the Americans had reneged, and said:

> No words of mine are needed to emphasise the deadly situation in which the whole world lies . . . the H-bomb carries us into domains which have never confronted practical human thought and have been confined to the realms of phantasy and imagination.

What was fantasy and what was reality? There was American pressure on the British to join in a military intervention in Vietnam, after the French defeat on 7 May at Dien Bien Phu. Churchill's refusal brought about talk of a 'British betrayal', and strains to the *quid pro quo* of the Special Relationship. Fears of a new Asian war were not unfounded; on 26 May an American admiral spoke of a 'campaign for complete victory' in Vietnam, including the use of nuclear weapons. A general described using atomic bombs to 'create a belt of scorched earth across the avenues of Communism to block the Asiatic hordes'. Dulles now said that he was 'very hopeful' that the British government would 'change their attitude'.

June 1954 was a period of particular uncertainty, with the Geneva talks on Vietnam being compared with those of Munich. Now it was the turn of American city populations to practise taking cover in air-raid shelters, while in Britain there was a revival of the Home Guard – recruitment was in progress at Wilmslow during the last week of May. The tension was as great in Europe as in Asia, with West German rearmament being the inflammatory issue. The rules had changed, and the past had changed its meaning. Not only the silver bars had been lost for good; other bridges had been destroyed, and new ones built in solid concrete. Now it was the turn of the U-boat men to be called back, while the hunt for spies and traitors was occupying their erstwhile enemies. It was on 2 June that the newspapers revealed that 'new man' at Princeton to be loyal but a 'security risk'. Robert Oppenheimer, guilty of wrong ideas and associations, was someone that no one could be certain about. And there was another special feature of the newspapers that Whitsun weekend. Faded, stilted, almost embarrassed tributes were being paid to the men who had landed on the beaches of Normandy, exactly ten years before.

Alan Turing was not an island, but a stray eddy in a sea of troubles. The coroner referred to the 'balance of his mind' and to him becoming 'unstable'. It was an image not remote from his own morphogenetic model of the moment of crisis. As the political temperature rose, his equilibrium would become more and more unstable. The smallest event could have been the trigger. One particular issue would have concentrated his demand for freedom on the one hand, and the implications of past promises on the other. Could he have gone abroad again in the summer of 1954 – when no one knew what might happen next, and in the midst of an official panic over homosexuality? The Foreign Office had been issuing stern memoranda on Soviet entrapment during the past year,[60] in parallel with an extension of 'positive vetting' on 31 March 1954, and fortified by the disclosures of the Russian defector Petrov. Meanwhile the Montagu trials had shown that fond British beliefs in velvet-glove government were not always to be sustained. There was always the possibility of another case being brought against Alan, manufactured out of an affair long in the past. This was one aspect of the wave of prosecutions now taking place, and one which threatened to drag down friends – even on the merest suspicions and flimsiest allegations. Even the newspapers, if he could bear to read them, could have told him this. He was in a corner. He had always been prepared to confine his fight to his own personal space – the space that others chose to allow him. But by now he was left no space at all.

E. M. Forster,[61] outdoing the King's heresy with grand bravura, had written in 1938 that if he were faced with the choice between betraying his country and betraying his friends, he hoped he would have the courage to betray his country. He would always put the personal above the political. But for Alan Turing, unlike Forster, or Wittgenstein, or G. H. Hardy, it was more than a theoretical question.

For him not only had the personal become the political, but the political was the personal. He had chosen and promised for himself in working for the government. The choice for him therefore was that between betraying one part of himself and betraying another part. And however much he wavered between these alternatives, there was a solid logic to the mind of security, one that could not be expected to take an interest in notions of freedom and development. He had no rights to such things, as he would have had to admit. He might have outwitted the Home Guard, but when it came to questions that mattered, there was no doubt that he had placed himself under military law. There was a war on; there was always a war on now.

Churchill had promised blood, toil, tears and sweat – and this was one promise which the politicians had kept. Half a million of Alan Turing's compatriots had been sacrificed ten years before, without much choice in their fate; to have the luxury of choice in matters of integrity and freedom was itself a great privilege. Only the 'heads in the sand' assumptions of 1938 had allowed him into such a position in the first place, and his position in 1941 was one for which many would have given all they had. Ultimately he could not have complained. The implications had proliferated, and arrived at a remorseless contradiction. It was his own invention, and it killed the goose that laid the golden eggs.

No one remotely mindful of such considerations could have wanted to make a fuss; and neither in any case could he speak of such things – that was the very point. Only in obscure clues and jokes could they emerge. In March 1954 he sent to Robin four last postcards. They were headed 'Messages from the Unseen World', an allusion to Eddington's 1929 book *Science and the Unseen World*. Robin kept only the last three, here shown.*

---

* It would be misleading to suggest that he had made any discovery in these jottings, but the underlying thoughts were in line with the developments of the 1950s and 1960s.

III. 'Arthur Stanley' is Eddington, and the first postcard alluded to cosmological questions. The 'light cone' is an important idea in relativity theory. Einstein's ideas were based on the concept of a point in space-time, this meaning a precise location in space at a precise instant of time. Imagining this as an instantaneous spark, the future 'light cone' of such a point is traced out by the expanding sphere of light from that spark.

By the 'Creation' he would mean the 'big bang'; it had been known since the 1920s that there were models of an expanding universe that agreed with Einstein's general theory of relativity, and in 1935 H. P. Robertson, whose lectures Alan attended at Princeton, had further developed the theory of them. Unfortunately the astronomers' observations of the galactic recession did not seem to be consistent with the Einstein theory, and only in the mid-1950s was the discrepancy eliminated. This was one reason why in 1948 H. Bondi, T. Gold and F. Hoyle had suggested a new theory involving 'continual creation' which eliminated the 'bang'. Alan might have heard Gold speak about it at the Ratio Club in November 1951. But apparently it did not deflect him from the earlier view, soon to be established much more firmly.

The emphasis on a description using light cones was not a trivial insight. Such an emphasis was emerging in a quite different way through the work of A. Z. Petrov in 1954, was taken up by H. Bondi and F. A. E. Pirani later in the 1950s, and entered very strongly into the ideas of Roger Penrose, who in the early 1960s formulated new ideas about space-time. In fact a 'Penrose diagram' of the universe would draw it as 'the interior of the Light Cone of the Creation'.

IV. Implicit here is the problem of physical determinism. Most physical laws, including Einstein's, are in the form of a differential equation, relating instantaneous rates of change to one another in such a way that in principle, given the state of a physical system at one time, it can be predicted at a later time by adding up the changes over the period. In the context of cosmology this begs the question of what the 'initial' state of the universe was – it was a very Eddingtonian suggestion that the study of the differential equations of physics could only be half the story. Here again, the question of the nature of the initial 'big bang' was to be of growing significance in the renaissance of relativity theory.

V. Again the allusion is to the problem of physical prediction – the wave functions somehow determining the events perceived as the pantomime of macroscopic life – and again the emphasis is on a description in terms of light rays. But the 'hyperboloids' suggest some quite novel geometric picture of his own, lost without trace.

VI. The reference to 'founts' suggests that he was thinking of describing the different elementary particles in terms of their corresponding symmetry groups – again in the mainstream of developments, although the picture as it unfolded in the 1960s was far more complicated than anyone would have known in 1954.

VII. He was certainly not the first to think that electric charge could be interpreted in terms of rotations, and his formula was too simple-minded. But 1954 saw the renewal of interest in 'gauge theories', which generalised this basic idea.

VIII. Often his letters closed with a brief line of personal comment, and this was surely the case here. There was certainly nothing new or speculative in scientific terms in this 'message', an allusion to the well-established Pauli Exclusion Principle. Back in 1929, when he read what Eddington had to say about the electron, Alan had noted the idea that the electrons of the universe had to be considered *en-masse*, not singly; the Pauli principle described an observed restriction on the collective behaviour which roughly speaking meant that no two electrons could be in the same place. Thus in each atom, the electrons would be stacked neatly in separate shells and orbits. As indeed he might have joked in 1929, it was like the House system that kept the boys from associating too freely. For their own benefit, of course: *Don't you see, Dr Turing, we have to do this for your own protection. . . .*

Messages from the Unseen World

? Is it perhaps what numbers ?

III The Universe is the interior of the Light Cone of the Creation

IV Science is a Differential Equation. Religion is a Boundary Condition

Arthur Stanley

V Hyperboloids of wondrous Light
Rolling for aye through Space and Time
Harbour those Waves which somehow Might
Play out God's holy pantomime

VI Particles are points

VII Charge $= \frac{e}{\hbar}$ avg of character of a 2π rotation

VIII The Exclusion Principle is laid down purely for the benefit of the electrons themselves, who might be corrupted (and become cheaper or demons) if allowed to associate too freely.

The old Empire was giving way to the institutions of Oceania. None of Alan Turing's friends saw this as a background that might be relevant to his death, nor saw him as playing the role of Casabianca after all. Not for about fifteen years would the various elements involved become mentionable at all, and even then no one could begin to put them together. There was no hushing-up operation in 1954 – it was not necessary, for no one thought anything nor asked any questions. The Wicked Witch of the West was caused no embarrassment, for the friends of Dorothy had nothing to go on. Few people on 7 June 1944, seeing the cycling civilian boffin, could have imagined a connection with the news of the great invasion: they did not need to know, nor want to know. Ten years later to the day, the links were literally unthinkable, and the death came as an individual hurt and loss, without suggesting any wider significance. Jung said:[62]

> Modern man protects himself against seeing his own split state by a system of compartments. Certain areas of outer life and of his own behaviour are kept, as it were, in separate drawers and are never confronted with one another.

Modern men had to protect themselves particularly carefully when they were confronted by Alan Turing, and they kept the compartments completely separate. So perhaps too did Alan Turing, when confronting his own situation.

Behind the singleminded Shavian figure that he cut, especially after the war, acting out in public a set of ideas with relentless intensity, and going to the stake like a modern Saint Joan, there had always been a more uncertain, contradictory person. Doubting Castle and Giant Despair had been his favourite passages in *The Pilgrim's Progress* as a small boy, and his part in the progress of mankind had been in keeping with them – the delectable mountains being few and far between. In particular there lay the uncertainty of all his relationships with institutions, neither fitting in, nor presenting a serious challenge. In this respect he shared something with many people deeply attracted to pure mathematics and science – never sure whether to regard social institutions as Erewhonian absurdities or as plain facts of life. Making a game out of anything, like G. H. Hardy (and like Lewis Carroll), he reflected the fact that mathematics could serve as protection from the world for one who was not so much blind to worldly affairs as only too sensitive to their horror. His offhand, self-effacing humour also shared something with the response of so many gay men to an impossible social situation: in some ways directing a bold, satirical defiance at society – yet ultimately resigned to it.

For Alan Turing these elements were aggravated by the fact that he never quite fitted into the roles of mathematician, scientist, philosopher or engineer – nor into the tail-end of the Bloomsbury set, nor indeed into any kind of set, even the Wrong Set. It was often a case of *Laughter in the Next Room* for Alan Turing, for people never knew whether to include him or not. Robin Gandy wrote[63] very soon after his death of how 'Because his main interests were in things and ideas rather than in people, he was often alone. But he craved for affection and companionship – too strongly, perhaps, to make the first stages of friendship easy for him . . .' And he was more alone than anyone could ever see.

A self-taught existentialist, one who had probably never heard of Sartre, he had tried to find his own road to freedom. As life became more complicated it became less clear where this should lead him. But why should it have been clear? This was the twentieth century, in which the pure artist felt called upon to become involved, and which was enough to make any sensitive person acutely nervous. He had done everything he could to restrict his involvement to the simplest sphere, as he had also tried to keep true to himself, but simplicity and honesty had not protected him from the consequences of that involvement: far from it.

The British university world was as well insulated from the twentieth century as anyone could hope for; and so often it saw his eccentricities, not his vision, offered vague tributes to cleverness, not serious criticism of his ideas, and remembered the bicycle stories rather than the great events. But although nothing if not an intellectual, Alan Turing never truly belonged to the confines of the academic world. Lyn Newman, who had the advantage of seeing that world at close quarters but from outside, articulated[64] more clearly than anyone else this lack of an easy identity; she saw him as 'a very strange man, one who never fitted in anywhere quite successfully. His scattered efforts to appear at home in the upper-middle class circles into which he was born stand out as particularly unsuccessful. He did adopt a few conventions, apparently at random, but he discarded the majority of their ways and ideas without hesitation or apology. Unfortunately the ways of the academic world which might have proved his refuge, puzzled and bored him . . .' There was an ambivalence in his attitude to what was, despite all its concomitant deprivation, a privileged upbringing: he jettisoned most of the paraphernalia of his class but in inner self-confidence and moral responsibility remained the son of Empire. There was a similar ambiguity in his status as an intellectual, not only in his disdain for the more trivial functions of academic life, but in the mixture of pride and negligence with which he regarded his own achievements.

There was another uncertainty in his attitude to the privilege of being a man in a male-dominated world. In most ways he took it entirely for granted. It was a weak point of King's liberalism that it rested upon wealth accumulated for the benefit of men alone, and he would have not been the person to question it. In conversation with Robin, who took a progressive line on the question of equal pay (the only issue, in this period, which kept feminism alive), Alan said simply that it would be unfair if women were off work having babies. Nor did he doubt that women would run round him clearing up the mess, and seeing to matters that he chose not to bother about himself. In conversation with Don Bayley at Hanslope he mentioned how he had been engaged, that he had realised it 'would not work' because of his homosexuality, but also remarked that if he were ever to marry, it would be to someone non-mathematical, who would look after his domestic needs – a conventional attitude much closer to his family's, and untrue to the way in which his friendship with Joan Clarke had developed. There was an unresolved contradiction here, at least at that stage of his life. He disliked the talking-down and triviality expected of men in 'mixed company' – and, no doubt, too, the pressure on him to display an erotic interest he did not feel – and largely avoided such social obligations. Yet when these constraints did not apply – especially, perhaps, with Lyn Newman, but to some extent with his own mother – he showed a mind more open to the other sex than that of many men to whom the word 'women' was a synonym for sexual possessions or distractions.

Nor did he ever do more for male supremacy than to share in its institutions. He never sought to justify his homosexuality in terms of preferring the superior male, for instance, and when set against the speech and writing of what everyone called the age of the common man, his comments were remarkably free of the stream of hostility, implicit or explicit, that most men felt entirely free to direct at the encroachments or pretensions of women. Alan certainly spoke of 'girls' doing the menial work, and by implication cast them in the role of 'slaves' at Bletchley; yet this was simply the way things were, and if anything he was just a little more conscious of the inequity than were others who took it entirely for granted. He did nothing to change it, but then he had never sought to change the world, only to interpret it.

He was not an Edward Carpenter, who saw a link between the low status of women and the stigma attached to his homosexuality. It would probably never have occurred to him that his own difficulties with the world were very akin to those suffered by women – as with the men's committee meetings held over his head, almost as if he were not there, and the way that people took little notice of what he had actually said or written, but remained obsessed by details of manners or appearance. Women had to learn to compensate for these indignities by making a special effort, but Alan Turing made no such attempt. He expected the male world to work for him, and was baffled when it did not.

He saw himself as a man doing a man's job in a man's world, and so far more conscious and definite, for him, was the fact that his love relations and power relations lay alike within that world. In this respect Alan Turing took most of the parts that society allowed – comic, tragic, pastoral, exile, outsider, in-between, and finally that of victim. But he had also risen above these roles, not only by avoiding the usual lying and cheating that went with them, but by doing the one thing that a homosexual should never, never do: becoming responsible for something that mattered. He also refused to be daunted by the unsympathetic ambience of the technical world. (Indeed, it was a kind of attempted love affair in which – as usual – he was firmly rebuffed.) His move to Manchester, for instance, might well have held a conscious rejection of the temptation to remain in the 'lovely backwater' of King's. Yet in that very determination he illustrated a problem which in the 1950s had hardly begun to emerge into consciousness: that in refusing social definition by a 'pansy' or 'aesthete' label, there lay the opposite danger of merely accentuating the accoutrements of 'masculinity'. His running, although it spoke of a search for wholeness, for another life earned independently of being 'a brain', and for relief from the aggressive feelings induced by a lifetime of banging his head on a wall, was perhaps touched by this. So too, perhaps, were his emotional reserve, his all-or-nothing response to difficulties, his insistence upon professional 'thinking' before off-duty 'feeling' – all influenced by a resolve not to be 'soft'. Yet soft machine he was.

The confusion and conflicts that underlay his apparently single-minded homosexual identity reflected the fact that the world did not allow a gay man to be 'ordinary' or indeed 'authentic'; to live simply, without making a fuss; to be truly personal, without taking a public stand. He, of course, was put particularly acutely on the spot. In 1938 Forster had spelt out the corollary to the claim for moral autonomy[65]: 'Love and loyalty to an individual can run counter to the claims of the State. When they do – down with the State, say I, which means that the State would down me.' But Forster never had to face this consequence, no more than Keynes was ever found out. It was Alan Turing, not as a King's intellectual, but as one of thousands of unfamous people, who had to resolve a moral crisis almost in silence, and almost alone. But even if the events of December 1951 had never led to that particular crisis, the contradictions might well have come to a head in some other way. There was no such thing as a 'simple' life for him, no more than there was a 'simple' science. Bletchley had proved G. H. Hardy wrong about pure mathematics; nothing was pure, and no one could be an island. Alan Turing might be Valiant-for-Truth, but even he had been led into the work of deception by science, and by sex into lying to the police.

The yellow brick road divided, and there were no signposts provided to say which was the true and which the imitation path. But the uncertainty in Alan Turing's life, the wavering between parts that struck observers most forcibly was seen not so much in terms of class, professional status, or gender, but in his oscillation between an 'adult' and 'child' role in life. To some people this was repulsive, to others an element of charm. While to some degree people used the word 'childish' to rationalise their surprise at a person who actually said what he thought, with little decoration or concealment, there was also something quite specifically odd about his manner, which became more noticeable in his late thirties at Manchester. A man with a quite powerful build, yet with the manners and movements of an 'undergraduate' or 'a boy', he was also disconcerting in his rapid changes of mood, between forcefulness and naiveté, bristling with silent fury, but then breaking out in earnest charm. 'Mercury' was Lyn Newman's image, linked with his running. It was an ambivalence with meanings at several different levels – an intellectual level in his refusal to be defined by his existing reputation, breaking instead into an entirely new sphere of work when approaching forty. And of course it held an erotic meaning, part of his response to the situation of homosexual men in general, in which the roles of seeker and sought were more fluid and diffuse than in heterosexual relationships. He had to keep moving; to keep on the go. These factors might indeed contribute a tension (though also a sheer fascination with life that others lost) as he became older. But beyond these meanings the boy-man quality of Alan Turing also reflected that most central question of his existence, one more special to himself. He had not wanted to 'come of age' at twenty-one, and as it transpired, he just avoided seeing the age of forty-two. He had never wanted to take on the *power* of adulthood, although he did not shirk all of its commitments. He was at the opposite pole from John von Neumann, although their brains touched so many common points. A master of committees, and a consultant to every American military organisation, with particular responsibility for the hydrogen bomb and the intercontinental ballistic missile, von Neumann in 1954 was a man of the world who dominated, and was not dominated by, his adopted country.* Alan Turing, in contrast, though born into the heart of an immensely confident administrative class, had only forced his ideas upon anyone when they were the alternative to catastrophic waste and folly. From the summer of 1933 that saw the mid-point of his life, to its end in the summer of 1954, he

---

* But there was a last parallel in that von Neumann also moved towards the problem of biological growth, although from a different point of view. His work was likewise left incomplete; he died of cancer on 8 February 1957.

had been engaged in a profound conflict between innocence and experience.

His contemporary, Benjamin Britten, who went the other way and withheld involvement, played out this theme in public after 1945. Alan Turing left almost nothing but those pages of a short story – pages which did, however, compress acute reflections upon his life. Describing himself taking his young man to the restaurant, he portrayed the scene thus:[66]

> . . . Upstairs Alec was taking off his overcoat; underneath as always he was wearing an old sports coat and rather unpressed worsted trousers. He didn't care to wear a suit, preferred the 'undergraduate uniform' which suited his mental age, and encouraged him to believe he was still an attractive youth. This arrested development also showed itself in his work. All men, who were not regarded as prospective sexual partners, were father substitutes to whom Alec had to be [*illegible*] showing off his intellectual powers. The 'undergraduate uniform' had no conscious effect on Ron. In any case his attention was now concentrated entirely on the restaurant and its happenings. Alec was enjoying himself now. Usually when he went to a restaurant he felt self-conscious, either for being alone or for not doing the right thing. . . .

As it happened, this was where the surviving pages came to an end – and at an appropriate point, for lonely consciousness of self-consciousness was at the centre of his ideas. But that self-consciousness went beyond Gödelian self-reference, abstract mind turning upon its abstract self. There was in his life a mathematical serpent, biting its own tail for ever, but there was another one that had bid him eat from the tree of knowledge. Hilbert once said that Cantor's theory of the infinite had created 'a paradise', from which mathematicians were not to be driven out. But Alan Turing lost that paradise, not because of what he thought but what he did. His problem lay in doing: doing or not doing the right thing.

No one in June 1954 perceived a symbolism in the apple that he ate, an apple filled with the poison of the 1940s. Without the context, the symbol had no meaning, and could no more be interpreted than the other tiny clues he left. He might even have had the symbolism in mind before the war, when he mentioned his suicide plan to his friend James Atkins.[67]* For it came just at the time when he had (in an equally off-hand fashion) mentioned to his mother his unsureness about the 'morality' of cryptography. And while she, the Stoney, was the believer in applied science, it was James who was the pacifist – they both had a position relevant to what for Alan Turing was a crucial change in his life, that of preparedness to know sin. He might have sensed that, for him, involvement with the world meant that he would be walking into danger all the time. He might have behaved as a child – as the child of the proud, impetuous and unlucky Turings, and the child of more worldly bridge-building Stoneys – but whether consciously or unconsciously, he was a child of his time.

His hints of self-revelation were so rare and cryptic, and showed such deep distaste for any kind of self-centred fuss, that all such questions must remain enigmatic. Another unanswerable question is that of how he finally perceived his great dream of computer intelligence, to which he had devoted the central part of his life. For although it was true, as Robin wrote, that he had given himself to ideas and things rather than to people, many of those things and ideas were the means by which he tried to approach the understanding of himself and other human beings, and to do so from first principles. That approach was one in which he had to regard social 'interference' as a secondary intrusion upon the individual mind. And while he had always conceded that this was a difficulty, in his last years he was taking a more active interest in other approaches to human life, in which interaction played a greater part. It was consistent with his general development that he told Don Bayley, in summer 1952, that mathematics was satisfying him less and less. Jung and Tolstoy were writers who placed the mind within a social and historical context, and there were Forster novels on his shelves when he died, novels in which the interplay of society and individual became a less machine-like play of ideas than that of Shaw, Butler and Trollope. Meanwhile social 'interference' in those last two years had played a peculiarly forceful part in his individual life. Could he have lost faith in the significance and relevance of his central ideas?

Combined with the more practical kind of disappointment,

---

* James Atkins himself had left the teaching of mathematics for music. He became a professional singer in 1949, and had a first Glyndebourne season in 1953.

regarding the incapacity of the Manchester computer (or indeed any computer of the time) to do justice to the scale of his vision, the post-war period had clearly seen an erosion of the confidence with which he emerged in 1945. On the other hand, he was not the person to give up ideas lightly, nor to allow the world to take them away from him. And neither was he the person to be disillusioned with science, by virtue of the fact that it had been turned upon himself; nor to abandon rationality because he had found himself at the receiving end of intelligence. His ruling passion for a concrete manifestation of the abstract, something that linked him with Gauss and Newton rather than with the pure mathematics of the twentieth century, inevitably took him into the application of science. Yet he showed no sign of intellectual illusions concerning the purposes of those applications. His remarks about computers had from the beginning been as ruthless as those of G. H. Hardy on mathematics. Never once had he suggested an application other than to pure research for its own sake, or to military purposes. He had never spoken of social improvement or economic welfare through science, and so had a position fortified against disillusion.

In 1946, making a brief reference to the American atomic bomb tests, he referred to 'the worst danger' being an 'anti-scientific reaction'. And however much assaulted by the application of science to 'organotherapy', for instance, he would not have questioned the structure of scientific knowledge itself. Indeed, he would have considered it extreme intellectual weakness to allow personal feelings to influence a view of scientific truth. He had often chided intellectuals for an 'emotional' rejection of the idea of intelligent machinery. It had been important for him to unshackle science from religious wishful thinking, and science for him remained independent of human purposes, judgments and feelings that were entirely irrelevant to the question of *what was so*. Edward Carpenter had called for a 'Rational and Humane Science', but in Alan Turing's book there was no reason whatever why the rational and the humane, the data and the instructions, should be correlated. His ruthless, raw view of science was something that Lyn Newman again captured with an image of him as 'the Alchemist' of the seventeenth century or before – recalling a time when science was not shrouded in titles and patronage and respectability, but was nakedly dangerous. There was a Shelley in him, but there was also a Frankenstein – the proud irresponsibility of pure science, concentrated in a single person. It was indeed that terrific concentration, combined as it was with an ability to throw out all that seemed irrelevant, and with a will to think about questions that everyone would have said were hopelessly difficult and complex, that was his secret. It was this process of abstracting a simple, clear principle, and then demonstrating its truth in some concrete way, that was his strength – it was this rather than the solving of problems within a given framework. But this kind of strength did not lend itself to some of the more subtle problems that were raised by his model of 'intelligence'.

He had, wrote Robin, 'a lack of reverence for everything except the truth', and his insistence on an uncompromising materialism was motivated by that obsession for keeping truth untainted by 'emotional' ideas about intelligence and consciousness. But in his concern to cut away the irrelevant he had brushed aside some fundamental questions about intelligence, communication, and language, questions which arose from the embodiment of brains in a human social world. This was not, however, so much the deficiency in *his* thought, as a reflection upon the method of science. His model of 'intelligence', using chess and mathematics as its paradigm, was one which simply reflected the orthodox view of science as the repository of objective truth. In the *Mind* article he had made it clear that he saw his model as one capable of absorbing all human communication, and this again reflected the positivist belief that science could elucidate human behaviour as it had already triumphed in the fields of physics and chemistry. The weak points of his argument were essentially the weaknesses of the analytical scientific method when applied to the discussion of human beings. Concepts of objective truth that worked so well for the prime numbers could not so straightforwardly be applied by scientists to other people.

As he himself explained in introducing the central morphogenetic idea, any kind of simplification was inevitably a falsification. If this was true in a discussion of the development of cells, it was all the more apposite a comment in respect of the development of human beings, whether of their 'intelligence' or of their yearning for communication, experience, and love. When science used human words about human beings, could it actually separate the 'data' from the 'instructions' of society? Could it 'observe' or 'experiment', or formulate a 'problem', independently of social institutions? Could its assessment of importance and significance of facts, however honestly noted, do other than reflect the imperatives of the dominant ideology? In the life sciences, the borderline between the spirit of truth and the *esprit de corps* was not as clear as it might seem in physics and chemistry. And this difficulty, that of separating fact from act, was very close to the weakness of his arguments for machine 'intelligence'.

This was a super-Gödelian problem, concerning the capacity of scientific language to jump outside the society in which it was embodied, and a problem to which Alan Turing's mind was not attuned. Nor, indeed, was the scientific mind of his time. Those who in the 1930s and 1940s wished to draw connections between social structures and scientific knowledge tended to be those determined to graft social systems onto science, or to derive them from science. Most notably the Nazi and Soviet ideologues were doing this, but Polanyi too, in opposing the influence of the mechanical marxism of the 1930s, was nudging science into a sophisticated revival of Christianity. He too was trying to push science around, wanting it to come up with answers that would fit in with a pre-existing religious and political philosophy. Such a thing was quite foreign to Alan Turing, who believed that he was keeping himself securely within the realm of experimental truth.

One person did, in contrast, investigate the capacity of language to separate the factual from the non-factual. But Wittgenstein's methods were such that hardly anyone could ever be sure what he meant. Alan Turing's approach was one that rode roughshod over Wittgenstein's questions in a search for the simple truth at the centre – but it had the virtue that the picture he drew was clear and plain, something that could in principle be tried out. As for the integration of a theory of logical problems, a psychological theory which had led him to 'the root of the problem' in his own unhappiness, Tolstoy's historical problem about the nature of individual action, Forster's questions about individuals and class consciousness – all this would be too much for anyone to encompass, and it was certainly not the way in which he worked or thought. At Bletchley he had worked on the central logical problems, finding bold and simple solutions, while a vast human organisation sprouted around him; it had not been his role to pull the whole complexity together.

He had clung to the simple amidst the distracting and frightening complexity of the world. Yet he was not a narrow man. Mrs Turing was right in saying, as she did, that he died while working on a dangerous experiment. It was the experiment called *life* – a subject largely inducing as much fear and embarrassment for the official scientific world as for her. He had not only thought freely, as best he could, but had eaten of two forbidden fruits, those of the world and of the flesh. They violently disagreed with each other, and in that disagreement lay the final unsolvable problem. In this sense his life belied his work, for it could not be contained by the discrete-state machine. At every stage his life raised questions about the connection (or lack of it) between the mind and the body, thought and action, intelligence and operations, science and society, the individual and history. But these were questions on which, except in the most special ways, he went out without a word of comment. Russell and Forster, Shaw and Wiener and Blackett held forth on such subjects; Alan Turing played the humble pawn.

He played the pawn, and ultimately he obeyed the rules. Alan Turing liked to consider himself as the heretic scientist, gloriously detached from the conventions of society in his quest for truth. But his heresy was directed only against the surviving fragments of a disintegrating religion, and the polite accommodations of the intellectual world. The hue and cry of philosophers seizing upon Gödel's theorem in defence of human freedom, as though minds could be regarded as static, isolated, academic intelligences, was to the real servitudes of the twentieth century what Lowes Dickinson had called Cambridge: 'a lovely backwater'. The mainstream was, in the 1920s, a case of[58]

> Jix and Churchill and Communists and Fascists and hideous hot alleys in towns, and politics, and that terrible thing called the 'Empire', for which everyone seems to be willing to sacrifice all life, all beauty, all that is worthwhile, and has it any worth at all? It's a mere power engine.

By the 1950s there was a new Empire, or rather two, each served by their respective scientists. Those great well-springs of the *Years of the Modern* – the liberation of the faculties of the individual, and the collective ownership of human resources – had been degraded into the liberalism of the Pentagon, and the socialism of the Kremlin. It was here that the important doctrines and heresies lay, not in the formalities of English class and Victorian religion.

In the 1930s, King's had held one central strength: Pigou and Keynes and Forster had not forgotten individual freedom when decrying the waste of *laissez faire*, and no more than Bertrand Russell did they fall for the glamour of the USSR. After Germany wrecked everything, and Hitler's curse lay upon the victors and the survivors, that stream of independent thought no longer had the same significance. But there was a fleeting moment after the war, in the period before Britain became Orwell's Airstrip One, when Forster could see the post-war world according to pre-war lights:[69]

> Owing to the political needs of the moment, the scientist occupies an abnormal position, which he tends to forget. He is subsidised by the terrified governments who need his aid, pampered and sheltered as long as he is obedient, and prosecuted under the Official Secrets Act when he has been naughty. All this separates him from ordinary men and women and makes him unfit to enter into their feelings. It is high time he came out of his ivory laboratory. We want him to plan for our bodies. We do not want him to plan for our minds . . .

Alan Turing *did* come out of his ivory laboratory, and in some ways came out further than Forster did. Nor was he the Bernal whom Forster was attacking in this passage, who believed that scientists should rule the world. But he did not offer a word on what turned out not to be an 'abnormal position' but rather the real orthodoxy of the 1950s: a dependence upon colossal machines. His work, while being as near to pacifism as military work could be, still had the effect of increasing state dependence upon machinery not only beyond the control, but even completely outside the knowledge, of those who paid for it. In this process Alan Turing remained a cipher.

In some ways the fears about scientists 'planning for our minds' were unjustified by events. Thus the plans for eradicating homosexuality by scientific means, like the claims of cybernetics, and like the other current dreams of truth drugs and lie detectors, were far too ambitious. They were certainly not a practical proposition

in the Britain of the 1950s.* Although academic research and
medical practice continued towards this goal, it never succeeded in
attracting the full support of government. Instead the issue of how
best to eliminate homosexuality was left as a juicy bone over which
the bulldogs of moral conservatism could fight with the forces of
technical progress. Meanwhile the growth of a new economy, in
which advertising, travel, leisure and entertainment made sexuality
ever more conscious an attraction, would undermine the conservative
and the medical models alike. There would even be room for the
notion of individual choice – an idea unheard in 1954. The state
never adopted such great plans, scientific or otherwise, to control the
behaviour of the whole population. There was an air of fantasy, ritual
display, clashing of symbols, about the whipped-up 'moral crisis'
of 1953–4. Instead the 1950s saw British government continue to

* In April and May 1954 the parliamentary debates were in terms of the idea (as the
outraged *Sunday Express* put it) that 'instead of the prison cell they should have the doctor's
clinic.' But more knowledgeable observers had recognised that talk of either punishing
or treating all homosexuals was quite unrealistic, and the publicity given to the Montagu
trials gave them the opportunity to point out the damage done to the repute of the judicial
system by a law so irregularly and partially enforced. A more practical policy was defined by
the *Sunday Times* in March; contrasting 'those things which must needs be legally tolerated'
with those which 'must be condemned and rooted out'. On 8 July, the Home Secretary
backed down and appointed J. F. Wolfenden, a public-school headmaster from 1934 to
1950, to chair a committee on the laws relating to homosexuality and prostitution. Thus
Alan Turing died just as a more central strand of British administration was reasserting
itself.

In fact his crime was of the kind that everyone agreed should continue to be the object
of state attention; meeting people in the street ('importuning'), and having an affair with
a working-class nineteen-year-old exemplified what was to be 'condemned and rooted
out'. But the number of prosecutions peaked in 1955, and then fell back until 1967. The
government failed to set up the special hospitals or camps suggested by the medical
profession, and the great panic dissipated rapidly after summer 1954. The most important
effect was that the silence was broken – a first BBC radio talk being allowed on 24 May. If
in fact it was the case of Alan Turing that had frightened the Churchill government out of
its wits, he also played a posthumous part in the defusing of the taboo.

He also died just before the international situation relaxed a little; at the Geneva
conference China agreed to the temporary partition of Vietnam. Meanwhile McCarthy's
star fell rapidly after he attacked the US Army and the CIA. Churchill went to Washington
on 24 June and repaired the Anglo-American rift. British military expenditure rose to a
dramatic peak in 1954 but thereafter declined until the mid-1960s. Everyone but Alan
Turing had a reprieve.

relinquish much of the civilian economy to the fray of international business, tempered by the traditions of class, tribe, religion, elections and so forth. Thus Winston Churchill set the people free.

This complex and contradictory future, rather than a 1930s vision of scientifically planned industry, or a 1950s fantasy of scientific mind-control, was the achievement of the new men. The old moral and social institutions, while persisting in form, would lose their absoluteness and all-embracing significance; even a bishop would soon be borrowing Carpenter's phrase and preaching a 'New Morality'. The lesson of the public schools, and of the no less grim training grounds of the lower orders, had been outdated in the 1920s and proved quite useless in crucial aspects of the Second World War, a fact learnt so reluctantly and so late. It was all the more pointless now that faith rested in the Wizard's mechanical Deterrent, with Heath Robinson contraptions of attack and counter-attack taking warfare further and further from human hands and minds, playing an unstable game in which every player was losing all the time, and in which the British government, not to be left out, had taken the lead in a proliferation of implications.

Alan Turing's split state prefigured the pattern of growth that he did not choose to live to see: a civilisation where the singing and dancing and mating – and the thinking about numbers – would be offered to a wider class, but one built around, and working to provide, methods and machines of inconceivable danger. And in his very silence, he typified the mainstream of scientific collaboration in this policy. It would soon be clear that suspicion of scientists' loyalties was just a passing difficulty; the arrogance of a few who thought they knew better than governments amounted to no more than the teething troubles in the establishment of the national security state. Who could see that in fact he had torn the curtain away and shown the fragile, erratic, embarrassing brain that lay behind the machine? For unlike Dorothy, he said not a word. He was no heretic: this was camouflage, although perhaps even he, who so rarely broke his promises, only just restrained himself at the end. Within his sphere he was the grand master; politically he was as he described himself in 1941: Churchill's obedient servant.

But he had never wanted to be the focus of the modern world's contradictions. It had been his trouble all along that, although driven by the desire to do something, he wanted to remain ordinary, to be left alone in peace. These were incompatible goals, and there was no consistency in him. Only in his death did he finally behave truly as he had begun: the supreme individualist, shaking off society and acting so as to minimise its interference. While *Nineteen Eighty-Four*, which so impressed him, held references to science and intelligence which made a telling counterpoint to his own ideas, there was a level on which Orwell was saying something very similar to himself. Orwell would have cared little for the legacy of Bletchley Park, another development of the Ministry of Truth, nor for the computers being built by people utterly heedless of questions such as those of *whose* intelligence was to be mechanised, and what it was for. He would have loathed the King's social culture, and the strand in Alan Turing that shared something with Edward Carpenter's erotic short-cut to socialism. Yet there was an underlying common bond: that of the few cubic centimetres within the skull which were all that could ever be called one's own, and which were to be defended at all costs against the ravages of the world. For all *his* contradictory elements, Orwell did not lose faith in the capacity of Oldspeak to convey the truth; and his dream of the plain-speaking Englishman was close to Alan Turing's simple model of the mind – the vision of a science independent of human error.

Bleak visionaries alike, they came from an England less lush than Cambridge, and breathed a cold mountain air from which fainter hearts drew back. They contradicted each other, for much of what Alan Turing wanted – both in science and sex – could hardly be described in Oldspeak, while George Orwell's idea of truth required a connection of the mind with the world that the Turing machine did not have, and the Turing brain did not entirely want. Neither thinker could do justice to the whole, and nor could the whole complex person that was Alan Turing keep true to his simple ideas. Yet in reaching the *niente* of his *Sinfonia Antartica*, he kept as close to his vision as the exigencies of the world allowed. Never content with the academic problems of dots and brackets, he found a purer end than Winston Smith.

With so few messages from the unseen mind to work on, his inner code remains unbroken. According to his imitation principle, it is quite meaningless to speculate upon his unspoken thoughts. *Wovon man nicht sprechen kann, darüber muss man schweigen.* But Alan Turing could not possess the philosopher's detachment from life. It was, as the computer might put it, the unspeakable that left him speechless.

Postscript

As I lay with my head in your lap camerado,
The confession I made I resume, what I said to you
    and the open air I resume,
I know I am restless and make others so,
I know my words are weapons full of danger, full of death,
For I confront peace, security, and all the settled laws,
    to unsettle them,
I am more resolute because all have denied me than
    I could ever have been had all accepted me,
I heed not and have never heeded either experience,
    cautions, majorities, nor ridicule,
And the threat of what is call'd hell is little or nothing to me,
And the lure of what is call'd heaven is little or nothing to me ;
Dear camerado! I confess I have urged you onward
    with me, and still urge you, without the least
    idea what is our destination,
Or whether we shall be victorious, or utterly quell'd and defeated.

Alan Turing's body was cremated on 12 June 1954 at the Woking
Crematorium. His mother, brother, and Lyn Newman attended the
ceremony. The ashes were dispersed in the gardens at the same place
as those of his father. There is no memorial.

Author's Note

For a figure in world history, there is very little source material from which to reconstruct a picture of Alan Turing – few original documents, and little in the way of published commentary. Secrecy and embarrassment of various kinds are partly responsible, but there is a paucity of information even where taboo subjects are not involved. The early development of the ACE, for instance, is covered incompletely by the surviving records – and some of the most interesting exist only thanks to unofficial individual initiative. The ACE represented a major act of public enterprise, and the events of 1946–9 largely determined the shape taken in Britain by what was soon to be seen as a second industrial revolution. Had cooperation between government, industry and brain power been continued in peace as in war, the future of the British economy might have been very different. But no special effort was made to record the course of decisions made, nor has the subject subsequently attracted the interest of historians or journalists or political theorists. And what is true of the ACE as a whole is even more true of Alan Turing's personal part.

One must recognise, however, that Alan Turing did not conduct his life as that of a figure in world history: he tried as best he could to continue the life of an ordinary mathematician. And mathematicians (as compared with literary or political figures, entertainers or spies) do not usually expect to be heard of or written about, whatever their contribution. They do not really expect other people to know what mathematics is, and are generally happy to be left alone. When judged by mathematical standards, one could not say that for a figure of his stature there has been any particular deficiency in records or neglect in reputation.* Pathetically small by worldly standards, the

---

* For mathematicians he has immortality through the expression *Turing machine*. Many people must have used his name without any conception of his historical existence – the nearest thing to the life as a disembodied spirit that he once pondered on. Going even further, modern papers sometimes employ the usage of *turing machine*. Sinking

corpus of biographical material is still substantial in comparison with that of others in his profession.

Taking first the question of what was written of him after his death, over the next twenty years or so, there were of course obituaries: Max Newman in The *Times*, Robin Gandy in *Nature*, Philip Hall in the King's College *Annual Report*, and various more minor tributes. Newman followed by writing the Biographical Memoir to which Alan, as a Fellow of the Royal Society, was entitled. One of the fuller and more penetrating of the series, it treated his life and work from the point of view of a pure mathematician. The Second World War thus appeared as an interruption to his work on logic and the theory of numbers. The subject of his war work had to go unmentioned but so, with ruthless consistency, was the subject of practical computers relegated to a few lines. This analysis embodied the outlook of an intellectual tradition to which Alan Turing had certainly half-belonged, but it was not the whole story.

One person was not satisfied with this assessment, and sensed that some other kind of recognition was due. This was Mrs Turing, who in 1956 embarked upon writing a biography of her son – an extraordinary development by any standards, in which a seventy-five-year-old Guildford lady, not hitherto notable for literary or social confidence, and knowing almost nothing about science, was left to piece together some of the debris from the wreck of the modern world's dream. Her Victorian values still unshaken, she retained a strong belief in the idea that Alan's work had been and would be for the benefit of humanity.

Her slim volume appeared in 1959. There was perhaps a better book in Sara Turing, one that would have been a genuine *memoir* which placed the death alongside the other mysteries (as they were to her) of what her son had spent his life doing: something that could have pointed tellingly to the twentieth-century separation of science from ordinary life, and to the efforts both he and she had made to overcome that separation – though not succeeding. But this was not what she did: her book was in the form of a biography

---

without a capital letter into the collective mathematical consciousness (as with the *abelian group*, or the *riemannian manifold*) is probably the best that science can offer in the way of canonisation.

and written with an apparent emotional detachment which was in itself remarkable, considering the frightful circumstances which had inspired it.

One reason that Alan's mother could cast herself as a detached observer was that in so many ways she *was* writing about a stranger. The reader was not to know, but there was very little in her narrative of his early life – and the period up to 1931 absorbed a third of her account – which did not come from surviving letters and school reports. She was surely trying to prolong in her mind the lately increased *rapport* with her son, by projecting it back into a past life of which she knew little – not an inkling, for instance, of Christopher Morcom's significance for her son's development. The objective stance then obliged her to set out his scientific career – another impossibility. Alan had once compared the work of writing programs to imitate intelligent behaviour as like being set to write an account of 'family life on Mars'. Mrs Turing had set herself a task of nearly equal difficulty; rather as a computer might be programmed to write sentences of grammatical form, she was able to make a jigsaw of the titles of his papers, bits cribbed from the extant obituaries, comments solicited from other people, and newspaper cuttings. Yet she had little conception of what it meant.*

Her position of weakness was accentuated by an extraordinarily obsequious attitude to anyone of rank or office, which meant that by implication she put her son at the level of a promising sixth-former. Indeed her whole book read much like a school report. The flow of tributes bore witness to the fact that she was still having to convince herself that he had turned out satisfactorily after all,

---

* But she did this very carefully, and only when trying to draw connections did the facade of intelligence fail. Even so, the jigsaw puzzling had more the effect of omitting and diminishing much of what he had done, rather than suggesting that he had done something more than was in fact the case. There was just one point where a definitely misleading statement resulted: cribbing a phrase by Philip Hall which in fact referred to the zeta-function machine, and setting it in a different context, she led readers to deduce that Alan had set about constructing a *universal* machine before the war. This no doubt encouraged A. Cave Brown in *Bodyguard of Lies* to suggest that he had done so in order to decode the Enigma. This suggestion then appeared as fact in J. Rohwer, *The Critical Convoy Battles of 1943*. Thus one sees a myth being made. The trouble is that truth does not reside in strings of symbols on the page; the business of interpreting them correctly requires experience.

and indeed to her astonishment that there was a world in which he had actually been admired. Undercutting him again and again, *Computable Numbers* was good because Scholz had been impressed by it, his interest in the brain was significant because Wiener and Jefferson approved. . . . Alan might have seen this assessment as a fate worse than death, although it was partly the outcome of his own failure to promote himself.

Yet there was one point his mother grasped that better-informed people could be too sophisticated to see, namely that in 1945 he had set out to build a computer. She stuck to this at a time when everything surrounding the subject was still suffused in embarrassment. And more generally, she displayed an amazing tenacity and nerve in tackling the male institutions from which she was excluded, and in refusing to be daunted by the polite evasiveness that she met. For of course there were two areas that were out of bounds – what he had done in the war and his homosexuality. A number of people felt they could not contribute anything whatever with honesty unless the unmentionable were mentioned – and of course it was *not* mentioned, no more than in any of the other written pieces. With the war she got a little further, being allowed to say that 'he was one of a team whose joint work was an important factor in our winning the war' – a hint which was as strong as anything that appeared in the next ten years.* Tiptoeing among the minefields – and perhaps anyone but her would have found it impossible to continue – she did, ultimately, stand up for him as few others would.

What was sad was not that she failed to do what was manifestly beyond her powers; it was that although she had her own insights and stories they did not, in fact, add up to an understanding or even a formulation of the particular enigma that Alan had posed for her. At the end she closed with a comparison of what she had written with the *Lives of the Saints* – Alan would have derided this attempt to get him through the eye of the needle, and yet had she pursued with any seriousness the subtle conflict of the 'pure' and the 'applied' she could quite legitimately have expressed it in religious language and

---

* Though Geoffrey O'Hanlon had spilt the beans in 1954, writing in his obituary of Alan in *The Shirburnian* that 'During the war he was engaged in breaking down enemy codes. . . . His work was hush-hush, not to be divulged even to his mother.'

found something quite special to say. But here she offered nothing: for examination marks, government service, and the building of machines appeared alike as undifferentiated Good Things. The absence of any question mark around the place of science was perhaps one reason why her book, pitiful by the standards that would be expected of a literary or political biography, attracted the gentle praise of critics. It was a holiday for a world trying to forget Dr Strangelove. At last, it seemed, here was a scientist untouched by the traumas of the 1940s and 1950s! There was a half-truth in this notion, there being something of the 1880s in Alan Turing as well as in his mother; but again it was hardly the whole story.

During the 1960s, and into the 1970s, Newman and Sara Turing were the sources upon which various encyclopedia entries, potted biographies and popular articles drew. But by the turn of the decade there were small sprouts of a different kind of comment pushing their way through the Stoney ground. One factor was simply the expansion and greater sophistication of what had become 'computer science', slightly modifying the status of computers as altogether *infra dig.* for a mathematician. It was in 1969 that Donald Michie published the NPL report *Intelligent Machinery*, he himself being concerned to set a lead in British developments. He commented in this period on how the prevailing attitude was that ideas about machine intelligence were a diversion from serious work; but the 1970s ushered in a greater appreciation of the computer as a universal machine, concerned with any and every form of logical manipulation, and not necessarily at work on arithmetical calculations. This general development encouraged a clearer understanding of what Alan Turing had envisaged from the beginning.*

---

* Perhaps, however, the magic of the 1940s could never be recaptured amidst the dull white heat of the 1960s technological revolution. And perhaps the closest evocation of the Turing spirit in this period was not in science, but in science fiction – a cinematic *Back to Methuselah*. The date of *2001: A Space Odyssey* was presumably taken from the fifty-year prophecy in the *Mind* paper, which was specifically cited in Arthur C. Clarke and Stanley Kubrick's book. Their picture of HAL was essentially based on Turing ideas; and, quite uncannily, their plot had HAL destroyed by 'the logic of the planners', whose 'twin gods of Security and National Interest meant nothing to HAL. He was only aware of the conflict that was slowly destroying his integrity – the conflict between truth, and concealment of truth.'

It was also in 1969, as the computer came of age, that it was first noted, in articles by Mike Woodger and R. Malik, that Alan Turing had emerged from the war with a practical knowledge of electronics. This fact, quite at odds with the prevailing stereotype of 'the logician', as he appeared in H. H. Goldstine's standard academic account of *The Computer from Pascal to von Neumann* in 1972, took some time to be assimilated; so likewise did the place of the ACE in computer history. The compilation by B. Randell of classic papers documenting the origins of digital computers (see note 5.23) relegated mention of the ACE to its bibliography, but the mini-boom of computer history did not entirely pass it by: in 1972 the original report was issued by the NPL, and it received a first serious review in 1975.

Meanwhile the turn of the decade had also seen the purpose of Bletchley Park become mentionable, although the first outright claim for its strategic significance came only in 1974 with F. W. Winterbotham's *The Ultra Secret*. This book made no mention of Alan Turing, but in the same year A. Cave Brown's imaginative work *Bodyguard of Lies* contained many sentences in which the word 'Turing' appeared, sometimes in conjunction with words such as 'machine' and 'Bombe'. The floodgates were opened. Meanwhile Jack Good and Donald Michie had published certain disclosures concerning the electronic machinery at Bletchley. Drawing these developments together, the inquiries of B. Randell, motivated partly by the question of understanding Alan Turing's part in the origin of the computer, enjoyed some success. His revelations of the Colossus technology did in fact reflect the achievement of Newman and Flowers rather than anything directly attributable to Alan Turing, but it meant that a first serious glimpse of the gigantic scale of operations had been given. Much of this, together with other mid-1970s disclosures, was brought together in a BBC television programme, one of a series on *The Secret War*, broadcast early in 1977.

Since 1969 was also the year of gay liberation, it meant that another change in how it was possible to think about Alan Turing occurred during the 1970s. This was not the result of the Wolfenden reforms – which, although delayed through the strenuous efforts of Viscount Montgomery and others, had passed into law as the Sexual

Offences Act in 1967. Setting an 'age of consent' at twenty-one, this rationalisation and modernisation of the law ensured that the Turing crime remained a crime.* It was rather the brief renaissance of American liberalism that permitted a reversal of the concept of 'problem': seeing society as a problem for the individual rather than *vice versa*. In its way this development created the conditions for a rediscovery of Alan Turing's life as much as did the other openings of the new decade: not just in that it was possible to say he was a homosexual, but in appreciating his pride, his stubbornness, and the moral force he brought to bear as a very private, reserved, shy man who nonetheless insisted that this was not a matter for hiding.

Taking these developments together, the 1970s therefore made it possible for someone to see *who Alan Turing was*, in a way that no one in his lifetime (but he himself) could have done. It so happened that I was placed in such a position as to be struck by all these developments. The name of Alan Turing first impinged on me in the summer of 1968 – essentially a reflection of the burgeoning of computer science, for I was reading about cybernetics and Turing machines as a mathematics student. In fact I did not choose to work

---

* This is not always appreciated. Thus a typical view emerges in *The Mighty Micro* by Christopher Evans (Gollancz, 1979), which was perhaps the first widely disseminated mention of the trial:

> . . . he came to a tragic end. A solitary individual who confided little in other people, he was also a practising homosexual at a time when homosexuality was viewed as a criminal offence. Somehow he brushed with the law – the sad, sorry details are hard, and perhaps unnecessary to come by – and one evening, depressed and disillusioned, he retired to his room and bit into an apple laced with potassium cyanide.

But in 1977 Lord Halsbury had led the defeat of an attempt to reduce the 'age of consent' to eighteen – and any kind of street encounter, whether or not involving money, is still illegal. Prosecutions for homosexual offences, furthermore, have roughly *trebled* since 1967, and currently run at twice the 1952 level. State control of sexuality has changed since the 1950s, but the elements of youth and of street life are 'viewed as' crimes as much now as then. More directly pertinent to the theme of *The Mighty Micro* is that the computer has facilitated a far more comprehensive process of 'brushing with the law', with 'sad, sorry details' filed away for the police, employers and security officials quite regardless of whether crimes have been committed. The process of positive vetting, in particular, is supposed to prevent a homosexual ever again having access to anything of state importance. Whatever the arguments, these are not issues of the past. Nor is the 'detail' of chemical castration a dead question. But even prophets of the 'information explosion' can exhibit a wish not to know.

in this field, moving instead into mathematical physics, learning about relativity and quantum mechanics from Roger Penrose as a postgraduate student from 1972 onwards, and trying to make a contribution to the Penrose theory of twistors.*

But in 1973 the name of Alan Turing impressed itself on me again, this time through another compartment of my life. I was then a member of a group that had formed within the London Gay Liberation Front to write a pamphlet criticising the medical model of homosexuality. One of the other members, David Hutter, had heard something of the end of the Turing story from Nick Furbank. Knowing nothing of his secret work, and believing that his death could be accounted for by the effect on his work of the hormone treatment, we included a paragraph which used this idea to illustrate our theme. Thus after twenty years there was a first public squeak of protest.

Lurking for years at the back of my mind, with a sense that I ought to have found out more about what happened, this story suddenly leapt to the fore again on 10 February 1977. On that day, while having lunch with Roger Penrose's research group at Oxford, there was a conversation about the famous *Mind* article which took me back to my earlier interest in Turing ideas – and then quite independently, another about the BBC programme on Bletchley Park, which had been broadcast the previous evening. Roger Penrose commented on its fleeting mention of Alan Turing; he had long since heard talk of him as a man 'hounded to his death' but just recently there had been rumours of a man who had 'deserved an earldom'. Nothing was clear or connected, and it was to be three years before I could compose a coherent interpretation of what had happened – but it was enough for me to sense

> A war O soldiers not for itself alone,
> Far, far more stood silently waiting
>   behind, now to advance in this book.

---

* Twistors, as geometric objects which allow the formulation of a different description of space and time from the usual one, are not quite 'Hyperboloids of wondrous Light'. But they are based on light rays and an objective of twistor theory is the unification of general relativity and quantum mechanics. So the last Turing messages have a special resonance for me. And the privilege of working in so dramatically original a theory, and studying under such a person, has been a vital ingredient in my approach to Alan Turing.

It had to be done, and it was the right moment to start. A first step was that of collecting extant publications, much as described above. But of course it was necessary to get much closer to my subject. So I turn now to the question of original Turing papers, where my first acknowledgement of a debt must be to Mrs Turing. According to her own book she once told Alan that she was putting things by for the use of a future biographer, and he grunted a gruff assent. Certainly she took the trouble to keep letters from her son's schooldays onwards, used them for her book, and then in 1960 deposited them as a small archive in the library of King's College, Cambridge.* She added to these seventy-seven letters a number of ancillary items, such as certain of the correspondence that arose in writing her book. A few other items went to Sherborne School.

Mrs Turing died at the age of ninety-four on 6 March 1976, and so never knew of me, nor indeed of what her son had done for the battle of the Atlantic. Before she died, however, she endowed an A. M. Turing Trust, to be chaired by Donald Michie, by then Professor of Machine Intelligence at Edinburgh University. It so happened that in 1977, just as I was first making enquiries, the Trustees had arranged for the deposit of all the surviving Turing papers in the King's College archive. These papers had been kept since 1954 by Robin Gandy, himself now a distinguished mathematical logician at Oxford; but in 1977 they were being sorted and catalogued by Jeannine Alton of the Contemporary Scientific Archive Centre, also at Oxford. That these Turing trails had focussed on Oxford was coincidental, but helpful in my early efforts to grasp my subject. I should like at this point to record particular thanks to Robin Gandy, Donald Michie, and Jeannine Alton, together with other members of the A. M. Turing Trust, who have helped me to marshal support and resources for my endeavour from the start. Others, since 1977, have also played parts of vital significance for this book; but I retain a special gratitude for

---

* Some have *not* been preserved; thus conspicuously absent is any announcement of his engagement in 1941. Mrs Turing also performed some half-hearted censorship on the letters, which like all incomplete suppression merely draws attention to the anxieties of the censor. Only one sentence, probably a reference to the engagement, is actually rendered illegible. I also feel a distinct debt to Lyn Newman for a subtle moral encouragement: in her introduction to Mrs Turing's book she called it 'a source-book for a future biographer', which is exactly how I have used it.

those who were prepared to assist me at a time when I had nothing but a shaky idea. I should add that the use I have made of this assistance is, of course, entirely my own responsibility; the interpretation I offer of material made available to me is likewise my own.

The enlarged Turing archive in King's College, at first sight thin, has indeed provided a backbone of documentary material. Here an acknowledgement is due to Alan Turing himself ; he kept nothing in the way of form letters, or the general accretion of correspondence overtaking those rising on the academic scale. But he took care to retain markers of most of the main points in his intellectual life. Indeed he had kept the star globe and the zeta-function machine gear wheels, although these were disposed of after his death. So interested in learning and development, he certainly cared about his own past.

So much, then, for what was already collected together in 1977. My own enquiries have since elicited further documents from a number of private and public sources, and meanwhile the growing interest in computer history has had the effect of giving me the benefit of recent Turing-related work by other researchers. Yet still the documentary evidence, in itself, would not add up to a portrait of Alan Turing. Only by meeting so many people who knew him could this picture have been drawn. Here again a first acknowledgment must go to my subject himself, who left behind a fund of goodwill on which I have repeatedly drawn. But in describing this aspect of my work, the word for what I have gained is less 'information' than 'experience'.

The work of compiling this string of symbols on my typewriter has been very different from anything performed by a computer, and different too from my mathematical work in the way that it has required so much interference with other people's lives. If this book is truly a biography – a writing of *life*, not a collection of facts – then it is because people have been prepared to allow that interference, and to entrust me with words and ideas that still have living force. The act of disturbing what was a strange kind of Sleeping Death had in fact involved some difficult – and often rather moving – moments. Thus Mr Arnold Murray, in conveying his recollections to me, was also shedding an albatross that had hung around his neck for twenty-five years. For he had returned to Manchester in 1954, only to be confronted by the news of Alan Turing's death. He was made to feel that he was to blame; and being both particularly vulnerable, and entirely unaware of a larger context, accepted a profound and unmentionable guilt. His success as a musician in the 1960s, and his movement into married life, could not resolve a trauma which had to wait until 1980 for enlightenment.

This one example must suffice to indicate why, in a number of cases, my gratitude to people who have helped me goes well beyond the terms of a conventional acknowledgement. In some cases this will be abundantly clear from the text; in others the true nature of the debt will be invisible. Nor is the following list even complete; but I should like to thank these and other contributors to my picture of Alan Turing for all that they have done:

J. Anderson, James Atkins, Don Atkinson, Bob (once Augenfeld), Patrick Barnes, John Bates, S. G. Bauer, Donald Bayley, R. Beadon, G. Black, Victor F. Beuttell, Matthew H. Blamey, R. B. Braithwaite, R. A. Brooker, W. Byers Brown, Mary Campbell (née Wilson), V. M. Cannon Brookes, David Champernowne, A. Church, Joan (née Clarke), F. W. Clayton, John Croft, Donald W. Davies, A. S. Douglas, Roy Duffy, D. B. G. Edwards, Ralph Elwell-Sutton, D. B. Eperson, Alex D. Fowler, T. H. Flowers, Nicholas Furbank, Robin Gandy, A. Gleason, A. E. Glennie, Harry Golombek, Jack Good, E. T. Goodwin, Hilla Greenbaum, Philip Hall, FRS, Arthur Harris, David Harris, Kenneth Harrison, Norman Heatley, Peter Hilton, F. H. Hinsley, Peter Hogg, N. E. Hoskin, H. D. Huskey, Neville Johnson, R. V. Jones, FRS, W. T. Jones, T. Kilburn, FRS, Leo Knoop, Walter H. Lee, Sir James Lighthill, FRS, R. Lockton, D. C. MacPhail, Malcolm MacPhail, Sir William Mansfield Cooper, A. V. Martin, P. B. C. Matthews, FRS, W. Mays, P. H. F. Mermagen, J. G. L. Michel, Donald Michie, Sir Stuart Milner-Barry, Rupert Morcom, Arnold Murray, D. Neild, E. A. Newman, M. H. A. Newman, FRS, John Polanyi, FRS, F. V. Price, J. W. S. Pringle, FRS, M. H. L. Pryce, FRS, David Rees, FRS, B. Richards, T. Rimmer, K. V. Roberts, Norman Routledge, David Sayre, Claude E. Shannon, Christopher Steed, Geoff Tootill, J. D. Trustram Eve, W. T. Tutte, Peter Twinn, S. Ulam, J. S. Vine, A. G. D. Watson, Mr and Mrs R. V. B. Webb, W. Gordon Welchman, A. C. Wesley, Patrick Wilkinson, J. H. Wilkinson, FRS, Cicely Williams (née Popplewell), R. Wills, Mike Woodger, Shaun Wylie.

It would be impossible to list all those who, besides these first-hand witnesses, have helped me by answering enquiries, commenting on draft, and in many other ways. But I should like to mention:

A. O. Childs (Sherborne School), J. E. C. Innes (Old Shirburnian Society), V. Knowles (Manchester University), Simon Lavington (Department of Computer Science, Manchester University), David Leigh *(The Guardian)*, Julian Meldrum (Hall-Carpenter Archives, London), J. E. Taylor (National Archives, Washington), Christopher Andrew, Duncan Campbell, Martin Campbell-Kelly, Peter Chadwick, Stephen Cohen, Cy Deavours, Robin Denniston, Fisher Dilke, D. Dunnill, James Fleck, Steven Hicks, David Hutter, David Kahn, Peter Laurie, Sir Bernard Lovell, FRS, J. Maunder, Roger Penrose, FRS, Felix Pirani, Brian Randell, Jeffrey Weeks.

One other person, however, has played a decisive role in translating my ideas into the practical form of a book. This is Piers Burnett, who was responsible for commissioning me to write this book, and has seen me through all its difficulties, reading and advising on numerous drafts. Originally my contract was with André Deutsch Ltd, of which Piers Burnett was a director, and which advanced £5500 for the book. (Certain difficulties arising in 1981 on the completion of my work, made a transfer necessary. The first British edition was published by Burnett Books, in association with the Hutchinson Publishing Group.) I have had no grant or subsidy from any other source. While this was by publishing standards a generous investment, and one which made the difference between the existence and the non-existence of this book, it did not easily cover my two years of full-time work from 1978 to 1980. Everything had to be done in a Turingesque shoestring style. There may be some virtues in this constraint, but it has certainly placed me in debt to many friends and friends of friends. My archive research and interviews in North America, for instance, would have been well-nigh impossible if I had not been given so many places to stay. At home I have also demanded great patience from those (particularly Peter Chadwick and Steve Hicks) who have had to share the tension and anxiety without the satisfactions.

Besides Piers Burnett there have been certain other people who somehow could see what I saw in this work, and who supplied a kind of moral support without which I could hardly have continued. One of these people commented very helpfully, at a critical stage, that I ought to leave at least something for others to find out! Certainly there are gaps, and trails I never followed. There will also be errors, though none I hope too serious, both of commission and omission. Perhaps, if the micro-electronic revolution brings about the end of the printed book, it will become possible to undertake continuous revision of published work. Meanwhile, a line must be drawn around an imperfect creation, in the knowledge that Turing scholarship has yet far to go. I shall transfer the documentary material I have accumulated to the Contemporary Scientific Archive Centre, for deposit at King's College, Cambridge; and I shall likewise pass on corrections and additions that I may receive in the future. Perhaps as important, however, to the continuation of Turing studies, is the fact that as the world changes so will its perception of who Alan Turing was. Even while writing this book the word 'computer' has changed its social meaning: a universal machine with the scale and speed of the ACE he envisaged is now sitting on my desk, hardly bigger than my hand. The algorithms embodied in the Bombes now amount to no more than a few lines of BASIC. Personal interaction with a private computer, with its little battles over storage and displays and checking, is now an experience common to hundreds of thousands of people. There is no knowing what this may lead to, but it has changed our perception of the past. If genetic engineering makes advances comparable to those of information technology, then again, his later work may appear in quite a new light.

Alan Turing presumably thought that eventually a machine would be capable of writing a book such as this. In his 1951 radio talk, set against the opening of the Festival of Britain, he commented that 'It is customary . . . to offer a grain of comfort, in the form of a statement that some peculiarly human characteristic could never be imitated by a machine. I cannot offer any such comfort, for I believe that no such bounds can be set.' There was an element here of *épater les bourgeois*, especially when he gave as example of a peculiarly human quality, that of being 'influenced by sex appeal'. Yet he was quite serious in describing the 'almost certain' advent of intelligent machines as a development 'which can give us anxiety' of a kind much more urgent than that of the Darwinian fright of a century before 'that we might be superseded by the pig or the rat.'

In actual fact I should gladly have surrendered my technical problems to a machine – and a word processor would have saved weeks of chopping and gluing – but these have not been the real difficulties in the composition of this book. More pressing, to give just one very pertinent example, has been the need to overcome the twentieth-century chasm between scientific thought and human life – moreover to resist the strong view held in certain quarters that my book should actually reinforce that separation. I have had to live, and even to struggle a little, for my vision; to put this point of view I have had to *act* this point of view.

One event of particular interest for me while working on these ideas was the appearance of Douglas Hofstadter's Gödel, Escher, Bach in 1979. It put my work in a Tangled Loop, for central to his book is its exploration of a topic I have brushed aside – the significance of Gödel incompleteness and Turing undecidability for the concept of Mind. I do not myself believe that these results, concerning as they do infinite, static, undisturbed logical systems, have any direct consequence for our finite, dynamic, interacting brains. Far more significant, in my view, is the limitation of human intelligence by virtue of its social embodiment – and this is a problem relegated to a marginal place in Hofstadter's work as in so many other accounts, though I have placed it at the centre of my own. The study of Alan Turing's life does not show us whether human intelligence is limited, or not limited, by Gödelian paradoxes. It does show intelligence thwarted and destroyed by its environment. But why, as Alan Turing might have put it, should machine intelligence be any less constrained by worldly reality? Indeed, there seems every reason to suppose that the clever machine will accommodate itself to the crazy demands of the political system in which it is embodied. In the academic sanctuary it is too easy to concentrate on infinitely more theoretical considerations.

For this reason my anxieties are rather different from those Alan Turing referred to: what worries me most is not whether we say machines are 'thinking' or not, but the place of such 'thought' in the body politic. Given the state of our modern Heartbreak House, I fear a victory not of anyone's intelligence, but of the pig and the rat.

Notes

These notes are not intended to supply a complete catalogue of sources for every statement. They cover only (1) the sources of direct quotations (2) the specification of documents and publications mentioned in the text (3) a fairly complete list of known documents with first-hand information about AMT, and (4) some source-critical comments and other points that require a discussion outside the time-frame of the text. I have not annotated sources which are fully specified within the text itself, and I have not regarded AMT's letters home (as held in the King's College archive) as all requiring explicit identification. I have not employed the academic formula of 'Private Communication' to indicate material gained from interviews: this seems to add nothing useful, and the reader will have in any case to trust me as the historical journalist with what I offer in the way of new biographical material. These notes must also serve as an inadequate bibliography; a full discussion of the literature surrounding AMT's work would go far beyond the scope of the present book. The same applies to 'further reading', although here I make one exception: *Mathematics Today* (ed. L. A. Steen, Springer Verlag, 1978).

I have used the following abbreviations throughout:

EST: The biography *Alan M. Turing*, by Sara Turing (Heffers, Cambridge, 1959).
KCC: The archive of letters and other documents relating to AMT held in the library of King's College, Cambridge.

Esprit de Corps

1.1. *The Lay of the Turings*, composed in about 1850 by the Rev. Henry Mackenzie, Bishop of Nottingham and son-in-law of the seventh Baronet. Perfect bad verse. A less romantic genealogy is detailed in Burke's Baronetcy.

1.2. H. D. Turing's daughter, Penelope Turing, wrote an autobiography *Lance Free* (1968).

1.3. Julius Turing's Record of Service is in the India Office Library, London.

1.4. The Stoney genealogy is given in Burke's *Irish Family Records*.

1.5. *The Road to Wigan Pier*, Part Two (Gollancz, 1937).

1.6. In an unpublished autobiography, *The Half Was Not Told Me*.

1.7. Quoted in *EST* from a letter written to Mrs Turing after AMT's death.

1.8. The original edition was entitled *A Child's Guide to Living Things* (Doubleday, Page & Co., New York, 1912).

1.9. Mrs Turing deposited in *KCC* sixteen letters of AMT from Hazelhurst, six from Sherborne. The first two of these, as quoted here, did not actually bear the year '1923'. This, however, was what Mrs Turing guessed in her annotations, and it is consistent with Sunday being letter-writing day at Hazelhurst, as seems to have been the case.

1.10. As note 1.6

1.11. Mrs Turing's own words in *EST*.

1.12. A. B. Gourlay, *A History of Sherborne School* (Sawtells, Sherborne, 1971).

1.13. *The Western Gazette*, 14 May 1926.

1.14. Alec Waugh, *The Loom of Youth* (Richards Press, 1917). Alec Waugh was at Sherborne from 1911 to 1915.

1.15. Nowell Charles Smith, *Members of One Another* (Chapman & Hall, 1913). A book of sermons for the years 1911–13. It may be somewhat anachronistic to quote from the pre-war period, but from all accounts, very little had changed in 1926.

1.16. Quoting from here onwards the comments made on AMT's school reports. These were donated by Mrs Turing to the library of Sherborne School.

1.17. Letter to the author from Mr D. B. Neild, 23.12.78.

1.18. A. H. T. Ross compiled an extensive book of revealing reminiscences, *Their Prime of Life* (Warren & Sons, Winchester, 1956). The 'House Letter' of 1928 is entirely typical of its style and content.

1.19. Quoting from a letter to the author of Mr M. H. Blarney, 9.7.78.

1.20. As note 1.12.

1.21. Letter to the author from Canon D. B. Eperson, 16.1.78.

1.22. The popular account was the English translation of Albert Einstein, *Relativity: The Special and the General Theory* (tr. R. W. Lawson, Methuen, 1920). Unfortunately it is not clear how and when the discovery of relativity fitted into AMT's development. The Memo Book containing his notes, held in *KCC*, carries Mrs Turing's assertion that the notes were written for her at Christmas 1927 – an astonishingly early date in view of the unanimous verdict of the schoolmasters that he could not express himself. The notebook contains calendars for 1928 and 1929 printed at the back, and so would presumably have been on sale at that Christmas. But this early date would not be consistent with the supposition that he got the statement of the geodesic law of motion from Eddington's book, which appeared only in 1928. So as a working compromise I have set my account in the context of late 1928. It may be that this does an injustice to him, and understates the contrast between his intellectual development and the general Sherborne appreciation of it. There is no other piece of evidence until the reference to relativity by Christopher Morcom in his letter of 19 August 1929. This reads as though they had at least spoken together of the subject. For another point of corroboration, see note 1.27. A related question is that of how AMT found the Einstein and Eddington books – here credit must be due to the Sherborne librarian or some other helpful hand. It is a good reminder of how incomplete our knowledge must sometimes be.

1.23. The quoted passages come from the letters and notes that AMT wrote for Mrs Morcom in 1930 and 1931 (see page 60, and note 1.26).

1.24. This report is held together with the school reports at Sherborne. Mrs Turing has annotated it as being either 1929 or 1930, and my placing it in 1929 is only a matter of guesswork.

1.25. A. H. T. Ross (note 1.18) makes specific mention of the danger of accepting holiday invitations from boys in other houses. Curiously enough he was writing his remarks on 'Problems' and 'Tone' in the spring and summer of 1954, with the result that they are interspersed with comments on the impact of the Montagu trials and news of AMT's death.

1.26. AMT kept the letters he had from Christopher Morcom, and other souvenirs (see page 53). In 1931 Mrs Morcom copied out the letters, and then the originals, which AMT kept all his life, were returned after AMT's death. The Morcom family also kept the letters written by AMT just before and then after Christopher's death. I am deeply indebted to Mr Rupert Morcom for making all these and other family documents available.

1.27. There are no letters in *KCC* between those of May 1926 and this one. The paraphrase is of a passage on page 215 of Sir Arthur Eddington, *The Nature of the Physical World* (Cambridge University Press, 1928). This is good evidence that by this time he had absorbed Eddington's account of relativity, which comes well before this discussion of the new quantum mechanical picture of matter.

The Spirit of Truth

2.1. This letter is not in *KCC*. Another loss is that of the letters AMT had both from his mother and his father at this time. According to *EST* he also kept these all his life. This deprives us in particular of a glimpse of the relationship between father and son. Mrs Turing had her say later, but in this as in so many other ways Mr Turing's part has just been wiped out.

2.2. Quoted in *EST* from a letter written to Mrs Turing by A. J. P. Andrews after AMT's death.

2.3. Letter to the author from Major L. Knoop, 24.1.79.

2.4. No diary has survived from this or any other part of AMT's life.

2.5. Letter to the author from Mr Patrick Barnes, 12.2.79.

2.6. This was the 1922 edition of *Mathematical Recreations and Essays* (Macmillan).

2.7. A short biography of Alfred W. Beuttell (1880–1965) was commissioned and published privately by Victor Beuttell in 1971, under the title *The Man Who Made Linolite*.

2.8. *The Shirburnian*, *36*, page 113.

2.9. Here and elsewhere I have drawn upon C. Reid, *Hilbert* (George Allen & Unwin; Springer Verlag, 1970) for quotations.

2.10. As note 3.3.

2.11. The paper was in *J. Lond. Math. Soc.* 8 (1933). Champernowne's result concerned 'normal numbers', an application of the study of the 'real number' system as it had developed since the late nineteenth century. A 'normal' number was defined as one whose decimal expansion contained the ten digits equally and evenly distributed in a certain precise sense. It was already known that if a real number were picked 'at random', then there was a probability of one hundred per cent that it would be 'normal'. Yet no actual example of a 'normal number' was known until Champernowne produced one. AMT took some interest in the question later. There was a connection with his interest in randomness, but also a similarity to the concept of computability. For a 'random' real number has a probability of one hundred per cent of being uncomputable, but it requires some effort to produce, as he did, an example of an uncomputable number. *KCC* contains a letter from G. H. Hardy to AMT on 'normal numbers', undated but presumably of the later 1930s.

2.12. It was undated, but written out on Clock House notepaper. This places it as composed on one of his visits. Mr Rupert Morcom writes that he believes it was written before 1933, and the handwriting style would support this belief. My guess is that 1930 is too early for the McTaggart reference, and that the style is more consistent with AMT's wider-ranging intellectual life at Cambridge. These considerations all point to 1932. But certainly AMT could have thought in terms very like these at any time since 1929 or so, and the date of this piece of writing is not too significant.

2.13. Quoting from the English translation of Laplace's *Essai sur les probabilités* (Dover edition, 1951).

2.14. In his obituary of AMT in the *Shirburnian*, 1954.

2.15. Quoted in *EST* from a letter written to her by Geoffrey O'Hanlon.

2.16. A. W. Beuttell, 'An Analytical Basis for a Lighting Code', in *The Journal of Good Lighting*, January 1934.

2.17. I am grateful to Professor W. T. Jones for bringing this passage to my attention in describing the impression AMT made on him in 1937. (See page 173). Keynes' talk on *My Early Beliefs*, given in 1938, was published after his death as one of *Two Memoirs* (Rupert Hart-Davis, 1949).

2.18. *The Autobiography of G. Lowes Dickinson*, published posthumously (Duckworth, 1973).

2.19. *New Statesman and Nation*, 4 February 1933. The progressive journal here used the medical model for homosexuality.

2.20. J. S. Mill, *On Liberty* (1859). I owe to Robin Gandy the identification of AMT as 'a J. S. Mill man'. In fact I have chosen to set AMT against less business-like and competitive libertarians, but certainly this essay contains many points of contact with AMT's outlook and convictions.

2.21. *Maurice*, written in 1913, was published after E. M. Forster's death in 1971.

2.22. The passage quoted was actually written by Shaw in 1944, but it only condensed the comment in Shaw's Preface to *Back to Methuselah* of 1920.

2.23. Bertrand Russell's *Introduction to Mathematical Philosophy* (George Allen & Unwin, 1919) did not deal with the background in geometry, but started with the problem of giving meaning to the Peano axioms. However, I have included mention of Hilbert at this point, in order to lend greater unity to the discussion.

2.24. The minutes are held in the University Library, Cambridge.

2.25. *The Times*, 10 November 1933. But if the mathematicians had delivered a politically advantageous formula, they had surrendered little in private content. The phrase 'a mixture of logic and intuition' was unexceptionable (compare AMT's remarks *apropos* of the ordinal logics in 1938); and the work of Gödel had just recently served to delineate the limitations of deductive logic.

2.26. The standard work for this course was Whittaker and Robinson, *The Calculus of Observations*, 1924.

2.27. Lindeberg, *Math. Zeitschrift 15* (1922).

2.28. AMT would have offered about six advanced courses for the Schedule B examination. Unfortunately the records of the Faculty of Mathematics do not seem to show what these were.

2.29. AMT's fellowship dissertation, *On the Gaussian Error Function*, remained unpublished. The original typescript is held in *KCC*.

2.30. As note 2.9.

2.31. An English translation of Gödel's paper is in *The Undecidable*, ed. Martin Davis (Raven Press, New York, 1965).

2.32. This was Hardy's Rouse Ball Lecture for 1928, published in *Mind*, 1929, as 'Mathematical Proof'.

2.33. AMT's paper was 'Equivalence of Left and Right Almost Periodicity', *J. Lond. Math. Soc. 10* (1935).

2.34. J. von Neumann, *Trans. Amer. Math. Soc. 36* (1934).

2.35. For a modern biographical study, with many points of contact with this book, see Steve J. Heims, *John von Neumann and Norbert Wiener* (MIT Press, 1980).

2.36. AMT also corresponded with von Neumann. In *KCC* there is an isolated letter from von Neumann to 'My dear Mr Turing', dated 'December 6' without year. It concerns a theorem about topological groups proposed to him by AMT. The year is most likely 1935; von Neumann's letter contains a reference to the mailboat, so this could not be 1936 or 1937. By 1938 AMT's research interests had moved away from this field. My search through the von Neumann papers in the Library of Congress did not reveal any more of this correspondence.

2.37. AMT's great paper, quoted here, was 'On Computable Numbers, with an Application to the Entscheidungsproblem', *Proc. Lond. Math. Soc.* (2), *42* (1936). It is reprinted in *The Undecidable* (as note 2.31).

2.38. Did AMT think in terms of constructing a universal machine at this stage? There is not a shred of direct evidence, nor was the design as described in his paper in any way influenced by practical considerations. Yet in his obituary of AMT in *The Times*, Newman wrote: 'The description that he then gave of a "universal" computing machine was entirely theoretical in purpose, but Turing's strong interest in all kinds of practical experiment made him *even then* interested in the possibility of actually constructing a machine on these lines.' (My italics.) Newman did not repeat this claim in his Royal Society memoir, in which the practical side was so much played down, although there he commented on how bold an innovation it had been to bring 'paper tape' into symbolic logic. Both comments reflected the impact made by AMT's concreteness upon a classical pure mathematician, but like the other obituary writers, Newman was concerned to delineate AMT's mental unorthodoxy, rather than to document anything in the history of technology. We have nothing more to go on. My own belief is that the 'interest' must have been at the back of his mind all the time after 1936, and quite possibly motivated some of his eagerness to learn about engineering techniques. But as he never said or wrote anything to this effect, the question must be left to tantalise the imagination.

New Men

3.1. A. Church, 'A Note on the Entscheidungsproblem', in *J. Symbolic Logic*, 1 (1936), reprinted in *The Undecidable* (note 2.31). The paper of a year earlier was 'An Unsolvable Problem of Elementary Number Theory', *Amer. J. Math.* 58 (1936), presented 19 April 1935.

3.2. The first letter in what was to be a more copious flow of correspondence during the Princeton period, in which AMT managed while away to think of something to say every three weeks or so. There are only eighteen letters in *KCC* for the five academic years 1931 to 1936, but twenty-eight for the two Princeton years. This frequency was never resumed, a total of nine letters home representing the remaining sixteen years of his life.

3.3. G. H. Hardy, *A Mathematician's Apology* (Cambridge University Press, 1940).

3.4. When Mrs Turing came to write her biography, she found herself better informed about AMT's environment at Princeton than anywhere else, thanks to his letters. Though largely transcribing the information in these, she added one story which does not derive from a *KCC* letter: 'Though prepared to find democracy in full flower, the familiarity of the tradespeople surprised him; he cited as an extreme case the laundry vanman who, while explaining what he would do in response to some request of Alan's, put his arm along Alan's shoulder. "It would be just incredible in England."' Perhaps there was something of an 'alas!' in AMT's remark, which would not have fitted in with Mrs Turing's ideas about tradesmen.

3.5. Two postcards from Scholz, dated 11 February and 15 March 1937, are in *KCC*.

3.6. A remark quoted in the review of von Neumann's contributions to the 'Theory of Games and Mathematical Economics', by H. W. Kuhn and A. W. Tucker, *Bull. Amer. Math. Soc. 64* (1958).

3.7. Published posthumously in *The Undecidable* (note 2.31).

3.8. Post's paper is reprinted in *The Undecidable*.

3.9. Letter to the author from Dr A. V. Martin, 26.1.78.

3.10. See note 8.67.

3.11. The brief logic paper was in *J. Symbolic Logic, 2* (1937). The work related to that of Baer was in *Compositio Math. 5* (1938). The other group theory paper was in *Ann. Math. (Princeton) 39* (1938).

3.12. A copy of von Neumann's letter is held in AMT's file at the Department of Mathematics, Princeton University. Formal recommendation of AMT came from the Vice-Chancellor of Cambridge University on 25 June.

3.13. One letter from Bernays to AMT, dated 24 September 1937, is in *KCC*. AMT's correction note appeared in *Proc. Lond. Math. Soc. (2) 43* (1937). There were other mistakes and inconsistencies in the specification of the universal machine, some of them detailed by Post in a 1947 paper (reprinted in *The Undecidable*, as note 2.31).

3.14. *J. Symbolic Logic, 2* (1937).

3.15. As described in *The Undecidable*, page 71.

3.16. J. B. Rosser, *J. Symbolic Logic 2* (1937).

3.17. Letter from A. E. Ingham dated 1 June 1937 in *KCC*.

3.18. The following account draws heavily on H. H. Edwards, *Riemann's Zeta Function* (Academic Press, New York, 1974), which also discusses AMT's contributions.

3.19. S. Skewes, *J. Lond. Math. Soc. 8* (1933). There is a letter in *KCC* from Skewes, dated 9 December 1937, with a brief expression of interest in AMT's ideas.

3.20. A. G. D. Watson, 'Mathematics and its Foundations', in *Mind 47* (1937).

3.21. AMT was right. During the war Gerard Beuttell made important contributions to the design of instruments to estimate the visual range by measuring the scattering of light within a small enclosed space. (*J. Scientific Instruments, 26* (1949)). He died on a meteorological reconnaissance flight over the north Atlantic in early 1945.

3.22. Letter to the author from Dr M. MacPhail, 17.12.77.

3.23. It was in service until 1960, then being supplanted by a digital computer, and may now be seen in the Liverpool City Museum.

3.24. Letter from E. C. Titchmarsh in *KCC*.

3.25. The original PhD thesis is held in the mathematics library at Princeton

University; it was published as 'Systems of Logic based on Ordinals' in *Proc. Lond. Math. Soc. (2) 45* (1939), and reprinted in *The Undecidable*.

3.26. Letter to the author from Professor S. Ulam, 16.4.79.

3.27. C. Andrew, 'The British Secret Service and Anglo-Soviet Relations in the 1920s, Part I' in the *Historical Journal, 20* (1977).

3.28. Hinsley I (see note 3.31), page 10.

3.29. Hinsley I, page 20.

3.30. Administrative files relating to GC and CS are held at the Public Record Office in FO 366.

3.31. F. H. Hinsley *et al.*, *British Intelligence in the Second World War.* Volume I (1979), Volume II (1981). Published by HMSO as an official war history.

3.32. FO 366/978.

3.33. Hinsley I, page 54.

3.34. As note 3.27.

3.35. Hinsley I, page 53.

3.36. Hinsley I, page 54.

3.37. From records of the Faculty of Mathematics, Cambridge University.

3.38. Parts of the revised *Encyklopädie* appeared in December 1939, but Scholz's section on the foundations of mathematics, including the reference to AMT's work, had to wait until August 1952.

3.39. A transcript compiled from notes taken by others attending the lectures has been published as *Wittgenstein's Lectures on the Foundations of Mathematics, Cambridge 1939*, ed. Cora Diamond (Harvester Press, 1976). The quoted dialogue comes from lectures 21 and 22. It is perhaps a pity that the most extensive verbatim record of AMT should be concerned with a discussion which was not central to his concerns, and where he was not in his element. AMT sometimes liked to give the impression that he had scored off Wittgenstein at some point, but if so the evidence is not to be found in this transcript. In fact he showed a curious diffidence, one feature being that despite long discussions about the nature of a 'rule' in mathematics, AMT never offered a definition in terms of Turing machines.

3.40. This is in *KCC*. It was corrected and completed by A. M. Cohen and M. J. E. Mayhew, *Proc. Lond. Math. Soc. (3) 18 (*1968). Using AMT's approach they reduced the 'Skewes number' to $10^{10^{529.7}}$ But in 1966 R. S. Lehman had by another method reduced the bound to the comparatively miniscule value of 1.65×10^{1165}.

3.41. His paper 'A Method for the Calculation of the Zeta-function' appeared only in 1943, in *Proc. Lond. Math. Soc. (2) 48*.

3.42. Quoting from a copy of part of the letter made by Mrs Turing and deposited in *KCC*. My guess is that she omitted some reference to the function of the proposed machine as a cipher generator, not knowing whether this would be a transgression of secrecy.

3.43. Minutes of the Council of the Royal Society.

3.44 The blueprint, initialled 'D.C.M.', is in *KCC*.

3.45 Hinsley I, page 51.

The Relay Race

4.1. Letter and list of names in FO/366/1059, which contains no further reference to AMT.

4.2. M. Muggeridge, *The Infernal Grove* (Collins, 1973).

4.3. Pre-eminently H. F. Gaines, *Elementary Cryptanalysis*, 1939. Only at the end of the 1970s did a serious technical discussion of specific modern cipher systems begin to appear.

4.4. I am grateful to the staff of the National Archives, Washington, for bringing this material to my attention. In late 1940 the German raider *Komet* made several captures of British merchant ships and took this code and cipher material. This then found its way into German archives captured after the war.

4.5. There is an account of Polish Enigma work in the appendix to J. Garlinski, *Intercept* (Dent, 1979). A fuller and better version is given by M. Rejewski, 'How Polish Mathematicians Deciphered the Enigma', in *Annals of the History of Computing 3* (1981). This would seem to be the definitive account, ending much confusion and speculation in earlier discussions.

4.6. Hinsley I, page 490, itself quoting from the Polish claim at the time.

4.7. Hinsley I, page 492.

4.8. Letter to the author from Professor R. V. Jones, 7.2.78, expanding upon a passage of his *Most Secret War* (Hamish Hamilton, 1978).

4.9. The following account of the Bombe is a simplified version of Gordon Welchman, *The Hut Six Story* (McGraw Hill, New York; Allen Lane, London, 1982). It is worth noting Welchman's comment: 'We thought very little, in these hectic days, of who should take credit for what.' AMT would have thought least of all, though he did say that he thought Welchman's idea had been the important one. Establishing priority and originality is hard enough in open work, let alone when considering ideas kept secret for over forty years. I hope that the departure from truth, in this and other passages suffering from the same difficulty, is not too great. The more important point lies in the fact that pre-war cryptology, fossilised and isolated by secrecy, was transformed as soon as any contemporary mathematical mind was brought to bear on the subject.

4.10. Hinsley I, page 493. The account in B. Johnson, *The Secret War* (BBC, London, 1978), identifies AMT as the 'emissary', following a statement made to BBC researchers by General Bertrand before his death. This seems rather unlikely as he was working on the Bombe, not the sheets, and as this was not really a job for a 'man of the Professor type'. But it might be so – I have found no further evidence one way or the other. *EST* has a story concerning AMT being sent abroad, a mix-up over papers, and managing for a day with 'a few francs', but this could be taken to fit the 1945 mission (page 391).

4.11. P. Beesly, *Very Special Intelligence* (Hamish Hamilton, 1977), which gives the Admiralty side of the story.

4.12. Hinsley I, page 103.

4.13. Hinsley I, page 336.

4.14. Hinsley I, page 163.

4.15. F. W. Winterbotham, *The Ultra Secret* (Weidenfeld & Nicolson, 1974), which gives a secret-service view.

4.16. P. Beesly, as note 4.11.

4.17. Hinsley I, page 109.

4.18. Hinsley I, page 144.

4.19. Hinsley I, page 336.

4.20. I. J. Good, 'Studies in the History of Probability and Statistics XXXVII. A. M. Turing's Statistical Work in World War II', in *Biometrika* 66 (1979), which this description of AMT's ideas follows closely. Further details are given in a note by Good appended to the article by M. Rejewski (note 4.5).

4.21. Quoting from I. J. Good's lecture at the National Physical Laboratory, 1976; this has since been published in slightly revised forms in several places, the most accessible being as a paper 'Pioneering Work on Computers at Bletchley' in the misleadingly entitled volume *A History of Computing in the Twentieth Century*, eds N. Metropolis, J. Howlett and G. C. Rota (Academic Press, New York, 1980).

4.22. Quotation is from Beesly, as note 4.11, although I follow Hinsley in stating the capture to have been planned and not an accident.

4.23. Messages as translated into English at the time, and taken from the first few pages of the gigantic PRO file DEFE 3/1.

4.24. Hinsley I, page 337.

4.25. Beesly, as note 4.11, pages 57, 97.

4.26. Hinsley I, pages 273–4.

4.27. Quoted in *EST*. There he appeared anonymously (presumably because working for GCHQ) as a colleague who later proved a 'staunch friend' – Mrs

Turing's only concession to the events of 1952.

4.28. Hinsley I, page 296.

4.29. R. Lewin, *Ultra Goes to War* (Hutchinson, 1978), page 183.

4.30. Obituary of A. C. Pigou by D. G. Champernowne in *Roy. Stats. J. A122* (1959).

4.31. As note 4.2.

4.32. Dorothy Sayers, *The Mind of the Maker* (Methuen, 1941). AMT referred to reading it in the first wartime letter to his mother, in August 1941 (see note 5.8), saying 'You should read it when you come.' The quoted passage is the one he himself quoted in 1948 (see page 475).

4.33. Princeton records show that von Neumann gave a popular lecture on the game of poker on 19 March 1937. It would be very surprising if AMT had not attended it. He did not, in his discussions with Jack Good, draw a connection between his chess programs and game theory – nor indeed with the machines of *Computable Numbers*. But I have assumed that he had a general acquaintance with game theory, just as he could hardly have forgotten his own 'machines'. I have also given space to game theory for another reason: AMT certainly showed an interest in it later, and often pointed out examples of strategies in everyday life.

4.34. AMT's letters to Newman are in *KCC*. They are undated but can mostly be placed by passing references to events.

4.35. This essay, 'The Reform of Mathematical Notation and Phraseology', remained unpublished. The typescript is in *KCC* with other unpublished work on type theory. Excerpts are included in a historical paper by R. O. Gandy, 'The Simple Theory of Types', in *Logic Colloquium 1976*, eds. R. O. Gandy and J. M. E. Hyland (1977).

4.36. AMT's joint paper with M. H. A. Newman was 'A Formal Theorem in Church's Theory of Types', in *J. Symbolic Logic* 7 (1942).

4.37. AMT's paper appeared in the same 1942 volume of the *Journal of Symbolic Logic*. The two 'forthcoming' papers, 'Some Theorems about Church's System' and 'The Theory of Virtual Types', never appeared. But in 1947 (see page 446 and note 6.34) he submitted a further paper on type theory which represented a revision of work done at this period.

4.38. Hinsley I, page 338.

4.39. Beesly, as note 4.11, page 164. I have inserted 'September' for Beesly's 'November' to make it consistent with the Hinsley account.

4.40. Letter quoted from Hinsley II, page 655.

4.41. Hinsley II, page 657.

4.42. Quoting from B. Randell, *The Colossus*, an account written from the engineering side. First published as a University of Newcastle-upon-Tyne report in 1976 this is now available in the Metropolis volume (see note 4.21).

4.43. Peter Hilton was speaking informally during a conference session of *Reminiscences of Logicians*, published as a section of *Algebra and Logic*, Springer Mathematical Notes 450; ed. J. Crossley, 1975.

4.44. As note 4.21.

4.45. Hinsley II follows earlier writers in calling the German enciphering machine a *Geheimschreiber*. But my understanding is that there was more than one type of machine covered by this generic term, and that the photograph of a Siemens machine in B. Johnson, *The Secret War* (as note 4.10) is not actually of the one that was deciphered as Fish.

4.46. As note 4.43.

4.47. Stories taken from *EST*. With mind-boggling *sang froid* she added:.'The idea of "Prof." being nearly arrested caused much amusement in his department.'

4.48. As note 4.43.

4.49. Hinsley II, page 56.

4.50. I am grateful to the State Department for supplying copies of documents relating to AMT's entry to the US in 1942. They are of a purely routine administrative nature. They account for all the references to AMT in the general index to State Department files held in the National Archives, Washington. In contrast, there are no corresponding British documents. There is a reference in the index to Foreign Office correspondence for 1942 to 'Turing: sea travel facilities to Washington: finances.' But the relevant file has been 'weeded': destroyed.

4.51. FO/371/32346.

4.52. As note 4.2.

4.53. I owe this reference to Dr G. DiVita. It was on 14 February 1941. Of course the report gave no indication of the scale and modernity of operations, but it is curious to see any mention whatever of 'Nazi codes' being broken, when comprehensive secrecy on the subject lasted for a quarter of a century after Nazi Germany's demise.

4.54. For AMT's pre-war voyages I was able to draw upon the Board of Trade passenger lists. But none exist for the wartime period, so the evidence here is indirect. The State Department information (see note 4.50) shows that he was admitted at New York on 13 November 1942. Information from the Naval Historical Center, Department of the Navy, Washington D. C. shows that this was the day that the *Queen Elizabeth* arrived.

The fast passenger liners converted to troop transporters being the normal means of ferrying high-level personnel, I have assumed this to settle the question. But it is confused by the statement of Mrs Turing that his passage west was very overcrowded, and that he was the only civilian on board apart from a couple of children. Here, surely, she was mistaken. It was the *eastbound* passages which were desperately crowded. The *Queen Elizabeth* carried a mere 557 passengers west, most of them civilians, to return in March with 10,261 troops. See also note BP 11, for evidence regarding AMT's eastbound voyage.

Bridge Passage

BP1. This is the most substantial anecdote from the wartime period in *EST*, and a rare example of where Mrs Turing's deference to officials took second place to an authentic Alan Turing tone of voice. Apart from some early childhood details, this is also almost the first place in *EST* where Mrs Turing's personal recollection emerges. My guess is that it was the hint of AMT making an important mission to America that made her take more notice of him.

BP2. The index to Foreign Office correspondence for 1943 contains on page 428 a reference to complaints (themselves 'weeded') about insufficient accreditation. This might have included AMT's – but in any case adds a few decibans to an otherwise unlikely-sounding story.

BP3. Beesly, as note 4.11, pages 152, 161.

BP4. The date of arrival, like subsequent dates and details of AMT's period in New York, derives from contemporary Bell Laboratories personnel records made available to me. But with AMT's visit making such an impression on his mother, there are a few odd details of reminiscence in *EST*, based presumably on his cryptic replies to her interrogation. There was 'some hold-up about his job, which involved a useless period of idling in New York' – very likely the two weeks or so before 19 January, and due to 'clearance' arrangements, just the thing to annoy him. Mrs Turing's own assertion about the purpose of his visit to America was that 'he probably saw something of the progress of computing machinery in the States.' But probably AMT said 'Oh, seeing some of their machines, Mother', and the word 'computing' was Mrs Turing's guesswork. She also wrote 'He seems to have taken the opportunity to visit Princeton' – he could certainly easily have stopped off on one of his several journeys between New York and Washington. Her oddest comment was in reference to the mix-up upon his arrival: 'Even on Ellis Island he would have found something of interest, perhaps more than he found at Washington.' While this again reflected the way that everyone in secret work had to spin out the trivial and play down the serious, there is a hint here that while AMT could hardly say, 'Well, Mother, I was handing over to the Americans all the work we've been doing for the last three years,' he allowed something of this to come through.

BP5. *National Service in War and Peace (1925–1975)*, the second volume of *A History of Engineering and Science in the Bell System*, Bell Telephone Laboratories, 1978. I have taken my account of the Vocoder and the X-system directly from this source. The X-system was 'one of the starting points of the digital transmission age that followed', despite being until 1975 'an unmentionable system'.

BP6. PRO file CAB 79/25. The Memorandum referred to has not been released. I am indebted to David Kahn for this reference.

BP7. Minute dated 27 April 1943, in CAB 79/27.

BP8. C. E. Shannon, 'Communication in the Presence of Noise', *Proc. I.R.E.* (1948) is annotated: 'Original manuscript received by the Institute, July 23 1940.' His paper 'Communication Theory of Secrecy Systems', *Bell System Technical Journal*, 1949, a very rare example of cryptology treated from a post-1930 standpoint in the open literature, was originally 'A Mathematical Theory of Cryptography', a Bell confidential report dated 1 September 1945.

BP9. C. E. Shannon, 'A Symbolic Analysis of Relay and Switching Circuits', *Trans. Amer. I.E.E. 57* (1938). According to the Bell *History*, it was as a result of this that 'the design of relay circuits changed rapidly from being a somewhat esoteric art to being a science, and it became possible to teach it as an engineering discipline.'

BP10. W. S. McCulloch and W. Pitts, 'A Logical Calculus of the Ideas Immanent in Nervous Activity', *Bull. Math. Biophys. 5* (1943). Their paper contained no reference to *Computable Numbers*, but in a discussion after a lecture of von Neumann (the same as in note 6.57), McCulloch mentioned that it was AMT's paper that had inspired their ideas. See von Neumann's *Collected Works* (Pergamon, 1963), volume V, page 319.

BP11. The evidence regarding AMT's eastbound voyage is less clear than for that of November 1942. According to *EST* 'he returned in a destroyer or similar naval vessel and experienced a good tossing on the Atlantic.' But I think that she was mistaken here; it is hard to believe that the 'top cryptanalyst' would have been entrusted to a destroyer when a fast independent troop transporter was available. Instead, I think her recollection of him being the only civilian on board a crowded ship (see note 4.54) must in fact refer to *this* voyage. Then it fits (apart from the 'couple of children') with the information (from the Naval Historical Center, Washington) about the *Empress of Scotland*. This sailing, furthermore, was the only independent eastbound sailing in the rest of March. The week's delay may be accounted for by the fact that this was when the convoy battle was at its height. Since Mrs Turing was certainly fallible – an example being in her annotations to *KCC*, which incorrectly stated Jack Crawford's death to have occurred before AMT's visit in 1938 – I have based the narrative on this not quite conclusive evidence.

BP12. I am grateful to Richard Plant for pointing out this reference in H. Heiber, *Reichsführer!* (Deutsche Verlags-Anstalt, Stuttgart, 1968).

BP13. E. M. Forster, 'Post-Munich', 1939, reprinted in *Two Cheers for Democracy* (Edward Arnold & Co., 1951).

Running Up

5.1. Volume I of *Allied Communications Intelligence and the Battles of the Atlantic*, report SRH 009 declassified by the National Security Agency, available at the National Archives, Washington.

5.2. Kapitän H. Bonatz, head of the *B. Dienst*, quoted in M. Middlebrook, *Convoy* (Allen Lane, 1976).

5.3. PRO file FO 850/171 contains a memorandum of May 1945, from the Cypher Policy Board to the Foreign Office, with instructions for use of the Typex. It explains that 'When enciphering on the Typex machine, the enciphered version of a letter can never be the letter itself. This sometimes makes it possible to assign with absolute accuracy even a small number of words known or estimated to be in a message to the actual letters of the cypher version . . .', and gives procedures for burying addresses and other stereotyped beginnings and endings amidst nonsense, inserting extra letters between and within words, and so forth. These were just the procedures which, if correctly applied, would render Enigma transmissions immune to decipherment. One cannot tell whether the existence of such a memorandum means that the lesson had, or had not, been learnt by British operators within the six years of war.

5.4. The year between March 1943 and March 1944 is the least well documented of AMT's life. Certainly there is clear evidence that both before and after this period he was engaged as a cryptographic consultant, and it is reasonable to suppose that this was also true during this year of catching-up. There arc surely some interesting facts yet to emerge concerning his interaction with this phase of the war, although my impression is that there was nothing engaging him with the intensity that characterised the earlier period. The other dark year of his life was, of course, his last. There might in fact be some connection between them, since if he continued to do top level work in examining Anglo-American communication systems in preparation for D-Day, he would have had access to new American machine systems and much else still important in the years after the Second World War. And who would have known how much he knew? As one apparently given access to American establishments on Churchill's personal authority, he was quite outside the usual service systems.

5.5. Frank Clare, *The Cloven Pine* (Secker & Warburg, 1943).

5.6. B. Randell, *The Colossus*, as note 4.42.

5.7. The Explanatory Caption attached to the photographs of the Colossus in PRO file FO 850/234 claims a direct link with Babbage and *Computable Numbers:*

Babbage's work in 1837 first established the logical principle of digital computers. His ideas were developed further in Turing's classical paper in 1936. The COLOSSUS machine produced by the Department of Communications of the British Foreign Office, and put into operation in December 1943, was probably the first system to implement these principles successfully in terms of contemporary electronic technology. . . . The requirement for the machine was formulated by Professor M. H. A. Newman, and the development was undertaken by a small team led by T. H. Flowers. A. Turing was working in the same department at that time, and his earlier work had its full influence on the design concept.

I assume that 'the logical principle . . .' means 'conditional branching'. Although this makes sense as retrospective comment, it is not my impression that this analysis was formulated at the time, and still less that there is some document in GCHQ dating from 1943 with references to Babbage or to *Computable Numbers*. The first two sentences have rather the function of giving a suitably imposing rationale for 'declassification'. The reference to AMT in the last sentence also seems to me to be misleading, except in the extremely general sense that he had done so much for the mechanisation of processes before Newman arrived. The essential part that AMT played in this development was in providing a statistical theory: not the machine, but the purposes for which it would be used.

5.8. The third of only three wartime letters in *KCC*. The first, written in August 1941 while at Portmadoc, gave a few details of the holiday and a reference to the Dorothy Sayers book; the second, later in 1941, mentioned a week in Cambridge and meeting David Champernowne ('Didn't find any others I know except the old fogies'), slight bombing at Shenley, and a possible visit to Rossall to see to Bob's future. I am grateful to Canon H. C. A. Gaunt for finding the dates of the Lake District holiday in A. C. Pigou's diaries. These, incidentally, show that this was the only visit AMT made there except that in 1948.

5.9. CAB 80/41. I found no further references to AMT in these or the corresponding American files.

5.10. FO/850/256.

5.11. Shannon had included it in the paper submitted in 1940 (see note BP8). Professor I. J. Good has written to the author: 'The "sampling theorem" . . . is not due to Shannon although it is often attributed to him. It dates back at least to E. T. Whittaker, *Proc. Roy. Soc. Edin. 35* (1915).'

5.12. As note 4.21.

5.13. J. Ramsbottom, *Edible Fungi*, 1943.

5.14. These pages are in *KCC*. They begin in mid-sentence, and are bereft of the necessary definitions, so do not make much sense. But the underlying problem addressed by AMT was clearly that of finding 'exceptional' rotor wirings permutations with some symmetrical feature, leaving a non-randomness that the cryptanalyst could exploit. Such wirings would have to be avoided when constructing an Enigma-type machine. The pages also give a strong impression of the high-powered algebraic and statistical work he had done on rotor machines.

5.15. Quoting from *The War Speeches of the Rt. Hon. Winston S. Churchill* (Cassell, 1951–2).

5.16. According to *EST* he was actually 'offered a lectureship' at Cambridge in 1945, and M. H. A. Newman's *Biographical Memoir* also states this. But the records of the Faculty Board lend no support to this claim. Most likely he was speaking to his mother of continuing as a Part III lecturer, just as he would have been in 1940 but for the war.

5.17. In *Mind*, 1950 (see page 522 and note 7.34).

5.18. In the ACE report (note 6.1).

5.19. In *Intelligent Machinery* (note 6.53). These three quotations express ideas so fundamental, and so characteristic of his thought, that I believe the anachronism of setting them in summer 1945 is justified.

5.20. Quoting from Mrs Turing's words in *EST*. This is explicitly given as a nugget of recollection, even if given a maternal gloss of 'service'. (I have inserted 'machine' for her word 'computer', since in this context there is no distinction whatever in meaning, and I do not wish to introduce the word prematurely.)

5.21. Reprinted fully in *Faster than Thought* (note 8.25).

5.22. Quoting from letters in the von Neumann archive, Library of Congress, Washington DC.

5.23. The relevant passage is in the extract included in *The Origins of Digital Computers*, ed. B. Randell (Springer Verlag, Berlin, 1973).

5.24. This letter was to Col. L. B. Simon at the Ballistics Research Laboratory.

5.25. Frankel's letter was written to B. Randell in 1972, in connection with the latter's detective work 'On Alan Turing and the Origins of Digital Computers'. This paper, quoting Frankel's letter, appeared in *Machine Intelligence* 7 (Edinburgh University Press, 1972). See also note 5.26.

5.26. Randell's 1972 work (note 5.25) arose from the fact that the EDVAC report was supposed by everyone to be the *fons et origo* of the digital computer. In trying to see how the ACE 'fitted in', he came across an assertion by Lord Halsbury, writing in 1959 as managing director of the NRDC, that one of the most important events in the evolution of the modern computer was 'of course the meeting of the late Doctors Turing and von Neumann during the war' (*Computer Journal*, I, 1959).

Randell continued to stress this question of a meeting, but my own conclusion is that whether or not they happened to meet (and I have found no more evidence of a meeting than Randell did), Halsbury was mistaken in thinking it important. The story of AMT and von Neumann is that of two utterly different personalities, in different social environments, but drawn to parallel problems within the movement of mid-century science. Either figure was perfectly capable of assembling the necessary ideas for the digital computer out of the conjunction of Hilbertian rationalism and Second World War technology. Both did, responding in slightly different ways according to their circumstances. There is no gap on either side that needs to be explained by a meeting, or some other conspiracy theory of history. Much the same applies to the question of when and how AMT discovered Babbage's work: it would have fascinated and encouraged him, but was ultimately irrelevant.

Mrs Turing got the picture exactly right when she wrote that his aim was 'to see his logical theory of a universal machine, previously set out in *Computable Numbers*, take concrete form in an actual machine.' Since she knew nothing of *Computable Numbers* but that a German professor had commended it, this was certainly not her own analysis; Newman could have guided her (see note 2.38) but her statement was more definite than anything that Newman had written. Most likely it was simply what AMT himself had told her again and again, trying to explain that all the logic she had thought so useless in the 1930s had come to something practical after all. The connection between 1936 and 1945 was also perfectly clear during AMT's time at the NPL. It was only later that this simple and direct truth was forgotten, to the extent that in 1972 Randell, writing his historical paper, could see 'no obvious connection' between the Universal Turing Machine and the ACE; and mentioned the ACE report only in terms of its relationship to the EDVAC report. It is astonishing how difficult people have found it, both in AMT's own time and since, to accept that he could both think of something abstract, and set out, without making any particular fuss, to make it concrete. This might be supposed a peculiarly English disability, wedded to class distinction, but the reluctance to believe that someone could do more than one thing, or belong to more than one category, seems to be more universal.

5.27. Sir George Thomson, describing Sir Charles Darwin in a *Biographical Memoir of the Royal Society*, 1963.

5.28. Here and frequently in the following narrative I draw on the collected minutes and reports of the Executive Committee of the NPL.

5.29. *Nature*, 7 April 1945.

5.30. Memorandum by J. R. Womersley, 'ACE Project – Origin and Early History', dated 26 November 1946, in DSIR 10/385. According to *EST*: 'On submission to the Government of the outline of his design for such a [universal] machine he was taken on to the staff . . . in October, 1945.' While AMT might have given some kind of verbal description to Womersley, no formal 'submission' appears in the records and most likely Mrs Turing was going on the memory of AMT's ACE report being formally accepted a few months later.

5.31. Lyn Newman, in her introduction to *EST*.

5.32. Edward Carpenter's autobiography, *My Days and Dreams* (George Allen & Unwin, 1916).

5.33. Forster's article in *Tribune* was reprinted in *Two Cheers for Democracy* (as BP. 13), where it followed a 1942 essay on Tolstoy voicing similar thoughts: 'Do you yourself believe in simplicity as a cure for our present troubles? And, if so, how do you think simplicity can be worked in a world that has become industrialised? Tolstoy's outlook was agricultural: he never realised the implications of the machine.'

5.34. Quoting from Angus Calder, *The People's War* (Jonathan Cape, 1969).

Mercury Delayed

6.1. AMT's report was headed only 'Proposed Electronic Calculator', and did not use the name 'ACE'. But he used the name as soon as the report was discussed, and so to simplify matters I have called it 'the ACE report'. A copy of the report, though without the pages of diagrams, is in DSIR 10/385, the main file covering the ACE development from 1946 to 1948. A complete version was issued in a limited edition by the Division of Computer Science, National Physical Laboratory, in April 1972 as report Com. Sci. 57. A first analysis of it was made in 1975 by B. E. Carpenter and R. W. Doran, later appearing in *Computer Journal 20* (1977).

6.2. The fragments consist of just four typescript pages. They survive only because he used the reverse sides as scrap paper on which to explain some circuit theory to Mike Woodger in 1947.

6.3. This was his lecture to the London Mathematical Society, 20 February 1947 (see page 448). The typescript (until now unpublished and unquoted) is in *KCC*. I have employed it here at the cost of jumping out of sequence, as I could hardly improve upon his own exposition of the ideas announced more starkly in the original ACE report.

6.4. In January 1947 J. V. Mauchly pointed out the idea that 'one set of instructions' could 'modify another set'. Paper reprinted in *The Origins of Digital Computers* (as note 5.23), page 366.

6.5. Mike Woodger has told me of a reference in NPL files, since destroyed, to the ACE report being in existence by the end of 1945. But in any case it was ready in time for Womersley to compose his own report by 13 February 1946; this was essentially 1945 work.

6.6. In the papers of Mike Woodger.

6.7. Paper E.881 in the NPL Executive Committee records.

6.8. Both letters are in DSIR 10/385.

6.9. J. H. Wilkinson has described his association with AMT and the ACE project in an article in *The Radio and Electronic Engineer*, July 1975, in the *Pioneers of Computing* oral history, ed. C. Evans, Science Museum, London, 1975, and in a paper 'Turing's work at the National Physical Laboratory . . .' in the Metropolis volume (see note 4.21).

6.10. DSIR 10/385.

6.11. Letter in the von Neumann archive, Library of Congress.

6.12. Minutes of the Council of the Royal Society, 1946.

6.13. S. S. Snyder, *Influence of US Cryptologic Organisation on the Digital Computer Industry.* Declassified NSA report SRH 003, 1977, available at the National Archives, Washington DC.

6.14. H. H. Goldstine, *The Computer from Pascal to von Neumann* (Princeton University Press, 1972), mentions this visit and the 'third version' in passing on page 218. My study of the Goldstine archive at Hampshire College, Amherst, Mass., failed to locate the 'third version'.

6.15. Letter in the von Neumann archive, Library of Congress. My search revealed only one reference to AMT in von Neumann's letters beside this glancing comment on the thesis of 'finitely many mental states'. It was in a letter to Wiener of 26 November 1946: 'I did think a good deal about self-reproductive mechanisms. I can formulate the problem rigourously [*sic*], in about the style in which Turing did it for his mechanism. I can show that they exist in this system of concepts . . .'

6.16. H. Hotelling, *Ann. Math. Stat. 14* (1943).

6.17. I. J. Good's book was not published until 1950. In the meantime Shannon's theory of communication had emerged from wartime secrecy in 1948, and Good was able to add a few comments to his text remarking upon the similarity of Shannon's concepts to those of 'weight of evidence'.

6.18. A. Wald, *Sequential Analysis*, 1947. In *KCC* there is a manuscript by AMT headed 'Sequential Analysis' and outlining the ideas: as with the algebraic work (note 5.14) he might well have felt there should be something in his papers to reflect the mathematical substance of his work. (But Wald's theory was used in R. B. Braithwaite's discussion of scientific method, and AMT later found it coming into Robin Gandy's work on the logic of science; so war work was not his only point of contact.)

6.19. D. Gabor, *J. Inst. Elect. Eng. 93* (1946).

6.20. *The Times*, 1 November 1946.

6.21. *Nature*, 20 April 1946 and 12 October 1946.

6.22. Hartree on 7 November, Darwin on 13 November 1946.

6.23. *The Electrician*, 8 November 1946.

6.24. *Surrey Comet*, 9 November 1946.

6.25. *The Listener*, 14 November 1946 (page 663). A photograph (page 672) claimed to show an earnest engineer 'wiring one of the sections of the automatic computer' at the NPL; but it was later revealed (page 755) to be nothing of the kind.

6.26. TRE documents (see note 6.27) show that F. C. Williams was supplied with the ACE report only in October 1946, and so could not have read of the 'regeneration' principle there. It was not, apparently, an obvious idea: Williams' account in the *Pioneers of Computing* oral history (note 6.9) explains that it was some time before 'the penny dropped'. No one at the time, nor since, seems to have noticed that AMT thought of it earlier; just one example of the refusal of people to believe he could do anything practical.

6.27. These are not NPL minutes but TRE documents, quoted and discussed by S. H. Lavington in *Electronics and Power*, November 1978, and then in his *Early British Computers* (Manchester University Press, 1980).

6.28. The ensuing correspondence between M. V. Wilkes and Womersley, and AMT's reaction to it, has been taken from a copy held in the papers of Mike Woodger.

6.29. As note 6.6.

6.30. The lectures described Versions V, VI and VII of the ACE design. Only the first two and part of the last were actually given by AMT. Hartree's notes of the last two lectures are held in the Hartree archive, Christ's College, Cambridge;

and photocopies of these are in *KCC*. The whole lecture course, however, was written up by T. H. Marshall for a report, *The Automatic Computing Engine*, for the Mechanical and Optical Instruments Branch, Military College of Science, Shrivenham. This was dated February 1947.

6.31. Remarks by Professor M. V. Wilkes in a covering note to the Wilkes-Womersley correspondence (see note 6.28), dated 7 February 1977.

6.32. *Daily Telegraph*, 27 December 1946; *Evening News*, 23 December 1946.

6.33. *Proceedings of a Symposium on Large-Scale Digital Calculating Machinery*, published as volume XVI of the Annals of the Computation Laboratory, Harvard, 1948.

6.34. AMT's paper was 'Practical Forms of Type Theory', in *J. Symbolic Logic*, 13 (1948). There are extensive drafts in *KCC* (see also note 4.37).

6.35. H. H. Goldstine (note 6.14) refers to this visit on pages 191, 219, 291. AMT's results on matrix inversion were more general than those of von Neumann and Goldstine, though the latter appeared first *(Bull. Amer. Math. Soc. 53,* 1947). AMT's paper (note 6.47), when it appeared in 1948, described the relation thus: 'In the meantime another theoretical investigation was being carried out by J.v. Neumann, who reached conclusions similar to those of this paper for the case of positive definite matrices, and communicated them to the writer at Princeton in January 1947 before the proofs given here were complete.'

6.36. The MCS report (see note 6.30) contains a reference to this reflection problem, suggesting that this was what was on his mind at the time. The patent (number 694,679) was filed only in 1952. Another patent (718,895), in the joint names of Turing, Woodger and Davies, and covering aspects of the ACE design, was filed in 1951. These were the only patents bearing AMT's name. Both were taken out by the National Research Development Corporation and at the NPL were regarded light-heartedly. No benefit accrued to the individuals named.

6.37. Note dated 14 August 1946 in DSIR 10/275.

6.38. As note 5.21.

6.39. I have assumed that he was clear on this point all along – after all, it was what he himself had proved in 1936! He must very early have faced the question as to how it was that his universal machine, *without* using program modification, could be set up to simulate the progress of a 'learning machine'. I have here quoted his best-expressed answer to this question from the 1950 *Mind* article (note 7.34), although he also discussed it, not quite so clearly, in the 1948 report (note 6.53). Not everyone was clear on this point; thus Goldstine (as note 6.14, page 266) supposed that program modification would extend the range of possible operations.

6.40. As note 6.6.

6.41. Letter undated. Mermagen had asked him to give a talk at Radley and

AMT replied characteristically that he would do so when he had 'lantern slides and possibly even an instructional film, which would make it more fun.'

6.42. C. G. Darwin, *The Next Million Years* (Rupert Hart-Davis, 1952).

6.43. Diagrams of this work, dated 2 March 1947, survive in Mike Woodger's notebook, as do details of Huskey's 'Test Assembly'.

6.44. As note 6.6.

6.45. *The Times*, 28 August 1947.

6.46. Indulging my Carpenter parallel again: words he used just before his own move from Cambridge to the North in the 1870s. From S. Rowbotham and J. Weeks, *Socialism and the New Life* (Pluto Press, 1977), page 35.

6.47. AMT's paper was 'Rounding-off Errors in Matrix Processes', in *Quart. J. Mech. App. Math. I* (1948), appearing in Russian translation in *Uspek. Matem. Nauk. (NS)* 6 (1951). It required NPL permission for him to publish. *KCC* contains a letter from Sir Charles Darwin to AMT, dated 11 November 1947, acknowledging the copy submitted to him for approval. '. . . I must say that I read it through with some attention and interest, but spent most of the time cursing you for giving me such a perfectly smudgy copy to read. Next time I hope somebody else and not myself [will] be the sufferer, but I think the best plan would be to get some better carbon paper.'

6.48. *Progress Report on the Automatic Computing Engine*, Mathematics Division, National Physical Laboratory, April 1948. This internal report, classified as 'confidential', contained extensive examples of programming for the ACE design as it then stood. Progress on each of the current British projects was also summarised by H. D. Huskey on his return to the United States in *Math. Tables and Other Aids to Computation*, *21*, page 213 (1948); this included a brief critique of AMT's plans for the ACE. For a recent account of the programming techniques developed at the NPL, see M. Campbell-Kelly, 'Programming the Pilot ACE . . .', in *Annals of the History of Computing*, *3* (1981).

6.49. Minutes of the Senate Committee, Manchester University, 22 March 1948.

6.50. The minutes of the Moral Science Club give no more than the title of the talk, which was presented in S. Toulmin's rooms.

6.51. The discussion in J. von Neumann and O. Morgenstern, *Theory of Games and Economic Behaviour*, 1944, had approximated the game of poker by treating the cards as having a continuous range of values. AMT's work differed only in that it considered the cards as a discrete set. This manuscript, and that of his analysis of the game of Psychology, are in *KCC*. He used the reverse sides of papers used in the King's College Choir School examinations.

6.52. As note 6.6.

6.53. The original typescript is in *KCC*. It was published in *Machine Intelligence 5*, eds B. Meltzer and D. Michie (Edinburgh University Press, 1969). Unfortunately this edition is marred by misprints, in particular the date 8/7/48 appearing as '8 August 1947'.

6.54. AMT's attitudes contrast particularly sharply with the male competitiveness described so acutely, if unconsciously, in J. Watson, *The Double Helix*, 1968.

6.55. AMT's letters have not survived, but his programs were copied into G. C. Tootill's notebook. The long division routine is dated 8 July 1948.

6.56. Described in more detail in an article on computer chess in *Personal Computing*, January 1980.

6.57. A lecture 'The General and Logical Theory of Automata', rendered as a paper in 1951, and included in volume V of von Neumann's *Collected Works* (Pergamon, 1963).

The Greenwood Tree

7.1. F. C. Williams in the *Pioneers of Computing* oral history (see note 6.9).

7.2. Letter in the von Neumann archive, Library of Congress.

7.3. In his *Programmers' Handbook* (note 7.7), page 4.

7.4. F. C. Williams, 'Early Computers at Manchester University', in *The Radio and Electronic Engineer*, 1975.

7.5. Quoting from a progress report made by M. H. A Newman, considered by a Manchester University committee (which 'Mr Turing attended by invitation') on 15 October 1948.

7.6. Lyn Newman's introduction to *EST*.

7.7. His *Programmers' Handbook* was a duplicated document of over 100 pages, dated March 1951. It was rapidly superseded by new versions thereafter.

7.8. A slightly revised version of an account written by her in 1969, and quoted by M. Campbell-Kelly, 'Early Programming Activity at the University of Manchester', in *Annals of the History of Computing, 2* (1980). This paper gives detailed examples of the programming work.

7.9. In an appendix to the *Programmers' Handbook*, giving an account of the prototype machine and the work done on it.

7.10. The design survives as an appendix dated 21 November 1949 to an 'Informal Report on the Design of the Ferranti Mark I Computing Machine', in the papers of G. C. Tootill.

7.11. There might well have been other popular articles on this theme, but I have simply taken the one noted by Mrs Turing. My research in the Wiener archive at MIT did not bring to light any correspondence with AMT or comment on the 1947 visit; most likely it had no great significance for either of them. For a more serious and much more sympathetic account of Wiener's ideas see the study by Steve Heims (as note 2.35).

7.12. In *Faster than Thought*, page 323 (see note 8.25).

7.13. *British Medical Journal*, 25 June 1949.

7.14. *The Times*, 11 June 1949.

7.15. This letter, Newman's letter, and photographs of the prototype computer, all appeared in *The Times* on 14 June 1949.

7.16. Proceedings published in duplicated form by the Mathematical Laboratory, Cambridge, 1950. From a technical point of view AMT's paper was a first 'program proof', anticipating ideas of the 1960s. It has recently been reproduced, annotated and reviewed by F. L. Morris and C. B. Jones, *Annals of the History of Computing* 6 (1984).

7.17. M. V. Wilkes, *Computers Then and Now*, the 1967 Turing Lecture of the [American] Association for Computing Machinery.

7.18. Instead of following the policy adopted in 1946, the Pilot ACE was used as a working computer, and duplicated as a commercial version, DEUCE, by English Electric. It may now be seen in the Science Museum, London. It went there in 1958 when superseded at the NPL by a larger machine called 'ACE'. At the opening day the then Superintendent of the appropriate NPL Division declared 'Today, Turing's dream has come true.' But the 1958 ACE was a tardy anachronism: it had retained mercury delay lines in the age of magnetic core store, and even vacuum tubes in the era of the transistor. This was not his dream.

7.19. AMT described the computer work in his paper 'Some Calculations of the Riemann Zeta function', in *Proc. Lond. Math. Soc. (3) 3* (1953). Giving as much detail as he did about the base-32 coding and the running of the machine was itself a highly characteristic Turing touch, one not at all to be expected in a pure mathematical paper by a more conventional person. I have inserted the word 'prototype' to avoid a tiresome confusion caused by the fact that he called the prototype the 'Mark I', and then the 1951 'Mark I', the 'Mark II'. The names used in my text are those that prevailed. Although AMT did not get very far himself with the computations, his method was sound and was applied by D. H. Lehmer in 1955-6 to check that the first 25,000 zeroes of the zeta function all lie on the critical line.

7.20. Quoted from the article in *Mind* (note 7.34). There it formed part of his argument that a determinate system need not necessarily be predictable in practice; a machine need not behave in a recognisably 'mechanical' way.

7.21. Unpublished account by David Sayre, 1969. He adds 'One does not expect to find in one man both the most admirable intellect one had met *and* a person of the rarest human quality, but Turing was such a man, at least for me.'

7.22. D. Sayre, 'Some Implications of a Theorem Due to Shannon', *Acta Cryst. 5* (1952). But Dr Sayre writes: 'a more important paper by the Japanese crystallographers S. Hesoya and M. Tokonami in 1967 comes, I think, much closer to what Turing had in mind.'

7.23. *Symposium on Information Theory, London Papers*. Report of proceedings published by Ministry of Supply, 1950; re-issued by the Institute of Radio Engineers, 1953. The proceedings contain other comments by AMT and also note his unpublished work on chess-playing machines.

7.24. C. E. Shannon, 'Programming a Computer for Playing Chess', *Phil. Mag. Ser. 7, 41* (1950).

7.25. Correspondence and notes relating to the Ratio Club are held by Dr J. A. V. Bates, at the National Hospital for Nervous Diseases, London.

7.26. W. Ross Ashby, *Design for a Brain* (1952), and W. Grey Walter, *The Living Brain* (1953).

7.27. As note 4.43.

7.28. AMT's paper was 'The Word Problem in Semi-groups with Cancellation', *Ann. Math. (Princeton) 52* (1950). This was reviewed, clarified, and careless misprints corrected, by W. W. Boone, *J. Symbolic Logic 17* (1952).

7.29. The letter from von Neumann is in *KCC*. My search in the von Neumann archive did not uncover any reply from AMT.

7.30. Quoting from M. Polanyi, *Personal Knowledge* (Routledge & Kegan Paul, 1958), pages 397, 40. Polanyi's weighty volume was based on his Gifford Lectures for 1951–2.

7.31. K. Popper, 'Indeterminism in Quantum Physics and in Classical Physics', *Brit. J. Phil. Sci.*, 1950.

7.32. Quoted by Polanyi, as note 7.30, page 20.

7.33. Professor W. Mays, Department of Philosophy, Manchester University, has made available to me rough notes of this Discussion.

7.34. The 1950 *Mind* article has been reprinted in several anthologies, most recently *The Mind's I*, eds. D. Hofstadter and Daniel C. Dennett (Basic Books, New York; Harvester, Brighton, 1981).

7.35. Letter in *KCC*. The index to the Russell papers has no mention of AMT except for the receipt in 1937 of *Computable Numbers*.

7.36. See note 7.58.

7.37. Edward Carpenter, *Civilisation its Cause and Cure*, first published 1889, quoted here from 1921 edition (George Allen & Unwin); the chapter 'Modern Science: a Criticism'.

7.38. In the January 1952 radio discussion (page 567 and note 7.61).

7.39. In the May 1951 radio talk (note 7.56).

7.40. R. V. Jones, *Most Secret War* (as note 4.8), page 522.

7.41. He had a place in *Who's Who* after the 1951 election to the Royal Society.

7.42. As discussed by N. Bohr, *Nature 131* (1933), page 457.

7.43. C. H. Waddington, *Organisers and Genes*, 1940.

7.44. P. Weiss, *Quart, Rev. Biol.*, 1950.

7.45. As note 7.30, pages 339, 356, 400.

7.46. Here quoting from AMT's paper 'The Chemical Basis of Morphogenesis', *Phil. Trans. Roy. Soc. B 237* (1952).

7.47. *Ibid.*

7.48. Actually two books: G. Bentham, *Handbook of the British Flora*, revised by Sir J. D. Hooker and A. B. Rendle, 1947, and A. R. Clapham, T. G. Tutin, E. F. Warburg, *Flora of the British Isles*, 1952.

7.49. Published as J. Z. Young, *Doubt and Certainty in Science*, 1951.

7.50. Writing after AMT's death to Mrs Turing, and as quoted in *EST*.

7.51. Letters in *KCC*.

7.52. Not in *KCC*, but quoted in *EST*.

7.53. Writing after AMT's death to Mrs Turing, and as quoted in *EST*.

7.54. Quoting from J. A. Symonds, *Shelley* (Macmillan, 1887).

7.55. A manuscript of the result, and of Whitehead's letters, are in *KCC*.

7.56. Transcript in *KCC*. The BBC has not preserved the recording of this, nor of the January 1952 broadcast. Nor does any other tape-recording seem to have survived, so that AMT's unusual voice is lost to posterity.

7.57. Letter in *KCC*.

7.58. This talk, 'Intelligent Machinery, a Heretical Theory' was given to the '51 Society' at Manchester, presumably in or after 1951. The typescript is in *KCC*. It was reprinted in *EST*.

7.59. The proceedings were printed by Ferranti Ltd. They also record AMT's comments on other talks during the course of the conference.

7.60. Manuscript in *KCC*. Only three pages survive, of which the section quoted here is from the first, and that quoted on page 654 is from the third. In between his story diverges from what actually happened in December 1951, bringing in different characters and locations. I have taken this to confirm my impression from other sources that AMT 'knew the score' in Manchester already; this was not his first Manchester pick-up, although it might well have been the first time that he invited someone home as a boyfriend. For this reason I have included a transitional passage on page 539. AMT's story is also concerned to give equal space to 'Alec', *i.e.* himself, and 'Ron', and so contains phrases in which he imagines himself as seen by a hard-up youth: '. . . Didn't seem to be very well dressed. What an overcoat! . . . No, he was having a furtive look. Just a bit shy. . . . Seemed to be quite a toff after all. You could tell by the way he talked . . .'

7.61. Transcript in *KCC*.

7.62. Lecture 30 of the 1939 course (see note 3.39).

On the Beach

8.1. Documents relating to the case are held at Chester Record Office. They include the statements made by AMT and Arnold Murray, and the police account of what was said on the evening of 7 February 1952. See also note 8.17.

8.2. *Sunday Pictorial*, 25 May, 1 June, 8 June 1952. The series reflected the fact that there had been correspondence on homosexuality in the *British Medical Journal* since a conference in September 1947; this in turn took off from more theoretical agitation of the 1930s and a government *Report on the Psychological Treatment of Crime* of 1939.

8.3. Page 166 of G. Westwood (actually a pseudonym, for Michael Schofield), *Society and the Homosexual* (Gollancz, 1952).

8.4. A. C. Kinsey *et al.*, *Sexual Behaviour in the Human Male* (W. B. Saunders, Philadelphia & London, 1948) page 261.

8.5. *Alderley Edge and Wilmslow Advertiser*, 29 February 1952.

8.6. J. W. S. Pringle, 'The Origin of Life', in no. VII of the *Symposia of the Society for Experimental Biology*, 1953.

8.7. I am grateful to Professor W. Byers Brown for diary entries which give these details. Curiously, Prigogine later forgot about the discussions of AMT's ideas at Manchester. In his paper (with G. Nicolis and A. Babloyantz), in *Physics Today*, November 1972, Prigogine included a historical passage (which can also stand here to indicate a Nobel prize winner's assessment of the significance of AMT's work): 'The development of irreversible thermodynamics of open systems by the Brussels school had, by the 1950s, led to the investigation of non-linear processes. . . . It was only then that we noticed a remarkable paper by A. M. Turing (1952) who had actually constructed a chemical model showing instabilities. His work had previously escaped our attention because it dealt with the more specific subject of formation of morphogenetic patterns. The work we have undertaken since then has demonstrated the relation of this type of behaviour to thermodynamics as well as its wide applicability to biology.'

8.8. The discussions were fully minuted in an internal Nuffield Foundation report, kindly made available to me.

8.9. Blair Niles, *Strange Brother* (Liveright, New York, 1931).

8.10. S. J. Glass, H. J. Duel and C. A. Wright, 'Sex Hormone Studies in Male Homosexuality', *Endocrinology 26* (1940).

8.11. S. J. Glass and R. H. Johnson, 'Limitations and Complications of Organotherapy in Male Homosexuality', *J. Clin. Endocrin.*, 1944.

8.12. C. W. Dunn, *J. Amer. Med. Ass. 115*, 2263 (1940).

8.13. A. Karlen, *Sexuality and Homosexuality* (Macdonald, London, 1971), page 334.

8.14. F. L. Golla and R. Sessions Hodge, 'Hormone Treatment of the Sexual Offender', *The Lancet*, 11 June 1949.

8.15. D. E. Sands, 'Further Studies on Endocrine Treatment in Adolescence and Early Adult Life', *J. Mental Science*, January 1954.

8.16. As note 8.3, pages 69, 70.

8.17. In contrast to the committal proceedings, which are fully documented (note 8.1), the Quarter Sessions trial records are limited to bare statements of the charges and judgments, and the report in the *Alderley Edge and Wilmslow Advertiser*, 4 April 1952. Many questions thus remain unanswered. Was there a psychiatric report? Who suggested the hormone treatment, and what claims were made of it; at what point did AMT learn of it and agree to it? Did the Home or the Foreign Office intervene, and if so how? Unfortunately it is not even possible to discover how unusual the probation condition was: there are no statistics available for the administration of 'organotherapy'.

8.18. As note 8.15. The cited paper by S. Zuckerman was in *Ciba. Found. Coll. Endocrin. 3* (1952).

8.19. Writing in her introduction to *EST*.

8.20. Quotation from Part Two of the Epilogue to *War and Peace*, tr. Rosemary Edmonds (Penguin, 1957).

8.21. On 24 April 1978, the author was shown a list including a number of Scandinavian and Greek addresses, which happened to be among AMT's unpublished work on morphogenesis. This address list (in AMT's handwriting) has since been 'lost'. This 'loss' occurred at the Atomic Weapons Research Establishment, Aldermaston.

8.22. C. W. Wardlaw, 'A Commentary on Turing's Diffusion-Reaction Theory of Morphogenesis', *The New Phytologist 52* (1953). An article by the mathematician H. S. M. Coxeter, in *Scripta Math. 19* (1953), referred briefly to the Fibonacci numbers in phyllotaxis, and to the expected appearance of a paper by AMT on how the numbers arise in the course of plant growth.

8.23. Held in *KCC*.

8.24. Much material on the draughts and love-letter programs are held in the Christopher Strachey archive, Bodleian Library, Oxford. But the quoted love-letter is one that the public knew about: according to S. Lavington, *A History of Manchester Computers* (National Computer Centre, Manchester, 1975) it appeared in the 1955 edition of *Pears Cyclopaedia*.

8.25. *Faster than Thought*, ed. B. V. Bowden (Pitman, 1953). The glossary entry, even if hard on AMT in a tongue-in-cheek way, was in fact typical of the felicitous editing and commentary by Vivian Bowden, relieving the otherwise mundanely technical descriptions. It also reprinted the whole of Lady Lovelace's memoir on the Analytic Engine.

8.26. Translated from French and published as J. Piaget, *Logic and Psychology* (Manchester University Press, 1953).

8.27. Photocopies of this and the subsequent letters from AMT to Robin Gandy are in *KCC*.

8.28. C. G. Jung, 'Approaching the Unconscious', in the collection edited by him, *Man and his Symbols* (Aldus Books with W. H. Allen, 1964).

8.29. This was a Symposium on Automatic Digital Computation, held at the NPL from 25 to 28 March 1953. AMT gave no talk. Notes made by Mike Woodger show that he commented on applications of computers to pure mathematics after a talk by J. C. P. Miller of Cambridge – mainly on the zeta function but also mentioning problems in algebraic topology.

8.30. *The Shirburnian*, 1953.

8.31. The attribution to 1953 may here be wrong. This fragment of a letter in *KCC* was headed only 'May'; it might have been May 1954. It was a note of apology for not visiting the Newmans (who retained their house in a village outside Cambridge) when on a visit to Cambridge two weeks or so earlier. 'I found such a round of gaieties had been arranged for me that it was quite impossible to get out to see you.' Nothing else of this correspondence has survived; my suspicion is that it probably held the most revealing and sophisticated psychological comment that he ever put into letters. But also, of course, an area where AMT's life could not be separated from the privacy of others. Mrs Newman died in 1973.

8.32. Letter to the author from Professor J. Polanyi, 6.10.78.

8.33. Minutes of the University Council show that this had been decided by January or February 1953.

8.34. As note 8.21.

8.35. Fritz Peters, *Finistère* (Gollancz, 1951).

8.36. Rodney Garland, *The Heart in Exile* (W. H. Allen, 1953).

8.37. Depositions and post-mortem report are in *KCC*. The coroner's remarks were reported in the local newspaper on 18 June 1954, in the *Daily Telegraph* on 11 June.

8.38. Mrs Turing left the golden teaspoon in *KCC*.

8.39. The last, at least, in the series in *KCC*.

8.40. Letter in *KCC*.

8.41. Unpublished work in *KCC*. The second paper, as drafted, fell into three parts: I. Geometrical and descriptive phyllotaxis. II. Chemical theory of morphogenesis. III. A solution of morphogenetical equations for the case of spherical symmetry. This last part was the work of Bernard Richards. The passage quoted, the 'Outline of Development of the Daisy', did not fall within any of these; it was in a mass of less coherent ancillary material which worked out in more detail some specific examples of the 'chemical theory'.

8.42. P. S. Novikov, *Doklady Akad. Nauk. SSSR (N.S.) 85* (1952).

8.43. A few pages of this work survive in *KCC*; not enough to see where it might have led. He seems to have been interested in reformulating the connection between spinors and vectors, ideas probably inspired by Dirac.

8.44. P. A. M. Dirac, *Nature 139*, 323 (1937).

8.45. Robin Gandy wrote to Newman very shortly after the death, and described these ideas. The letter is in *KCC*. The problem to do with quantum-mechanical 'observation', that 'a watched pot never boils', is still a live question. I owe Philip Pearle for a recent reference: Ahanarov *et al.*, *Phys. Rev. D 21*, 2235 (1980).

8.46. One negative point deserves mention: rumours floated about after AMT's death that a second charge had been brought, but these had absolutely no foundation. One other curiosity reflects ambiguously upon his relationship with the state at the end of 1952, which was rather before the real crackdown began. On 28 November 1952 AMT wrote to Sir John Stopford, the Vice-Chancellor, saying that he had been 'invited by the Foreign Office to give a number of lectures at five German universities' in the spring, and that he would 'very much like to go' for the two weeks required. Perhaps this was a form of compensation for the loss of his consultancy work – but it is surprising that he should actually have been encouraged into the treacherous fields of post-war Germany. In the event he did *not* go after all. Permission was, of course, granted by the university, but on 22 January AMT wrote: 'I have however now cancelled my tour, as I found myself unable to undertake the work that would be involved.' Was this the real reason? There is obviously much more as yet unknown, and it can only be noted that British official silence regarding 'security' is total .

8.47. *Interim Report submitted to the Committee on Expenditures in the Executive Departments by its Subcommittee on Investigations pursuant to S. Res. 280 (81st Congress)*. Reprinted in D. W. Cory, The *Homosexual in America*, (Greenberg, New York, 1951).

8.48. Alan Moorehead, The *Traitors* (Hamish Hamilton, 1952).

8.49. In contrast, British policy only emerged ten years later, on account of a case where in late 1953 the vetters had *not* made a very searching investigation, and where a press campaign forced the government to admit that it had employed a homosexual in the diplomatic service. The quotations are from the *Report of the Tribunal appointed to Inquire into the Vassall Case and Related Matters*, 1963. There is a very interesting contemporary account in fictional form of these issues in Rodney Garland, *The Troubled Midnight* (W. H. Allen, 1954).

8.50. Here and elsewhere in this passage I draw on Peter Wildeblood, *Against the Law* (Weidenfeld & Nicolson, 1955).

8.51. *Hansard*, Parliamentary Debates (Commons) *521*, pages 526 and 1297.

8.52. *Hansard*, Parliamentary Debates (Lords), *187*, pages 737–767.

8.53. Published in the *Journal of Mental Science*, April 1954.

8.54. K. Sand and H. Okkels, in *Endokrinologie 19* (1938).

8.55. *Hansard*, Parliamentary Debates (Commons) *526*, page 1866.

8.56. Quoting this report from Peter Wildeblood's book.

8.57. R. S. Cline, *Secrets, Spies and Scholars* (Acropolis Books, Washington DC, 1976). Cline wrote as a retired deputy director of the CIA.

8.58. This was certainly just the kind of thing that the CIA was supposed to know about, but apparently it did not. According to a letter of 29.11.79 to the author, the CIA has no records concerning AMT.

8.59. As note 8.39.

8.60. As note 8.49.

8.61. In *What I Believe*, 1938, reprinted in the anthology *Two Cheers for Democracy*, as note BP 13.

8.62. As note 8.28.

8.63. An appreciation written by Robin Gandy and submitted to *The Times* to follow the more formal obituary by Max Newman, but not published. It is in *KCC*.

8.64. Her introduction to *EST*.

8.65. As note 8.61.

8.66. See note 7.60.

8.67. The letter itself did not survive, although the fact that AMT did write to James Atkins in early 1937 is established by AMT's reference in a letter home to his having sent an offprint of *Computable Numbers* to his friend. I have drawn on James Atkins' recollection of the 'apple' and 'electrical wiring' mentioned in the letter, a recollection as clear and distinct and unsolicited as any that have come my way. The sceptical reader may wonder whether this was not a projection of the news of 1954 into a memory of 1937. Here again it is as clear as anything I have been told that James Atkins never knew that an apple had actually featured in AMT's death, until he heard of it from me. It had not been mentioned in the *Daily Telegraph*, where he had read the report in 1954, and he had no knowledge that Mrs Turing or anyone else had written about it.

8.68. As note 2.18.

8.69. In *The Challenge of our Time*, a 1946 radio broadcast, published in *Two Cheers for Democracy*, as note BP 13. A rather more polemical critique of Forster was given by the author and David Hutter in *With Downcast Gays* (Pomegranate Press, London, 1974; Pink Triangle Press, Toronto, 1977).

Acknowledgements

The reproduction of Alan Turing's letters and papers has been made possible by the kind permission of the Fellows and Scholars of King's College, Cambridge, the A. M. Turing Trust, Robin Gandy, and P. N. Furbank. I am grateful to Donald Michie, Chairman of the A. M. Turing Trust, and to Peter Croft, Librarian of King's College, Cambridge, for facilitating the granting of this permission. I am grateful also to the Headmaster of Sherborne School for permission to reproduce those records held at Sherborne.

Transcripts of Crown Copyright records in the Public Record Office appear by permission of the Controller of Her Majesty's Stationery Office. So do the extracts from *British Intelligence in the Second World War*. I am grateful also for specific permission to reproduce photographs of the Colossus and the Delilah.

Private letters and papers are reproduced by permission of the many individuals who have kindly made them available.

I am grateful to the London Mathematical Society, and to Harvester Press, for permission to reproduce long extracts. The authors and publishers of other quotations are acknowledged in the Notes above.

Index

Page header